EVAPOTRANSPIRATION

Edited by **Leszek Łabędzki**

Evapotranspiration
Edited by Leszek Łabędzki

Published by InTech

Janeza Trdine 9, 51000 Rijeka, Croatia

Publishing Process Manager Iva Lipovic

Technical Editor Teodora Smiljanic

Cover Designer InTech Design Team

Additional hard copies can be obtained from orders@intechopen.com

Evapotranspiration, Edited by Leszek Łabędzki
p. cm.
ISBN 978-953-307-251-7

For EU product safety concerns :
IN TECH d . o . o . , Prolaz Marije Krucifikse Kozulić 3 , 51000 Rijeka , Croatia ,
info@intechopen . com or visit our website at intechopen . com .

Meet the editor

Leszek Labedzki graduated in water management and hydrology from the Technical University of Warsaw, Poland, in 1978. From the beginning of his research activity until today he has been working in the Institute of Technology and Life Sciences at Falenty, Poland, where he has been an assistant, senior assistant, lecturer, associate professor and professor. His research interests are agrometeorology and agricultural water management, comprising wide problems of water conditions in soils and agricultural areas, especially evapotranspiration, irrigation water requirements, water circulation in the soil-plant-atmosphere system and water distribution in agricultural areas. He teaches water management and operation of irrigation systems at the University of Technology and Life Sciences in Bydgoszcz, Poland.

Contents

Preface

Farmers, agriculture advisers, extension services, hydrologists, agrometeorologists, water management specialists and many others are facing the problem of evapotranspiration, i.e. the actual water use from the surface covered with plants. The book "Evapotranspiration" gives an example of the contribution of science to the solving of this problem.

As evapotranspiration is a very complex phenomenon, comprising different aspects and processes (hydrological, meteorological, physiological, soil, plant and others) and depending on many factors, the angles under which evapotranspiration can be viewed and analysed are various. That is why the book presents the results of an international effort to analyze the process of evapotranspiration from different points of view.

In recent years there have been many publications concerning evapotranspiration, issued both as books and as scientific articles in different journals. This book is published with two primary intentions in mind. The first concerns its multinational character, reflecting the views, opinions and study results of teams from different regions of the world. The second concerns the multidisciplinary character of the presented analyses and approaches which cover a broad range of scientific expertise, committed to improve our knowledge of this complex phenomenon and especially the links between diverse disciplines of knowledge interested in evapotranspiration.

The book consists of three major parts. The first part consists of chapters relating different concepts, methods of measurements, predicting and modelling evapotranspiration. The second part deals with different remote sensing techniques for measuring various parameters used for calculating evapotranspiration. The last and the largest part of the book contains chapters presenting the applications of different methods of measurements, calculations and analyses of evapotranspiration as well as reporting case studies.

The book does not try to cover all possible angles, points of view, methods and theoretical considerations of the subject matter. We hope that this book can be seen as an attempt to present a broad body of experience contained by reporting different views of the authors and the results of their studies. We also hope that this book will stimulate the researchers to formulate new questions regarding evapotranspiration, to find new solutions for already known questions, to search better and more reliable methods of calculating or measuring evapotranspiration. Although many researchers have been engaged in evapotranspiration for many years, although from time to time the opinions can be heard that everything is known about it, reading the chapters in this book, it can be stated that there is still a lot to be made in this field.

The book is intended for professional scientists and engineers who are active in meteorology, climatology, hydrology, agronomy and other environmental fields, and who wish to study the phenomenon of evapotranspiration. I believe that the book will be useful and helpful for all those with an interest in evapotranspiration, both for the researches for whom evapotranspiration is a focus of their activity as well as for those who have more practical approach and use evapotranspiration as an input in other fields of their activity.

It was a pleasure for me and the publisher to accompany and support the fruitful international cooperation in the preparation of this book. I wish to thank all the authors for their contributions. During the editing process many people helped, directly or indirectly. MSc Iva Lipovic deserves special thanks for extending immense support throughout the entire period of publishing of this book.

January 2011

Prof. Dr. Leszek Łabędzki
Institute of Technology and Life Sciences
Poland

Part 1

Concepts, Methods and Modelling

1

Water Vapor Flux in Agroecosystems Methods and Models Review

Víctor Hugo Ramírez-Builes[1] and Eric W. Harmsen[2]
[1]*National Coffee Research Center-Cenicafé*
[2]*University of Puerto Rico*
[1]*Colombia*
[2]*USA*

1. Introduction

The water vapor flux in agroecosystems is the second largest component in the hydrologycal cycle. Water vapor flux or evapotranspiration (ET) from the vegetation to the atmosphere is a widely studied variable throughout the world. ET is important for determining the water requirements for the crops, climatic characterization, and for water management. The estimation of ET from vegetated areas is a basic tool to compute water balances and to estimate water availability and requirements. During the last sixty year several methods and models to measure the water flux in agroecosystems have been developed. The aim of this chapter is to provide a literature review on the subject, and provide an overview of methods and models developed which are widely used to estimate and/or measure ET in agroecosystems.

Evapotranspiration constitutes an important component of the water fluxes of our hydrosphere and atmosphere (Conroy *et al.*, 2003), and is a widely studied variable throughout the world, due to it applicability in various disciplines, such as hydrology, climatology, and agricultural science. Pereira *et al.*, (1996) has reported that the estimation of ET from vegetated areas is a basic tool for computing water balances and to estimate water availability and requirements for plants. Measurement of ET is needed for many applications in agriculture, hydrology and meteorology (Suleiman & Crago, 2004). ET is a major component of the hydrologic water budget, but one of the least understood (Wilson et al., 1992). ET permits the return of water to the atmosphere and induces the formation of clouds, as part of a never-ending cycle. ET also permits the movement of water and nutrients within the plant; water moving from the soil into the root hairs, and then to the plant leaves.

ET is a complicated process because it is the product of the different processes, such as evaporation of water from the soil, and water intercepted by the canopy, and transpiration from plant leaves. Physiological, soil and climatic variables are involved in these processes. Symons in 1867 described evaporation as "*...the most desperate art of the desperate science of meteorology*" (Monteith, 1997). The first vapor flux measurements were initiated by Thornthwaite and Holzman in 1930s, but that works was interrupted by World War II (Monteith, 1997). In the late 1940s Penman (1948) published the paper "*Natural Evaporation from open Water, Bare Soil and Grass*" in which he combined a thermodynamic equation for the surface heat balance and an aerodynamic equation for vapor transfer. The "Penman equation" is one of the most widely used equations in the world. The equation was later modified by Monteith (1965; 1981) and is widely known as the "*The Penman-Monteith Model*". It is also necessary to introduce a review of the work of Bowen, who in 1926 published the relationship between the sensible and latent heat fluxes, which is known as the "*Bowen ratio*". Measurement of the water vapour flux became a common practice by means of the "*Bowen ratio energy balance method*" (Tanner, 1960).

Allen et al. (1998) separated the factors that affect the ET into three groups: *a) Weather parameters,* such as radiation, air temperature, humidity and wind speed. The evapotranspirational power of the atmosphere is expressed by the reference crop evapotranspiration (ET_o) as the Penman-Monteith (FAO-56), or using direct measurements of pan evaporation data (Doorenbos & Pruit, 1977), or using other empirical equations; *b) Crop factors* such as the crop type, variety and developmental stage should be considered when assessing the ET from crops grown in large, well-managed fields. Differences in resistance to transpiration, crop height, crop roughness, reflection, ground cover and crop rooting characteristics result in different ET levels in different types of crops under identical environmental conditions. Crop ET under standard conditions (ET_c) refers to excellent management and environmental conditions, and achieves full production under given climatic conditions (equation 2); and *c) Management and environmental conditions* (ET_{cadj}). Factors such crop water stress, soil salinity, poor land fertility, limited applications of fertilizers, the presence of hard or impermeable soil horizons, the absence the control of disease and pest and poor soil management may limit the crop development etc., and reduce the ET, (ETc_{adj} equation 3).

One of the most common and fairly reliable techniques for estimating ET_0 is using evaporation pan data when adjustments are made for the pan environment (Grismer et al. 2002) using the pan evaporation and the pan coefficient (K_p).

$$ET_o = K_p . E_p \tag{1}$$

Where E_p is the pan evaporation (mmday^{-1}), and K_p is the pan coefficient, and depends on location, so it is important to know or calculate this coefficient before calculating the ET_o. Allen et al., (1998) gave a methodology to know or calculate *Kp,* and is essentially a correction factor that depends on the prevailing upwind fetch distance, average daily wind speed, and relative humidity conditions associated with the siting of the evaporation pan (Doorenbos & Pruitt 1977)

2. Crop water flux using single crop coefficients- The FAO approach

The United Nation Food and Agricultural Organization (FAO) is also well knew as the "*Two steps method*", which is very useful for single crops and when "reference" conditions are available (i.e., no crop water stress). In this case, crop evapotranspiration (ET_c) can be estimated using equation 2 (Doorenbos & Pruitt 1977; Allen at al., 1998):

$$ET_c = K_c . ET_o \tag{2}$$

where K_c is the coefficient expressing the ratio between the crop and reference ET for a grass surface. The crop coefficient can be expressed as a single coefficient, or it can be split into two factors, one describing the affect of evaporation and the other the affect of transpiration.

As soil evaporation may fluctuate daily, as a result of rainfall and/or irrigation, the single crop coefficient expresses only the time-average (multi-day) effects of crop ET, and has been considered within four distinct stages of growth (see. FAO,56. Allen et al., 1998). When stress conditions exist, the effects can be accounted for by a crop water stress coefficient (K_s) as follows:

$$ET_{cadj.} = K_s . K_c . ET_o \quad (3)$$

2.1 Crop coefficients

Although a number of ET_c estimation techniques are available, the crop coefficient (K_c) approach has emerged as the most widely used method for irrigation scheduling (Hunsaker et al., 2002). As ET is not only a function of the climatic factors, the crop coefficients can include conditions related to the crop development (K_c), and non-standard conditions (K_s).
The K_c is the application of two concepts: a) crop transpiration represented by the basal crop coefficient (K_{cb}), and b) the soil evaporation K_e (Allen et al., 1998) as follow:

$$K_c = K_{cb} + K_e \quad (4)$$

K_c is an empirical ratio between ET_c and ET_o over grass or alfalfa, based on historic measurements. A K_c curve is constructed for an entire crop growing season, and which attempts to relate the daily water use rate of the specific crop to that of the reference crop (Hunsaker et al., 2002).
The FAO paper # 56 (Allen et al., 1998) provided detailed instructions for calculating these coefficients. For limited soil water conditions, the fractional reduction of K_c by K_s depends on the crop, soil water content, and magnitude of the atmospheric evaporative demand (Doorenbos &Pruitt, 1977).
The value for K_c equals K_{cb} for conditions where, first, the soil surface layer is dry (i.e., when $K_e = 0$) and, second the soil water within the root zone is adequate to sustain the full transpiration (non-stressed conditions, i.e., $K_s = 1$). When the available soil water of the root zone becomes low enough to limit potential ET_c, the value of the K_s coefficient is less than 1 (Allen et al., 1998; Hunsaker, 1999, Hunsaker et al., 2002).
The soil evaporation coefficient accounts for the evaporation component of ET_c when the soil surface is wet, following irrigation or rainfall (Allen et al., 1998; Hunsaker et al. 2002). When the available soil water of the root zone become low enough, crop water stress can occur and reduce ET_c. In the FAO-56 procedures, the effects of water stress are accounted for by multiplying K_{cb} (or K_c) by the water stress coefficient (K_s).

$$K_c . K_s = (K_{cb} . K_s + K_e) = ET_c/ET_o \quad (5)$$

Where $K_s < 1$ when the available soil water is insufficient for the full ET_c and $K_s = 1$ when there is no soil water limitation on ET_c. Thus, to determine K_s, the available soil water within the crop zone for each day needs to be measured or calculated using a soil water balance approach (Hunsaker et al., 2002).
The estimation of K_e using the FAO-56 method, requires the use of the soil field capacity (FC), the permanent wilting point (PWP), total evaporable water (TEW), the fraction of the soil surface wetted (f_w) during each irrigation or rain, and the daily fraction of the soil surface shaded by vegetation (f_c), or conversely the unshaded fraction ($1-f_c$). Hunsaker et al., (2002) reported an exponential relation between $1-f_c$ and height of the Alfalfa crop.

The measurement of K_e and K_{cb} can be made by performing a *daily water balance*, and use of the following equations from FAO Paper 56 (Allen et al., 1998).

$$ET_c = (K_{cb}+K_e)\ ET_o \tag{6}$$

$$K_{cb} = (ET_c/ET_o)-K_e \tag{7}$$

The soil evaporation (E) can be calculated using the equation (8)

$$E = K_e\ ET_o;\ (8) \text{ and } K_e \text{ is equal to } K_e = E/ET_o \tag{9}$$

The soil evaporation (E) can be measured using the water balance (equation 10)

$$E = D_{e,i-1} - (P_i - RO_i) - \frac{I_i}{f_w} + \frac{f_i}{f_{ew}} + Tew_{,i} + DP_{e,i} \tag{10}$$

where: $D_{e,i-1}$ is the cumulative depth of evaporation following complete wetting from the exposed and wetted fraction of the topsoil at the end of day i-1 (mm), P_i is the precipitation on day i (mm); RO_i is precipitation runoff from the surface on day i (mm), I_i is the irrigation depth on day i that infiltrates into the soil (mm), E_i is evaporation on day i (i.e., E_i=K_e/ET_o) (mm), $T_{ew,i}$ is depth of transpiration from the exposed and wetted fraction of the soil surface layer on day i (mm), f_w is fraction of soil surface wetted by irrigation (0.01-1), and f_{ew} is the exposed and wetted soil fraction (0.001-1).

The ratio of reference evaporation to reference transpiration depends on the development stage of the leaf canopy expressed as "δ" the dimensionless fraction of incident beam radiation that penetrates the canopy (Cambell &Norman, 1998; mentioned by Zhang et al. 2004).

$$\delta = \exp(-c.LAI) \tag{11}$$

where c is the dimensionless canopy extinction coefficient, and therefore evaporation and transpiration can be calculate how:

$$E_o = \delta.ET_o \tag{12}$$

$$T = (1-\delta).ET_o \tag{13}$$

Hunsaker, (1999) found that ET_c in cotton was higher when the crop was submitted to high depth of irrigation (820-811mm) that when have low depth of irrigation level (747-750mm), similar to the $K_{cb}K_s$ curves, obtaining higher values than the treatment with high frequency (i.e.; $K_{cb}K_s$ =1.5, 90 days after planting) than the low frequency (i.e.; $K_{cb}K_s$ = 1.4, 90 days after planting).

2.2 Limitations in the use of K_c

Katerji & Rana, (2006) reviewed recent literature related to K_c and found differences of ± 40% between K_c values reported in the FAO-56 paper (Allen et al., 1998) and the values experimentally obtained, especially in the mid growth stage. According to the authors, these large differences are attributable to the complexity of the coefficient K_c, which actually integrate several factors: aerodynamic factors linked to the height of the crop, biological factors linked to the growth and senescence of the surfaces leaves, physical factors linked to evaporation from the soil, physical factors linked to the response of the stomata to the vapour pressure deficit and agronomic factors linked to crop management (distance between rows, using mulch, irrigation system, etc.). For this reason K_c values need to be evaluated for local conditions.

The variation in crop development rates between location and year have been expressed as correlations between crop coefficients and indices such as the thermal base index, ground cover, days after emergence or planting, and growth rate (i.e., Wright & Jensen, 1978; Hunsaker, 1999; Brown et al., 2001; Nasab et al., 2004; Hanson & May 2004; Madeiros et al., 2001; Madeiros et al., 2005; and Ramírez, 2007). The K_c is well related with the growing degree grades-GDD and with the fraction of the soil cover by vegetation (f_c) (**Fig. 1**), and depends on the genotype and plant densities (Ramirez, 2007). The equations for two common beans genotypes and two plant densities are:

The equations based on CGDD and f_c for common bean genotype Morales with 13.6 plants.m^{-2} are:

$$K_c = -3x10^{-6} CGDD^2 + 0.0033CGDD - 0.053; R^2 = 0.76; p < 0.0001 \tag{14}$$

$$K_c = -1.4019 f_c^{\ 2} + 2.5652 f_c - 0.2449; R^2 = 0.70; p < 0.0003 \tag{15}$$

The equations based on CGDD and fc for common bean genotype SER 16, with 6.4 plants.m^{-2} are:

$$K_c = -3x10^{-6} CGDD^2 + 0.0034CGDD - 0.0515; R^2 = 0.60; p < 0.0001 \tag{16}$$

$$K_c = -0.6726 f_c^{\ 2} + 1.90086 f_c - 0.2560; R^2 = 0.60; p < 0.0032 \tag{17}$$

2.3 Water stress coefficient (K_s)

The soil water stress coefficient, K_s, is mainly estimated by its relationship to the average soil moisture content or matric potential in a soil layer, and it can usually be estimated by an empirical formula based on soil water content or relative soil water available content (Jensen et al., 1970, cited by Zhang et al., 2004).

The K_s is an important coefficient because it indicates the sensitivity of the crop to water deficit conditions, for example corn grain yield is especially sensitive to moisture stress during tasselling and continuing through grain fill. Roygard et al., (2002) observed that depletion of soil water to the wilting point for 1 or 2 days during tasselling or pollenization reduced yield by 22%. Six to eight days of stress reduced yield by 50%.

Allen et al. (1998), presented the following methodology for estimating K_s:

$$K_s = \frac{TAW - Dr}{TAW - RAW} = \frac{TAW - Dr}{(1-p)TAW} \tag{18}$$

where TAW is total available water and refers to the capacity of the soil to retain water available for plants (mm), Dr is root zone depletion (mm), RAW is the readily available soil water in the root zone (mm), p is the fraction of TAW that the crop can extract from the root zone without suffering water stress.

$$TAW = 1000(\theta_{FC} - \theta_{WP}) Z_t \tag{19}$$

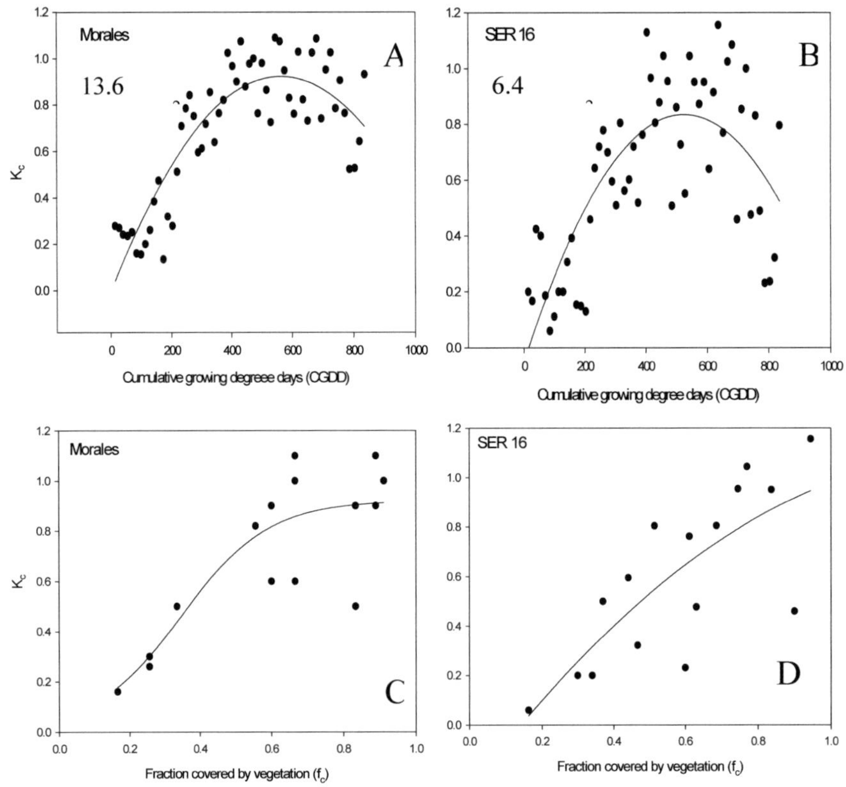

Fig. 1. Crop coefficients (K_c) as related to cumulative growing degree days (CGDD) and fraction covered by vegetation (f_c) for two common bean genotypes: **A**. Morales CGDD vs K_c, **B**. SER 16 CGDD vs K_c, **C**. Morales f_c vs K_c, **D**. SER 16 f_c vs K_c. The curves were fitted from growth periods V1 to R9 (Data from: Ramirez, 2007). *(These data's were obtained under the project sponsored by NOAA-CREST (NA17AE1625), NASA-EPSCoR (NCC5-595), USDA-TSTAR-100,USDA Hatch Project H-402, and University of Puerto Rico Experiment Station)*

where θ_{FC} is the water content at field capacity ($m^3.m^{-3}$), θ_{WP} is the water content at wilting point ($m^3.m^{-3}$), and Z_t is the rooting depth (m)

$$RAW = pTAW \tag{20}$$

Allen et al. (1998) give values to different crops (FAO, 56.p163). Roygard et al. (2002) and Zhang et al. (2004), reported that K_s is a logarithmic function of soil water availability (Aw), and can be estimated as follow.

$$K_s = \ln(Aw+1)/\ln(101) \tag{21}$$

and Aw is calculated according to the equation

$$Aw = 100\left(\frac{\theta_a - \theta_{wp}}{\theta_{FC} - \theta_{wp}}\right) \tag{22}$$

where θ_a is average soil water content in the layers of the root zone depth.
An example of the relationships between K_s and available soil water changes, estimated as a root zone depletion, is presented by Ramirez (2007). The root zone depletion (D_r), can be calculated using the water balance equation:

$$D_{r,i} = D_{r,i-1} - (P - RO)_i - I_i + ET_{c,i} + DP_i \tag{23}$$

where $D_{r,i}$ is the root zone depletion at the end of day i; $D_{r,i-1}$ is water content in the root zone at the end of the previous day, i-1; $(P\text{-}RO)_i$ is the difference between precipitation and surface runoff on day i; I_i is the irrigation depth on day i; $ET_{c,i}$ is the crop ET on day i and DP_i is the water loss from the root zone by deep percolation on day i; all the units are in mm. The root zone depletion associated with a K_s = 1.0 (i.e., no water stress), was up to 10 mm for a root depth between 0 to 20 cm, and up to 15 mm for a root depth of 0 to 40 cm in common beans. Fifty percent of the transpiration reduction was reached for D_r = 22 mm and 25 mm for the common bean genotype Morales and genotype SER 16, respectively. Transpiration ceased completely (K_s = 0) when Dr = 37 mm and 46 mm, respectively, for Morales and SER 16 (Ramírez, 2007).

3. Lysimeters as a direct water vapour flux measurement

The word '*lysimeter*' is derived from the Greek root 'lysis,' which means dissolution or movement, and 'metron,' which means to measure (Howell, 2005). Lysimeters are tanks filled with soil in which crops are grown under natural conditions to measure the amount of water lost by evaporation and transpiration (Allen et al., 1990). A lysimeter is the method of determining ET directly. The lysimeter are tanks buried in the ground to measure the percolation of water through the soil. Lysimeter are the most dependable means of directly measuring the ET rate, but their installation must meet four requirements for the data to be representative of field conditions (Chang, 1968).
Requirement 1. The lysimeter itself should be fairly large and deep to reduce the boundary effect and to avoid restricting root development. For short crops, the lysimeter should be at least one cubic meter in volume. For tall crops, the size of the lysimeter should be much larger.
Requirement 2. The physical conditions within the lysimeter must be comparable to those outside. The soil should not be loosened to such a degree that the root ramification and water movement within the lysimeter are greatly facilitated. If the lysimeter is unclosed on the bottom, precaution must be taken to avoid the persistence of a water table and presence of an abnormal thermal regime. To ensure proper drainage, the bottom of an isolated soil column will often require the artificial application of a moisture suction, equivalent to that present at the same depth in the natural soil (Coleman, 1946)
Requirement 3. The lysimeter will not be representative of the surrounding area if the crop in the lysimeter is either taller, shorter, denser, or thinner, or if the lysimeter is on the periphery of no-cropped area. The effective area of the lysimeter is defined as the ratio of the lysimeter ET per unit area of the surrounding field. The values of this ratio, other than unity, are caused by the in homogeneity of the surface. The maintenance of uniform crop height and density is not an easy task in a tall crop, spaced in rows. If the surface is indeed inhomogeneous, there is no adequate way to estimate the effective area from tank area overlap corrections or plant counts.

Requirement 4. Each lysimeter should have a "guard-ring" area around it maintained under the same crop and moisture conditions in order to minimize the clothesline effect. In arid climates, Thornthwaite in 1954, suggested that a "guard-ring" area of ten acres may or may not be large enough. Where several lysimeters are installed in the same field, the "guard-ring" radius may have to be about ten times the lysimeter separation (Chang, 1968).
Lysimeters surrounded by sidewalks or gravel will not provide reliable data, nor will lysimeters planted to a tall crops if it is surrounded by short grass, or planted to grass and surrounded by a tall crop. Differences in growth and maturity between the lysimeters plants and surrounding plants can result in significant differences in measured ET in and outside the lysimeter (Pruitt & Lourence, 1985; mentioned by Allen et al., 1990). The lysimeters are divided basically in two types: Weighing and Non-weighing, each of which are described below.

3.1 Non-Weighing Lysimeters

Also called *Drainage lysimeters,* they operate on the principle that ET is equal to the amount of rainfall and irrigation water added to the system, minus percolation, runoff and soil moisture changes. Since the percolation is a slow process, the drainage lysimeters is accurate only for a long periods for which the water content at the beginning exactly equals that at end. The length of such a period varies with the rainfall regime, frequency and amount of irrigation water application, depth of the lysimeters, water movement, and the like. Therefore, records of drainage lysimeters should be presented only in terms of a long-period more than one day (Chang, 1968), and they are not useful for estimating hourly ET.
Allen et al. (1990) discusses two types the non-weighing lysimeters: a) *non- weighing constant water-table type,* which provides reliable data in areas where a high water table normally exists and where the water table level is maintained essentially at the same level inside as outside the lysimeters; b) *Non-weighing percolation type,* in which changes in water stored in the soil are determined by sampling or neutron methods or other soil humidity sensors like TDR, and the rainfall and percolation are measured.
General principles of a drainage lysimeter: Provisions are made at the bottom of the lysimeter container to collect and measure volumetrically the deep percolation. Precipitation is measured by rain gauge(s). *Evapotranspiration* is considered as the difference among *water applied, water drainage and soil water change.* (Teare et al., 1973; Xingfa et al., 1999)
When filling-in a lysimeter, the soil dug out from the lysimeter's pit is replaced in the container, special precautions are needed to return the soil to its original status by restoring the correct soil profile and compacting the soil layers to the original density. It is desirable to have a similar soil state inside the lysimeter relative to the outside. However if the roots are well developed and nutrients are available, and as long as the water supply to the roots is unrestricted, dissimilar soil will not give significant variation in water use and yield, provided other conditions are similar.(Xingfa et al. 1999).
Although disturbed soil in filled-in lysimeters does not pose serious problems in ET measurement, the soil can affect plant growth. Breaking up the soil, will change soil structure, aeration, and soil moisture retention characteristics. The lysimeters should provide a normal rooting profile. It should be large enough to render the effect of the rim insignificant. It can give relatively large errors in the ET measurement if the container is small. However, the greater the lysimeters area, the more costly and complicated the installation and operation becomes. (Xingfa et al., 1999).

Installation and walls: The wall can be different materials: reinforced concrete, polyester reinforced with steel, fiberglass or plastic. The installation proceeds in the following steps: Excavation (e.g. 1m*1m*1.2m) in the experimentatal site. Each layer of soil (e.g. 0-30 cm, 30-60cm and 60-100 cm) is separated. Once the excavation it completed, the lysimeter is placed in the excavated hole with 4 wooden boards outside. Before repacking the soil layers, make a V-shaped slope at the bottom and place a 25 mm inside diameter perforated PVC pipe (horizontal). There should be a screen material placed around the perferated pipe to avoid the soil particles from entering the pipe. Connect an access tube (25 mm PVC), approximately 1 m long (vertical). Cover the horizontal pipe with fine gravel approximately 3-5 cm thick. Fill the container with the excavated soil where each layer is repacked inside the lysimeter to match the original vertical soil state. (Xingfa et al. 1999).

3.2 Weighing Lysimeters

A weighing Lysimeter is capable of measuring ET for periods as short as ten minutes. Thus, it can provide more additional information than a drainage lysimeter can. Problems such as diurnal pattern of ET, the phenomenon of midday wilt, the short-term variation of energy partitioning, and the relationship between transpiration and soil moisture tension, can be investigated only by studying the records obtained from a weighing lysimeter. (e.g.; Chang, 1968; Ritchie & Burnett, 1968; Takhar & Rudge, 1970; Parton et al., 1981; Steiner et al., 1991; Allen et al., 1998; Loos et al., 2007; von Unold &Fank, 2008)

Weighing lysimeters make direct measurements of water loss from a growing crop and the soil surface around a crop and thus, provide basic data to validate other water vapor flux prediction methods (e.g.; Dugas et al., 1985; Prueger et al., 1997; López-Urrea et al., 2006; Vaughan et al., 2007). The basic concept of this type of lysimeter is that it measures the difference between two mass values, the mass change is then converted into ET (mm) (Malone et al., 1999; Jhonson et al., 2005).

During periods without rainfall, irrigation and drainage, the ET rate is computed as indicated by Howell (2005), as:

$$ET = \lfloor A_l[(M_i - M_{i-1})/A_l]/A_f \rfloor / T_i \qquad (24)$$

where ET is in units of (mm.h^{-1} or Kg.m^2) for time interval i; M is the lysimeter soil mass, (Kg); A_l is lysimeter inner tank surface area (m^2); A_f is lysimeter foliage area (mid wall-air gap area) (m^2); T is the time period (h). The ratio A_f / A_l is the correction factor for the lysimeter effective area. This correction factor assumes the outside and inside vegetation foliage overlap evenly on all of the sides or edges. If there is no overlap, as occurs in short grass, the A_f / A_l=1.0 (Howell , 2005).

Weighing lysimeters provide the most accurate data for short time periods, and can be determined accurately over periods as short as one hour with a mechanical scale, load cell system, or floating lysimeters (Allen at al., 1990). Some weighing lysimeters use a weighing mechanism consisting of scales operating on a lever and pendulum principle (Harrold & Dreibelbis (1951), mentioned by Malone et al. (1999)). However, some difficulties are very common like: electronic data logger replacement, data logger repair, load cell replacement, multiflexor installation etc. (Malone et al., 1999).

The measurement control in these lysimeters are important because of the following issues: a) re-calibration requirements, b) measurement drift (e.g., slope drift, variance drift), c) instrument problems (e.g., localized non-linearity of load cell, load cell damage, data logger

damage), d) human error (e.g.; incorrectly recording data during calibration) and e) confidence in measurement results (Malone et al. 1999).
A load cell is a transducer that coverts a load acting on it into an analog electrical signal. The electrical signal is proportional to the load and the relationship is determined through calibration, employing linear regressions models (mV/V/mm water), and it is used to determine mass changes of a lysimeter over the period interest (e.g. day, hour, etc.).
The lysimeter characteristics can be different, for example: Malone et al., (1999) built a lysimeter of the following form: 8.1 m^2 in surface area and 2.4 m depth, the lysimeter is constructed without disturbing the soil profile and the underlying fracture bedrock. The soil monolith is supported by a scale frame that includes a 200:1 lever system and a counterweight for the deadweight of the soil monolith. The gap between the soil in the lysimeters and the adjacent soil is between 5.1 cm and 7.0 cm except at the bottom slope where the runoff trough is located, this same author has given instructions for achieving a good calibration for this type of lysimeter.
Tyagi et al., (2000) in wheat and sorghum used two rectangular tanks, an inner and outer tank, constructed from 5-mm welded steel plates. The dimensions of the inner tank were 1.985 x 1.985 x 1.985 m and those of the outer tank were 2.015 x 2.015 x 2.015 m. The lysimeters were situated in the center of a 20-ha field. The size ratio of the outer tank to the inner tank is 1.03, so the error due to wall thickness is minimal. The effective area for crop ET was 4 m^2. The height of the lysimeter rim was maintained near ground level to minimize the boundary layer effect in and around the lysimeter. The lysimeter tank was suspended on the outer tank by four load cells. The load cells were made out of the steel shear beam type with 40,000-kg design load capacity. The total suspended mass of the lysimeter including tank, soil, and water was about 14,000 kg. This provided a safety factor of 2.85. The high safety factor was provided to allow replacement of a load cell without the danger of overloading and also to account for shock loading. A drainage assembly connected with a vertical stand and gravel bedding to facilitate pumping of drainage water was provided. The stand pipe also can be used to raise the water table in the lysimeter.
To calculate the ET using Lysimeter, we need to employ the soil water balance (SWB) equation:

$$ET = R + I - P - Rff +/- \Delta SM \quad (25)$$

Where: R is the rain, I is the irrigation, Rff is the runoff, and +/-ΔSM soil moisture changes, all in mm.
The size of the Lysimeter is an important element to be considered in water vapor flux studies with this method. For example, Dugas & Bland (1989) evaluated small lysimeters (<1.0 m^2) and reported significant differences in the ET estimations, basically associated with the differences in the leaf area index (LAI) inside the lysimeters, which differed among lysimeter, this problem can be addressed using LAI corrections.

3.3 Calibration of the weighing Lysimeter

Seyfried et al.,(2001) made a weighing lysimeter calibration by placing known weights on the lysimeters and then recording the resultant pressure changes. The weights used in that study were as follows: 19.9 kg for supportive blocks placed on the lysimeter, 43.4 kg for the tank which contained the weights, and then twenty-four 22.7 kg sacks of rock added in four-sack increments. The weight of each sack corresponded to about a 13 mm addition of water;

so that weight increments were equivalent to ~52 mm and the total range was~360 mm of water. Measurements were made both as weight was added and removed.
The main arguments against the use of weighing lysimeters for monitoring water balance parameters and measuring solute transport parameters in the soil and unsaturated zone has been the discussion of potential sources of error, such as the well known oasis effect, preferential flow paths at the walls of the lysimeter cylinders due to an insufficient fit of soil monoliths inside the lysimeters, or the influence of the lower boundary conditions on the outflow rates (Fank, 2008).

4. The micrometeorological methods

For many agricultural applications, micrometeorological methods are preferred since they are generally non-intrusive, can be applied on a semi-continuous basis, and provide information about the vertical fluxes that are occurring on scales ranging from tens of meter to several kilometres, depending the roughness of the surface, the height of the instrumentation, and the stability of the atmosphere surface layer. Meyers and Baldocchi (2005) have separated micrometeorological methods into four categories: 1) eddy covariance, 2) flux-gradient, 3) accumulation, and 4) mass balance. Each of these approaches are suitable for applications that depend on the scalar of interest and surfaces type, and instrumentation availability. Some of these methods are described in the following sections of this chapter.

4.1 Humidity and temperature gradient methods

Movement of energy, water and other gases between field surface and atmosphere represent a fundamental process in the soil-plant-atmosphere continuum. The turbulent transport in the surface boundary layer affect the sensible (H) and latent (λE) heat fluxes, which along with the radiation balance, govern the evapotranspiration and canopy temperature (Ham and Heilman, 2003).
Monteith and Unsworth (1990) presented the functional form of the gradient flux equation, and which has been applied by Harmsen et al. (2006), Ramírez et al. (2008) and Harmsen et al. (2009):

$$ET = \left(\frac{\rho_a \cdot c_p}{\gamma \cdot \rho_w}\right) \cdot \frac{\left(\rho_{vL} - \rho_{vH}\right)}{\left(r_a + r_s\right)} \tag{26}$$

where ρ_w is the density of water, ρ_v is the water vapor density of the air, ρ_a is the air density, γ is the psychometric constant, c_p is specific heat of air, r_a and r_s are aerodynamic and bulk surfaces resistances (all these variables are discussed in detail below). L and H are vertical positions above the canopy (L: low and H: High positions), for example in small crops like beans or grass, possible values of L and H could be 0.3 m and 2 m above the ground, respectively.
Harmsen et al. (2006) developed an automated elevator device (ET Station) for moving a temperature and relative humidity sensor (Temp/RH) between the two vertical positions (**Fig.2**). The device consisted of a plastic (PVC) frame with a 12 volt DC motor (1/30 hp) mounted on the base of the frame. One end of a 2-m long chain was attached to a shaft on the motor and the other end to a sprocket at the top of the frame. Waterproof limit switches

were located at the top and bottom of the frame to limit the range of vertical movement. For automating the elevator device, a programmable logic controller (PLC) was used which was composed of "n" inputs and "n" relay outputs. To program the device, a ladder logic was used which is a chronological arrangement of tasks to be accomplished in the automation process. The Temp/RH sensor was connected to the elevator device, which measured relative humidity and temperature in the up position for two minutes then changed to the down position where measurements were taken for two minutes. This process started each day at 0600 hours and ended at 1900. When the elevator moves to the up position it activates the limit switch which sends an input signal to the PLC. That input tells the program to stop and remain in that position for two minutes. At the same time it activates an output which sends a 5 volt signal to the control port C2 in the CR10X data logger in which a small subroutine is executed. This subroutine assigns a "1" in the results matrix which indicates that the temperature and relative humidity corresponding to the up position. At the end of the two minute period, the elevator moves to the down position and repeats the same process, but in this case sending a 5 volts signal to the data logger in the control port C4, which then assigns a "2" in the results matrix. All information was stored in the weather station data-logger CR-10X (Campbell Scientific, Inc) for later downloading to a personal computer.

Fig. 2. Automated elevator device developed for moving the Temp/RH sensor between the two vertical positions. **A.** Temp/RH sensor in down position and **B.** Temp/RH sensor in up position. Measuring over common bean (*Phaseolus vulgaris* L.). (*These data's were obtained under the project sponsored by NOAA-CREST (NA17AE1625), NASA-EPSCoR (NCC5-595), USDA-TSTAR-100,USDA Hatch Project H-402, and University of Puerto Rico Experiment Station*)

4.2 The Bowen-ratio energy balance method

The basis for this method is that the local energy balance is closed in such a way that the available net radiative flux (Rn) is strictly composed of the sensible (H), latent (λE), and ground heat (G) fluxes, other stored terms such as those related to canopy heat storage and photosynthesis are negligible (Meyers and Baldocchi, 2005).

This method combines measurements of certain atmospheric variables (temperature and vapour concentration gradients) and available energy (net radiation and changes in stored thermal energy) to determine estimates of evapotranspiration (ET) (Lloyd, 1992). The method incorporates energy-budget principles and turbulent-transfer theory. Bowen showed that the ratio of the sensible- to latent-heat flux (β) could be calculated from the ratio of the vertical gradients of temperature and vapour concentration over a surface under certain conditions.

Often the gradients are approximated from air-temperature and vapour-pressure measurements taken at two heights above the canopy. The Bowen-ratio method assumes that there is no net horizontal advection of energy. With this assumption, the coefficients (eddy diffusivities) for heat and water vapour transport, kh and kw, respectively, are assumed to be equal. Under advective conditions, kh and kw are not equal (Verma et al., 1978; Lang et. al., 1983 cited bay Tomilson, 1997) and the Bowen-ratio method fails to accurately estimate ET.

Based on the assumption that Kh and Kw are equal, and by combining several terms to form the psychometric constants, the Bowen-ratio take the form to the equation 28. Although the theory for this method was develop in the 1920s by Bowen (Bowen, 1926), its practical applications has only been possible in recent decades, due to the availability of accurate instrumentation (Payero et al., 2003). The Bowen ratio initial concept is shown below:

$$\beta = \frac{PC_p Kh \frac{dT}{dz}}{\lambda \varepsilon Kw \frac{de}{dz}} \tag{27}$$

If it is assumed that there is no net horizontal advection of energy, equ. 27 can be simplified as shown below:

$$\beta = \frac{PC_p \frac{dT}{dz}}{\lambda \varepsilon \frac{de}{dz}} \tag{28}$$

where P is the atmospheric pressure (kPa), C_p is the specific heat of air (1.005 J/g°C), ε is the ratio molecular weight of water to air = 0.622 and λ is the latent heat of evaporation (Jg^{-1}). Once the Bowen ratio is determined, the energy balance (equ. 29) can be solved for the sensible-heat flux (H) and latent-heat flux (λE).

$$Rn = \lambda E + H + G \tag{29}$$

where R_n Rn is the net radiation, λE is the latent-heat flux, H is the sensible-heat flux and G is the soil-heat flux ($W.m^{-2}$).

$$H = \beta \lambda E \tag{30}$$

$$\lambda E = \frac{(R_n - G)}{(1+\beta)} \tag{31}$$

The latent heat flux can be separated into two parts: the evaporative flux E ($g\ m^{-1}$ day-1) and the latent heat of vaporization λ (Jg^{-1}), which can be expressed as a function of air temperature (T-°C) (λ = 2,502.3-2.308 T). The latent-heat of vaporation (λ) is defined as the amount of energy required to convert 1 gram of liquid water to vapour at constant temperature T. Sensible-heat flux (H) is a turbulent, temperature-gradient driven heat flux

resulting from differences in temperature between the soil and vegetative surface and the atmosphere.
The soil-heat flux (G) is defined as the amount of energy moving downward through the soil from the land surface, caused by temperature gradient. This flux is considered positive when moving down through the soil from the land surface and negative when moving upward through the soil toward the surface (Tomilson, 1994). The soil heat flux is obtained by measuring two soil heat flux plates below the soil surface at 2 and 8 cm, soil moisture at 8 cm, and soil temperature at 6 cm between the two soil heat flux plates (Campbell Scientific, Inc. 1998).
Because the soil-heat flux is measured below the soil surface, some of the energy crossing the soil surface could be stored in, or come from, the layer of soil between the surface and flux plate located closest to the surface, for this reason a change in storage term, S is added to the measured heat flux (equ. 33). (Cambell Scientific, Inc. 1991):

$$S = \left[\frac{\Delta Ts}{\Delta t}\right] d\rho_b \left(C_s + \left(WC_w\right)\right) \tag{32}$$

where S is the heat flux going into storage (Wm^{-2}), Δt is the time interval between measurement (sec), ΔTs is the soil temperature interval between measurement, d is the depth to the soil-heat-flux plates (0.08m), ρ_b is the bulk density of dry soil, C_s is the specific heat of dry soil (840 J/Kg°C), W is the water content of soil (kg the water/ Kg the soil) and C_w is the specific heat of water (4,190 J/Kg°C). The soil heat flux (G) at the surface is obtained by including the effect of storage between the surface and depth, *d*, using equation 11.

$$G = \left(\frac{FX1 + FX2}{2}\right) + S \tag{33}$$

where FX_1 is the soil-heat flux measured 1 (Wm^{-2}), FX_2 is the soil-heat flux measured 2 (Wm^{-2}).
One of the requirements for using the Bowen-ratio method is that the wind must pass over a sufficient distance of similar vegetation and terrain before it reaches the sensors. This distance is referred to as the fetch, and the fetch requirement is generally considered to be 100 times the height of the sensors above the surface (Campbell, 1977). More detail about determination of the minimum fetch requirement is presented later in this chapter.
Hanks et al. (1968), described by Frank (2003), reported $\lambda E/Rn$ of 0.16 for dry soil conditions and 0.97 for wet soil conditions; On the other hand he found $\lambda E/Rn$ to be lowest in grazed prairie, suggesting that defoliation changes the canopy structure and energy budget components, which may have contributed to increase water loss through evaporation compared with the non-grazed prairie treatment. Hanson and May (2004), using the Bowen Ratio Energy Balance Method to measure ET in tomatoes, found that ET rates decreased substantially in response to drying of the soil surface.
Perez et al. (2008) proposed a simple model for estimating the Bowen ratio (β) based on the climatic resistance factors:

$$\beta = \frac{\Delta + \gamma}{\Delta} \cdot \frac{1 + S}{1 + C} - 1 \tag{34}$$

$$C = \frac{\gamma . r_i}{\Delta . r_a} \tag{35}$$

$$S = \frac{\gamma}{\Delta + \gamma} . \frac{r_c}{r_a} \tag{36}$$

where r_c is the canopy resistance (s m^{-1}) based on the "big leaf" concept, and r_a is the aerodynamic resistance (s m^{-1}). These resistance factors are described in detail in the next section. The factor r_i is the climatological resistance as reported by Montheith (1965):

$$r_i = \frac{\rho_a C_p VPD}{\gamma (R_n - G)} \tag{37}$$

where ρ_a is the air density at constant pressure (Kg.m^{-3}), C_p is the specific heat of moist air at constant pressure (1004 J.Kg^{-1} $^0C^{-1}$), VPD is the vapour pressure deficit of the air (Pa), γ is the pychrometric constant (Pa. $^0C^{-1}$) and R_n and G are in W.m^{-2}. For homogeneous canopies, the effective crop surface and source of water vapour and heat is located at height d + z_{oh}, where d is the zero plane displacement height and z_{oh} is the roughness length governing the transfer of heat and vapour (Allen *et al.* 1998).

4.3 The Penman-Monteith Method

The important contribution of Monteith and Penman`s original equation was the use of resistances factors, which was based on an electrical analogy for the potential difference needed to drive unit flux systems that involve the transport of momentum, heat, and water vapour (Monteth and Unsworth, 1990; Monteith, 1997). The resistances have dimensions of time per unit length, as will be described later. This methodology calculates the latent heat flux using the vapour pressure deficit, the slope of the saturated vapour-pressure curve and aerodynamic resistance to heat, and canopy resistance in addition to the energy-budget components of the net radiation, soil heat flux, and sensible heat flux. Field measurements of air temperature, relative humidity, and wind speed are needed to determine these variables (Tomilson, 1997).

Equation 38 describes the Penman-Monteith (P-M) method to estimate the λE (Allen at al., 1998, Kjelgaard and Stockle, 2001)

$$\lambda E = \frac{\Delta s (R_n - G) + \rho_a C_p \frac{VPD}{r_a}}{\Delta s + \gamma \left(1 + \frac{r_s}{r_a} \right)} \tag{38}$$

where λE, R_n, and G in Wm^{-2}, VPD is vapour pressure deficit (kPa), ΔSis slope of saturation vapor pressure curve (kPa $°C^{-1}$) at air temperature, ρ is density of air (Kgm^{-3}), C_p in J $Kg^{-1}°C^{-1}$, γ in kPa $°C^{-1}$, r_a is aerodynamic resistant (s m^{-1}) r_s surface resistance to vapour transport (s m^{-1}).

According to Monteith (1997), the appearance of a wind-dependent function in the denominator as well as in the numerator implies that the rate of evaporation calculated from the P-M model is always less dependent on wind speed than the rate from the corresponding Penman equation when other elements of climate are unchanged. In general, estimated rates are usually insensitive to the magnitude of r_a and the error generated by neglecting the influence of the buoyancy correction is often small. In contrast, the evapotranspiration rate is usually a strong function of the surfaces resistance (r_s).

Kjelgaard and Stockle (2001) say the surface resistance (r_s) parameter in the P-M model is particularly difficult to estimate due to the combined influence of plant, soil and climatic factors that affect its value. The magnitude of the stomatal resistance can be estimated in principle from the number of stomata per unit leaf area and from the diameter and length of pores, which is difficult and therefore rarely measured; therefore, the stomatal resistance is usually calculated from transpiration rates or estimated gradients of vapour concentration (Monteith, 1997).

Knowing the value of the a***erodynamic resistance (r_a)*** permits estimation of the transfer of heat and water vapour from the evaporating surface into the air above the canopy. The aerodynamic resistance for a single leaf to diffusion through the boundary layer surrounding the leaf, within which the transfer of heat, water vapour, etc., occurs, proceeds at a rate governed by molecular diffusion. Provided the wind speed is great enough and the temperature difference between the leaf and air is small enough to ensure that transfer processes are not affecting by gradients of air density, the boundary layer resistance depends on air velocity and on the size, shape, and altitude of the leaf with respect to the air stream. In very light wind, the rates of transfer are determined mainly by gradients of temperature and therefore by density, so that the r_a depends more on the mean leaf-air temperature difference than on wind speed. According to Thom (1975), the r_a for heat transfer can be determined by:

$$r_{ah} = \frac{\rho C\rho(T_s - T_a)}{H} \tag{39}$$

At the field level, r_a for homogeneous surfaces, such as bare soil or crop canopies, there is a large-scale analogous boundary layer resistance, which can be estimated or derived from measurements of wind speed and from a knowledge of the aerodynamic properties of the surface as is described later (Monteith, 1997). The r_a can be determined given values of roughness length (Z_o) and zero plane displacement height (d), that depend mainly on crop height, soil cover, leaf area and structure of the canopy (Massman, 1987; Perrier, 1982; Shaw and Pereira, 1982 cited by Alves et al. 1998),

$$r_a = \frac{Ln\left[\frac{(Z_m - d)}{Z_{om}}\right] Ln\left[\frac{(Z_h - d)}{Z_{oh}}\right]}{K^2 u_z} \tag{40}$$

where Z_m is height of wind measurements (m), Z_h is height of humidity measurements (m), d is zero displacement height (m), Z_{om} is roughness length governing momentum transfer of heat and vapour (m) is 0.123h, Z_{oh} is roughness length governing transfer of heat and vapour (m) is $0.1Z_{om}$, K is the von Karman`s constant (0.41), u_z is win speed at height z.

This equation is restricted for neutral stability conditions, i.e., where temperature, atmospheric pressure, and wind speed velocity distribution follow nearly adiabatic conditions (no heat exchange). The application of the equation for short time periods (hourly or less) may require the inclusion of corrections for stability. However, when predicting ET_o

in the well watered reference surface, heat exchange is small, and therefore the stability correction is normally not required (Allen et al., 1998).
Alves et al. (1998) state that though this is the most used expression for r_a, in fact it is not entirely correct, since it assume a logarithmic profile from the source height ($d + Z_{oh}$) with increasing z in the atmosphere, using the concept to the "big leaf", equ. 40 can be modified as follows:

$$r_a = \frac{Ln\left(\frac{z-d}{h_c - d}\right) Ln\left(\frac{z-d}{Z_{om}}\right)}{K^2 u_z} \tag{41}$$

where h_c is the height of the crop canopy.
According to Tollk et al. (1995), the r_a to momentum transport in the absence of buoyancy effects (neutral stability) follows the equation:

$$r_{am} = \ln[(Zi - d)/Z_{om}]^2 / k^2 u_z \tag{42}$$

Under adiabatic conditions, the equations must be corrected using the Richardson number for stability correction, assuming similarity in transport of heat and momentum, yielding:

$$r_{ah} = r_{am}(1 + 5R_i) \tag{43}$$

The Ri for stability conditions is considered when ($-0.008 \le R_i \le 0.008$) and is calculated by

$$R_i = [g(T_a - T_s)(Z - d)] / T_{av}.u^2{}_z \tag{44}$$

where g is the acceleration of the gravity (9.8 m.s^{-2}), T_a is the air temperature (K), T_c is the plant canopy temperature (K), T_{av} is the average temperature taken as $((T_a+T_c)/2)$. The advantage of the R_i over other stability corrections is that it contains only experimentally determined gradients of temperature and wind speed and does not depend directly on sensible heat flux (Tolk et al., 1995).
The ***bulk surface resistance (r_s)*** describes the resistance of vapour flow through transpiring crop leaves and evaporation from the soil surface. Where the vegetation does not completely cover the soil, the resistance factor should indeed include the effects of the evaporation from the soil surface. If the crop is not transpiring at a potential rate, the resistance depends also on the water status of the vegetation (Van Bavel, 1967; Allen et al., 1998), and for this case they proposed the use of the following approximate:

$$r_s = \frac{r_L}{LAI_{active}} \tag{45}$$

where LAI_{ctive} is 0.5 times the leaf area index (m^2 of leaf per m^2 of soil), and r_L is bulk stomatal resistance, which is the average resistance of an individual leaf, and can be measured using an instrument called a porometry, the first stomatal readings were developed by Francis Darwin who develop his horn hygrometer (Turner, 1991).
The r_L readings are highly variable and depend on several factors, such as: crop type and development stage, the weather and soil moisture variability, the atmospheric pollutants and the plant phytohormone balance (Turner, 1991). Typically to determine minimum r_L

using a porometer, fully expanded, sunlit leaves near to the top of the canopy are surveyed during maximum solar irradiance (approximately solar noon under cloudless conditions) and low VPD periods (Kjelgaard and Stockle, 2001). This "standard" value from literature or porometer measurements are hereafter identified as r_{Lmin}. In addition, r_L has been shown to increase with increasing VPD and/or reduced solar irradiance (R_s). Adjustment factors for VPD (f_{VPD}) and R_s (f_{Rs}) were empirically derived and used as multipliers of r_{Lmin}.
The dependence of r_L on VPD can be represented by a linear function (Jarvis, 1976) as

$$f_{VPD} = a + bVPD \tag{46}$$

where a and b are linear regression coefficients, and f_{VPD} is equal to 1 (no adjustment) for VPD $\leq$ a threshold value, which can be taken as 1.5 kPa. The same authors presented a calibrated form of equation 46 for corn as, $f_{VPD} = 0.45+0.39(VPD)$.
Kjelgaard and Stockle (2001) presented a modified form of the adjustment factor:

$$f_{Rs} = \frac{R_{s\max}}{C_2 + R_s} \tag{47}$$

where R_s and $R_{s\,max}$ are the actual and maximum daily solar irradiance (MJ m^{-2} day^{-1}) and C_2 is a fitted constant.
Taking the maximum of the adjustment factors for VPD and Rs, r_{Lmin} is modified to give the r_L (Kjelgaard and Stockle, 2001):

$$r_L = r_{L\min} \max\{f_{VPD}, f_{Rs}\} \tag{48}$$

where f_{VPD} and f_{Rs}, are equal to or larger than 1.
Alves et al. (1998) indicate that the surface resistance term (r_s) has been the most discussed in the literature. Several components to be considered here include: a) The resistances to water vapour at the evaporating surfaces: plants and their stomates (r_s^c) and soil (r_s^s); b) the resistance to vapour transfer inside the canopy from these evaporating surfaces up to the "big leaf" (r_s^a). The resistance r_s^c, can be approximated using equ. 50.

$$r_s^c = \frac{\left(\sum_{i=1}^{n} \frac{1}{r_{stj}}\right)^{-1}}{LAI} \tag{49}$$

Where r_{st} is the single leaf stomatal resistance (sm^{-1}), n is a leaf number.
The bulk surface resistance can also be calculated using the inversion of the Penman-Monteith equation with incorporation of the Bowen ratio as follow (Alves et al. (1998) and Alves and Pereira, 2000):

$$r_s = r_a\left(\frac{\Delta s}{\gamma}\beta - 1\right) + \frac{\rho_a C_p VPD}{\gamma \lambda E} \tag{50}$$

Accurate prediction of r_s requires a good estimate of the Bowen ratio (β). Ramirez (2007) has used the following inversion form of the Penman-Monteith equation to obtain estimates of r_s:

$$r_s = r_a.\frac{\left[\frac{\Delta(R_n - G) + \rho_a C_p\left(\frac{VPD}{r_a}\right)}{\lambda E} - \Delta - \gamma\right]}{\gamma} \tag{51}$$

Similarly these authors, analysing the resistance concepts, concluded that the r_s of dense crops cannot be obtained by simply averaging stomatal resistance because the driving force (vapour pressure deficit) is not constant within the canopy.

Saugier (1977) addressed canopy resistance (r_c), stating that it is normally a mixture of soil and plant resistances to evaporation. If the top the soil is very dry, direct soil evaporation may be neglected and r_c is approximately equal to the leaf resistance (r_L) divided by the LAI. Baldocchi *et al.* (1991), indicated that the inverse of the 'big-leaf' model (eg., inverse of the P-M model) will be a good estimate of canopy resistance or surface resistances if certain conditions are met. These conditions include: i) a steady-state environment; ii) a dry, fully developed, horizontally homogeneous canopy situated on level terrain; iii) identical source-sink levels for water vapour, sensible heat and momentum transfer, and negligible cuticular transpiration and soil evaporation.

Szeicz and Long (1969) describe a profile method to estimated r_S as,

$$r_s = \frac{\rho_a.C_p, VPD}{\gamma.\lambda E} \tag{52}$$

These methods can be used in the field when the rate of evapotranspiration is measured by lysimeters or calculated from the Bowen ratio energy balance method, and the temperature, humidity and wind profiles are measured within the boundary layer simultaneously.

Ortega-Farias et al. (2004), evaluated a methodology for calculating the canopy surface resistance ($r_{cv} \approx r_s$) in soybean and tomatoes, using only meteorological variables and soil moisture readings. The advantage of this method is that it can be used to estimate λE by the general Penman-Monteith model with meteorological reading at one level, and without r_L and LAI measurements.

$$r_s = \frac{\rho_a.c_p.VPD}{\Delta.(R_n - G))}.\frac{\theta_{FC} - \theta_{WP}}{\theta_i - \theta_{WP}} \tag{53}$$

Where θ_{FC} and θ_{WP} are the volumetric moisture content at field capacity (fraction) and wilting point (fraction), respectively, and θ_i is a volumetric soil content in the root zone (fraction) measured each day.

Kamal and Hatfield (2004), used the equation 50 to determine the surface resistance in Potato: and stated that the canopy resistance (r_c in $s.m^{-1}$; "mean stomatal resistances of crops"), can be determined by dividing the r_s by the effective LAI as defined by other authors such as Hatfield and Allen (1996) and for well watered crops, r_c can be can be estimated using equation 54.

$$r_c = \frac{0.3LAI + 1.2}{LAI} r_s \tag{54}$$

Kjelgaard and Stockle (2001) discussed the estimation of canopy resistance (r_c) from single-leaf resistance (r_L) (equation 55), as originally proposed by Szeicz and Long (1969):

$$r_c = \frac{r_L}{LAI_{active}} \tag{55}$$

Kamal and Hatfield (2004) divided the surface resistance (r_s) used in the P-M model into two components, and conceptualized an excess resistance (r_o) in series with the canopy stomatal resistance. This excess resistance was linked to the structure of the crop, particularly crop height.

$$r_s = r_c + r_o \tag{56}$$

Pereira et al. (1999) stated that the surface resistance (r_s) is the sum of two components: one corresponding mainly to the stomatal resistance (r_{st}), the other to the leaf boundary layer and turbulent transfer inside the canopy (r_{ai}) (equation 57), thus, surface resistance is not a purely physiological parameter:

$$r_s = r_{st} + r_{ai} \tag{57}$$

Stomatal resistance can take values from 80 $s.m^{-1}$ to 90 $s.m^{-1}$ as a common range for agricultural crops suggested a value of 100 $s.m^{-1}$ for most arable crops (Monteith, 1981). The **table 1** lists mean average values for various crops under well water conditions.
The r_L is strongly dependent on the time of day (basically due to the temporal nature of climatic conditions), for the soil moisture content and by the genotype. **Fig. 3A**, shows how larger differences in r_L occur, with and without drought stress, after 9:00 am until late in the afternoon, and the most critical point is at 13:00 hours when the highest VPD occurred. For this reason, when this variable (r_L) is not measured, appropriate parameterisation is required for good water flux or ET estimation, especially under drought stress conditions. In **Fig. 3C**, it is possible to see in a common bean genotype under drought stress conditions, lower r_L as compared with less drought resistance during several days with drought stress.
Perrier (1975), as reported in Kjelgaard and Stockle (2001), conceptualized the excess resistance (r_o) as a linear function of crop height and LAI:

$$r_o = ah_c + bLAI \tag{58}$$

where a and b are constants. For corn, Kjelgaard and Stockle (2001) parameterized equ. 58 as follows: $r_o = 16.64h_c + 0.92LAI$.
Canopy resistance can also be determined from leaf or canopy temperature since it is affected by plant characteristics, eg. Leaf area index (LAI), height, and maturity. Soil factors (Available soil water-ASW, and soil solution salinity) and weather factors (R_n and wind speed) also affect the canopy resistance.
Montheith (1965) showed that transpiration rate physically depends on relative changes of surface temperature and r_a, and concluded that r_a depends on the Reynolds number of the air and can be determined from wind speed, the characteristic length of the plant surface, and the kinematic viscosity of the air. An increase in r_c for Wheat was caused by a decrease in total leaf area, by an increase in the resistance of individual leaves due to senescence, or by a combination of both effects; in Sudan grass, r_c increased with plant age and a decrease in soil moisture. Van Bavel (1967) studied Alfalfa throughout an irrigation cycle and found

that canopy resistance increased linearly with decreasing soil water potential. Kamal and Hatfield (2004) found an exponentially inverse relationship between canopy resistance and net radiation, and a linear inverse relationship between canopy resistance and available soil water.

Cover crops	**r_L**	**Source**	**Cover crops**	**r_L**	**Source**
	s/m			s/m	
Corn	200	Kirkham et al. (1985)	Cassava	714 Between 476 to 1428	Oguntunde (2005). This data under limited soil water conditions.
Sunflower	400	Kirkham et al. (1985)	Eucalyptus	200-400	Pereira and Alves (2005)
Soybean and potato	350	Kirkham et al. (1985)	Maple	400-700	Pereira and Alves (2005)
Sorgum	300	Kirkham et al. (1985)	Crops-General	50-320	Pereira and Alves (2005)
Millet	300	Kirkham et al. (1985)	Graim sorghum	200	Pereira and Alves (2005)
Aspen	400	Pereira and Alves (2005)	Soybean	120	Pereira and Alves (2005)
Maize	160	Pereira and Alves (2005)	Barley	150-250	Pereira and Alves (2005)
Alfalfa	80	Pereira and Alves (2005)	Sugar beet	100	Pereira and Alves (2005)
Clipped grass (0.15 m)	100-150	Pereira and Alves (2005)	Clipped and Irrigated grass (0.10-10.12m)	75	Pereira and Alves (2005)
Common beans	170-270	Ramirez et al. (2007)	Sorghum	192	Stainer et al. (1991)
Corn	264	Ramirez and Harmsen (2007). Unpublished data.	Andes Tropical Forestry	132	Ramirez and Jaramillo (2008). (Calculated)
Coffee	149	Ramirez and Jaramillo (2008). (Calculated)	Coffee	150	Angelocci et al. (1983)
Wheat	134	Howell et al. (1994)	Corn	252	Howell et al. (1994)
			Sorghum	280	Howell et al. (1994)

Table 1. Average values of the stomatal resistance (r_L) for several crops.

The Drainage and Irrigation Paper-FAO 56 (Allen et al., 1998) recommends the Szeicz and Long (1969) method for calculating r_s (equation 55), in which an average of r_L for different positions within the crop canopy, weighted by LAI or $LAI_{effective}$ is used. This method seems to give good results only in very rough surfaces, like forest and partial cover crops with a dry soil (Monteith, 1981). Alves et al. (1998) concluded that r_s of dense crops cannot be obtained by simply averaging stomatal resistance (r_L) because VPD, which is the "driving force", is not constant within the canopy. Alves and Pereira (2000) have stated *"The PM model can be used to predict ET if accurate methodologies are available for determining the r_s that take into account the energy partitioning"*.

In addition to the lack of r_s values for crops, questions have been raised relative to the appropriateness of using the PM model for partial or sparse canopies because the source/sink fluxes may be distributed in a non-uniform manner throughout the field (Ham and Heilman, 1991; Kjelgaard et al., 1994; Farahami and Bausch, 1995; Ortega-Farias et al.; 2006). Adequate parameterization of the surface resistance makes the P-M model a good estimator of ET (i.e., Saugier and Katerji, 1991; Rana et al., 1997a; Alves and Pereira, 2000; Ortega-Farias et al., 2004).

Ramirez (2007), reported that the daily ET estimation with the P-M model with r_s based on r_L and $LAI_{effective}$ gave a good estimation in two common bean genotypes with variable LAI, without and with moderate drought stress for both years (2006 and 2007).

Ramirez et al. (2008) an reported inverse relation between r_a and r_s and r_L in beans (*Phaseolus vulgaris* L), as well as those reported by Alves and Pereira (2000) (**Fig. 4**), which implies that with low r_a (windy conditions), the r_L (and therefore r_s) increases. The Alves and Pereira (2000) study did not measure the r_L, rather the r_s was estimated based on micrometeorological parameters.

Disparities in the measured r_s using the P-M inverse model arise from: a) imperfect sampling of leaves and the arbitrary method of averaging leaf resistance over the whole canopy, b) from the dependence of r_s on non-stomatal factors such as evaporation from wet soil or stems, or others and c) the complex aerodynamic behaviour of canopies (Monteith, 1995).

Lower LAI index (LAI <1.0) and drought stress also affect the precision in the r_s estimation (eg., Ramirez, 2007). Use of the $LAI_{effective}$ when LAI < 1.0 is not necessary and tends to overestimate the r_s and under-estimate the ET. Katerji and Perrier (1985) found for LAI >1.0 a good agreement between measurement values of evapotranspiration over alfalfa crops using the energy balance method, and values calculated with P-M equation using variable r_s.

Katerji and Perrier (1983) proposed to simulate r_s using the following relation:

$$\frac{r_s}{r_a} = a\frac{r^*}{r_a} + b \tag{59}$$

where **a** and **b** are linear coefficients that need empirical determination, r^*(s.m^{-1}) is a climatic resistance (Katerji and Rana, 2006) giving by:

$$r^* = \frac{\Delta + \gamma}{\Delta\lambda} \cdot \frac{\rho C_p VPD}{(R_n - G)} \tag{60}$$

Table 2 presents values of a and b for several crops.

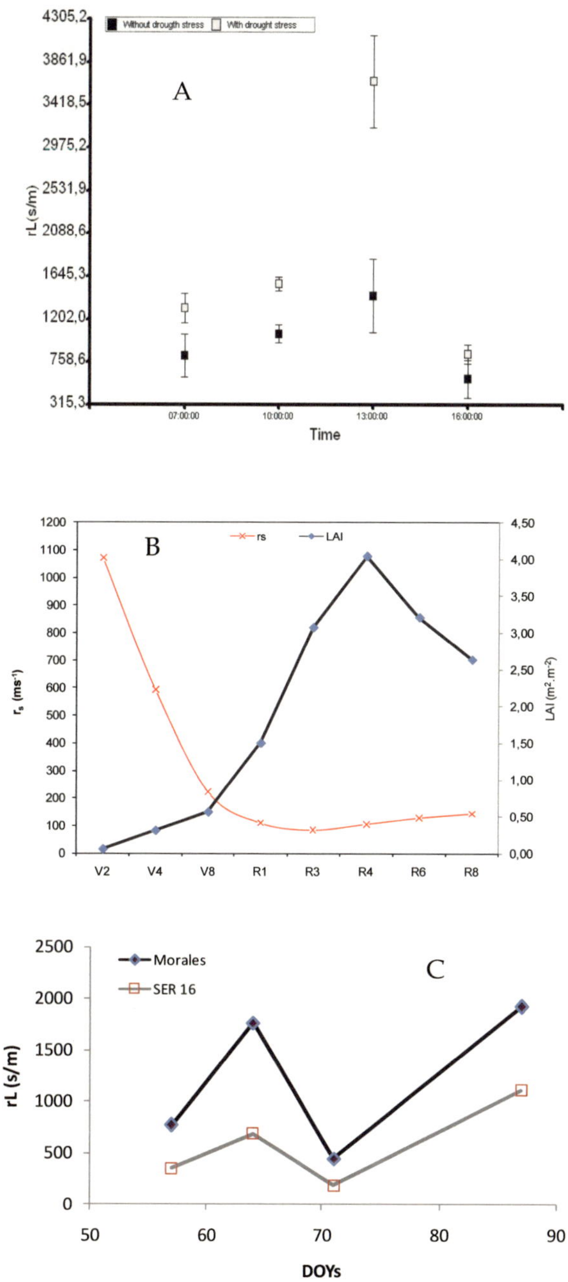

Fig. 3. Relationship between **A.** Changes in the stomatal resistance during the day with and without drought stress in *Phaseolus vulgaris* L. genotype 'Morales'. B. Surfaces resistance and Leaf area index, and **C.** Stomatal behaviour represented in stomatal resistance (r_L) under drought stress conditions for two common bean genotypes -'Morales' lest drought tolerant and 'SER 16' drought stress tolerant.

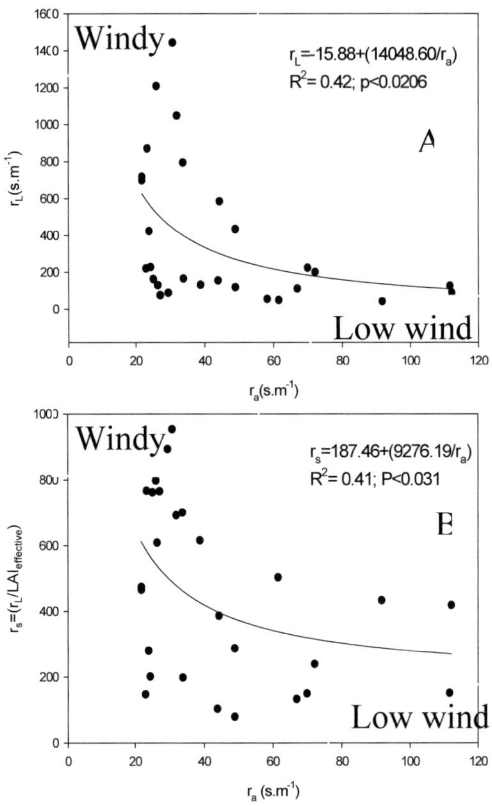

Fig. 4. Aerodynamic resistance (r_a) as a function of: **A**. Stomatal resistance (r_L) and **B.** Measured surface resistance: $r_s = r_L/LAI_{effective}$, **(Datas from: Ramirez, 2007)**.

Crop	a	b	Source
Grass	0.16	0.0	Katerji and Rana (2006)
Tomato	0.54	2.4	Katerji and Rana (2006)
Grain sorghum	0.54	0.61	Katerji and Rana (2006)
Soybean	0.95	1.55	Katerji and Rana (2006)
Sunflower	0.45	0.2	Katerji and Rana (2006)
Sweet sorghum	0.845	1.0	Katerji and Rana (2006)
Grass (Tropical climate)	0.18	0.0	Gosse (1976) in Rana et al. (1997)a
Grass (Mediterranean climate)	0.16	0.0	Rana et al. (1994)
Alfalfa	0.24	0.43	Katerji and Perrier (1983) in Rana et al (1997)a
Sorghum	0.94	1.1	Rana et al. (1997)b
Sunflower	0.53	1.2	Rana et a.l (1997)b

Table 2. Coefficients a and b for several crops

The Penman-Monteith model is considered as a 'single-layer' model, Shuttlewoth and Wallace (1985) developed a 'double-layer' model, relying on the Penman-Monteith model concept to describe the latent heat flux from the canopy (λT) and from the soil (λE) as follows:

$$\lambda T = \frac{\Delta(R_n - R_{ns}) + \rho Cp \dfrac{VPDo}{r^c{}_a}}{\Delta + \gamma\left(1 + \dfrac{r^c{}_s}{r^c{}_a}\right)} \tag{61}$$

$$\lambda E = \frac{\Delta(R_{ns} - G) + \rho Cp \dfrac{VPDo}{r^s{}_a}}{\Delta + \gamma\left(1 + \dfrac{r^s{}_s}{r^s{}_a}\right)} \tag{62}$$

where R_{ns} is the absorbed net radiation at the soil surface, $r^c{}_a$ is the bulk boundary layer resistance of the canopy elements within the canopy, $r^c{}_s$ is the bulk stomatal resistance of the canopy, $r^s{}_a$ is the aerodynamic resistance between the soil and the mean canopy height, $r^s{}_s$ is the surfaces resistance of the soil and VPD_o is the vapour pressure deficit at the height of the canopy air stream.

4.4 The double-layer Shulttleworth-Wallace model

The Shulttleworth-Wallace Model (S-W) assumes that there is blending of heat fluxes from the leaves and the soil in the mean canopy airflow at the height of the effective canopy source (Shulttleworth and Wallace,1985). The full expression of the Shulttleworth-Wallace Model (S-W) model is presented by Zhang *et al.* (2008) as follow:

$$\lambda ET = \lambda E + \lambda T = C^S_{SW} PM^S_{SW} + C^P_{SW} PM^P_{SW} \tag{63}$$

$$PM^S_{SW} = \frac{\Delta A_{SW} + \left[\left(\rho CpD - \Delta r^s_a\right)\left(A_{SW} - A^s_{SW}\right)/\left(r^a_a + r^s_a\right)\right]}{\Delta + \gamma\left[1 + r^s_s/\left(r^a_a + r^s_a\right)\right]} \tag{64}$$

$$PM^P_{SW} = \frac{\Delta A_{SW} + \left[\left(\rho CpD - \Delta r^P_a A^s_{SW}\right)/\left(r^a_a + r^P_a\right)\right]}{\Delta + \gamma\left[1 + r^P_s/\left(r^a_a + r^P_a\right)\right]} \tag{65}$$

$$C^s_{SW} = \frac{1}{1 + \left[R^S_{SW} R^a_{SW} / R^P_{SW}\left(R^S_{SW} + R^a_{SW}\right)\right]} \tag{66}$$

$$C^P_{SW} = \frac{1}{1 + \left[R^P_{SW} R^a_{SW} / R^S_{SW}\left(R^P_{SW} + R^a_{SW}\right)\right]} \tag{67}$$

$$R^S_{SW} = (\Delta + \gamma) r^s_a + \gamma r^s_s \tag{68}$$

$$R^P_{SW} = (\Delta + \gamma) r^P_a + \gamma r^P_s \tag{69}$$

$$Ra_{SW}^{P} = (\Delta + \gamma) r_a^a \tag{70}$$

where λE is the latent heat flux of evaporation from the soil surfaces (W/m2), λT the latent heat fluxes of transpiration from canopy (W/m2), r_s^p the canopy resistance (s/m), r_a^p the aerodynamic resistance of the canopy to in-canopy flow (s/m), r_s^s the soil surfaces resistance (s/m), r_a^a and r_a^s the aerodynamic resistance from the reference height to in-canopy heat exchange plane height and from there to the soil surface (s/m), respectively, A_{sw} A_{SW}^s are the total available energy and the available energy to the soil (W/m2), respectively and defined as follow:

$$A_{sw} = R_n - G \tag{71}$$

$$A_{SW}^s = R_{nsw}^{\ s} - G \tag{72}$$

where $R_{nsw}^{\ s}$ is the net radiation fluxes into the soil surface (W/m2), and can be calculated using the Beer's law as follow:

$$R_{nsw}^{\ s} = R_n . \exp^{(-c.LAI)} \tag{73}$$

where c is the extinction coefficient of light attenuation (e.g.; Sene, 1994 indicate c=0.68 for fully grown plant, c=0 for bare soil; Zhang *et al.*, 2008 use 0.24 for vineyard crops).
The ***surfaces resistance*** is calculated as follow:

$$r_s^p = \frac{r_{st\,min}}{LAI_{effective} \Pi_i F_i(X_i)} \tag{74}$$

where $r_{st\ min}$ is the minimal stomatal resistance of individual leaves under optimal conditions. $LAI_{effective}$ is: equal to LAI for LAI ≤2.0; LAI/2 for LAI ≥4.0 and 2 for intermediate values of LAI, X_i is a specific environmental variable, and $F_i(X_i)$ is the stress function with $0.0 \leq F_i(X_i) \geq 1.0$ (from: Jarvis, 1976).

$$F_1(S) = \left(\frac{S}{1100}\right)\left(\frac{1100 + a_1}{S + a_1}\right) \tag{75}$$

$$F_2(T) = \frac{(T - T_L)(T_H - T)^{(TH - a_2)/(a_2 - T_L)}}{(a_2 - T_L)(T_H - a_2)^{(TH - a_2)/(a_2 - T_L)}} \tag{76}$$

$$F_2(D) = e^{-a_3 D} \tag{77}$$

$$F_4(\theta) = \begin{bmatrix} 1 - -if - - - - - \theta \geq \theta_F \\ \dfrac{\theta - \theta_W}{\theta_F - \theta_W} - if - -\theta_F < \theta < \theta_W \\ 0 - -if - - - - - \theta \leq \theta_W \end{bmatrix} \tag{78}$$

where S is the incoming photosynthetically active radiation flux (W/m^2), T is the air temperature (°K), θ_F is the soil moisture at field capacity (cm^3/cm^3), θ_w is the soil moisture at wilting point (cm^3/cm^3), and θ is the actual soil moisture in the root zone. (cm^3/cm^3). T_H and T_L are upper and lower temperatures limits outside of which transpiration is assumed to cease (°C) and are set at values of 40 and 0°C (e.g.; Harris et al., 2004; Zhang et al., 2008). The a1, a2, and a3 are derived by multi-variate optimization, and are 57.67,25.78 and 9.65, respectively (Zhang et al., 2008).

The ***aerodynamic resistances*** r_a^a and r_a^s are calculated from the vertical wind profile in the field and the eddy diffusion coefficient. Above the canopy height, the eddy diffusion coefficient (K) is given by:

$$K = ku^*(z-d) \tag{79}$$

where u* is the wind friction velocity (m/s), k is the van-Karman constant (0.41), z is the reference height (m), and d the zero plane displacement (m). The exponential decrease of the eddy diffusion coefficient (K) through the canopy is given as follow:

$$K = k_h.\exp\left[-n\left(1-\frac{z}{n}\right)\right] \tag{80}$$

where k_h is the eddy diffusion coefficient at the top of the canopy (m^2/s), and n is the extinction coefficient of the eddy diffusion. Brutsaert (1982) cited by Zhang et al., (2008) indicate that n =2.5 when h_c <1 m; n = 4.25 when h_c>10 m, linear interpolation could be used for crops with h between those values. k_h is determined as follow.

$$k_h = ku^*(h_c-d) \tag{81}$$

The aerodynamic resistance r_a^a and r_a^s are obtained by integrating the eddy diffusion coefficients from the soil surface to the level of the "preferred" sink of momentum in the canopy, and from there to the reference height (Shutlleworth and Gurney, 1990, mentioned by Zhang et al., 2008) as follow:

$$r_a^{\ a} = \frac{1}{Ku^*}\ln\left(\frac{z-d}{h_c-d}\right)+\frac{h_c}{nk_h}\left[\exp\left[n\left(1-\frac{zo+d}{h_c}\right)\right]-1\right] \tag{82}$$

$$r_a^{\ s} = \frac{h_c\exp^{(n)}}{nk_h}\left[\exp\left(\frac{-nz_o'}{h_c}\right)-\exp\left[-n\left(\frac{z_0+d}{h_c}\right)\right]\right] \tag{83}$$

The bulk boundary layer resistance of canopy is calculated as follow:

$$r_a^{\ p} = \frac{r_b}{2LAI} \tag{84}$$

where r_b is the mean boundary layer resistance (s/m) (e.g.; Brisson et al., 1998, recommend use 50 s/m).

The **soil surface resistance** r_s^s is the resistance to water vapour movement from the interior to the surface of the soil, and is strongly depending of the water content (θ_s), and is calculated using the Anandristakis et al. (2000) expression:

$$r_s^s = r_{s\ \min}^s f(\theta_s) \tag{85}$$

where θ_s is soil volumetric water content (cm³/cm³), and $r_{s\ \min}^s$ is the minimum soil surfaces resistance, that correspond with the soil field capacity (θ_{FC}) and is assumed equal to 100 s/m (e.g.; Camilo and Gurney, 1986; Zhang et al., 2008).

The $f(\theta_s)$ is expressed according with Thompson (1981) and mentioned by Zhang et al. (2008) as follow:

$$f(\theta_s) = 2.5\left(\frac{\theta_{FC}}{\theta_s}\right) - 1.5 \tag{86}$$

4.5 Clumping model

The Clumping model is based in the Shulttleworth-Wallace model, this model separate the soil surfaces into fractional areas inside and outside the influence of the canopy, and include the fraction of canopy cover (*f*) in his calculation. Brenner and Incoll (1997) and Zhang et al. (2008) express the model as follow:

$$\lambda E = \lambda E^s + \lambda E^{bs} + \lambda T = f\left(C_c^s PM_c^s + C_c^p PM_c^p\right) + (1-f)C_c^{bs} PM_c^{bs} \tag{87}$$

where λE^s is the latent heat of evaporation from soil under the plant (W/m²); λE^{bs} is the latent heat of evaporation from bare soil (W/m²); *f* is the fractional vegetative cover and the other terms are expressed as follow:

$$PM_c^p = \frac{\Delta A_c + \left[\dfrac{\left(\rho CpD - \Delta r_a^p A_c^s\right)}{r_a^a + r_a^p}\right]}{\Delta + \gamma\left[1 + \dfrac{r_s^p}{r_a^a + r_a^p}\right]} \tag{88}$$

$$PM_c^s = \frac{\Delta A_c + \left[\dfrac{\left(\rho CpD - \Delta r_a^s A_c^p\right)}{r_a^a + r_a^s}\right]}{\Delta + \gamma\left[1 + \dfrac{r_s^s}{r_a^a + r_a^s}\right]} \tag{89}$$

$$PM_c^{bs} = \frac{\Delta A_c^{bs} + \left[\dfrac{(\rho CpD)}{r_a^a + r_a^{bs}}\right]}{\Delta + \gamma\left[1 + \dfrac{r_s^{bs}}{r_a^a + r_a^{bs}}\right]} \tag{90}$$

$$C_c^s = \frac{R_c^{bs} R_c^p \left(R_c^s + R_c^a\right)}{\left[R_c^s R_c^p R_c^{bs} + (1-f)R_c^s R_c^p R_c^a + fR_c^{bs} R_c^s R_c^a + fR_c^{bs} R_c^p R_c^a\right]} \tag{91}$$

$$C_c^p = \frac{R_c^{bs} R_c^s \left(R_c^p + R_c^a\right)}{\left[R_c^s R_c^p R_c^{bs} + (1-f)R_c^s R_c^p R_c^a + fR_c^{bs} R_c^s R_c^a + fR_c^{bs} R_c^p R_c^a\right]} \tag{92}$$

$$C_c^{bs} = \frac{R_c^s R_c^p \left(R_c^{bs} + R_c^a\right)}{\left[R_c^s R_c^p R_c^{bs} + (1-f)R_c^s R_c^p R_c^a + fR_c^{bs} R_c^s R_c^a + fR_c^{bs} R_c^p R_c^a\right]} \tag{93}$$

$$R_c^s = (\Delta + \gamma) r_a^s + \gamma r_s^s \tag{94}$$

$$R_c^p = (\Delta + \gamma) r_a^p + \gamma r_s^p \tag{95}$$

$$R_c^{bs} = (\Delta + \gamma) r_a^{bs} + \gamma r_s^{bs} \tag{96}$$

$$R_c^a = (\Delta + \gamma) r_a^a \tag{97}$$

Where A_c, A_c^p, A_c^s and A_c^{bs} are energy available to evapotranspiration, to the plant, to soil under shrub and bare soil (W/m2) respectively, r_a^{bs} the eddy diffusion resistance from in-canopy heat exchange plane height to the soil surface (s/m), r_s^{bs} the soil surfaces resistance of bare soil (s/m).

The ***Available energy*** for this model, the net radiation (R_n) is divided is divided into net radiation in the plant (R_n^p) and the net radiation in the soil (R_n^s). If the energy storage in the plant is assumed to be negligible, then:

$$R_{nc}^s = R_n \exp^{(-CLAI/f)} \tag{98}$$

$$R_{nc}^p = R_n - R_{nc}^s \tag{99}$$

$$A_c^s = R_{nc}^s - G^s \tag{100}$$

$$A_c^{bs} = R_n - G^{bs} \tag{101}$$

$$A_c^p = R_{nc}^p \tag{102}$$

Where R_{nc}^p and R_{nc}^s are the radiation absorbed by the plant and the radiation by the soil (W/m2) respectively, G^s and G^{bs} are the soil heat flux under plant and bare soil (W/m2) respectively, C is the extinction coefficient of light attenuation according for Sene (1994) is equal to 0.68 for fully grown plant.

The resistance for the bare soil surfaces r_s^{bs} can be calculated equally as in the S-W model, mentioned before. The aerodynamic resistance between the bare soil surface and the mean surfaces flow height (r_a^{bs}) can be calculated assuming that the bare soil surface is totally unaffected by adjacent vegetation so that is aerodynamic resistance equal to r_a^b and defined for:

$$r_a^b = \ln \frac{\left(\frac{Z_m}{Z'_o}\right)^2}{k^2 U_m} \tag{103}$$

Where Z_m is the mean surface flow height (m), and could be assumed equal to $0.75h_c$, and u_m is the wind speed at the Z_m (m/s).

According with Zhang et al. (2008), the aerodynamic resistance (r_a^{bs}) varies between r_a^b and r_a^s as *f* varies from 0 to 1, and the functional relationship of this change is not know.

4.6 Combination model

Theoretical approaches to surface evaporation from the energy balance equation combined with sensible heat and latent heat exchange expressions give the following definition for actual evapotranspiration (Pereira et al., 1999).

$$ET = \frac{\Delta}{\Delta+\gamma}\left[(R_n - G) + \frac{\rho Cp}{\Delta} Hu(VPDa - VPDs)\right] \tag{104}$$

Where R_n - G = available energy (MJ/m2) for the canopy, comprised of net radiation, R_n and the soil heat flux, G; $H(u)$ = exchange coefficient (m/s) between the surface level and a reference level above the canopy but taken inside the conservative boundary sublayer; *VPDs* and *VPDa* (kPa) = vapour pressure deficits (VPD) for the surface level and the reference level, respectively; ρ= atmospheric density (kg/m3); *Cp* = specific heat of moist air (J/kg7C); Δ = slope of the vapour pressure curve (Pa/°C); and γ= psychrometric constant (Pa/°C).
To obtain evapotranspiration with (104) the most difficult term to estimate is *VPDs*, representing the vapour pressure deficit at the evaporative surface. If *VPDs* can be associated with a surface resistance term (r_s), then ET can be calculated directly from the flux equation:

$$ET = \frac{\rho Cp}{\gamma}\frac{VPDs}{r_s} \quad (87) \tag{105}$$

and

$$r_a = \frac{1}{Hu} \tag{106}$$

r_a can be calculate using the equations discussed later. Two main solutions can be defined from (104) using climatic data:
1. The case of full water availability corresponding to saturation at the evaporative surface. Then *VPDs* = 0 and r_s becomes null. Eq. (104) then gives the maximum value for ET, the potential evaporation (EP), which depends only on climatic driving forces:

$$EP = \frac{\Delta(R_n - G) + \rho CpF(u)VPDa}{\Delta\lambda} \tag{107}$$

In where F(u) = 1/r_a. The combination the equations can get:

$$ET = \frac{EP}{(1 + \frac{\gamma}{\Delta+\gamma}\frac{r_s}{r_a})} \tag{108}$$

2. The case for equilibrium between the surface and the reference levels corresponds to V*PDs* = V*PDa*. In this case, the evapotranspiration is referred to as the equilibrium evaporation (*Ee*).

$$Ee = \frac{\rho Cp}{\gamma}\frac{VPDa}{r_e} \tag{109}$$

Where r_s was renamed r_e, termed the equilibrium surface resistance, indicating that the term, in this case, represents the surface resistance for equilibrium evaporation. The value for r_e depends predominately on climatic characteristics although these characteristics are influenced by R_n and G of the vegetative surface. For purposes here, the r_e term can be called the climatic resistance for the surface.

$$r_e = \frac{\rho Cp}{\gamma}\frac{\Delta+\gamma}{\Delta}\frac{VPDa}{R_n - G} \tag{110}$$

EP can be estimate:

$$EP = Ee\left(1 + \frac{\gamma}{\Delta+\gamma} + \frac{r_e}{r_a}\right) \tag{111}$$

and ET can be estimate using:

$$ET = \frac{EP_{(36)}}{\left(1 + \frac{\gamma}{\Delta+\gamma}\frac{r_s}{r_a}\right)} \tag{112}$$

4.7 Pristley and Taylor model

Pristley and Taylor (1972), propose to neglected the aerodynamic term and fix the radiation term by introducing a dimensionless coefficient (α).

$$ET = \alpha\frac{\Delta}{\Delta+\gamma}(R_n - G) \tag{113}$$

where ET is water flux under references conditions (well watered grass) in mm.day^{-1}; R_n and G are net radiation and soil heat flux respectably in mm.day^{-1}; Δ and γ in kPa.°C^{-1}. The term α is given as 1.26 for grass field in humid weather conditions, and was adopted by Pristley and Taylor (1972) for wet surfaces; however α is ranging between 0.7 to 1.6, over various landscapes (Flint and Childs, 1991).
According with Zhang et al. (2004), the term α can be calculated as follow:

$$\alpha = \frac{\lambda E(\Delta+\gamma)}{\Delta(R_n - G)} = \frac{\Delta+\gamma}{\Delta(1+\beta)} \tag{114}$$

Also de α term sensible at the soil moisture changes (Eg.; Grago and Butsaert, 1992; Grago 1996; and Zhang et al., 2004), that relation can be estimated using a models like:

$$\alpha = k\left[1 - \exp\left(-c\frac{\theta - d}{\theta_{FC}}\right)\right] \tag{115}$$

where k, c and d are parameters of the model, θ is the actual volumetric soil moisture content (cm^3.cm^{-3}) and θ_{FC} is the volumetric moisture content at field capacity (cm^3.cm^{-3}).

4.8 Eddy covariance method

The eddy covariance method is, in general, the most preferred because it provide a direct measure of the vertical turbulent flux across the mean horizontal streamlines, provided by fast sensors (~10 Hz) (Meyers and Baldocchi, 2005). Realizing the limitation of the Thornthwaite-Holzman type of approach, Swinbank (1951) cited by Chang (1968) was the first to attempt a direct measurement by the so-called eddy correlation technique. The method is based on the assumption that the vertical eddy flux can be determined by simultaneous measurements of the upward eddy velocity and the fluctuation in vapour pressure. Actually is a routinely technique for direct measurement of surfaces layer fluxes of momentum, heat, and traces gases (CO_2, H_2O, O_3) between the surfaces and the turbulent atmosphere (Massmam, 2000).

This system recognizes that the transport of heat, moisture, and momentum in the boundary layer is governed almost entirely by turbulence. The eddy correlation method is theoretically simple using an approach to measure the turbulent fluxes of vapour and heat above the canopy surface. The eddy correlation fluxes are calculated and recorded in a 30 min or less temporal resolution. Assuming the net lateral advection of vapour transfer is negligible, the latent heat flux (evapotranspiration) can be calculated from the covariance between the water vapour density (ρ_v) and the vertical wind speed (w).

$$\lambda E = \lambda \overline{w' \rho_v'} \qquad (116)$$

where λE is the latent heat flux (W m^{-2}), λ is the latent heat of vaporization (J kg^{-1}), ρ_v' is the fluctuation in the water vapour density (kg m^{-3}), and w' is the fluctuation in the vertical wind speed (m s^{-1}). The over bar represents the average of the period and primes indicate the deviation from the mean values during the averaging period. According with Weaver (1992) the eddy correlation method depends on the relations between the direction of air movement near the land surface and properties of the atmosphere, such as temperature and humidity.

The sensible heat flux can be calculated from the covariance of air temperature and the vertical wind speed.

$$H = \rho_a C_p \overline{w'T'} \qquad (117)$$

Where H the sensible heat flux (W m^{-2}), ρ_a the air density (kg m^{-3}), C_p the specific heat of moist air (J kg^{-1} $^oC^{-1}$) and T' the fluctuation in the air temperature (^oC).

The fine wire thermocouples (0.01 mm diameter) are not included in the eddy correlation system. The air temperature fluctuations, measured by the sonic anemometer, are corrected for air temperature fluctuations in estimation of sensible heat fluxes. The correction is for the effect of wind blowing normal to the sonic acoustic path. The simplified formula by Schotanus et al. (1983) is as follows:

$$\overline{w'T'} = \overline{w'T_s'} - 0.51\left(\overline{T + 273.15}\right)\overline{w'q'} \qquad (118)$$

Where $w'T'$ is rotated covariance of wind speed and sonic temperature (m ^oC s^{-1}), T is air temperature (^oC) and q is the specific humidity in grams of water vapour per grams of moist air.

Two Eddy covariance systems are used to measure the water vapour fluxes, the open path and close path. According with Anthoni et al. (2001) the Open-path eddy covariance

systems require corrections for density fluctuations in the sampled air (Massman and Lee, 2002; Webb et al., 1980) and in general closed-path system require incorporation of a time lag and corrections for the loss of high frequency information, due to the air being drawn through a long sampling tube (Massman and Lee, 2002; Moore, 1986).
The most common correction in the eddy covariance system is described by Wolf et al. (2008) as: i) Coordinate rotation, ii) Air density correction, and iii) Frequency -dependent signal loss.
Estimation of turbulent fluxes is highly dependent on the accuracy of the vertical wind speed measurements. Measurement of wind speed in three orthogonal directions with sonic anemometer requires a refined orientation with respect to the natural coordinate system through mathematic coordinate rotations (Sumner, 2001). The vector of wind has three components (*u, v, w*) in three coordinate directions (x, y, z). The z-direction is oriented with respect to gravity, and the other two are arbitrary. Baldocchi et al. (1988) provide procedures to transform the initial coordinate system to the natural coordinate system. Described in details by Sumner (2001), the coordinate system is rotated by an angle η about the z-axis to align *u* into the x-direction on the x-y plane, then rotated by an angle θ about the y-direction to align w along the z-direction. The results force $\overline{v}$ and $\overline{w}$ equal to zero, and $\overline{u}$ is pointed directly to the air stream. When θ was greater than 10 degrees, the turbulent flux data should be excluded based on the assumption that spurious turbulence was the cause of the excessive amount of the coordinate rotation.

$$\cos\theta = \sqrt{\frac{\left(u^2 + v^2\right)}{\left(u^2 + v^2 + w^2\right)}} \tag{119}$$

$$\sin\theta = \frac{w}{\sqrt{\left(u^2 + v^2 + w^2\right)}} \tag{120}$$

$$\cos\eta = \frac{u}{\sqrt{\left(u^2 + v^2\right)}} \tag{121}$$

$$\sin\eta = \frac{v}{\sqrt{\left(u^2 + v^2\right)}} \tag{122}$$

The latent heat and sensible heat fluxes are computed from the coordinate rotation-transformed covariance:

$$\left(\overline{w'\rho_v'}\right)_r = \overline{w'\rho_v'}\cos\theta - \overline{u'\rho_v'}\sin\theta\cos\eta - \overline{v'\rho_v'}\sin\theta\sin\eta \tag{123}$$

$$\left(\overline{w'T_s'}\right)_r = \overline{w'T_s'}\cos\theta - \overline{u'T_s'}\sin\theta\cos\eta - \overline{v'T_s'}\sin\theta\sin\eta \tag{124}$$

After the coordinate rotation, the final latent heat flux can be estimated from equation (116) plus the following correction of air density (C_{air}) (Webb et al., 1980) and correction of oxygen (CO_2) (Tanner and Greene, 1989).

$$C_{air} = \frac{\overline{\rho_v H}}{\rho C_p (T + 273.15)} \lambda \tag{125}$$

$$C_{O2} = \frac{FK_o \overline{H}}{K_w (T + 273.15)} \lambda \tag{126}$$

Where F is the factor used in krypton hygrometer correction that accounts for molecular weights of air and oxygen, and atmospheric abundance of oxygen, equal to 0.229 g $^oC\,J^{-1}$, K_o is the extinction coefficient of hygrometer for oxygen, estimated as 0.0045 $m^3\,g^{-1}\,cm^{-1}$, K_w is the extinction coefficient of hygrometer for water, from the manufacture is 0.149 $m^3\,g^{-1}\,cm^{-1}$.
With the measured four flux components from the energy balance equation, the energy balance should be closed, however, this is not practically the case. A tendency to underestimate energy and mass fluxes has been a pervasive problem with the eddy covariance technique (Ham and Heilman, 2003). Ham and Heilman (2003) reported clousure of 0.79 for priarine locations and 0.96 for forest. Ramirez and Harmsen (2007-Data without publication) indicate 0.71 for grass and 0.75 for corn.

- The errors in eddy covariance method could be associated with:

1. Accuracy of the R_n and G measurements (errors are often 5 to 10%).
2. The length scale of the eddies responsible for transport (if is larger, the frequency response and sensor separation error may have been smaller)
3. Sensor separation and inadequate sensor response (can generate 15% of Underestimation of λE by (Ham and Heilman, 2003) and 10% reported by Laubach and McNauhton (1999).
4. Ham and Heilman (2003) conclude *"The inherent tendency to underestimate fluxes when using eddy covariance may be linked to the errors caused by sensor separation and inadequate frequency response of the sensors. The correction proposed by Massman and Lee (2002) is difficult to implement for the non-specialist because they require calculation of cospectra using high-frequency (10Hz) data, and also is required expertise experience to interpret the cospectra properly"*

The "energy balance closure" is corrected using the Bowen ratio (Kosugi and Katsuyama, 2007) as follow:

$$H = \beta * \lambda E \tag{127}$$

$$\lambda E = Rn - G - H \tag{128}$$

Where: β and λE came from eddy covariance system, R_n and G are measured.
The Massman analytical formulae for spectral corrections to measuremed momentum and scalar fluxes for eddy covariance systems. Massman (2000) develop an analytical method for frequency response corrections, based in the Horst's (1997) approach develops as follow:
For *Stable atmospheric conditions* ($0<\zeta\leq 2$).

a. Fast-response open path system

$$\frac{Flux_m}{Flux} = \left[\frac{ab}{(a+1)(b+1)}\right]\left[\frac{ab}{(a+p)(b+p)}\right]\left[\frac{1}{(p+1)}\right]\left[1 - \frac{p}{(a+1)(a+p)}\right] \tag{129}$$

b. Scalar instrument with 0.1-0.3s response

$$\frac{Flux_m}{Flux}=\left[\frac{ab}{(a+1)(b+1)}\right]\left[\frac{ab}{(a+p)(b+p)}\right]\left[\frac{1}{(p+1)}\right]\left[1-\frac{p}{(a+1)(a+p)}\right]\left[\frac{1+0.9p}{1+p}\right] \quad (130)$$

c. Unstable atmospheric conditions (ζ≤0)

$$\frac{Flux_m}{Flux}=\left[\frac{a^\alpha b^\alpha}{(a^\alpha+1)(b^\alpha+1)}\right]\left[\frac{a^\alpha b^\alpha}{(a^\alpha+p^\alpha)(b^\alpha+p^\alpha)}\right]\left[\frac{1}{(p^\alpha+1)}\right]\left[1-\frac{p^\alpha}{(a^\alpha+1)(a^\alpha+p^\alpha)}\right] \quad (131)$$

where the subscript m refers to the measurement flux, $a=2\pi\int x\tau_h$; $b=2\pi\int x\tau_b$; $p=2\pi\int x\tau_c$, and τ_h and τ_b are the equivalent time constant associated with trend removal (τ_h) and block averaging (τ_b). For relatively broad coespectra with relatively shallow peaks, such as the flat terrain neutral/stable flat terrain coespectrum α=0.925, and for sharper, more peaked coespectra, such as the stable terrain coespectra α=0.925 (Kaimal et al., 1972). These approximations are clearly easier to employ than numerical approaches and are applicable even when fluxes are so small as to preclude the use of *in situ* methods. Nevertheless, this approach is subject at the next conditions: i) horizontally-homogeneous upwind fetch, ii) the validity of the co-espectral similarity, iii) sufficiently long averaging periods, and preferably, iv) relatively small corrections (Massman, 2000).

4.9 The infrared surface temperature methods

The surface temperature has also been used for the estimation of the sensible heat flux (H) using the resistance model (Alves et al., 2000)

$$H=\rho.Cp\frac{To-Ta}{r_a} \quad (132)$$

Where ρ is air density (Kgm^{-3}), Cp specific heat at constant pressure (Jkg^{-1}°C^{-1}), To is the temperature at surface level (°C), Ta is the temperature at the reference level (°C), and r_a is the aerodynamic resistance to heat flux between the surface and the reference level (sm^{-1}), the latent heat flux (λE) can be computed as the residual term in the energy balance.

$$\lambda E=Rn-G-H=Rn-G-\rho.Cp\frac{To-Ta}{r_a} \quad (133)$$

Alves et al. (2000) say the radioactive surface temperature has a several drawbacks. Thermal radiation received by the instrument can originate from the leaves but also from de soil, and the measurement can be highly dependent on crop cover, inclination of radiometer and sun height and azimuth, especially en partial cover crops, the first one lies in the use of an adequate value of r_a .

Where d is zero plane displacement height (m), Z_{oM} and Z_{OH} are the roughness lengths (m) for momentum and heat respectively, k is the von Karman constant, u_z is the wind speed (ms^{-1}) at the reference height z (m), and ψM and ψH are the integrated stability functions for describing the effects of the buoyancy or stability on momentum transfer and heat between the surface and the reference level.

The necessary instruments are: Wind speed and direction sensor at (0.85 and 1.46m), psychrometer at the same height that wind sensor, a net radiometer placement a 1.5 m and infrared thermometer perpendicular to the rows the crop, and positioned at an angle of 60° below horizontal to view the top leaves of the plants at 0.40 m distance. (Alves et al., 2000) Sensible heat flux, H is calculated with the flux applied to levels Z_1 and Z_2.

$$H = \rho Cp \frac{T1 - T2}{[ra]_1^2} \tag{134}$$

$[ra]_1^2$ is the aerodynamic resistance to heat flux between the two levels, and is computed using the equation

$$[ra]_1^2 = \frac{\ln\left(\frac{Z2 - d}{Z1 - d}\right)}{ku^*} \tag{135}$$

with u* the friction velocity, obtained in the process of determining aerodynamic parameter d and Z_{oM} from the win profile measurements.
The air temperature at the surface level (To) is calculate using:

$$To = Ta + \frac{Hr_a}{\rho Cp} \tag{136}$$

The stability conditions can be calculated using the Richardson number

5. Fetch requirements

The air that passing over a surface is affected by the field surfaces feature (Rosenberg et al., 1983); the minimal fetch requirement can be estimated based on the thickness of the internal boundary layer (δ in m) and a roughness parameter (Z_o in m) considering the minimal and maximal crop height during the grown season. The δ can be calculated using the relation proposed by Monteith and Unsworth (1990).

$$\delta = 0.15.L^{4/5}.Z_o^{1/5} \tag{137}$$

where L is the distance of traverse (fetch) across a uniform surface with roughness Z_o. The Z_o for crops is approximately one order of magnitude smaller than the crop height h, and can be calculated according with Rosenberg et al. (1983) as follow:

$$\text{Log}_{10} Z_o = 0.997 \log_{10} h - 0.883 \tag{138}$$

As a factor of safety a height to fetch of 1:50 to 1:100 is usually considered adequate for studies made over agricultural crop surfaces (Rosenberg et al., 1983, Allen et al., 1998) but may be too conservative and difficult to achieve in practice. Alves et al. (1998) obtained full profile development using a 1:48 fetch relation in Wheat and lettuce. Heilman et al. (1989) found that for Bowen-Ratio estimates a fetch 1:20 was sufficient over grass, and Ham and Heilman (1991) and Frithschen and Fritschen (2005) obtained similar results.

6. Stability correction

The gradient method need a stability correction, one of the most used is the Monin-Obukhov stability factor (ζ) described by (Rosemberg et al., 1983; Campbell, 1985; Prueger and Kustas;2005).

$$\xi = \frac{(-k.z.g.H)}{\left(\rho_a.C_p.T_a.u*^3\right)} \tag{139}$$

where k is von Karman's constant, z is height of wind and air temperature measurements (m), g is the gravitational constant (9.8 $m.s^{-2}$), H = β.λE, T_a is air temperature (°K), u* is the friction velocity given by Kjelgaard et al. (1994) without the stability correction factor:

$$u* = \frac{k.u_z}{\ln\left(\frac{z-d+Z_{om}}{Z_{om}}\right)} \tag{140}$$

flux with a negative sign for ζ indicate unstable conditions and needs to be exclude, in flux under unstable conditions the λE is over R_n (**Fig.5a**), when the flux with negative ζ are exclude, λE is low that R_n (**Fig.5b**).
Payero et al. (2003) indicate that fluxes with incorrect sign and β ≈ -1 should not be considered when estimated the energy balance components by the energy balance Bowen ratio method. The negative ζ are corresponded with negative β (**Fig.6**).
The Richardson number (Ri) represented by the equation 44, also is well know as stability factor (e.g.; Alves et al., 2000; Tolk, et al., 1995) and represent the ratio of the buoyancy – "thermal effect" to mechanical –"wind shear" (Prueger and Kustas, 2005). Negative values indicate instability conditions where surfaces heating enhances buoyancy effects, and positive Ri values indicate a stable conditions where temperature near the surfaces are cooler than away from the surfaces.

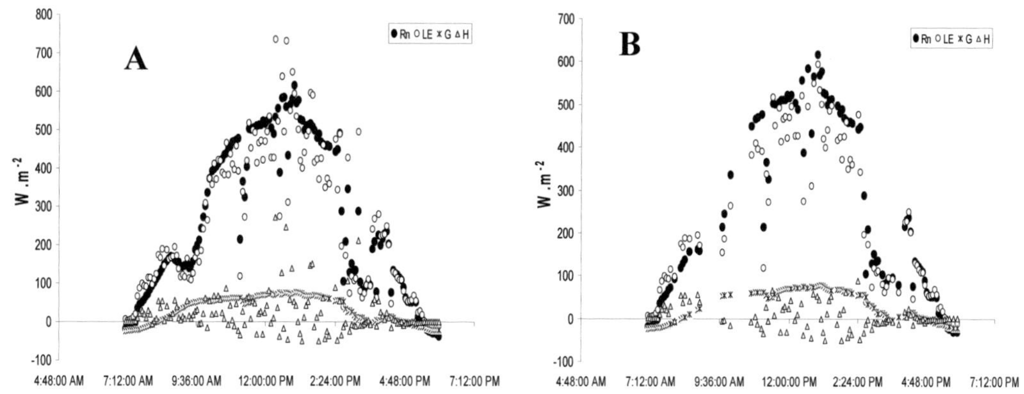

Fig. 5. Energy balance components measured by Bowen ratio method in grass **A.** without stability correction and **B.** with stability correction.

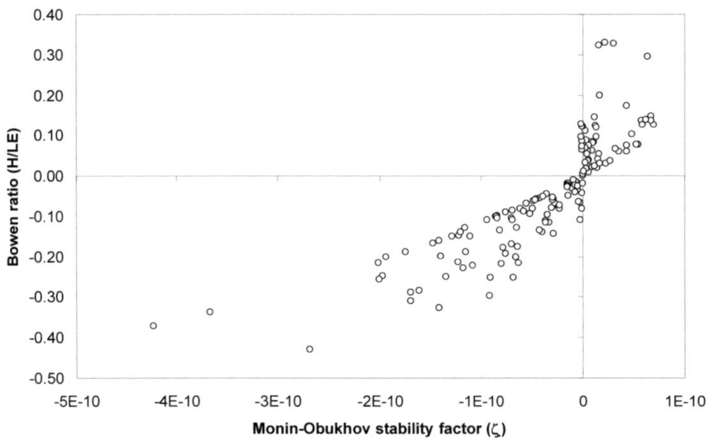

Fig. 6. Relationship between Bowen ratio (β)) and the Monin-Obukhov stability factor

7. Acknowledgements

Much of the data presented in this chapter were obtained under the project sponsored by NOAA-CREST (NA17AE1625), NASA-EPSCoR (NCC5-595), USDA-TSTAR-100, and USDA Hatch Project H-402.

We are also grateful with the University of Puerto Rico Experiment Station and the National Coffee Research Center (Cenciafé-Colombia) for providing support to finish this document.

8. References

Alves, I.; Perrier, A. and Pereira, L. S. (1998). Aerodynamic and Surface Resistente of Complete Cover Crops: How Good is the "Big Leaf"?. *American Society of Agricultural Engineers.* 41(2):345-351

Alves, I.; Frontes, J.C. and Pereira, L.S. (2000). Evapotranspiration Estimation From Infrared Surface Temperature. I: The Performance of the Flux Equation. *American Society of Agricultural Engineers.* 43(3):591-598.

Alves. L. and L.S. Pereira. (2000). Modeling Surface resistance from climatic variables?. *Agricultural. Water. Management.* 42:371-385.

Allen G.R.; Cuenca H.R.; Jensen E.M.; Blatchley K.R. & Erpenbeck M.J. (1990). Evapotranspiration and irrigation water requirements. *American Society of Civil Enginneers.* New York (USA).332p.

Allen, G.R.; Pereira, S.L.; Raes, D. & Smith, M. (1998). Crop evapotranspiration: Guidelines for computing crop water requerements. *Food and Agrocultural Organization of the United Nations* (FAO).56. Rome.300p.

Anandristakis, M., A.Liakatas., P. Kerkides., S. Rizos., J.Gavanonis. and A. Poulavassilis. (2000). Crop water requirements model tested for crops grown in Grece. *Agricultural. Water Management.*42:371-385.

Angelocci, L.R.; BruninI, O. and Magalhaes, A.C.(1983). Variacao de resisténcia estomática á difusao de vapor da água associada ao estado de energia da água na folha em cafeeiros jovens. Universidade Federal de Viçosa. 18p In: REUNION y Simpósio Relaciones Água-Planta, 9. Viçosa (Brasil), Julho 25-28.

Anthoni, A.; Unsworth, M.; Law, B.; Irvine, J.; Baldochi, D.; Kolle, O. Knohl, A. and Schulze, E.D. (2001). Comparison of open-path and Closed-path eddy covariance system. Max-Planck Institute for Biogeochemie, Jena, D-07745, Germany,2p.

Baldocchi, D. Hicks, B. and Meyers, T. P. (1988). Measuring biosphere-atmosphere exchanges of biologically related gases with micrometeorological methods. *Ecology* 69(5): 1331-1340.

Baldocchi, D.; Luxmoore, R.J. and Hatfield, J.L. (1991). Discerning the forest from the trees: an assay on scaling canopy stomatal conductance. *Agricultural and Forest Meterology.* 54:197-226.

Bowen, I.S. (1926). The ratio of heat losses by conduction and by evaporation from any water surface. *Physical Review.* 27:779-788.

Brenner, A.l. and Incoll, L.D. (1997). The effect of clumping and stomatal response on evaporation from sparsely vegetation shrunblands. *Agricultural and Forest Meterology.* 84: 187-205.

Brisson, N.; Itier, B.; L'Hotel, J.C. and Lorendeau, J.Y. (1998). Parameterisation of the Shuttleworth-Wallace model to estimate daily maximum transpiration for use in crop models. *Ecologycal Modeling.*107:159-169.

Brown, P.A.; Mancino, C.F.; Joung, M.H.; Thompson, T.L; Wierenga P.J. & Kopec, D.M. (2001). Penman Monteith crop coefficients for use with desert turf system. *Crop Science.* 41:1197-1206.

Campbell Scientific, INC (1991-1998). Bowen Ratio Intrumentation Instruction Manuel.23p. IN: www.Campbell.com

Campbell, G.S. (1977). An introduction to environmental biosphysics. Springer-Verlag, New York, New York, USA.

Campbell, G.S. (1985). Soil physics with basic transport models for soil-plant systems. *Developments in soil Science* No. 14. Elsevier, ed.150p.

Camilo, P.J. and Gurney, R.J. (1986). A resistance paramenter for bare soil evaporation models. *Soil Science.* 141:95-106.

Chang, J.H. (1968). Climate and Agriculture and Ecological Survey. University of Hawaii. Ed. Aldine Publishing Company. Chicago. (USA).304p.

Coleman, E.A. (1946). A laboratory of lysimetric drainage under controlled soil moisture tension. *Soil Science.* 62:365-382.

Conroy, J.W.; WU.J. & Elliot, W. (2003). Modification of the evapotranspiration routines in the WEPP model: Part I. ASAE Annual International Meeting. Las Vegas Nevada, USA: 27-30 July. 16p. In: http://www.pubs.asce.org/WWWdisplay.cgi?8801815

Doorenbos, J. & Pruitt, W.O. (1977). Guidelines for predicting crop water requirements. Food and Agricultural Organization of the United Nations (FAO). Publication No. 24. Rome. 300p.

Dugas, W.A.; Dan, R. & Ritcchie, J.T. (1985). A weighing Lysimeter for evapotranspiration and root measurements. *Agronomy Journal.* 77:821-825.

Farahami, H.J. and Bausch, W.C. (1995). Performance of evapotranspiration models for maize-bare soil to close canopy. *Trans. ASAE.* 38:1049-1059.

Fank, J. (2008). Monolithic field Lysimeter-a precise tool to close the gap between laboratory and field scaled investigations. *Geophysical Research Abstracts*. Vol 10. 8p.

Flint, A.L. and Child, S.W. (1991). Use the Pristley-Taylor evaporation equation for soil water limited conditions in a small forest clearcut. *Agricultural and Forest Meterology*.56:247-260.

Frank, A.B. (2003). Evapotranspiration from Northern Semiarid Grasslands. *Agronomy Journal*. 95:1504-1509.

Fritschen, J.L. and Fritschen, L.C. (2005). Bowen ratio energy balance method. In: *Micrometeorology in Agricultural Systems*, Agronomy Monograph no, 47. American Society of Agronomy, Crop Science Society of America, Soil Science Society of America, 677 S. Segoe Rd., Madison, W153711, USA: 397-405.

Grago, R.D. and Brutsaert, W. (1992). A comparison of several evaporation equations. *Water Resources Research*.32:951-954.

Grago, R.D. (1996). Comparison of the evaporative fraction and the Pristley-Taylor α for parameterisation day time evaporation. *Water Resources Research*.32:1403-1409.

Grismer, M. E.; Orang, M.; Snyder, R. & Matyac. R. (2002). Pan Evaporation to Reference Evapotranspiration Conversion Methods. *Journal of Irrigation and Drainage Engineer*. 128(3):180-184.

Ham, M.J. and Heilman, L.J. (1991). Aerodynamic and surface resistance affecting energy transport in a sparse crop. *Agricultural and Forest Meterology*.53:267-284.

Ham, M.J. and Heilman, L.J. (2003). Experimental Test of Density and Energy-Balance Corrections on Carbon Dioxide Flux as Measured Using Open-Path Eddy Covariance. *Agronomy Journal*. 95:1393-1403.

Hanson, R.B. & May, D.M. (2004). Crop coefficients for drip-irrigated processing tomatoes. In: ASAE/CSAE Annual International Meeting. Fairmont Chateau Laurier, The Westing, Government Centre, Ottawa, Ontario, Canada.12p.

Harmen, W.W.; Ramirez, B. V.H.; Gonzalez, E.J; Dukes, D. M. and Jia, X. (2006). Estimation of short-term actual crop evapotranspiration. Proceedings of the 42nd *Annual Meeting of the Caribbean Food Crops Society*, July 9-14, 2006, San Juan, Puerto Rico.12p.

Harmsen, E.W.; Ramirez, B.V.H.; Dukes, M.D.; Jia. X.; Perez, A. L.; Vasquez, R. (2009). A ground-base method for calibrating remotely sensed temperature for use in estimating evapotranspiration. *WSEAS Transactions and Environment and Development*.1(5):13-23.

Hatfield, J.L. and Allen, G.R. (1996). Evapotranspiration estimates under deficient water supplies. *Journal of Irrigation and Drainage. Engineer*. 122:301–308.

Harris, P.P., Huntingford, C., Cox, P.M., Gash, J.H.C. and Malhi, Y. (2004). Effect of soil moisture on canopy conductance of Amazonian rainforest. *Agricultural and Forest Meteorology*. 122:215-227.

Heilman, J.L.; Brtittin, C.L. and Neale, C.M.U. (1989). Fetch requirements from Bowen ratio measurements of latent and sensible heat fluxes. *Agricultural and Forest Meteorology*.. 44: 261-273.

Howell, T.A. (2005). Lysimetry. In: *Soil and the Environment*. Edited by Elsevier: 379-386.

Howell, T.A.; Steiner, J.L.; Scheider, A.D.; Evett, S.R. and Tolk. A.J. (1994). Evapotranspiration of irrigated winter wheat, sorghum and corn. *ASAE Paper* No.94-2081, ASAE, St. Josep, MI

Horst, T.W. (1997). A simple formula for attenuation of eddy covariance flux measurements. *Boundary Layer Meteorology*. 82:219-233.

Hunsaker, D.J. (1999). Basal Crop Coefficients and Water Use For Early Maturity Cotton. *American Society of Agricultural Engineers*.42(4):927-936.

Hunsaker, D.J.; Pinter, P.J.Jr. & Cai, H.(2002). Alfalfa Basal Crop Coefficients for FAO-56 Procedures in the De*sert Regions of the Southwestern U.S. American Society of Agricultural Engineers*.45(6):1799-1815.

Jarvis, P.G. (1976). The interpretations of the variations in leaf water potential and stomatal conductance found in canopies in the field. *Phil Transaction Royal Society London Bulletin*. 273:593-610.

Jhonson, R.S.; Williams, L.E.; Ayars, J.E. & Trout. T.J. (2005). Weighing lysimeter aid study of water relations in trees and vine crop. *California Agriculture*.59(2):133-136.

Kamal, H.A. and Hatfield, L.J. (2004). Canopy Resistance as Affected Soil and Meteorological Factors in Potato. *Agronomy Journal*. 96:978-985.

Kaimal, J.C.; Wyngaard, J.C.; Izumi, Y. and Cote, O.R. (1972). Deriving power spectra from a three-component sonic anemometer. *Journal Applied Meteorology*. 7:827-837.

Katerji, N. and Perrier, A. (1983). Modelling the real evapotranspiration in Alfalfa crop. Role of the crop coefficients. *Agronomie*.3(6):513-521 (in: French).

Katerji, N. and Perrier, A. (1985). Determinations of the canopy resistance to the water vapour diffusion of the six canopy cover. *Agricultural Meteorology* 34:105-120 (in: French).

Katerji, N. and Rana, G. (2006). Modelling evapotranspiration of six irrigated crops under Mediterranean climate conditions. *Agricultural and Forest Meterology*.138:142-155.

Kirkham, M.B.; Redelfs, M.S.; Stone, L.R. and Kanemasu, E.T. (1985). Comparison of water status and evapotranspiration of six row crops. *Field Crops Research*.10:257-268.

Kjelgaard, J.F. and Stokes, C.O. (2001). Evaluating Surface Resistance for Estimating Corn and Potato *Evapotranspiration with the Penman-Monteith Model. American* Society of *Agricultural Engineers*.44(4):797-805.

Kjelgaard, J.K.; Stockle, C.O.; Villar Mir, J.M.; Evans, R.G. and Campbell, G.S. (1994). Evaluation methods to estimate corn evapotranspiration from short-time interval weather data. *Transaction of ASAE*. 37(6): 1825-1833,

Kosugi, Y. and Katsuyama. M. (2007). Evapotranspiration over Japanese cypress forest II. Comparison of the eddy covariance and water budget methods. *Journal of Hydrology*.334:305-311.

Laubach, J. and McNauhton (1999). A spectrum-independent procedure for correcting eddy flux measurements with separated sensors. *Boundary Layer Meteotology*. 89:445-467.

Lloyd, W.G. (1992). Bowen-Ratio measurements. Evapotranspiration measurements of native vegetation, Owens Valley, California. U.S Geological Survey. *Water-Resources Investigations Report* 91-4159:5-18

López-Urrea, R.; Martin de Santa Olalla, F..; Febeiro, C. & Moratalla. A. (2006). Testing evapotranspiration equiations using lysimeter observations in a semiarid climate. *Agricultural Water Management*.85(1-2):15-26.

Loos, C.; S. Gainyler.; E. Priesack. (2007). Assessment of water balance simulations for large-scale weighing lysimeters. *Journal of Hydrology*.335(3-4):259-270.

Madeiros, G.A.; Arruda, F.B.; Sakai, E. & Fujiwara, M. (2001). The influence of the crop canopy on evapotranspiration and crop coefficients of beans (*Phaseolus vulgaris* L.). *Agricultural Water Management* 49:211-234.

Malone, R.W.; Stewardson, D.J.; Bonta, J.V. & Nelsen, T. (1999). Calibration and Quality Control of the Coshocton Weighing Lysimeters. *Transaction of the ASAE/American Society of Agricultural Enginieer*.42(3):701-712.

Massman, W.J. (2000). A simple method for estimate frequency response corrections for eddy covariance systems. *Agricultural and Forest Meteorology*.104:185-198.

Massman, W.J. and Lee, X. (2002). Eddy covariance flux corrections and uncertainties in long term studies of carbon and energy exchange. *Agricultural and Forest Meteorology*.113:121-144.

Meyers, P.T. and Baldocchi, D.D. (2005). Current micrometeorological flux methodologies with applications in agriculture. In: *Micrometeorology in Agricultural Systems*, Agronomy Monograph no, 47. American Society of Agronomy, Crop Science Society of America, Soil Science Society of America, 677 S. Soe Rd., Madison, W153711, USA: 381-396.

Monteith, L.J. (1965). Evaporation and the environment. *Symp.Soc.Exper.Biol*.19:205-234.

Monteith, L.J. (1981). Evaporation and surface temperature. *Quarterly Journal of the Royal Meteorology Society*. 107(451):1-27.

Monteith, J.L. (1995). Accommodation between transpiring vegetation and the convective boundary layer. *Journal of Hydrology*. 166:251-263.

Monteith, , L.J. (1997). Evaporation Models. In: *Agricultural Systems Modelling and Simulation*. Edited by Robert M. Peart and R. Bruce Curry. University of Florida, Gainesville, Florida: 197-234.

Monteith, J. L. and Unsworth, M. H. (1990). Principles of Environmental Physics, 2nd ed., *Edward Arnold Publisher*. 291 pgs.

Moore, C.J. (1986). Frequency response corrections for eddy correlation systems. *Boundary-Layer Meteorology*, 37: 17-35.

Nasab, B.S.; Kashkuli, H.A.; Khaledian, M.R. (2004). Estimation of Crop Coefficients for Sugarcane (Ratoon) in Haft Tappeh of Iran. ASAE/CSAE Annual International Meeting. Fairmont Chateau Laurier, The Westing, Government Centre Ottawa, Ontario, Canada. 1-4 August.10p.

Oguntunde, G.P. (2005). Whole-plant water use and canopy conductance of cassava under limited available soil water and varying evaporative demand. *Plant and Soil*. 278:371-383.

Ortega-Farias, S.O.; Antonioletti, R. and Brisson, N. (2004). Evaluation of the Penman-Monteith model for estimatig soybean evapotranspiration. *Irrigation Science* 23:1-9.

Ortega-Farias, S.O.; Olioso, A.; Fuentes, S.; Valdes, H. (2006). Latent heat flux over a furrow-irrigated tomato crop using Penman-Monteith equation with a variable surfaces canopy resistance. *Agricultural Water Management.* 82:421-432

Parton, W.J.; Lauenroth, W.K. & Smith, F.M. (1981). Water loss from a shortgrass steppe. *Agricultural Meteorology*. 24:97-109.

Payero, J.O.; Neale C.M.U.; Wright J.L.and Allen. R.G. (2003). Guidelines for validating bowen ratio date. *American Society of Agricultural Enginieers*. 46(4):1051-1060.

Penman; H.L. (1948). Natural Evaporation from open water, bare soil and grass. *Proceedings of the Royal Society of London. Series A, Mathematical and Physical Sciences*. 193(1032):120-145.

Penman, H.L. (1956). Evaporation: An introductory survey. *Netherlands Journal Agricultural Science*.1:9-29, 87-97,151-153.

Pereira S. L.; Perrier, A.; Allen G, R. & Alves, I. (1996). Evapotranspiration: Review of Concepts and Future Trends. In; Evapotranspiration and Irrigation Shceduling. Proceedings of the International Conference. November 3-6. San Antonio, TEXAS:109-115.

Pereira S. L.; Perrier, A.; Allen G, R.; Alves, I. (1999). Evapotranspiration: Concepts and Future Trends. *Journal Irrigation and Drainage Engineer*.125(2):45-51.

Pereira S.L.; & Alves, I. (2005). Crop water requirements. In: *Soil and Enviroment*. Edited by Elsevier: 322-334.

Perez, P.J.; Castellvi, F. and Martinez-Cob. A. (2008). A simple model for estimation the Bowen ratio from climatic factors for determining latent and sensible heat flux. *Agricultural and Forest Meteorology*. 148:25-37.

Priestley, C.H.B. and Taylor, R.J. (1972). On the assessment of the surfaces heat and evapotranspiration using large-scale parameters. *Monthly Water Review*. 100(2):81-92.

Prueger, H.J.; Hatfield, J.L.; Aase, J. K. & Pikul Jr. J.L. (1997). Bowen-Ratio comparison with lysimeter evapotranspiration. *Agronomy Journal*. 89(5): 730-736.

Prueger, H.J. and Kustas, P.W. (2005). Aerodynamic methods for estimating turbulent fluxes. In: *Micrometeorology in Agricultural Systems*, Agronomy Monograph no, 47. American Society of Agronomy, Crop Science Society of America, Soil Science Society of America, 677 S. Segoe Rd., Madison, W153711, USA:407-436.

Ramírez, B.V.H. (2007). Plant-Water Relationships for Several Common Bean Genotypes (*Phaseolus vulgaris* L.) with and Without Drought Stress Conditions. M.Sc. Thesis, Agronomy and Soils Department, University of Puerto Rico - Mayaguez Campus. 190 pages.

Ramirez, B.V.H.; Harmsen, W.E. and Porch, G.T. (2008). Estimation of Actual Evapotranspiration Using Measured and Calculated Values of Bulk Surfaces Resistance. *Proceedings of the World Environmental and Water Research Congress*. Honolulu (Hawaii). 10p.

Ramírez, B.V.H. and Jaramillo, R.A. (2008). Modifications in the Superficial Hydrology associated at the cover changes in the Colombian´s Tropical Andes. In: XIV Colombian Soil Society Meeting.13p. (in Spanish).

Rana, G.; Katerji, N.; Mastrorilli, M.; El Moujabber,M El. and N. Brisson. (1997a). Validation of a model of actual evapotranspiration for water stressed soybeans. *Agricultural and Forest Meteorology*.86:215-224.

Rana, G.; Katerji, N.; Mastrorilli, M. and El Moujabber, M. (1997b). A model for predicting actual evapotranspiration under soil water stress in a Mediterranean region. *Theorical Applied. Climatolology*. 56:45-55.

Ritchie, J.T. & Burnett, E. (1968). A precision weighing lysimeter for row crop water use studies. *Agronomy Journal*. 60(5):545-549.

Rosenberg, J. N.; Blad, B.L. and Verma, S.B. (1983). Microclimate; The Biological Environment. A Wiley-Interscience Publication, Jhon Wiley and Sons. Edt. 495p.

Roygard, F.R.; Alley, M.M.& Khosla, R. (2002). No-Till Corn Yield and Water Balance in the Mid-Atlantic Coastal Plain. *Agronomy Journal*. 94:612-623.

Saugier, B. (1977). Micrometeorology on Crops and Grasslands. Environmental Effects on Crop Physiology. *Proceeding of a Symposium held at Long Ashton Research Station Uviversity of Bristol.* Academic Press:39-55.

Saugier, B. and Katerji, N. (1991). Some plant factors controlling evapotranspiration. *Agricultural and Forest Meteorology*.54:263-277.

Sene, K.J. (1994). Parameterisation for energy transfer from a sparse vine crop. *Agricultural and Forest Meteorology*.71:1-18.

Seyfried, M.S.; Hanson, C.L.; Murdock, M.D. & Van Vactor. S. (2001). Long-term lysimeter database, Reynolds Creek Experimental Watershed, Idaho, United States. *Water Resources Research*.37(11):2853-2856.

Schotanus, P.; Nieuwstadt, F. T. M. and de Bruin, H. A. R. (1983). Temperature measurement with a sonic anemometer and its application to heat and moisture fluxes. *Boundary-Layer Meteorology* 50: 81-93.

Shuttleworth, W.J.; and Wallace, J.S. (1985). Evaporation from sparse crops-An energy combination theory. *Quarterly Journal of the Royal Meteorology Society*..111:839-855.

Steiner, J.L.A.; Howell, T.A. & Schneider, A.D. (1991). Lysimeter evaluation of daily potential evapotranspiration models for grain sorghum . *Agronomy Journal*.83(1):240-247.

Suleiman, A. & Crago, R. (2004). Hourly and Daytime Evapotranspiration from Grassland Using Radiometric Surface Temperature. *Agronomy Journal* .96:384-390.

Sumner, D. M. (2001). Evapotranspiration from a cypress and pine forest subjected to natural fires in Volusia County, Florida, 1998-99. U. S. *Geological Survey Water-Resources Investigations Report* 01-4245. Washington, D.C.: U.S. Geological Survey.

Szeicz, C., and Long, L.F. (1969). Surface resistance of crop canopies. *Water Recourses Research*. 5(8):622-633.

Takhar, H.S. & Rudge, A.J. (1970). Evaporation studies in standard catchments. *Journal of Hydrology*.11(4):329-362.

Tanner, C.B. (1960). Energy balance approach to evapotranspiration from crops. *Soil Science. Society of American Proceedings*. 24(1):1-9.

Tanner, B. D., and Greene, J. P. (1989). Measurements of sensible heat flux and water vapour fluxes using eddy correlation methods. *Final report to U.S. Army Dugway Proving Grounds*. DAAD 09-87 D-0088.

Teare, L.D.; H. Schimmulepfenning.; and R.P.Waldren. (1973). Rainout shelter and drainage lysimeters to quantitatively measure drought stress. *Agronomy Journal*. 65(4):544-547.

Thom, A.S. (1975). Momentum, mass and heat exchange of plant communities. In: *Vegetation and the atmosphere*: 57-109 Edit by Monteith. J.L Academic Press, New York.

Tolk, A.J.; Howell, A.T.; Steiner, L.J. and Krieg, R.D. (1995). Aerodynamic Characteristics of Corn as Determined by Energy Balance Tecniques. *Agronomy .Journal* .87:464-473.

Tomilson, A.E. (1994). Instrumentation, Methods, and Preliminary Evaluation of Evapotranspiration for a Grassland in the Arid Lands Ecology Reserve, Beton Country, Washington, May-Ostober 1990. U.S. Geological Survey. *Water-Resources Investigations Report* 93-4081 and Washington State Department of Ecology.32p

Tomilson, A.E. (1997). Evapotranspiration for Three Sparce-Canopy Sites in the Black Rock Valley, Yakima Country, Washington, March 1992 to October 1995. U.S. Geological Survey. *Water-Resources Investigations Report* 96-4207 and Washington State Department of Ecology.88p.

Turner, C.N. (1991). Measurement and influence of environmental and plant factors on stomatal conductance in the field. *Agricultural and Forest Meteorology*. 54:137-154.

Tyagi, N.K.; Sharma, D.K. & Luthra, S.K. (2000). Evapotranspiration and Crop Coefficients of Wheat and Sorghum. *Journal of Irrigation and Drainage Engineer*.126(4):215-222.

van Bavel, C.H.M. (1967). Changes in canopy resistance to water loss from alfalfa induced by soil water depletion. *Agricultural and Forest Meteorology*. 4:165–176.

Vaughan, P.J.; Trout, T.J. & Ayars, J.E. (2007). A processing method for weighing lysimeter data and comparison to micrometeorological ETo predictions. *Agricultural. Water Management*. 88(1-2):141-146.

Von Unold, G. & Fank, J. (2008). Module design of field lysimeter for specific application needs. *Water Air Soil Pollut: Focus* 8:233–242

Weaver, L.H. (1992). Eddy-Correlation measurements at sites C and F. Evapotranspiration Measurements of Native Vegetation, Ownes Valley. California. U.S. Geological Survey. *Water-Resources Investigations Report* 91-4159:25-33.

Webb, E. K.; Pearman, G. I. and Leuning, R. (1980). Correction of flux measurements for density effects due to heat and water vapour transfer. *Quarterly Journal of the Royal Meteorology Society*.. 106: 85-100.

Wilson, D.; Reginato, R. & Hollet, J.K. (1992). Evapotranspiration measurements of native vegetation, Ownes Valley, California. U.S. Geological Survey. *Water Resources Investigations Report* 91-4159:1:4 .

Wolf,A.; Saliendra, N.; Akshalov, K.; Johnson, D.A. and Laca, E. (2008). Effect of different eddy covariance balance closure and comparisons with the modified Bowen ratio System. *Agricultural and Forest Meteorology*.148:942-952.

Wright, J.L. & Jensen, M.E. (1978). Development and evaluation of evapotranspiration models for irrigation scheduling. *Trans ASAE*. 21(1):88-96.

Xingfa, H.; Viriyasenakul, V. & Dechao, Z. (1999). Design, Construction and Installation of Filled-in Drainage Lysimeter and Its Applications. Proceeding of 99 *International Conference on Agricultral Engineering*. Beijing, China, December:II-162-II-167.

Zhang, Y.; Liu, C.; Yu, Q.; Shen, Y.; Kendy, E.; Kondoh, A.; Tang, C. & Sun, H. (2004). Energy fluxes and Pristley-Taylor parameter over winter wheat and maize in the North China Plaint. *Hydrology Process.*18 :2235-2246.

Zhang, B.; Kang, S.; Li, F. and Zhang. L. (2008). Comparison of three evapotranspiration models to Bowen ratio-energy balance method for vineyard in an arid desert region of northwest China. *Agricultural and Forest Meteorology.*148:1629-1640.

2

Conceptual Elements and Heuristics from Complexity Paradigm Suitable to the Study of Evapotranspiration at the Landscape Level

Claudia Coronel, Omar Tapia Silva, Gilberto Hernández, José Manuel Madrigal, Edgar Rosales, Alejandro Toledo, Mauricio Galeana, Alejandra López Caloca and José Luis Silvan
Centro de Investigación en Geografía y Geomática "Ing. Jorge L. Tamayo"
Universidad Autónoma Metropolitana, Iztapalapa, Distrito Federal
México

1. Introduction

Due to the importance of the impacts from anthropogenic process on global change, there has been a growing interest in predicting climate variability on seasonal scales. In order to accomplish this, better knowledge about and modeling of earth surface energy exchanges and evaporation is required trough diverse scales, from local to continental extents. Recognizing this need, we state as general objective of this chapter to advance towards a conceptual and methodological framework that sustain a better representation of evapotranspiration at landscape scale. Along this chapter are discussed the reasons, why traditional schemes based on classical science fail in order to explain, emergent, convergent and highly dynamic behaviors observed in evapotranspiration process. On the application of more integrative approaches for modeling, the search for organizing principles regulating function response on evapotranspiration systems, is required. Emphasis on the structure-function relation, adaptation in social-ecological systems and the development of methodological strategies themselves appear to be the most relevant elements to the incorporation of resilience as an analogical model and of the complexity paradigm as a conceptual framework for the modeling of the evapotranspiration process.

2. Opportunities and challenges for modeling evapotranspiration from the classical perspective and the analogical model approach

2.1 Review of evapotranspiration models from hydrology and climatology

Over the past three decades, there has been a growing interest in predicting climate variability on seasonal and interannual scales due to the recognition of the importance of anthropogenic processes on global change. This awareness process has led to the development of new experimental technologies and modeling for generalizing local and regional observations of mass and energy exchanges on the terrestrial surface. Such generalizations represent valuable data for general circulation models (Shuttleworth & Gash, 2005; Taylor et al., 1998). Major developments on evapotranspiration estimation include semi-distributed and semi-empirical models semi-distributed and semi-empirical models that represent energy or water balances on the land surface-atmosphere interface and incorporate local and regional effects of land cover heterogeneity on evapotranspiration (Norman et al., 1995; Xu et al., 1996; Bastiaanssen et al., 1998; Sandholt et al., 2002; Allen et al., 2005; Courault et al., 2005; Garatuza-Payan & Watts, 2005; Neale et al., 2005; Senay et al., 2007). For a general view about mentioned models, see equation 2 for water balance.

$$Rn = H + \lambda LE + G_0 \quad (1)$$

$$SMt_1 = (SMt_0 + PRECt_1) - (EVAPt_1 + ROFFt_1) \quad (2)$$

In equation 1, from the total amount of available energy Rn (net radiation), a part is used to heat the air above the surface (H) (sensible heat flux), another part is used to heat the surface (G_0) (soil heat flux) and one more is used for changes in state (λLE) (latent heat flux). Water balance at hydrological unit, or water storage available in the soil, for a period t_1 (equation 2), resulting from the differential between the system input—given by the water storage available in the soil over a previous period, SMt_0, plus the effective precipitation, $PRECt_1$, minus the water interception produced by evapotranspiration, $EVAPt_1$, plus run-off, $ROFFt_1$.

Standing out among these efforts is the use of remote sensing and *in situ* auxiliary data to represent the variability of evapotranspiration (Baldocchi et al., 2001; Boegh et al., 2002; Schmugge et al., 2002). In the area of methods for integrating data from diverse sources are assimilation procedures that use the application of atmospheric and surface models supplied with satellite data calibrated with in-situ data (Crow & Wood, 2003; McCabe & Wood, 2006; Cleugh et al., 2007; Pan et al., 2008). In summary, these efforts have enabled the generation of knowledge about earth surface energy exchanges and improvements in its modeling. Nevertheless, there continues to be uncertainty regarding how land surface operate land surface-atmosphere interactions in different spatial and temporal contexts (Callaghan et al., 2004; McCabe et al., 2008). Global studies on seasonal climate variability (Dan & Ji, 2007) establish that "*the seasonal cycle of the exchanges of energy, mass and momentum between the land and atmosphere can be modified by vegetation through morphological change and carbon assimilation, and it plays a major role in the seasonal variations of local surface hydrology*". Meso-scale studies demonstrate significant differences in energy fluxes according to cover type through feedback mechanisms resulting in different cloud covers and differential rates in the development of the boundary convective layer (Taylor & Lebel, 1998; Mahrt et al., 2001; Clark et al., 2003). In turn, according to the temporal scale of observation, specific conductive processes are shown. For example, in zones with disperse vegetation the instantaneous evaporation rate is controlled by the rain pattern after precipitation events, while on a seasonal scale the evaporation patterns are highly correlated with the characteristics of the different types of cover (DOE, 1996).

From this perspective, it is crucial to incorporate into the modeling of evapotransporation (Et), multi-scale effects of local conductive processes that result in emergent behavior on a regional scale, and which cannot be represented in a simple manner. In terms of the current theory on biophysical processes, the emphasis has been on establishing the behavior of the variables from a perspective of processes and modeling and not as a specific and joint response of features or objects on the earth's surface. This means going beyond physical and complex models and an emphasis on calibration, and exploring those organizing rules that control the complexity of regional hydrology (Sivapalan, 2005; McDonnell et al., 2007). To answer the "why" questions, a coherent theory must be advanced that directly incorporates the pattern-process relations that constitute the landscape and lead to the organization of systems with specific functions susceptible to observation—in this case, the coupling of water, energy and nutrient cycles. These functions are known as functional traits and are the result of slow co-evolution or memory processes (Schröder, 2006; De Bello et al., 2010). For example, the height of plants and the size of seeds have evolved as a result of selective evolutionary pressure and constitute a climate footprint of the past. Other processes—such as phenology, soil moiture, snow and ice cover, and even biodiversity—constitute the terrestrial memory processes that contribute to the complexity observed in the components of the hydrologic cycle (Sivapalan et al., 2001, Shuttleworth y Gash, 2005).

All of these works agree that it is possible to significantly advance the knowledge about hydrologic processes—among them, evapotranspiration—through the recognition of regulatory principles that determine the rather complex behavior of the different components of the hydrologic cycle, especially in terms of the human role as a conductive element in very dynamic processes that alter the functioning of zones under management to such an extent and manner that exceeds our ability to control it. The following section explores the potential for incorporating conceptual elements of the complexity paradigm and those of landscape ecology in the modeling of evapotranspiration and other hydrologic processes.

2.2 Conceptual elements and heuristics from complexity paradigm and its application in modelling hydrologic processes

First, the conceptual elements, attributes and general methodologies of complexity, chaos and systems theories are presented. Then, a review is provided of the most relevant advances in the application of these theoretical frameworks to modeling hydrologic processes—among them, evapotranspiration.

In discussing a theory, the conceptual models of the theory itself and that represent the object of study can be used. With regard to the complexity paradigm, we shall discuss the concept of a complex system. This is composed of close interactions between its components in such a way that the events taking place significantly influence the probabilities of many types of subsequent events. Its functioning is the result of the confluence of multiple factors that interact in such a way that the system is not separable, but rather only semi-separable (Martin & Sunley, 2007).

Complex systems present key generic properties (Holland, 1995; Holland, 2002), such as:

a. Distributed representation: the functions and relations within the system are distributed on all of the scales with a high degree of connectivity. The paradox of scale is a property of complex systems.
b. Openess: the boundary between the system and its environment is not fixed or easy to identify. Dynamic and diffuse limits.
c. Chaotic behavior: Nonlinear dynamics among the variables that interact with the components of the system. (Contigences).
d. Emergences: The macro-scale structures and functioning arise from interactions among micro-scale components (building blocks).
e. Adaptive behavior and self-organization: the systems adapt their structures and dynamics in response to changes in the external environment or to internal changes through co-evolution mechanisms and critical situations.

These mechanisms are triggered in response to chaotic and highly unpredictable disturbances or messages. As a result of this large complexity framework, a series of approaches exist, such as Chaos Theory, Hierarchy Theory, Complex Adaptive Systems and Complex Social-ecological Systems (Valerie & Allen, 1996; Liu et al., 2007). The latter refers to integrated or coupled systems in which the population interacts with its natural components. Fundamental characteristics of Complex Social-ecological Systems are: the presence of nested hierarchies, or panarchies; their components interact reciprocally at diverse levels of organization as well as different spatial and temporal scales; their behavior is defined by the interactions among complex networks and; they present emergent properties (Cadenasso et al., 2006; Liu et al., 2007). See incises on figure 1.

Complex systems are composed of heterogeneous elements that interact, which means that the subsystems belong to the domains of very diverse disciplines. A multi-discipline presumes the integration of different disciplinary approaches (García, 2008) and viewing any problematic as a system whose elements are inter-defined, the study of which requires the coordination of disciplinary approaches that need to be integrated into one common approach (Holland, 2002). What integrates a multidisciplinary team that studies a complex system is a common conceptual and methodological framework derived from a shared concept of the science-society relation. This conceptual framework may be connected by guides and hypotheses that provide the threshold reasoning for what is known as heuristics, validation and approximation schemes that make up the conceptual model, or what is called the analogical model (Holland, 1995; Wu, 1999).

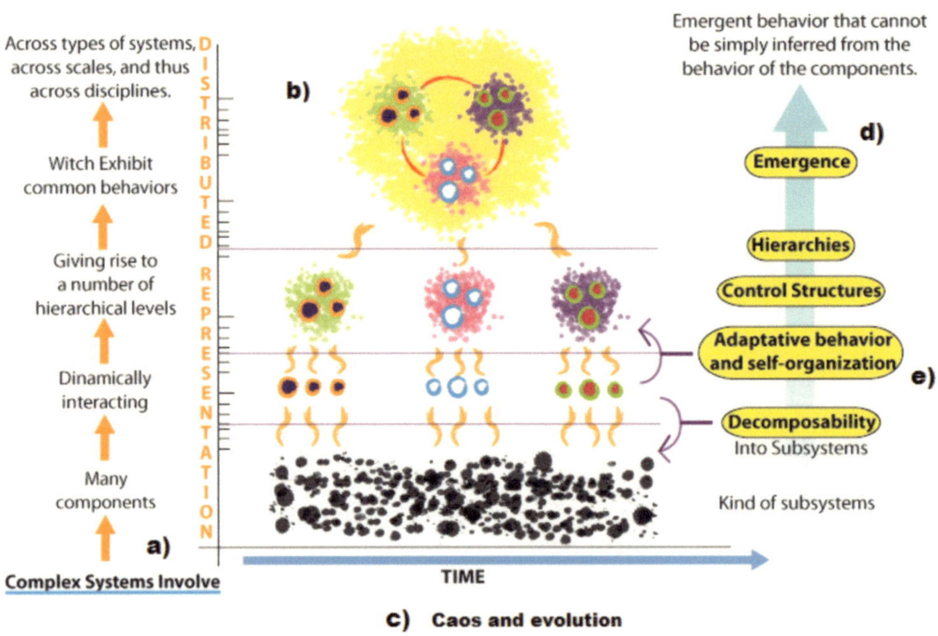

Fig. 1. Generic properties of complex systems. Modified from http://www.idiagram.com/examples/complexity.html

One last fundamental aspect of complex systems is the content of the information present in its components. The information contained in complex systems may be working information, the result of feedback from the components of the system due to a message or flow of energy, mass or language (Brooks et al., 1989). This type of information is any kind of phenotype which is the product of previous exposure to a message or specific stimulus and is the result of a process of learning and of co-evolution of the components of a system with respect to its environment (Wiener, 1954). This means that communication is a fundamental process of a system's self-organization; the final result is that the system both adapts to its medium and also modifies it (Günther & Folke, 1993).

Latent information is that which is contained in a system and is not expressed because it has not been fundamental to contributing to the collective functioning of the system and to facilitating its current adaptation. Latent information can become working information if a specific message is produced that triggers a result. Some of the components of the system have this function and are fundamental to its survival; these are called history tellers (Wiener, 1954). Complex systems are found in an enclave on "the edge of chaos," from which comes the notion of temporal-spatial contingency or convergence. Based on this point of view, Wiener develops his theory about cybernetics or message theory. His thesis is that the society can only be understood through the study of the communication messages and instruments that pertain to it.

With respect to the application of the conceptual elements of the complexity paradigm and the practical methodologies corresponding to this discipline as it pertains to the study of hydrologic processes, the importance of functional raits is recognized as a joint response of the elements of a landscape, an area subject to management or a hydrologic unit. In relation to the development of a regional hydrologic theory, organizing rules of emergence and connectivity are identified at the conceptual level, as well as the organization of hierarchies based on functional traits or self-similarity, and the hypothesis behind the optimization and structuring of hydrologic networks (Sivapalan, et al., 2001; McDonnell et al., 2007). Also recognized is the importance of contingencies, extreme values and trigger values—elements that come from chaos theory (Wiener, 1954; Gunther & Folke, 1993). Since these heuristics must be test and the diverse disciplines have a wide range of experience with them, the theory of the hydrology of catchments—from the perspective of functional traits—has to be the result of the integration of multidisciplinary conceptual frameworks and the design of observation schemes for the experimentation and testing of hypotheses. Said theory should also incorporate a more real or phenomenological representation of the processes.

With regard to a methodological proposal, some authors consider the integration of upward, downward approximations—although without theoretically offering a solution—and the definition of elemental units for the development of a classification scheme that supports the principles of scaling and based in functional aspects (Sivapalan, 2005; Schröder, 2006). Out of spatial modeling and Geomatics arise pattern discovery and the conceptualization of the geographic entities or components of a system, a methodological element indispensable to the development of classification and schemes of hierarchies in geographic processes (Haggett, 1965; Maceachren et al., 1999; Coronel, 2008). With the functional traits perspective, the social, cultural and economic nature of the managed areas or case studies is not made explicit, and although the definition of elemental units of analysis is recognized, it is not advanced in the conceptualization of hydrologic processes as the functioning of systems with complex behavior. Rather, this aspect is considered at the application level, with the definition of geospatial models for the characterization of run-off networks (Ruddell & Kumar, 2006) or in the simulation of scenarios for a management area (Parrott, 2009).

An alternate approach to functional traits is provided by biology, landscape ecology, urban planning and the very science of complexity. This approach stresses the importance of developing methodological practices specific to the science of complex systems and then applying them to the specifics of a problem or case study. Several authors consider scaling as a fundamental issue in the development of a theory that transcends classical scientific thought, but they do not propose a general theory of scaling which would contradict the complex paradigm (Günther & Folke, 1993; Haila, 2002; Walker et al., 2006). Instead, they suggest analogical models, including paradoxes, puzzle models, and the recognition of uncertainty as an attribute of complex processes. The authors mentioned above agree that scaling behavior requires the identification of critical scales in order to understand the dynamics of a system as a whole. The scale of relevance is defined by the process and its context; what is relevant is to apply scaling rules to investigation practices. This point of view is similar to that proposed in the last section of this chapter, called phenomenon driven analysis.

In order to put forth conceptual elements of a hydrologic theory, three definitions are proposed (Schröder, 2006), pattern, as structured spatial behavior that temporally differs significantly from a random process. Processes, understood as interactions among different objects in the environment. Function refers to the resulting response, or functioning, of a complete system based on its importance to maintaining the system itself. Besides these 3 basic elements, some additional elements that are suggested are simple fallible models and the science of place, or the temporal-spatial paradigm (McDonnell et al., 2007; Coronel, 2008).

A large part of what underlies these proposals to develop a catchment hydrologic theory is learning from simulated experimentation, such as in the functional traits approach and the more applied approach of the science of complexity through analogical models. This learning about the functioning of complex social-ecological systems emphasizes our ability to adapt, especially in the context of global change and the crisis of natural resources (Cutter et al., 2008; Castree, 2010).

The proposal hereins suggests moving toward a theory about the process of evapotranspiration through the application of analogical models while at the same time developing methodological practices. The strategy consists of capitalizing on multidisciplinary knowledge which offers integrating concepts related to eco-systemic capacities such as adaptability and transformability. There are important works that explore a similar strategy in a variety of hydrologic processes and highlight the importance of its adaptation. Among those are the nature of the heterogeneity of the earth's surface that drives the run-off networks and rain fields, applying the heuristics of self-organization and self-similarity (Gorenburg et al., 2001; Veitzer & Gupta, 2001), the temporal spatial variability associated with the process of evapotranspiration, and identifying diverse control factors in relevant scales and patterns (Albertson et al., 2001; Katul et al., 2001; Mahrt et al., 2001; Coronel et al., 2008). Terrestrial memory processes that are fundamental to long-term water balance are identified, among them vegetation and soil adjustments that relate energy and water balances and nutrient cycling (Post & Jones, 2001; Mackay, 2001; Coronel & Mora, 2008a; Coronel & Mora, 2008b). The application of analogical models can be seen in various works applied to the management of hydrologic basins from a more comprehensive perspective, such models have been used to demonstrate characteristics reproduced at different scales (Haila, 2002; Pickett et al., 2004; Gunderson et al., 2006).

In terms of an ecosystem's capacity to adapt and transform, this chapter proposes the explicit inclusion of the role of some of the complex social-ecological systems as triggers of chaotic processes or disturbances in their surroundings, without resulting in an increase in the system's self-organization (Folke & Kautsky, 1992). The capacity of regional systems, as well as the system under management, to absorb such disturbances at different temporal scales is reduced; that is, their resilience is affected. The emphasis on resilience as an integrating concept lies in its importance to adaptation, especially with regard to complex social-ecological systems which have a very short evolutionary history compared to the evolution of the natural systems that support them. In the next section, the capacities of ecosystems as resilience, adaptability and transformability are defined in greater detail, basic elements that contribute to the resilience of complex social-ecological systems are summarized, and the fundamental conceptual elements that define evapotranspiration systems are established.

3. Ecosystem capacities: from concepts to representation models

3.1 Defining resilience, adaptability and transformability

There is a established methodology to apply the concept of resilience as an analogical model, or metaphor for decision-making (Pickett et al., 2004). Most relevant when considering resilience in terms of analogy is making a direct connection between structure and function. The first step should be based on a clear and concise concept of resilience.

The concept of resilience was proposed in 1973 by Holling as a way to assist in understanding the dynamic of ecosystems (Gunderson, 2000). Resilience is "the persistence of relationships within the system and is a measure of the ability of those systems to absorb changes in state variables, driving variables and parameters, and still persist." Complementary to this, Holling defines stability as "the ability of a system to return to an equilibrium state after a temporal disturbance." Later interpretations use this definition of stability as part of the definition of resilience in considering it as a measurement of how fast a system returns to a state of equilibrium after a disturbance (resilience engineering) (Pimm, 1991; Gunderson, 2000).

While the concept of resilience refers to the structure and function relations that operate in ecosystems, concepts of adaptability and transformability, from the point of view of systems theory, refer to different degrees of change in social-ecological systems and are based on recognizing that humans influence resilience at different levels of organization. From Walker's definition of adaptability as "the capacity of social components in a system to manage ecological resilience" (Gunderson et al., 2006), and considering that natural systems—through their history of co-evolution with the environment—have also adapted and altered the environment (Sivapalan et al., 2001; McDonnell et al., 2007), we propose the term adaptability as the capacity of the components of a complex social-ecological system to manage its resilience. The above also focuses on the capacity that we have to maintain the goods and services that contribute to ecosystems or to even restore an undesirable regimen to a desirable one. Transformability "is the capacity to create a fundamentally new configuration of the system" (Walker et al., 2006). Social-ecological systems, as agro-ecosystems, are susceptible to failure in state policies and actions and the principal forces that result in the creation of new systems are crises in key resources and changes in social values (Gunderson et al., 1995).

There are some forms of interaction among different components of resilience—such as resistance, latitude, and precariousness—that occur in highly evolved systems by applying the principle of basins of attraction (Walker, et al., 2006). Resistance (R) refers to the ease or difficulty of a system to change into a new state of dynamic equilibrium; with regard to a basin of attraction, this corresponds to its depth (see figure 2). Latitude (L) refers to the amount of disturbance or change that can be exerted on a system without crossing the threshold value. This corresponds to the diameter of the basin and is related to the degree of variation in state variables that the system can manage. Precariousness (Pr) defines the current trajectory of the system in the basin of attraction and how close it is to a threshold value for its state variables. The position of the system (SEt) on the basin is indicated by the white dot.

In highly evolved systems, these thee aspects of resilience have co-adapted and are often strongly interrelated. Technician agricultural systems and fisheries have short co-evolutionary histories (Walker et al., 2004). Therefore, we cannot rely on such selected systems with appropriate feedback controls, and the likelihood of crossing thresholds is much higher.

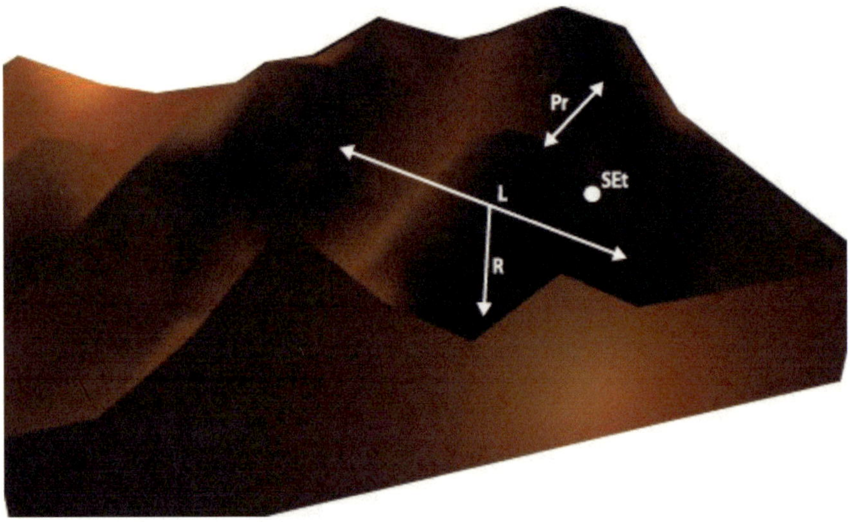

Fig. 2. Components of resilience as basin of attraction

There are more traditional examples of systems with a long history that show greater adaptation by permitting the natural recycling of nutrients and the accumulation of memory (Berkes & Folke, 1998). Panarchy, is the fourth aspect to be considered, since it involves a high degree of interdependence across scales. Losses, however, can be managed so as to be confined to smaller organizational scales, in order to minimize the loss of stability of the landscape (Ahern, 1999).

To continue building towards the application of resilience as an analogical model, heuristics from natural and social science perspectives that support this capacity need to be sought and discussed (Walker et al., 2006).

3.2 Heuristics behind resilience in complex social-ecological systems

An adaptability approach involves connecting structure and function, since the way in which processes and functions of systems are affected is mainly through changes in their structure. For there to be greater adaptability, as much should be known as possible about the resilience of managed systems. To this end, it is important to consider that slow variables control the resilience of the natural components of a social-ecological system while controlling the resilience of social components, whether by slow or rapid processes (Walker et al., 2006). Accordingly, it is important to know the scale domains of the driving variables of the resulting function of evapotranspiration systems (defined in section 3.3).

One possible cause of the loss of resilience is the lack of agreement among scales between social and ecological subsystems (Cumming et al., 2006). Some of the causes of this lack are population changes, food production, technology and transportation, human values and perceptions about nature. Over the long-term, a large part of this problem involves the development of flexible institutions that can adapt to the changing environment and the environmental experience being meaningful to social systems; that is, promoting more adaptable social systems (Szerszynsky, 2002). Several alternatives related to the above are presented in the following paragraphs.

The application of the analysis of resilience, or the application of the analogical model of resilience, changes the approach from one of maximizing sustainable productivity to one of managing adaptive resources and adaptive governance, the latter being the process of creating adaptability and transformability in complex social-ecological systems (Lebel et al., 2006; Vilardy, 2009).

An additional component for adaptability and co-responsible transformation is the creation of spaces for discussion that are essential for social networks to establish contact and promote learning (Gunderson et al., 2006). Several methodological and conceptual approaches exist that facilitate consensus in discussion spaces. Scenario planning—a bottom-up view—is a process of envisioning whether or not plausible transformations are desirable and bringing them into the social decision-making processes (Peterson et al., 2003). Participatory planning at regional or local scales by means of participatory cartography, mental maps or other techniques promote consensus about an issue in the geographic space (Martínez & Reyes, 2005; Romero & Farinós, 2006; Cataldo & Rinaldi, 2010) These approaches, based on social knowledge, are top-down methodological views and, together with bottom-up ones, contribute to an adaptation process. By making the ecological reality in which management conflicts operate explicit, social systems can then be adapted in order to implement adaptive governance.

It is worth stating a cautionary principle related to multi-objective management at the landscape scale, which is that because social-ecological systems have multiple interacting thresholds, they lead to different options for management regimes, only a few of which are feasible (Kinzing et al., 2006).

With regard to the behavior of social-ecological systems in general, certain evolutionary phases can be identified according to the adaptive cycle proposed by Holling in 1986. Said cycle undergoes abrupt changes involving creation or growth, conservation, release and reorganization phases. Although an adaptive approach has to be applied to all the phases of the adaptive cycle of a system, some authors especially emphasize the importance of the back loop, and in particular the flexible management needed to retain critical ecological resources (adaptive management), and the evolution of rules that influence resilience during self-organization (adaptive governance) (Abel et al., 2006; Gunderson et al., 2006). Maintaining critical resources and taking into account cross-scale linkages can increase the probability of desirable results during the release and reorganization phases; the possibility of this, depends on the systems in a large hierarchy being in a growth or stability phase and their having connections, such as memory and learning (heterogeneous landscapes can more efficiently stop the dispersion of fire and infestations at large scales) (Walker et al.,

2006; Peterson, 1998). Thus, a relevant task in the application of resilience as an analogical model is to characterize the system under management in terms of the phase of the adaptive cycle in which it is found at the moment in which management decisions are made.

The diversity of functions and responses are key elements to the resilience of social-ecological systems (Walker et al., 2006). The former is related to the performance, or functional effectiveness, of a system; the rule is the greater the diversity of functions, the more effectively a system will respond to disturbances. This does not occur in agricultural or pasture systems focused on maximizing the productivity of a few species and, thus, said systems will tend to be less resilient in the face of any disturbance. The diversity of responses refers to functional redundancy, that is, diverse groups fulfilling the same function. Apparent redundancy plays an important role in management and adaptive governance, while a focus on performance alone reduces apparent redundancy and, therefore, reduces resilience. In terms of functional diversity and the diversity of responses in different ecosystems, a direct connection has been established of the role of biodiversity in the functioning of complex social-ecological systems.

In summary, adaptability is determined by absolute and relative amounts of social, human, natural, manufacturing and financial capital, as well as by the institutional system and governance (Lebel et al., 2006). In terms of social capital, there are three determinants of adaptive capacity: multiple leaderships consolidated social networks and networks that favor the flow of information and trust in institutions (Gunderson et al., 2006). Another very important component of human capital is knowledge, such that the presence of different superimposed knowledge approaches will lead to consensus and shared goals for handling conflict and favoring adaptability (Reyes & Martínez, 2005). Adaptive learning is facilitated by open social networks that apply ecological knowledge across social and natural scales (Anderies et al., 2006). Adaptation and transformation depend on the social-ecological memory and are limited by legal and cultural factors, as well as key vulnerabilities that determine resilience (Abel et al., 2006). Closed institutional systems change only when forced by social or ecological crises, since their primary approach is being isolated, filtering information and maintaining prevalent theories and paradigms (Kinzing et al., 2006).

In terms of evapotranspiration systems, at the local level, systems tend to prefer management goals—whether urban or rural— such as agriculture or pasture production, forest management or conservation areas (Daily et al., 1997), each of which represent a basin of attraction with very different resource needs. The focus on maximizing the production of different components of the landscape can create conflicts due to the effects of certain systems over others. Added to this are effects at other scales that, together, result in the landscape having virtually no stability; for example, the problem of double exposure (Castree, 2010). The application of such a situation is shown with a case study in the agriculture zone of Bajío, in central Mexico where there are large agricultural areas that are inefficient in terms of the management of resilience. These zones are highly exposed to the production needs of the global market and all of the institutional, cultural and financial efforts are directed toward maintaining this production model, with a basin of attraction that is deep but highly precarious because of soil contamination, decreased aquifer recharge and excessive control of runoff (Soto & Soto, 1990; Alcocer et al., 2000; Maganda, 2004).

Once the general heuristics that determine resilience in complex social-ecological systems are stated and discussed, it is necessary to proceed toward the application of the concept as an analogical model. This requires translating the concept into schematic, structural and quantitative representations so as to apply it to specific real or hypothetical situations or issues. The model is the representation of the systems, its components, the spatial and temporal limits, the interactions among components and the possible degree of variation in the state variables and in the resulting function (Pickett et al., 1994; Pickett et al., 2004).

3.3 Toward the definition of evapotranspiration systems

We shall begin with the previous works that suggest the characterization of the functional response on evapotranspiration systems at different scales, while not necessarily defining what an evapotranspiration system would be. In hydrologic modeling, the human factor is not explicitly incorporated but rather is suggested through land surface-atmosphere interactions driven by changes in land cover. Water and nutrient (C and N) cycles are related through the optimization principle; that is, with the minimization of the use of water and the maximization of CO_2 exchange (Shuttleworth and Gash, 2005; Dan & Ji, 2007). Spatial-temporal variability implies the exchanges of water and energy between the land surface and the atmosphere, the context of which is fundamentally different that which regulates run-off processes. Evapotranspiration responds to the heterogeneity of all of the controlling factors, such as radiation, atmospheric conditions, soil properties, soil humidity, and vegetation conditions (Sivapalan et al., 2001).

In terms of energy balance, it has also been observed that there is a radiation threshold that can trigger changes in the adaptive cycle of plants, for example, speeding up the senescence and loss of biomass (Zavaleta et al., 2003). In the temporal context, at fine scales of seconds to minutes, energy fluxes are governed by atmospheric turbulence; and at intermediate scales of hours to days, by diurnal variations in radiation, average wind, atmospheric humidity and soil moisture (Katul et al., 2001). Natural temperate forest systems seem to successively evolve toward maintaining a stable level of water use efficiency, regulated by local control over temperature differences between the surface and air provided by rough layer, and by a quasilinear regional control over evapotranspiration provided by the vertical structure of the vegetation in such as way that the resulting function is a balance of water, energy and nutrient cycles, even under unlimited water conditions (Albertson et al., 2001, Katul et al., 2001) . There is no history indicating that these strategies for the optimization and stability of efficiency in the use of water operate in adaptive cycles, such as those of loss and destruction, but rather there appears to be a tendency toward the conservation of critical resources (water, nutrients, genetic pool) (Abel et al., 2006; Walker et al., 2006).

Interaction with the atmosphere is incorporated by the concept of the mixed layer, in which the influence of the heterogeneity of the surface becomes very small as the altitude increases. The altitude of the mixed layer reflects a gradual decrease in the influence of heterogeneity with altitude (Mahrt et al., 2001). Its effects depend on the spatial scale of the rough layer, with greater interaction when large areas of smooth surfaces are present (mono-crops, pastures and forests with little vertical structure) and practically zero effects when extensive zones with well-structured canopies are present. Interactions between the land surface and the atmosphere, that lack the effects of heterogeneity at small scales which are filtered, present patterns on the land surface that are driven by the middle wind (Sivapalan et al., 2001).

Another relevant factor is interaction with slow and seasonal processes—such as precipitation, water available in the soil and snow, the phenology of plants and the adaptive cycles in which systems are found. At scales of weeks to years, surface fluxes are governed by the seasonal variation in the wetness of the soil, the physiology the plants and the seasonal dynamics of Rn. Only at these scales are there interactions between run-off processes and evapotranspiration. Also studied is the role of the climate, soil and vegetation on the long-term water balance and the hydrologic equilibrium (through which run-off patterns, climatic controls and biogeochemical cycles in the soil cause an adjustment in the forest canopy) (Mackay, 2001). This means the recognition of the evolutionary adjustments of the ecosystem and soils to cycles in water, energy and soil nutrients and the atmosphere (C cycle). Other authors examine the role of forests, snow and soil in the capacity of some sub-basins to store and provide humidity at different spatial-temporal scales (Post & Jones, 2001); that is, the application of the notion of learning and function as a result of an evolutionary-adaptive process.

3.4 Elements that define an evapotranspiration system

The above works provide elements to define an evapotranspiration system. Such is the case for the rough layer, for which the principles of interaction or processes that control evapotranspiration are observed. A mixed layer can also be formed in which the boundary layer of the atmosphere exerts controls on state variables by humidity conditions in the atmosphere and wind (Albertson et al., 2001). The emergence of the mixed layer speaks to local controls (surface-air temperature difference, *d*t) and regional controls (vertical structure, LAI) on instantaneous and diurnal energy flows (see figure 3). The dotted line represents the altitude of the mixed layer, corresponding to the lower limit of the mixed layer in light gray. Under the mixed layer is the rough layer defined by the heterogeneity of the ground cover. The arrows indicate the system's inputs and outputs and general interactions among the components of the system—soil, cover, atmosphere—through matter and energy flows; the clouds emanating from the land cover represent surface flows driven by evapotranspiration and the effect of the mixed layer.

At temporal scales of days and weeks, significant interaction is observed in the adaptive and phenological cycles of the ecosystems, effective precipitation and available radiation (Shuttle worth & Gash, 2005; Coronel & Mora, 2006; Coronel & Mora, 2007). Soil moisture deserves special mention because it affects local and regional scales, first triggering the evaporation rate at the beginning of the humid season and then regulating the evapotranspiration rate during the different phases of the adaptive cycle by coupling the water and C cycles through principles of optimization and stability of efficiency in the use of water.

The vertical limits are defined by the rough layer and/or the mixed layer in the upper and lower limits by the A and B soil horizons that regulate the flows of nutrients and water available to plants. At the local scale, the horizontal limits would be defined by the heterogeneity of the surface, the proportion of area with vegetation cover, the lack of vegetation or bare soil, impermeable surfaces or water bodies. At the regional scale—due to complex interactions between the canopy and the atmosphere, as well as soil controls—the limits would be defined primarily by the rates of efficiency in the use of water, since this function emerges from the local and regional processes mentioned and, in the temporal context, it is dependent on the system's adaptive cycle, with controls exerted by the principles of optimization, stability and conservation according to the availability of critical resources. In summary, the limits of an evapotranspiration system are open, emergent, specific to the temporal-spatial context and highly interactive with the environment.

The resulting function of the evapotranspiration system, or efficiency in the use of water, summarizes the interactions among components in the vertical and horizontal planes and can trigger emergent responses under disturbance conditions depending on the co-evolution and memory of its components, the adaptive cycle in which it is found and the predominant management practices (notion of contingency or convergence). Although man is not a component of the system, human interaction with it through the intentional modification of different driving processes so as to control productivity can result in altering of adaptive cycle (for example, not permitting the reorganization phase to occur, which follows the release phase in annual agricultural crop systems, see figure 4). This does not consider the conservation of critical resources and does not favor diversity of functions and responses, processes that contribute to learning and the development of memory. There is excessive exportation of entropy into the environment with no increase in the self-organization by

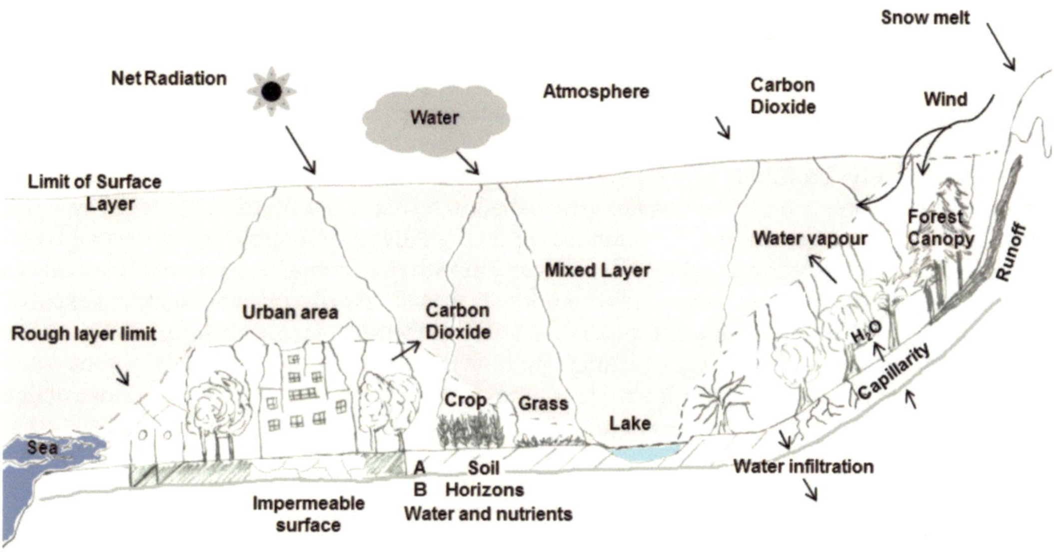

Fig. 3. Evapotranspiration system, components, limits and interactions

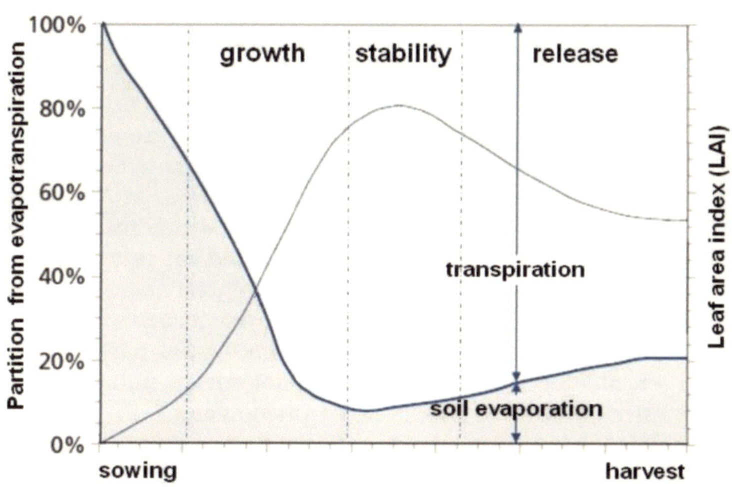

Fig. 4. Adaptive cycle for an annual crop system. Modified from FAO, (1990)

accumulating biomasa or recycling when significant controls are exerted on feedback mechanisms at local scales, such as the use of fertilizers and control measures for infestations and undergrowth (Günther & Folke, 1993). Although they satisfy certain societal needs, this type of local management affects a landscape's regional and/or slow stability processes, such as aquifer recharge, soil enrichment, landscape connectivity and biodiversity.

The development of generating heuristics of resilience and the emphasis on the structure-function connection in the modeling of evapotranspiration systems is fundamental to the adaptation of these systems, and thus it is important to validate them in terms of their role on driving configuration on evapotranspiration systems: panarchies, or interaction of processes across scales; emergences, results of feedback processes among the components of an evapotranspiration system; convergences or contingencies, primarily by management elements and meteorological disturbances and; changes in adaptive cycles. The emphasis on the scale of the landscape is established by its importance to its stability and to the strategy to decrease the effects at regional scales on the panarchy of evapotranspiration systems.

The next section discusses certain methodological strategies that can be implemented which would be greatly useful to demonstrate the four heuristics of resilience considered. In addition, information requirements are established in terms of the relation of these strategies with driving processes in the functional response of evapotranspiration systems. These options do not exhaust the possible alternatives but rather summarize some of the advances in terms of observation schemes, methodologies based on discovering temporal-spatial patterns and modeling of complex processes with bottom-up approaches. In addition, the relevance of communication and learning elements are discussed and the role of tools such as participatory planning, regional planning, and cybercartography as constructive methodological strategies for social adaptation.

4. Perspective for modeling evapotranspiration at application levels

4.1 Methodological strategies for the application of elements of complexity

Two approaches—bottom-up and top-down—stand out in the literature review of methodological strategies to incorporate the complexity paradigm at application levels (logical and computational) and epistemological levels.

The former consists of applying generic systemic models to the process of interest, differentiating components, interactions at local scales, and establishing feedback rules at regional scales. At the same time, state variables are defined, or even better, threshold values or control mechanisms that trigger emergent behavior are established (Costanza & Maxwell, 1991; Parrott & Kok, 2000). This approach seeks to simulate behavior at more regional scales, which later are compared with observations at the same scale, or the simulated behavior is analyzed in terms of the degree to which the presumed behavior is represented (test of hypothesis). These exercises are based on prior knowledge about the process in question and use conceptual and geographic representation models—especially if the linkage between structure and function needs to be incorporated. Examples of this approach are complex object-based models and the geospatial pattern discovery (Parrott, 2009; Maceachren et al., 1999; Coronel, 2008). Bottom-up modeling is based on a phenomenological strategy, or driven by phenomena; in which logical and geocomputational models inherit the behavior of observed or simulated processes.

The top-down application consists of producing analogical models, favoring the discussion of knowledge frameworks applied to the specific problems and the search for solutions and representation models according to heuristics discussed and agreed to at the conceptual level. The application of cybernetic communication processes, such as participatory cartography and regional planning, are examples that well-illustrate this approach (Romero & Farinos, 2006; Cataldo and Rinaldi, 2010). These facilitate learning about a region or area from the perspective of the processes that comprise it and promote social participation at diverse levels of organization. Thus, adaptation is facilitated by the creation of consensus, the strengthening of networks and the comprehensive vision of a territory.

Our proposal is to integrate both perspectives so as to take the modeling of evapotranspiration to representation and epistemological levels. This would involve reconsidering proposals that have been examined from the hydrology perspective, focusing on organizing principles, testing hypotheses by comparing case studies and characterizing functional traits. At the same time, geographic and representation models must be developed, based on which computational models would be implemented in order to simulate complex behaviors. In addition, it is necessary to promote the discussion of the application of analogical models other than those for basins of attraction, such as connectivity and self-similarity (McDonnell et al., 2007; Schröder, 2007), which are also relevant to the application of resilience for the modeling of evapotranspiration. Therefore, the use of cartographic contexts and models to assist learning and developing knowledge about the organizing principles of a landscape—as in the communication processes for geocybernetics and cybercartography—can benefit the adaptation and development of social knowledge networks (Reyes, 2005; Reyes et al., 2006).

In order to progress toward the spatial-temporal representation of evapotranspiration systems, it is suggested that geospatial patterns discovery be applied in order to show the validity of certain heuristics, such as emergences, convergences, panarchies and adaptive cycles (Haggett, 1965; Coronel, 2008). Examples of this are applications of said process for discovering patterns in order to demonstrate the emergence of seasonal spatial-temporal structures in *d*t shown below in the figure 5. Each picture represents one stational period; numbers in black are julian days of the year. 8 days scenes from MODIS surface temperature were used along with air temperature from a meteorological network.

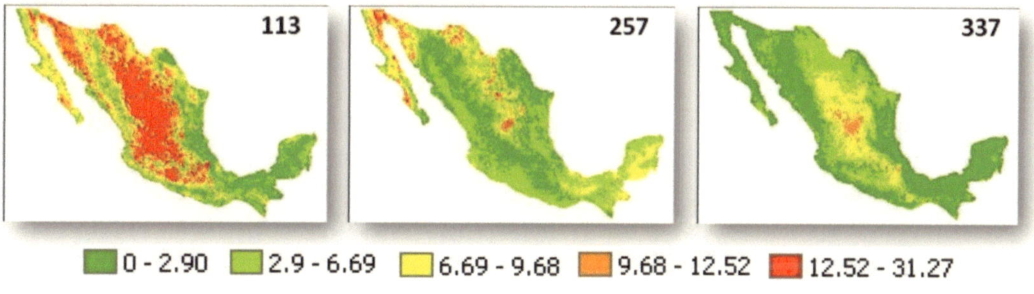

Fig. 5. Spatial-temporal structures on surface–air temperature differences *d*t (°C), in México, 2007

Some studies at the landscape scale have shown that the surface-atmosphere interactions observed in the surface layer are the result of local and regional processes of heterogeneity and cover roughness, as well as the effects of atmospheric variables at the meso-scale, such as air humidity and wind speed (Coronel et al., 2008; Albertson et al., 2001). This clearly indicates that an important feedback process occurs between local and regional scales (multi-scale effects in the panarchy of evapotranspiration systems). It is worth emphasizing that these works incorporate methodological elements to discover spatial-temporal patterns, including aggregation and disaggregation methods, geostatistics and exploratory spatial statistics, all of which seek to formulate the relations of association among processes.

In addition, documented cases exist of significant effects at the local and regional scales on hydrologic processes, such as increase in run-off and decrease in water recharge, which are directly caused by poor management of evapotranspiration systems based on a simple productivity approach. These effects are exacerbated by extreme regional precipitation events and by prolonged periods of drought or higher than normal air temperatures, which are considered convergences and are difficult to predict (Van Aalst, 2006; Cutter et al., 2008; Romero, 2010).

Lastly, it is possible to characterize the phases of the adaptive cycle in which an evapotranspiration system is found based on functional responses, such as the efficiency in the use of water, or the production of entropy in relation to an increase in self-organization, such as the application of primary productivity to the accumulation of biomass (Günther & Folke, 1993; Goetz et al., 2009).

4.2 The potential of observation schemes for incorporating organizing principles in evapotranspiration modeling

The design of observation schemes are largely based on the users' need for information about the phenomenon to be observed (Jensen, 2007). Thus, it is relevant to precisely identify the characteristics that a datum should present in order to most adequately represent a process. The literature recognizes that there are different levels of precision in data that should be considered for designing observation schemes: spatial and temporal precisions, contents of information or level of representation, geometric and thematic precisions (Schmugge et al., 2002; Couturier et al., 2009). Even more important is recognizing that prior knowledge about a process is what enables its spatial and temporal domain scales to be determined, in which the main variability of the process is observed.

Based on what has been discussed throughout this chapter, it is possible to recognize certain processes in evapotranspiration which directly relate to the previously discussed organizing principles: surface temperature difference, albedo (inversely proportional to the amount of energy that interacts with the surface or Rn) and primary productivity (related with the accumulation of biomass). It has been shown and discussed that these processes behave in a way that favors the specific behaviors of complexity that have been theoretically suggested. Thus, some of the potentials of remote sensing schemes are presented here, in terms of the representation of spatial and temporal behavior of these processes and their application to bottom-up procedures to show the validity of the emergences, convergences, panarchies and adaptive cycles, as heuristics of resilience.

The difference between surface temperatures should be represented with adequate resolution, given its effect at the local scale on sensible heat flow. Thus, advances in the characterization of surface temperature, with the use of remote sensing data in the thermal infrared portion (~ 10 μm wavelength), has been directed toward the discriminating covers according to their brightness temperature (Gillespie et al., 1998). In terms of the air temperature, meteorological observation networks are normally used and are efficient in representing the behavior of this variable, which has a more regional behavior (Coronel, 2008). Nevertheless, to represent the surface fluxes on a more regional scale, it has been shown that separating temperature according to cover does not contribute to the variation in latent flow and that *d*t tends toward middle or stable values for an area, as a control for the efficiency in the use of water through emergence behavior at the landscape scale (Albertson et al., 2001; Coronel & Mora, 2007). What has not yet been resolved is whether this behavior is a strategy of evapotranspiration systems in response to climatic variability.

Albedo represents the portion of total incident solar radiation that is reflected or is transmitted through the atmosphere and, finally, registered by sensors. Net solar radiation, available for energy balance, is the remaining proportion and, thus, because of its relevance, albedo should be adequately characterized. Albedo is correlated with erosion and decrease with soil moisture, and the density of vegetation is an attenuating factor in this process (Robinove et al., 1981). Albedo is also an important indicator of desertification processes (Xu et al., 2000). These findings suggest that albedo presents behavior that is regulated by local processes such as fractional vegetation cover, and increased by transformation on land cover. One way to study the effects of feedback between the atmosphere and surface flows (multi-scale interactions in panarchy and convergences) may be relating albedo at regional, continental and seasonal scales, with energy fluxes (Baldocchi et al., 2001; Brauman et al., 2007).

Primary productivity presents efficient control of water, carbon and nutrient cycles at local and regional scales. Along with albedo and *d*t, it should be observed and studied in detail and with precision, with an emphasis on those indicators of primary productivity that make up the vertical and horizontal structure of the canopy. Studies have been conducted of estimations of aerial biomass (leaves and branches) which have led to important debates about the main satellite components (Lu, 2006; Goetz et al., 2009). Lidar systems adequately represent the vertical structure of the canopy and, therefore, have great potential for estimating aerial biomass, although they present limitations in the observation of large areas due to maintenance and productions costs (Hyde et al., 2006). An alternative for mapping surface roughness is the integration of data from multiple sensors (Moghaddam et al., 2002; Coulibaly et al., 2008). This would enable characterizing both the horizontal structure of the cover (through optical sensors) as well as the vertical structure of the canopy (through radar and Lidar sensors). It is possible to find differences in the strategies for accumulation of biomass according to the adaptive cycle and to incorporate this information in complex models.

5. Conclusions and future considerations

Emphasis on the structure-function relation, adaptation in social-ecological systems and the development of methodological strategies themselves appear to be the most relevant elements to the incorporation of resilience as an analogical model and of the complexity paradigm as a conceptual framework for the modeling of the evapotranspiration process.

One of the most significant challenges for incorporating resilience in evapotranspiration systems is continuing to advance the definition of these systems, characterize their behavior and the criteria for organization that support classification systems, for the purpose of generating comparable information in different regions to advance learning and developing knowledge about evapotranspiration. This knowledge can be made available for regional planning in different areas, according to the information needs of decision-makers, within the application of participatory planning, and/or geocybernetics.

In modeling evapotranspiration systems at application and epistemological levels, important information needs exist regarding nutrients present in the soil (water, C and N) that strongly control evapotranspiration rates, and possible connections between biodiversity and the function response need to be explored through a functional group approach and by integrating indicators of this process, such as Beta biodiversity (Whittaker, 1972; Qian et al., 2009). Adequate schemes should be advanced for evaluating data of different variables associated with evapotranspiration—at spatial and/or biophysical levels of precision—with emphasis on the representation of driving processes and the testing of hypotheses through object-based models and the discovery of spatial-temporal patterns.

6. Acknowledgements

We have got valuable contribution and help on figure design by Cynthia Fabianna Galeana Pizaña and GabrielaLópez. Many thanks to Centrogeo by faciliting the work and supporting meetings and discussion spaces.

7. References

Abel, N., D. H. M. Cumming, & J. M. Anderies. (2006). Collapse and reorganization in social-ecological systems: questions, some ideas, and policy implications, Ecology and Society, *11,1, 17. ISSN:1708-3087*

Ahern, 1999, Spatial Concepts, Planning Strategies and future Scenarios: a Framework Method for Integrating Landscape Ecology and Landscape Planning. In: Landscape Ecological Analysis. Issues and Applications. *Klopatek & Gardner, New York. Springer-Verlag.175-201. ISSN:*

Ahl, V.; Allen, T. F. H. & Lerner, P. (1996). Hierarchy Theory: A vision, vocabulary and epistemology. Columbia University Press. ISBN: 0231084803 , USA

Albertson, J.D., Katul, G.G., & Wiber, P.., 2001. Relative importance of local and regional controls on coupled water, carbon, and energy fluxes, Adv Water Resour, *24, 1103-1118, ISSN:* 0309-1708

Alcocer, J., Escobar, E. & Lugo, A. (2000). Water use (and abuse) and its effects on the crater-lakes of Valle de Santiago, México. *Lakes & Reservoirs: Research and Management. Vol. 5, pp. 145-149, ISSN: 1440-1770.*

Allen R.G., Tasumi M., Morse A. & Trezza R., (2005). A Landsat-based energy balance and evapotranspiration model in Western US water rights regulation and planning. Irrigation and Drainage Systems, 19(3/4): pp. 251–268, ISSN: 0168-6291

Anderies, J. M., B. H. Walker, & A. P. Kinzig. (2006). Fifteen weddings and a funeral: case studies and resilience-based management. Ecology and Society, 11, 1, 21. *ISSN: 1708-3087*

Arai, T. & Kragic, D. (1999). Name of paper, In: Name of Book in Italics, *Name(s) of Editor(s), (Ed.), page numbers (first-last), Publisher, ISBN, Place of publication*

Baldocchi, D., Falge, E., Gu, L., Olson, R., Hollinger, D., Running, S., Anthoni, P., Bernhofer, C., Davis, K., Evans, F., Fuentes, J., Goldstein, A., Katul, G., Law, B.,Lee, X., Malhi, Y., Meyers, T., Munger, W., Oechel, W., K. T. Paw U, Pilegaard, K., H. P. Schmid, Valentini, R., Verma, S., Vesala, T., Wilson, K., & Steve W.(2001). FLUXNET: A new tool to study the temporal and spatial variability of ecosystem-scale carbon dioxide, water vapour and energy flux densities. Bulletin of the American Meteorological Society, 2001, 82, pp. 2415–2434, ISSN 1520-0477.

Bastiaanssen, W.G.M., Menenti, M., Feddes, R.A. & Holtslag, A.A. (1998). A remote sensing surface energy balance algorithm for land (SEBAL). International Journal of Hydrology 1998, pp. 212–213, ISSN: 2042-7808.

Berkes, F., & Folke, C. (1998) Linking Social and Ecological Systems Management Practices and Social Mechanisms for Building Resilience. ISBN: 0521591 1406, New York: Cambridge Univ. Press.

Boegh, E., Soegaard, H., & Thomsen, A. (2002). Evaluating evapotranspiration rates and surface conditions using Landsat TM to estimate atmospheric resistance and surface resistance. Remote Sensing of Environment. 2002, 79, pp. 329-343, ISSN:0034-4257.

Brauman KA, Daily GC, Duarte TK, & Mooney HA (2007) The nature and value of ecosystem services: an overview highlighting hydrologic services. Annual Rev Environ Resource , *32: 67–98, ISSN 1543-5938.*

Brooks, D., Collier, J., Maurer, B., Smith, J. & Wiley, E. (1989). Entropy and information in evolving biological systems, Biology & Philosophy, *Vol. 4, No. 4, 407-432, ISSN 0169-3867*

Cadenasso, M., Pickett, S. & Grove, J. (2006). Dimensions of ecosystem complexity: Heterogenety, connectivity, and history, Ecological Complexity 3, *1-12. ISSN 1476-945X.*

Callaghan, T.V., Björn, L.O., Chernov, Y., Chapin, T., Christensen, T.R., Huntley, B., Ims, R.A., Johansson, M., Jolly, D., Jonasson, S., Matveyeva, N., Panikov, N., Oechel, W., Shaver, G., Schaphoff, S., & S. Sitch (2004) Effects of Changes in Climate on Landscape and Regional Processes, and Feedbacks to the Climate System, Ambio, *Vol. 33, No. 7, Climate Change and UV-B Impacts on Arctic Tundra and Polar Desert Ecosystems (Nov., 2004), pp. 459-468*

Castree, N. (2010). Book review: Environmental change and globalization: double exposures. Progress in Human Geography, *34, 274-277. ISSN:0309-1325*

Cataldo, A. & Rinaldi, A. (2010). An onthological approach to represent knowledge in territorial planning science, Computers, Environment and Urban Systems, Vol. 34, No. 2, 117-132, Marzo 2010, ISSN: 0198-9715

Clark, D.B., Taylor, C.M., Thorpe, A.J., Harding, R.J. & Nicholls, M.E. (2003) The influence of spatial variability of boundary-layer moisture on tropical continental squall lines, Quartely Journal of the Royal Meteorological Society, *129, 1101-1121, ISSN: 1477-870X*

Cleugh, H.A., Leuning, R., Qiaozhen, M.,. & Running S.W. (2007). Regional evaporation estimates from flux tower and MODIS satellite data a CSIRO. Remote Sensing of Environment *106 (2007) 285–304, ISSN: 0034-4257.*

Constanza, R., & T. Maxwell, (1991) Spatial Ecosystem Modelling Using Parallel Processors, Ecological Modelling, *58, 159-183. ISSN:0304-3800*

Coronel, C. & F., Mora (2007) Spatial-temporal Patterns of Fractional Vegetation Cover and Land Surface Temperature for Monitoring the Water Balance, Proceedings of the 32th International Symposium of Remote Sensing of Environment, (junio 2007) Costa Rica .

Coronel, C. & F., Mora (2006) Dinámica de la fracción vegetal en la cobertura terrestre: Análisis subpixel en imágenes MODIS, Memorias del XII Symposium Internacional, Capitulo Colombia, pp. 79-85, ISBN: Centro Internacional de Alta Tecnología , SELPER, San Juan de Costa Rica

Coronel, C., (2008) Incorporación del proceso de descubrimiento de patrones geoespaciales en el estudio sobre componentes del balance hídrico, Master in Geomatics. Centrogeo, CONACYT, D.F., México.

Coronel, C., E. Rosales, F. Mora, A.A. López-Caloca, Tapia-Silva, F.-O, & G., Hernández (2008) Monitoring evapotranspiration at landscape scale in Mexico: Applying the energy balance model using remotely-sensed data, Proceedings of SPIE Europe - Remote Sensing, pp. 1-12, 71040H, University of Wales Institute, SPIE Europe, Cardiff, Wales, UK

Coulibaly, L., Migolet, P., Adegbidi, H., Fournier, R. and Hervet, E. (2008) Mapping aboveground forest biomass from Ikonos satellite image and multisource geospatial data using Neural Networks and Kriging interpolations. Geoscience and Remote Sensing Symposium, 2008. IGARSS 2008. IEEE International. *3: 298-301.*

Courault, D., Seguin, B., & A., Olioso (2005). Review on estimation of evapotranspiration from remote sensing data: From empirical to numerical modeling approaches. Irrigation and Drainage Systems, *19, 223-249, ISSN:1573-0654*

Couturier, S., Mas J.-F., Cuevas G., Benítez J., Vega-Guzmán A., & Coria-Tapia V. (2009). An accuracy index with positional and thematic fuzzy bounds for land-use/ land-cover maps. Photogrammetric Engineering and Remote Sensing, *75, 7, 789-805, ISSN: 0099-1112*

Crow, W.T., Wood, E.F.. (2003). The assimilation of remotely sensed soil brightness temperature imagery into a land surface model using Ensemble Kalman filtering: a case study based on ESTAR measurements during SGP97. Advances in Water Resources 26 (2003) pp. 137–149 ISSN 0309-1708.

Cumming, G., Cumming, D., & C. Redman. (2006). Scale mismatches in social-ecological systems: causes, consequences, and solutions. Ecology and Society *11(1):14. ISSN:1708-3087*

Cutter, S.L, Barnes, L., Berry, M., Burton, C., Evans, E., Tate, E., & Webb, J. (2008) A place-based model for understanding community resilience to natural disasters, Global Environmental Change, 18, 598–606, ISSN: 0959-3780

Daily, G.C., Alexander, S., Ehrlich, P.R., Goulder, L., Lubchenco, J., Matson, P.A., Mooney, H.A., Postel, s., Schneider, S.H., Tilman, D., & G.M. Woodwell (997) Ecosystem Services: Benefits Supplied to Human Societies by Natural Ecosystems, Issues in Ecology, No. 2, 1-16, ISSN 1092-8987

Dan L., Ji J, & Li Y. (2007) The interactive climate and vegetation along the pole-equator belts simulated by a global coupled model, Advances in Atmospheric Sciences, *24(2): 239-249 ISSN 0256-1530.*

de Bello, F., Lavorel, S, Dı´az, S., Harrington R., Cornelisse, J.H.C., Bardgett, R.D., Berg, M.P., Cipriotti, P., Feld, C., Hering, D., Martins da Silva, P Potts, S.G, Sandin, L., Sousa, J.P., Storkey, J., Wardle, D.A., & Harrison, P.A. (2010) Towards an assessment of multiple ecosystem processes and services via functional traits, Biodivers Conserv,. *Received: 31 March 2010*

DOE, (1996) Science Plan for the Atmospheric Radiation Measurement (ARM) Program, U.S. Department of Energy, Washington, DOE/ER-0670.

FAO. (1990). Evapotranspiración del cultivo, Guías para la determinación de los requerimientos de agua para los cultivos, FAO, ISBN 92-5-304219-2, USA

Folke, C. & Kautsky, N. (1992) Aqualcultural with its environment; prospects for sustanibility, Ocean and Coastal Management, 17, 1, pp. 5-24, ISSN: 0964-5691

Garatuza-Payan J. & Watts, C.J. (2005). The use of remote sensing for estimating ET of irrigated wheat and cotton in Northwest Mexico. Irrigation and Drainage Systems, *2005 19(3/4): pp 301–320, ISSN:1573-0654*

García R. (2008). Sistemas complejos: Conceptos, métodos y fundamentación epistemológica de la investigación interdisciplinaria, *Gedisa., ISBN: 8497841646 ISBN-13: 9788497841641, España.*

Goetz, S., Baccini, A., Laporte, N., Johns, T., Walker, W., Kellndorfer, J., Houghton, R., and Sun, M. (2009) Mapping and monitoring carbon stocks with satellite observations: a comparison review. Carbon Balance and Management *2009, 4:2.*

Gorenburg, I.P., McLaughlin, D.& Enterkhabi, D. (2001) Scale-recursive assimilation of precipitation data. Adv Water Resour, 24, 941-953, ISSN: 0309-1708

Gunderson L. H., C. S. Holling, and S. S. Light. (1995). Barriers and bridges to the renewal of ecosystems and institutions. Columbia University Press, ISBN 0-231-10102-3 New York, USA.

Gunderson, L. (2000). Ecological resilience –in Theory and application, . The human use of human beings, *Da Capo Press, ISBN 0-306-80320-8, Boston, USA*

Gunderson, L. H., S. R. Carpenter, C. Folke, P. Olsson, & G. D. Peterson. (2006). Water RATs (resilience, adaptability, and transformability) in lake and wetland social-ecological systems. Ecology and Society *11, ISSN: 1708-3087*

Günther, F., & C. Folke, (1993) Characteristics of Nested Living Systems, Journal of Biological Systems, *1, 3, pp. 257-274, ISSN:*

Haggett, P. 1965, Locational Analysis in Human Geography, *London Edward Arnold, ISBN:071315179X, London, UK*

Haila, Y. (2002) Scaling environmental issues: problems and paradoxes, Landscape and Urban Planning 61 2002 pp. 59–69 ISSN 0169-2046.

Helen A. Cleugh, Ray Leuning, Qiaozhen Mu, & Steven W. Running (2007). Regional evaporation estimates from flux tower and MODIS satellite data a CSIRO, Remote Sensing of Environment 106 (2007) 285–304, ISSN: 0034-4257

Holland, J.H. (1995). Hidden Order: How Adaptation Builds Complexity. *M.A. Addison-Wesley. ISBN: 1064-5462*

Holland, J.H. (2002). Complex adaptive systems and spontaneous emergence, In: Complexity and Industrial Clusters: Dynamics and Models in Theory and Practice (Contributions to Economics), Curzio, A.Q. and Fortis, M., pp. 25-34, Physica-Verlag HD, ISSN: 3790814717, Heidelberg

Holling, C. S. (1973). Resilience and stability of ecological systems. Annual Review of Ecology and Systematics, *4:1-23. ISSN: 1543-592X.*

Hyde, P., Dubayah, R., Walker, W., Bryant, J., Hofton, M. & Hunsaker, C. (2006) Mapping forest structure for wildlife habitat analysis using multi-sensor (LiDAR, SAR/ InSAR, ETM+, Quickbird) synergy. Remote Sensing of Environment, *102, 63-73.ISSN: 0034-4257*

Jensen, J. 2007. Remote Sensing of the Environment, An Earth Resource Perspective, Pearson Prentice Hall, ISBN 0-13-188950-8, NJ, USA

Katul, G., Lai, C.T., Schäfer, K., Vidakovic, B., Albertson, J., Ellsworth, D., & Oren, R. (2001) Multiscale analysis of vegetation surface fluxes: from seconds to years, Advances in Water Resources, *Vol. 24, 2001, 1119-1132, ISSN 0309-1708*

Kinzing, A. P., P. Ryan, M. Etienne, H. Allyson, T.Elmqvist, & B. H. Walker. (2006). Resilience and regime shifts: assessing cascading effects. Ecology and Society *11(1): 20. ISSN: 1708-3087*

Lebel, L., J. M. Anderies, B. Campbell, C. Folke, S. Hatfield-Dodds, T. P. Hughes, & J. Wilson. (2006). Governance and the capacity to manage resilience in regional social-ecological systems. Ecology and Society *11(1): 19. ISSN:1708-3087*

Liu, J., Dietz, Th., Carpenter, S., Alberti, M., Folke, C., Moran, E., Pell, Al., Deadman, P., Kratz, T., Lubchenco, J., Ostrom, E., Ouyang, Z., Provencher, W., Redman, Ch., Schneider, S. & Taylor, W. (2007). Complexity of coupled human and natural systems. Science. *317: 2007, pp. 1513-1516. ISSN 0036-8075 (print), 1095-9203 (online)*

Maceachren, Wachowicz, Edsall, Haug & Masters, 1999, Constructing knowledge from multivariate spatiotemporal data: integrating geographical visualization with knowledge discovery in database methods, Int. J. Geographical Information Science, *13, 4, 311–334, ISSN:*

Mackay, D.S. (2001) Evaluation of hidrologic equilibrium in a mountainous watershed: incorporating forest canopy spatial adjustment to soil biogeochemical process, Advances in Water Resources, *Vol. 24, 2001, 1211-1227, ISSN 0309-1708*

Maganda, MC. (2004). Disponibilidad de agua un riesgo construido. Vulnerabilidad hídrica y crecimiento urbano-industrial en Silao, Guanajuato (México). Tesis Doctorado en Antropología. CIESAS, Septiembre, 2004. Distrito Federal, México

Mahrt L., Vickers, D., & Sun, J. (2001) Spatial variations of surface moisture flux from aircraft data. Advances in Water Resources *24 (2001) pp. 1133-1141. ISSN: 0309-1708.*

Martin, R. & Sunley, P. (2007). Complexity Thinking and Evolutionary Economic Geography. Papers in Evolutionary Economic Geography. pp. 1—36, ISSN:

Martínez, E. & Reyes, M. (2005). Cybercatography and Society. In: Cybercartography: Theory and Practice, *Taylor, F. (Ed.), 99-122, Elsevier, ISBN 978-0-444-51629-9, Amsterdam, The Netherlands*

McCabe, M.F. & Wood, E. F. (2006) Scale influences on the remote estimation of evapotranspiration using multiple satellite sensors. Remote Sensing of Environment 105, (2006), pp. 271–285, ISSN: 0034-4257

McDonnell J.J., Sivapalan, M., Vaché, K., Dunn, S., Grant, G., Haggerty, R., Hinz, C., R. Hooper, Kirchner, J., Roderick, M. Selker, L.J., & Weiler, M. (2007). Moving beyond heterogeneity and process complexity: A new vision for watershed hydrology. Water Resources Research, *43, (july 2007), pp. 1-6, ISSN:0043-1397.*

Moghaddam, M., Dungan, J. and Acker, S. (2002) Forest Variable Estimaion From Fusion of SAR and Multispectral Optical Data. IEEE Transaction on Geoscience and Remote Sensing. *40:10, 2176-2187, ISSN:0196-2892*

Neale, C., Jayanthi, H.& Wright, J.L. (2005). Irrigation water management using high resolution airborne remote sensing, J. Irrigation and Drainage Systems, *19(3/4), 2005, pp. 321–336., ISSN:1573-0654*

Norman, J. M., W. P. Kustas, & K. S. Humes (1995). A two-source approach for estimating soil and vegetation energy fluxes from observations of directional radiometric surface temperature, Agric. For. Meteorol., *77, 263-293 ISSN:0168-1923.*

Parrot, L., (2009) Understanding and managing the landscape as a complex system: What can bottom-up modelling approaches contribute?., Applying Landscape science to Natural Resource Management. A Symposium of the Spatial Sciences Institute Biennial International Conference. *pp. 1-15, Adelaide Convention Centre, Adelaide, (october 2009), Australia*

Parrott, L. & Kok, R. (2000) Incorporating Complexity in Ecosystem Modelling, Complexity International, Vol. 7, (june, 2000) pp. 1-19, ISSN: 1320-0682.

Perry, Chris, (2005). Irrigation reliability and the productivity of water: a proposed methodology using evapotranspiration mapping, J. Irrigation and Drainage Systems, *19, 2005 pp 211-221, ISSN:1573-0654*

Peterson, G. D., C. R. Allen, & C. S. Holling. (1998). Ecological resilience, biodiversity, and scale. Ecosystems *1:6–18, SSN:1435-0629*

Peterson, G. D., G. S. Cumming, & S. R. Carpenter. (2003). Scenario planning: a tool for conservation in an uncertain world. *Conservation Biology* 17, 358–366, ISSN: 0888-8892

Pickett, S., Cadenasso, M., & Grove, J. (2004). Resilient cities: meaning, models and metaphor for integrating the ecological, socio-economic and planning realms. Landscape and urban planning, 69, 369–384, ISSN 0169-2046

Pickett, S., Jones, C., Kolasa, J., (1994). *Ecological Understanding: The Nature of Theory and the Theory of Nature.* Academic Press, San Diego, California. ISBN 0-12-554720-X

Pimm, S. L. (1991). *The balance of nature?*, University of Chicago *Press, ISBN: 9780226668307 Chicago, Illinois, USA*

Post, D.A. & Jones, J.A., 2001. Hydrologic regimes of forested, mountainous, headwater basin in New Hampshire, North Carolina, Oregon, and Puerto Rico. Adv Water Resour, *24, 1195-1210, ISSN:* 0309-1708

Qian, H., Kissling, D.W.., Wang, X., & Andrews, P. (2009) Effects of woody plant species richness on mammal species richness in southern Africa, Journal of Biogeography, 36, 1685–1697, ISSN: 1365-2699

Reyes, M. (2005). Cybercartography from a Modeling Perspective.In: Cybercartography: Theory and Practice, *Taylor, F. (Ed.), 63-98, Elsevier, ISBN:978-0-444-51629-9, Amsterdam, The Netherlands*

Reyes, M. & Martínez, E. (2005). Technology and Culture in Cybercartography. In: Cybercartography: Theory and Practice, *Taylor, F. (Ed.), 123-148, Elsevier, ISBN 978-0-444-51629-9, Amsterdam, The Netherlands*

Reyes, M., Taylor, F., Martínez, E. & López, F. (2006). Geo-cybernetics: A New Avenue of Research in Geomatics. Cartographica, *Vol. 41, No.1, 7-20, ISSN 0317-7173*

Reyes, M., Taylor, F., Martínez, E. López, F. (2006). Geo-cybernetics: A New Avenue of Research in Geomatics. Cartographica, Vol. 41, No.1, 7-20, ISSN 0317-7173

Riobinove, C.J.; Chavez, P.S.; Gehring, D.; & Holmgren, R. (1981)Arid land monitoring using Landsat albedo difference images. Remote Sens. Envrion, *11, 133-156. ISSN:0034-4257.*

Romero, J. & Farinós, J. (2006). Gobernanza territorial en España. Claroscuros de un proceso a partir del estudio de casos. *Universitat de Valèn*cia, ISBN 84-370-6542-9, Valencia España

Romero, L.P. (2010) Water in Mexico City: what will climate change bring to its history of water-related hazards and vulnerabilities?, Environment and Urbanization, 22; 157-178. ISSN:0956-2478

Ruddell, B. & Kumar, P. (2006). The Modelshed Geodata Model, In: Hydroinformatics, Kumar, P, Alameda, J., Bajcsy, P., Folk, M. & Markus, M. (Ed.), 81-99, CRC Press, ISBN 0-8493-2849-2, Boca Raton, FL, USA

Sandholt I, Rasmussen K, & Andersen J., (2002). A simple interpretation of the surface temperature/vegetation index space for assessment of surface moisture status, International Journal of Remote Sensing, *2002, 79(2): pp. 213-224. ISSN 0143-1161, Online ISSN: 1366-5901*

Schmugge T. J., Kustas W. P., Ritchie J. C., Jackson T. J. & Rango A., (2002). Remote sensing in hydrology. Advances in Water Resources, *25(2002), pp. 1367-1385 ISSN:0309-1708.*

Schröder, B. (2006). Pattern, process, and function in landscape ecology and catchment hydrology - how can quantitative landscape ecology support predictions in ungauged basins (PUB)? Hydrology and Earth System Sciences, *10, pp. 967-979. ISSN:1027-5606*

Senay, G.B., Budde, M., Verdin, J.P., & Melesse, A.M. (2007). A coupled remote sensing and simplified surface energy balance approach to estimate actual evapotranspiration from irrigated fields. Sensors, *2007, 7, pp. 979-1000, ISSN1424-8220.*

Shuttleworth, W.J., & J.H. Gash, (2005) The role of large-scale field experiments in water and energy balance studies, In: Encyclopedia of Hydrological Science, *M. G. Anderson Jhon Wiley & Sons, Ltd.*

Sivapalan, M. (2005) Pattern, Process and Function: Elements of a Unified Theory of Hydrology at the Catchment Scale, In: Encyclopedia of Hydrological Science, *M. G. Anderson Jhon Wiley & Sons, Ltd.*

Sivapalan, M., Kumar, P. & Harris, D. (2001). Nonlinear propagation of multi-scale dynamics through hydrologic subsystems. Advances in Water Resources, Vol. 24, No. 9, November, 2001, 935-940, ISSN 0309-1708

Soto, M. & Soto, F. (1990). Cambios recientes en el uso del suelo agrícola en la región del Bajío, Gto, Geografía y Desarrollo, *2, 5, 3–11. ISSN: 0187-6562.*

Szerszynsky B., (2002). Wild times and domesticated times: the temporalities of environmental lifestyles and politics, Landscape and Urban Planning, Vol. 61, No. 4, November 2002, 181-191, ISSN 0169-2046.

Taylor, C.M., & Lebel, T. (1998). Observational evidence of persistent convective – scale rainfall patterns. Monthly Weather Review, *126 (6), 1998, pp. 1597 -1607, ISSN: 1520-0493*

Van Aalst, M. (2006). The impacts of climate change on the risk of natural disasters. Disasters, Vol. 30, No. 1, 5-18, ISNN 0361-3666

Veitzer, S.A. & Gupta, V.K. (2001) Statistical self-similarity of width function maxima with implications to floods. Adv Water Resour, 24, 9, 955-965, ISSN: 0309-1708

Vilardy, Q.S.P. (2009). Estructura y dinámica de la ecorregión Ciénaga Grande de Santa Marta: una aproximación desde el marco conceptual de los sistemas socio-ecológicos complejos y la teoría de la resilencia Tesis de Doctorado Departamento interuniversitario de Ecología p.296 Universidad Autónoma de Madrid, 2009.

Walker, B. H., L. H. Gunderson, A. P. Kinzig, C., Folke, S. R. Carpenter, & L. Schultz. (2006). A handful of heuristics and some propositions for understanding resilience in social-ecological systems. Ecology and Society, *Vol. 11, No. 1: 13, ISSN 1708-3087.*

Walker, B., C. S. Holling, S. R. Carpenter, & A. Kinzig. (2004). Resilience, adaptability and transformability in social–ecological systems. Ecology and Society *Vol. 9, No. 2: 5, ISSN 1708-3087*

Whittaker, R. H. 1972. Evolution and measurement of species diversity. *Taxon, 21: 213-251.*

Wiener, N. (1954). The human use of human beings, Da Capo Press, ISBN 0-306-80320-8, Boston, USA

Wu, J. (1999). Hierarchy and Scaling: extrapolating information along a scaling ladder. Canadian Journal or Remote Sensing. *4, pp. 367-380. ISSN: 1712-7971.*

Xu, C-Y, J. Seibert & S. Halldin, (1996). Regional water balance modelling in the NOPEX area: development and application of monthly water balance models, Journal of Hydrology, *180, 1996, pp. 211-236, ISSN 0022-1694.*

Zavaleta, E.S., Thomas, B.D., Chiariello. N.R., Asner, G.P., Shaw, M.R. & Field, C.B. (2003). Plants reverse warming effect on ecosystem water balance. Proceedings of the National Academy of Sciences, *Vol.100, No 17, (august 2003),9892–9893, ISSN:0027-8424.*

Methods of Evapotranspiration Assessment and Outcomes from Forest Stands and a Small Watershed

Vladimir Cernohous[1], Frantisek Sach[1], Petr Kantor[2] and Vladimir Svihla[3]
[1]*Forestry and Game Management Research Institute, Research Station, Opocno*
[2]*Mendel University, Faculty of Forestry and Wood Technology, Brno*
[3]*Svihla, Fügnerova 809, Beroun*
Czech Republic

1. Introduction

The authors have used ET terms related to evapotranspiration terms which outcome from common equation of water budget. The components of water budget equation have been employed by the author prof. Kantor since 1970s. The terms are in compliance with terms of Encyclopedia of Forest Sciences (Burley et al., 2004) and with Committee of Hydrologic Impacts of Forest Management (National Academy of Sciences of the United States of America 2008).

All study and understanding begins with good definitions (Hewlett, 1982); therefore at first we present fundamental terms of forest hydrologic cycle taken into consideration for evapotranspiration of forest stands and small forested catchments (Bruijnzeel, 2004):

Precipitation (P) – rain, snow, and fog (occult precipitation)

Throughfall (Tf) – sum of direct throughfall reaching the forest floor without touching the canopy plus crown drip, once the storage capacity of the canopy has been filled

Stemflow (Sf) – part of precipitation travelling along the branches and trunks into the forest floor

Net precipitation (Pn) – sum of throuhgfall and stemflow [Pn = Tf + Sf]

Interception losses (Ei) – portion of the precipitation intercepted by the canopy and evaporated back to atmosphere – evaporation from a wet canopy [Ei = P – (Tf + Sf)]

Transpiration – water taken up by the roots and returned to the atmosphere via the process of transpiration (Et) – evaporation from dry canopy

Evaporation from the litter and soil surface (Es)

Evapotranspiration (ET) – total evapotranspiration equals to sum of Ei + Et + Es (in millimetres of water per time unit)

The interlocked character of the chief components of the hydrological cycle is summarized by the site or catchment water budget equation:

$$P = ET + Q + \Delta S + \Delta G \text{ (mm)} \tag{1}$$

where

Q is amount of drainage to deeper layers or streamflow,
ΔS change in soil water storage,
ΔG change in groundwater storage.
In forested catchments the hydrologic cycle thus in short involves precipitation, interception, evapotranspiration, overland flow, subsurface flow, groundwater flow and stream flow (**Fig. 1**).

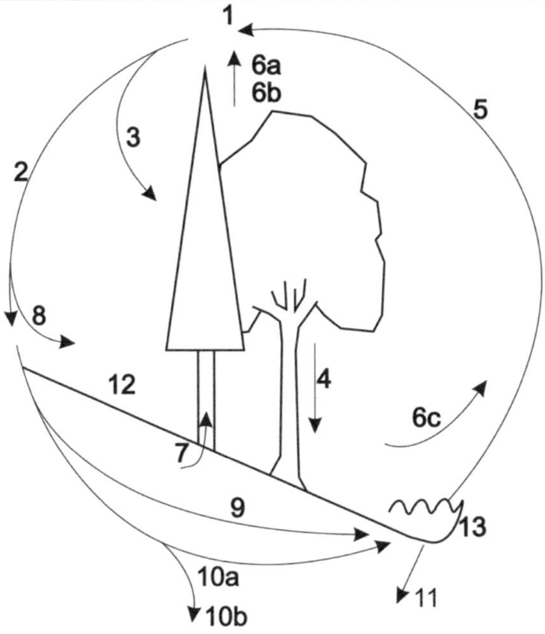

1	Atmosphere
2	Precipitation
3	Throughfall (direct throughfall and drip)
4	Stemflow
5	Evaporation
6a	Interception loss
6b	Transpiration of crown, canopy
6c	Evaporation from slope surface, forest floor and ground vegetation
7	Water uptake
8	Overland flow
9	Subsurface flow
10a	Groundwater flow
10b	Vertical percolation into the underlying bedrock and recharging deep groundwater
11	Stream flow
12	Slope
13	Stream

Fig. 1. Forest hydrologic cycle

On steeper hillslopes, which represent most of the topographical conditions of our hydrologic research areas in the Orlické hory Mts. and their piedmont, water infiltrating into soil is either stored in the soil or continues moving vertically to recharge local groundwater or flows as lateral subsurface flow; likewise comprised recent knowledge

Weiler & Mac Donnell (2004). In areas with a steep relief, the lateral subsurface flow (sometimes called interflow, throughflow, pipeflow, and subsurface stormflow) are very common in forest soils since the lateral hydraulic conductivity and the gravitational gradients are often high. Thus additional preferential flow pathways are present to enhance the downslope flow. A hypothetic hydrologic system for forest soil on a hillslope is presented in **Fig. 2** (by Thomas & Beasley, 1984).

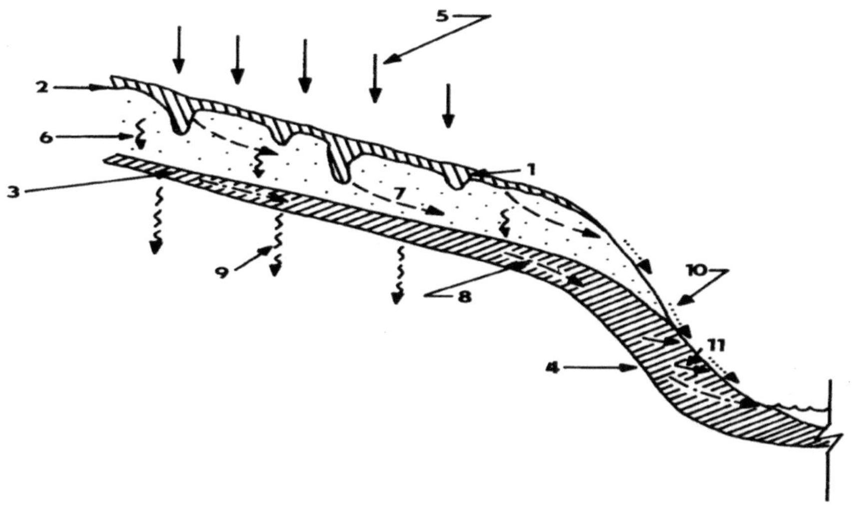

1- organic layer
2 - mineral soil surface
3 - saturated zone
4 - impeding layer
5 - net precipitation
6 - infiltration
7 - pipe flow
8 - seepage
9 - deep percolation
10 - overland flow
11 - return flow
(Thomas & Beasley, 1984)

Fig. 2. Hypothetic hydrologic system for forest soils on a hillslope

The system of research areas in piedmont and mountain region of the Orlické hory Mts., where the water regime of montane forest ecosystems and silvicultural methods are continuously investigated, is composed of three long-term experimental silviculture-hydrologic stations: Deštenská stráň - experimental balance areas on moderate WSW slope (alt. 890 m) to complex study water budget of Norway spruce and European beech stands during progress of forest reproduction, Česká Čermná - experimental runoff areas on steep south hillside of Dubovice (alt. 500 m) to study effect of regeneration felling (clear and shelterwood) on hillslope runoff and soil erosion, U Dvou louček - experimental watershed on SW slope with moderate gradient (alt. 880-950 m) to study effect of hydroameliorative treatment and conversion of forest ecosystem after salvage felling due to air pollution on water regime of the catchment.

For the main aims of various projects, which have involved optimization of tree species composition to ensure hydrologic services of forestry, technical and biological treatments to minimize overland flow and erosion from forest lands, reproduction methods decreasing risk of floods from rainstorms and snowmelt and water regime amendment of waterlogged forest soils to provide maximum retention and detention capability in mountain forests in protected region of natural water accumulation, the evapotranspiration studies play an important role. For the chapter of this book the ET has been described and assessed only in a growing period (May - October).

1.1 General characteristics of the area under study

The research areas are situated in the natural forest region of the Orlické hory Mts (24,000 ha) with their tops reaching 1,000–1,100 m alt. and in the natural forest area of the Foothills of the Orlické hory Mts (18,500 ha) with their tops between 500 and 800 m alt. From the geological point of view, the Orlické hory Mts belong to a crystalline complex with prevailing orthogneiss, while the fringe is formed by mica schist and paragneiss, the transition to the foothills is then formed by phyllites and amphibolites. Forest soils are moist and poor in minerals. Mountain brown soils (Cambisols) and humus podzols predominate (62% and 30% respectively). Soils affected by water occur less frequently (7%). The climate of the Orlické hory Mts ranges from cold in the uppermost part of the mountains to the temperate climate in their foothills. Mean annual air temperature ranges in the whole area from 4° to 7°C, in the growing season from 10° do 13°C. Precipitation is rather copious, with the frequent occurrence of fog (occult precipitation), in winter accompanied by frost deposit and glazed frost. Mean annual total precipitation ranges between 1,300 (mountains) and 800 mm (foothills), in the growing season from 750 to 410 mm. The number of days with snowfall is related to the altitude and ranges from 80 (mountains) to 40 (foothills). Duration of the growing season is also related to the altitude and ranges between 83 and 141 days. Western air flow predominates, NE destructive wind being dangerous for forest ecosystems. In general, the climate of the region is harsh with low temperatures, rich precipitation, frequent fogs and occasional destructive winds. Stream system is dense being characterized by a considerable slope and erosion activity. The greater part of the Orlické hory Mts and their foothills belongs to the basin of the Orlice River, the smaller one to the basin of the Metuje River. In 1956, a Protected Landscape Area occupying about 60% of the Orlické hory Mts natural forest region was proclaimed in the northern and central parts of the mountains. The Protected Region of the Natural Accumulation of Water of the same area was set up in the Orlické hory Mts in 1978.

The Orlické hory Mts and their foothills are situated between the 8th and 5th forest vegetation zones (FVZ). The 6th spruce/beech FVZ predominates (49% of the area) being followed by the 5th fir/beech FVZ (33%) and the 7th beech/spruce FVZ (16%). Acid edaphic categories (57% of the area) are mostly distributed while rich categories occupy only 24% of the area. Acid spruce/beech and acid fir/beech stands occupy the largest area, viz. 30 and 13%, respectively. Spruce management of higher altitudes (management sets of Nos. 51–57) occupies 80% of the area.

The Orlické hory Mts, particularly their uppermost parts, were in the last 15 to 20 years affected by air pollution disasters when extensive areas of damaged stands of all age categories were felled (thousands of hectares, today clear-cut areas and regenerated stands up to 20 years of age occupying about 3,200 ha). The nearly monotonous species composition (Norway spruce percentage about 90%) of the stands participated to a great extent in the extremely severe impacts of air pollution (particularly SO_2) on the local mountain ecosystems. Decrease in the biodiversity of forest ecosystems in the past period resulted in the marked reduction of their ecological and static stability. Subsequent enforced felling measures reduced their wood-producing as well as non-wood-producing functions/effectiveness. From the hydrological point of view it is important that extensive air pollution clear-cut areas have been created associated with the marked extension of the forest road network. The density of forest roads reaches about 51 m/ha, 1/3 of the figure being represented by logging roads with frequently unsatisfactory slope conditions and drainage. Regeneration of forest stands after the disasters is aggravated by difficult air pollution and ecotope conditions. It is particularly difficult to ensure forest ecosystem biodiversity. It is expected that richer tree species composition will be applied particularly through the increase in the percentage of broadleaved species (European beech, sycamore maple, Scotch elm, and soil-improving species) as well as silver fir.

1.2 Characteristics of experimental areas and watershed

1.2.1 The Deštenská stráň Hillside experimental area

The Deštenská stráň Hillside experimental area in the Orlické hory Mts serves for studying the water balance of spruce and beech ecosystems as representatives of two most important tree species of middle-altitude mountain locations in the Czech Republic. A couple of balance plots forms the research area. Both balance plots (40x30 m each) are 50 m apart being situated on a slope with WSW aspect and mean gradient 16° at an altitude of 890 m; their latitude and longitude determines 50°19′20′′ N and 16°21′45′ E respectively. Mean annual temperature is 4.9°C and mean annual precipitation 1200 mm.
From the forest site classification point of view, the spruce and beech stands belong to the most widespread forest type of the spruce/beech FVZ, to the forest type of an acid spruce/beech stand with Deschampsia (6K1). From the pedological point of view, both stands can be ranked among typical acid Cambisols of higher altitudes, sandy loam to loamy sand with the 50% mean admixture of skeleton, the proportion of which reaches 90 - 98% at a depth of 0.7–1.0 m (weathered parent rock being mica schist).
The study was started in autumn 1976. During the first five hydrological years (1 November 1976 to 31 October 1981), hydrological effectiveness of mature stands was studied. In winter 1981/1982, both stands were felled at a time and immediately (in spring 1982) the research plots were reforested again with Norway spruce and beech so that since 1 November 1982, the study of the water regime of both species could be continued under changed unfavourable air pollution/ecotope conditions.
In the course of the experiment, the following components of the water regime have been continuously studied by the same procedures and using the same measuring devices on both plots, both in mature stands and in newly established plantations/present young growths (Kantor, 1992, 1995): interception and transpiration of trees (in mature stands being determined numerically as the only unknown quantity in the equation of water balance), evaporation from the soil surface, interception and transpiration of ground vegetation, changes in soil moisture, overland flow, lateral water flow through the soil, water seepage to the subsoil (with the following subsurface flow), snowpack depth, density and water equivalent value, air temperature and humidity.

1.2.2 The U Dvou louček experimental watershed

The U dvou louček experimental watershed was established for the purpose of studying the problems of draining a waterlogged forested watershed situated on a mountain slope. The small forested watershed U Dvou louček occurs in the uppermost part of the Orlické hory Mts. The position of the watershed is determined by geographical coordinates 16° 30' 56'' E and 50° 13' 16'' N. The watershed is situated at 880–950 m altitude, mean altitude according to a hypsographic curve being 922 m.

The watershed is fan-shaped (substitute figure being a parabola with the side B length 820 m), form coefficient is 1.16, watershed divide length is 2,290 m (coefficient of the watershed divide division 1.13) and talweg length is 530 m. The area of the watershed amounts to 32.6 ha. The area exhibits variable slope, viz. in the lower part 7.5°, in the middle part 8.5° and in the upper part 4.3°. Mean slope calculated from the course of contours is 6.4°. The talweg inclination is 5.4°. The SW watershed aspect changes in border parts to the SE and W aspects. A watercourse draining the watershed is formed by two branches. The right branch exhibiting 5.9° inclination is 340 m long while the left branch having 5.3° inclination is 300 m long. One quarter of the watershed area is affected by high groundwater table (flowing water and spring area). Permanent waterlogging occurs on 5.5 ha, temporary waterlogging in the summer half-year does not exceed the area of 5 ha. Out of the period of soil profile waterlogging, volume soil moisture ranges from 30 to 60%. In the winter half-year, soil profile is saturated by water to the field moisture capacity in the whole watershed. The area of the full-grown European beech/Norway spruce stand (mean age 80 years) was 6.8 ha in 1991 (21% of the area) and till the year 1995 decreased to 5.7 ha (17%) due to the further disintegration of spruce ecosystems. The remaining area of the watershed is a clear-cut area resulted from air pollution salvage felling with a Norway spruce young plantation of various age (max. 15 years old). Road density in the watershed amounts to 62 $m.ha^{-1}$ being formed by skidding and hauling roads.

In terms of hydrogeology the watershed belongs to the crystalline complex of the Eagle Mts. with a low even medium crevice permeability of the rock and unstable penetration of the quaternary mantle. The rock consists namely of honeycombed muscovite gneiss of the Proterozoic Era and mica schist of the strontian series. Results of hydrodynamic tests in boreholes carried out in the watershed showed that gneiss in the hydrogeology structure of the watershed plays the role of a collector and the mica schist acts as a hydrogeology insulator. In the upper part of the watershed where gneiss appears the supply of ground water is dependent exclusively on atmospheric precipitation. Due to the good permeability of gneiss and gradient of the terrain the water continually flows off via the underground. In contrast, the area of the mica schist in the lower and middle parts of the watershed prevents the water to flow via the underground and it swells to the surface. Since the supply of water from the gneiss area is virtually non-stop, the area is continually waterlogged.

On the gneiss cambisol (brown forest soil) developed and partly also humus podzols; on the mica schist humus podzols, gleyey peat podzol and peat soils. The soil is loamy-sand, in some horizons clayey, extremely stony (20 to 50%) and non-homogeneous. From the point of view of the dynamics of soil water the soils are permeable, with a high infiltration capacity. The soil areas with gleyey podzolic soil and peat are oversupplied with water and have a high water level caused by a high inflow of water from the higher areas.

In the U Dvou louček watershed, hydrological and silvicultural research programmes are conducted. The hydrological programme includes watershed calibration in hydrological years 1991/92–1995/96 including recording hydropedological characteristics, manual implementation of the ecological draining measure in summer 1996 and the study of the effects of draining on further hydrology and hydropedology of the watershed. Draining measures aimed at the restoration of functions of the existing drainage system and the

interception of runoff from spring areas and areas with insufficient drainage were carried out on the area greater than 2 ha. The length of drainage ditches reached about 500 m. The silvicultural research programme follows up the implemented drainage. It deals with the improvement of the survival and growth of established spruce young plantations and increasing their biodiversity and ecological stability. In part of the drained area protected by fencing of an area of about 1 ha with three degrees of the soil profile moisture (moist, moister, moistest), European beech, sycamore maple and silver fir are interplanted into the spruce plantation and on unforested places since 1997. Various types of mound and ridge planting and amelioration using natural materials (basic rock meals) are applied.
To obtain input data, precipitation in the watershed is measured by 8 station rain gauges and 2 ombrographs connected with the automatic meteorological station. Other characteristics of the soil and air sphere are also determined as well as the dynamics of air pollution flow by a summation method. Vertical flow onto the bedrock using ten buried open lysimeters (deep infiltrometers) placed at a depth of 0.75 m with a total area of 1 m^2 and lateral subsurface flow from the interface of organic and mineral horizons and from the interface of more loose and more compacted mineral layers was also measured. Groundwater table is monitored by 52 water table measuring perforated pipes placed in two transects (perpendicularly to the contour and across ditches) and on four microplots differing in vegetation cover (peat moss, reed, hairgrass and mixed growth). Suction pressure of soil is measured by 288 tensiometers at a depth of 0.15–0.60 m in two elementary runoff-balance plots (10 x 10 m each) differing in moisture. On the elementary runoff-balance plots (ERBP) groundwater table is measured by 42 soil water level measuring perforated pipes built-in depth of 0.7 m. The flow (discharge) in the closing profile of watercourse was monitored by mechanical float water level recorder. Since 1996, the flow is recorded by manometric water level recorder with automatic data collection.

2. Method of ET calculation from water balance equation

2.1 ET calculated from complex of measured water balance components in a forest stand with only single unknown Et

2.1.1 Outline of method procedure

The method of ET assessment is performed on the Deštenská stráň Hillside experimental area. Investigations, from which the ET components for mature Norway spruce and European beech stands were obtained, we have done in a continuous sequence of five water years (from 1 Nov. 1976 to 31 Oct. 1981). Two components of vaporization process, i.e. interception losses (Ei) and evaporation from the litter and soil surface (Es) were ascertained directly; the tree canopy transpiration (Et) was estimated by calculation from water budget equation as the only unknown:

$$Et = P - Tf - Sf - Es - OlF - SslF - VF - \Delta S \text{ (mm)} \qquad (2)$$

where
P is precipitation, **Tf** throughfall, **Sf** stemflow, **Es** evaporation from forest floor and ground vegetation, **OlF** overland flow, **SslF** subsurface lateral flow, **VF** vertical flow (infiltration) to bedrock, and **ΔS** change in soil water storage.
From these components of water budget, the total evapotranspiration (ET) was possible to calculate as the sum of Ei + Et + Es (in millimetres of water per time unit).

Interception losses were calculated as the difference of open area precipitation P and net precipitation Pn = Tf + Sf. In both stands, throughfall (Tf) were measured by a series of 10 station ombrometers each of them with an orifice of 500 cm^2, located in spacing 4 m along a contour line. Stemflow (Sf) was collected from spiral collar attached to a tree trunk and conveyed by hose into gauge barrels; Sf was recorded from 3 trunks in the spruce stand and 11 trunks in the beech stand of all diameter classes.

Evaporation from forest floor and ground vegetation (Es) were found out using modified Popov's evaporimeter. It is a double cylinder container; the inner cylinder, its bottom is covered by wire mesh, is filled by an undisturbed soil monolith. Its circular cross section is 160 square centimetres and depth 0.2 m. The evaporimeter is inserted in the soil so that its upper edge may be levelled with surrounding soil surface. Evaporimeters were weighted in regular time intervals usually once a day. After precipitation, water infiltrated through evaporimeter was measured. From weight difference of the soil monolith the evaporation was calculated. In each of both balance plots 20 evaporimeters were installed. In the spruce stand, 6 evaporimeters were planted by bilberry (Vaccinium myrtillus) and six ones by hair-grass (Deschampsia flexuosa), two main dominant plants of ground vegetation. Evaporation from soil surface was estimated from weighing eight evaporimeters. In the beech stand, all twenty evaporimeters were used to determination of evaporation from soil surface.

The further necessary components of water balance were obtained by following procedures.

Precipitation (P) of open area was measured by 5 station rain gauges (ombrometers) with an orifice of 500 cm^2 on a clearcutting area of 150 x 90 m and in the middle of a gap 40 x 40 m in the immediate vicinity of the water balance plots.

Overland and subsurface lateral flow was measured on elementary runoff plots (ERP) each of 3.5 x 5 m in size. The lower edge of the runoff plot is presented by the timbering pit. Into the front wall, three gutters each with entering section are inserted and tamped beneath the undisturbed soil profile. The overland flow from a depth 0.05 m (horizon interface of LFH and A horizon) and subsurface lateral flow from depths of 0.25 and 0.7 m was collected.

Vertical flow (infiltration) through soil profile onto the subsoil at a depth of 0.7 m was measured in 3 lysimetric pits in a spruce stand and 3 lysimetric pits in a beech stand. The lysimetric pit of 4 m in length was equipped with 10 buried open lysimeters (deep infitrometers) with orifice of 0.3 x 0.3 m each.

Soil moisture was surveyed using a gravimetric method. Disturbed soil samples were taken from 4 soil horizons (characterizing a soil profile) on three places both in a spruce and in a beech stand. Sampling was done usually once a week during growing season in rainless days.

Measurements of water budget components during growing season were done in all days with measurable precipitation, i.e. usually 2 to 3 times in a week.

2.1.2 Results containing ET components

Interception losses (Ei) represent the most different component between spruce and beech stand. During five growing seasons (May - October) spruce canopy intercepted and later evaporated 150.1 mm, i.e. 20.6% of open area precipitation; beech canopy interception losses represented only 56.5 mm, i.e. 7.8% precipitation of open area. Significant lower Ei of beech stand was caused mainly by different values of stemflow. The spruce stand showed negligible stemflow 1.4% of growing period precipitation, while the beech one showed 18.8%, which presented substantial decrease of interception losses.

Interception losses of both stands were further decreased by occurrence of occult precipitation which increased near equally value of net precipitation in observed growing seasons by 40 to 70 mm, i.e. by 5 to 10% of summer precipitation. Practically the same ability of both stands to obtain occult precipitation is explainable on one hand by greater intercepting area of needle biomass, on the other hand by better conditions for precipitation travelling along the branches and trunks and lower storage capacity of the broadleaved tree.

Evaporation from forest floor and ground vegetation (Es) did not present any important difference between both stands during investigated growing seasons. They equal to 75.6 mm, i.e. 10.4% and 72.0 mm, i.e. 9.9% in the spruce and beech stand, respectively.

Transpiration (Et) represented the most important component of water cycle in the spruce and beech forest stands during observed growing seasons. Et of the spruce stand equalled to 182.4 mm, i.e. 25.1%, Et of the beech stand equalled to 176.3 mm, i.e. 24.3% of summer precipitation in average. Somewhat lower Et in the beech stand can be ascribed to shorter only five months lasted growing period of that tree species.

The values of further measured components necessary for calculation of transpiration from water budget equation of spruce and beech stand are presented in **Table 1** and **2**.

Total ET reached in the mature, fully-stocked Norway spruce stand during five investigated growing seasons 1977-1981 (summer water half-years, i.e. May - October) 408.1 mm, i.e. 56.1% of summer precipitation in average. Similarly characterized European beech stand vaporized 304.8 mm, i.e. 42.0%.

Both mature stands were clear-cut during winter 1981/1982 and both clearcuttings were reforested by the same tree species, i.e. by outplants of Norway spruce and European beech. The plantations influenced the water budget in years after reforestation only insubstantially. Both interception (Ei) and transpiration (Et) was negligible due to small needle and foliage biomass. Water regime in both established plantations was being successively influenced by evaporation from ground vegetation. In 5 years after logging weeds had been infesting near the whole surface of both balance plots. In vegetation cover the following species dominated: Rubus idaeus L., Carex sp., Avenella flexuosa (L.) Drejer, Calamagrostis arundinacea (L.) Roth, Deschampsia caespitosa (L.) Beauv., and Calamagrostis epigeios (L.) Roth. In summer mean dry matter of weed above ground biomass exceeded 3 metric tons per hectare.

Evaporation of ground vegetation (Es), i.e. its interception and transpiration ranged from 268 to 332 mm (i.e. from 40.8% to 67.7% of summer precipitation) during 5 growing seasons after clearcutting.

2.2 ET obtained from complex of partially measured water balance components and partially derived from tensiometric measurements in a small watershed with only single unknown E(t, s)

2.2.1 Outline of method procedure

The method of ET assessment is performed on the U Dvou louček experimental watershed (UDL). Two constructed elementary runoff-balance plots - ERBP (each of size 10 x 10 m and slope 6°) were used for ET determination by above mentioned method (viz heading). The first plot (named Nad cestou) is covered by 15 year-old Norway spruce, pole-stage stand of 6 m high; the second plot (named Pod cestou) is covered by a similar fully stocked spruce stand with 5% admixture of European beech. The soil type is Cambisol in the plot No. 1 and the humus podzol in the plot No. 2. Detail soil characteristics are shown in **Table 3**.

Month	Air temperature	Air humidity	Precipitation P	Interception Ei		Evaporation Es		Transpiration Et		Overland flow OlF		Subsurface lateral flow SslF		Vertical flow VF		Change in soil moisture ΔS	
	°C	%	mm	mm	%	mm	%	mm	%	mm	%	mm	%	mm	%	mm	%
May	8.2	76.5	67.5	21.6	32.0	13.9	20.6	35.4	52.4	0.1	0.1	0.2	0.3	15.3	22.7	-19.0	-28.1
June	12.3	79.2	115.3	33.2	28.8	17.8	15.5	31.9	27.7	0.6	0.5	1.2	1.0	27.8	24.1	2.8	2.4
July	11.4	84.6	212.0	36.5	17.2	13.5	6.4	34.9	16.5	1.3	0.6	3.2	1.5	114.2	53.9	8.4	3.9
Aug	12.4	84.1	133.8	28.1	21.0	15.3	11.4	31.5	23.6	0.9	0.7	2.6	1.9	59.2	44.2	-3.8	-2.8
Sept	9.2	87.1	100.3	13.0	13.0	8.8	8.8	30.5	30.4	0.1	0.1	0.7	0.7	42.2	42.0	5.0	5.0
Oct	5.8	85.9	98.2	17.7	18.0	6.3	6.4	18.2	18.6	0.4	0.4	0.7	0.7	52.9	53.9	2.0	2.0
Average Total	9.9	82.9	727.1	150.1	20.6	75.6	10.4	182.4	25.1	3.4	0.5	8.6	1.2	311.6	42.8	-4.6	-0.6

Table 1. Water balance of the mature Norway spruce stand over the growing periods from 1977 to 1981 (average values in particular months)

Month	Air temperature	Air humidity	Precipitation P	Interception Ei		Evaporation Es		Transpiration Et		Overland flow OlF		Subsurface lateral flow SslF		Vertical flow VF		Change in soil moisture ΔS	
	°C	%	mm	mm	%	mm	%	mm	%	mm	%	mm	%	mm	%	mm	%
May	8.1	75.9	67.5	10.5	15.6	12.6	18.7	29.9	44.3	0.5	0.7	0.4	0.6	32.3	47.8	-18.7	-27.7
June	12.3	78.7	115.3	13.6	11.8	16.4	14.2	38.1	33.1	1.0	0.9	0.8	0.7	42.0	36.4	3.4	2.9
July	11.9	84.3	212.0	9.5	4.5	13.1	6.2	36.3	17.1	3.6	1.7	3.5	1.7	137.5	64.8	8.5	4.0
Aug	12.7	83.2	133.8	11.6	8.7	15.5	11.6	36.1	27.0	2.1	1.6	1.8	1.3	69.9	52.2	-3.2	-2.4
Sept	9.6	86.3	100.3	2.0	2.0	8.7	8.7	28.3	28.2	0.9	0.9	0.6	0.6	55.3	55.1	4.5	4.5
Oct	6.0	85.2	98.2	9.3	9.5	5.7	5.8	7.6	7.8	0.7	0.7	0.4	0.4	72.5	73.8	2.0	2.0
Average Total	10.1	82.3	727.1	56.5	7.8	72.0	9.9	176.3	24.3	8.8	1.2	7.5	1.0	409.5	56.3	-3.5	-0.5

Table 2. Water balance of the mature European beech stand over the growing periods from 1977 to 1981 (average values in particular months)

Soil type	Depth (cm)	Soil porosity	Maximum capillary water capacity	Lento-capillary point	Retention capacity		
					Dynamic	Static	Total
Cambisol	4 - 15	87.53	75.87	40.26	11.66	35.61	47.27
	15 - 26	47.05	43.72	25.79	3.33	17.93	21.26
	26 - 42	25.33	21.72	14.98	3.61	6.74	10.35
	Average	45.78	40.07	23.64	5.71	16.43	22.14
Humus podzol	5 - 17	80.47	74.47	43.00	6.00	31.47	37.47
	17 - 25	50.52	44.10	36.16	6.42	7.94	14.36
	25 - 50	32.05	29.85	21.73	2.20	8.12	10.32
	Average	48.24	44.28	29.97	3.96	14.31	18.27

Table 3. The soil parameters of elementary runoff plots in the "U Dvou louček" watershed (volume percentage)

Notes: Maximum capillary water capacity by Novák (1954)

Lento-capillary point - soil water constant representing point of decreased availability for plants, i.e. soil moisture at pF 2.8

Retention capacity dynamic or static - volume of pores gravitational or capillary

Maximum capillary water capacity and total retention capacity of 0.6 m deep soil profile amounts to 236 mm and 129 mm for Cambisol and 266 and 111 mm for humus podzol.

Each of both ERBP was equipped by following instruments:

- 144 tensiometers in depths of 0.15, 0.3, 0.45, and 0.6 m with 36 repetitions in each depth
- 21 auger holes equipped by soil water level measuring perforated pipes with diameter of 22 mm built-in depth of 0.7 m.

Data coming from both ERBP were further completed by data from:

- 10 buried open lysimeters (deep infiltrometers) each of them with orifice 0.1 m^2 placed at the depth of 0.75 m below undisturbed forest soil layer
- 3 gutters for measurement of overland flow and 3 gutters for measurement of lateral shallow subsurface flow in common pit with whole catchment area of 3 x 2 m
- an automatic meteorological station and 8 additional ombrometers
- water level recorder with automatic data collection at closing profile of experimental watershed with automatic data recording

2.2.2 Description of ET model design

The scheme of the measurement assessment for the ET estimation is shown on the **Figure 3**. The ET is calculated as the single unknown from the following water budget equation:

$$P - \overbrace{Ei - Et - Es}^{ET} - Q(c,\ sc) - \Delta\ SW(a)\ -\ \Delta SW(g)\ - Q(g)\ = 0\ \text{(mm)} \tag{3}$$

where

P = Pn + Ei	precipitation of open area
Pn = P - Ei	net precipitation
Ei	interception losses of spruce pole-stage stand derived from relation of open area precipitation and net precipitation on the bases long-term measurements

Et + Es = E(t, s)	transpiration of forest stand and evaporation from forest floor and ground vegetation
Q(c, sc)	vertical flow (outflow or inflow) of capillary and semicapillary soil water
Δ SW(a)	change in content of capillary and semicapillary soil water in aeration (unsaturated) zone
Δ SW(g)	change in content of gravitation water in soil layer with depth of 0.6 m
Q(g)	discharge of gravitation water
E(t, s) = Et + Es	transpiration of spruce pole-stage stand and evaporation from forest floor and ground vegetation

The model (design, pattern) given by Figure 3 and Equation (3) takes in consideration two interlinking forest soil zones, i.e. zone gravitational and zone capillary including semicapillary. Soil moistures are derived from suction-pressure measurements using retention curves created by a laboratory. Vertical capillary and semicapillary outflow (or inflow) is determined on the basis of Darcy equation for an aeration (unsaturated) zone using measurement of suction pressures and responding coefficients of unsaturated hydraulic conductivity. These coefficients have been found out by treatment of volume soil samples using the vaporization method by Schindler with the instrument Ku-pFUGT Muenchebeck applying approximation of measured data by Van Genuchten equations in the programme RETC, version G.

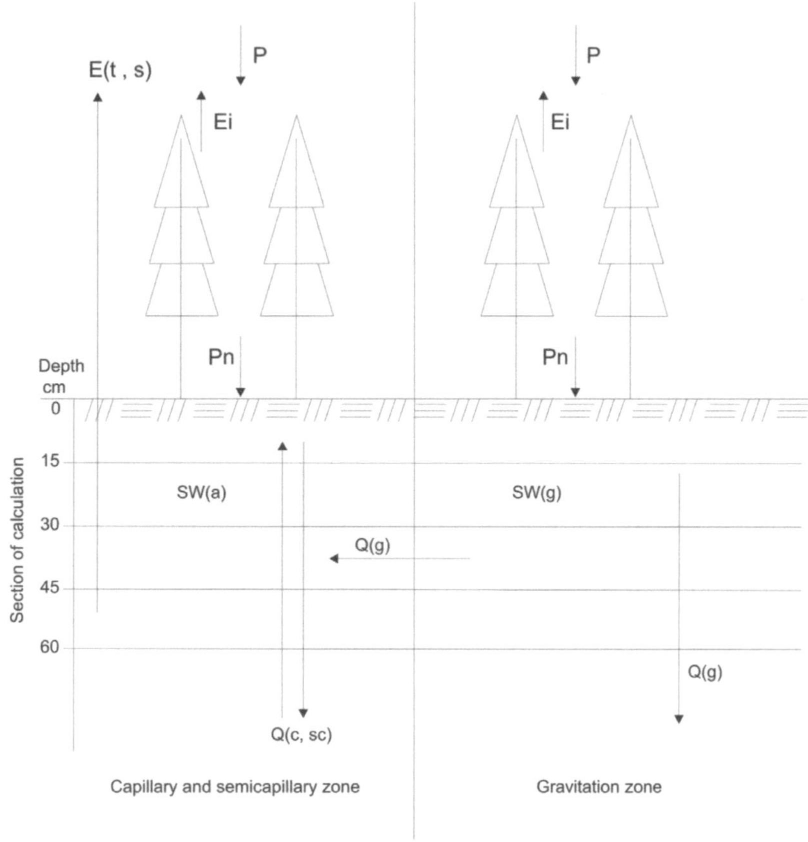

Fig. 3. Scheme of the measurement assessment for the E(t, s) estimation

Suction pressures for calculation of capillary, semicapillary soil water including its flow in partially saturated soil and hydraulic depths of gravitational water in auger holes are measured by tensiometers and perforated water-level pipes, respectively. These data are collected in 6 - 10 day intervals.

2.2.3 Results of model

Results of the model are presented for the growing period of moderately dry year 2008 (precipitation equalled to 77% of long-term mean) in **Tables 4** and **5**.

Overland flow (OlF) attains only negligible values. Therefore, these values are not taken in consideration at computing the model outcomes by the Equation (1).

Interception losses (Ei) of forest stand in common with E(t, s) results in estimating total evapotranspiration (ET) equal to 303 - 309 mm per evaluated time period of 148 days in the growing season. Interception equal to 22.9% of precipitation (P) is consistent with data from other research areas mentioned in the Orlické hory Mts.

Q(c, sc) represents ascending or descending vertical flow of capillary and semicapillary soil water in the aeration zone. The soil of both ERBP has not been during the whole growing period 2008 saturated. The soil moisture has not descended under lentocapillary point at pF 2.8 (point of decreased availability).

Gravitational water of auger holes and buried open lysimeters (subsoil infiltrometers) is located in irregular space network of gravitational pores. These devices serve for observing level of gravitational water and bring information on its dynamics. During dry period the instruments present vertical fluctuation of gravitational water; during stormflow they give level of gravitational water in gravitational pores in combination with its lateral and vertical movement. Volume portion of gravitational pores ranges between 3.3 - 11.7% in ERBP "Nad cestou" and 2.2 - 6.4% in ERBP "Pod cestou" in dependence on soil depth (cf. **table 3**). Gravitational pores create in forest soil irregular network spatially restricted. Flow in gravitational pores is evident and its proportion ranges between 28 - 30% of total discharge from the ERBP.

The total runoff of gravitational water during observed growing period amounted to depth of 30.7 mm (cf. **tables 4** and **5**). The value was estimated using similarity with hydrological balance model of the Divoká Orlice River basin (Horský 1970), where the experimental watershed is located. The time behaviour of gravitational water discharge was then derived in relation to time behaviour of runoff obtained from the closing profile of the "U Dvou louček" watershed. The procedure is substantiated by knowledge of hydrogeological survey that the portion of gravitational water vertically percolates into the underlying bedrock and recharges deep groundwater. At computing procedure the measurements from open subsoil infitrometers and auger holes were taken into account.

Transpiration of tree canopy and evaporation from forest floor and ground vegetation E(t, s) of depth 212 - 218 mm per evaluated time period of 148 days in the growing season, i.e. in average 1.4 - 1.5 mm per day, is consistent with measurements on other examined research areas in the Orlické hory Mts.

2.2.4 Conclusions

Estimation of total evapotranspiration (ET) equalled to 303 - 309 mm per evaluated time period of 148 days in the growing season 2008.

Period	Precipitation P	Interception Ei	Net precipitation Pn	Capillary and semicapillary outflow Q(c, sc)	Soil water content variation Δ SW(a)	Gravitational water variation Δ SW(g)	Gravitational water outflow Q(g)	E (t, s)
Year 2008	mm							
28. 5. – 3.6.	2.5	-1.0	1.5	-7.5	11.6	6.1	0.0	-11.7
4.6. – 10.6.	9.9	-4.2	5.7	-3.0	4.3	0.0	0.0	-7.0
11.6. – 17.6.	4.2	-2.9	1.3	-0.9	8.6	0.0	0.0	-9.0
18.6. – 24.6.	5.4	-2.8	2.6	0.8	7.8	0.0	0.0	-11.2
25.6. – 2.7.	37.7	-4.5	33.2	1.3	-5.6	0.0	-4.0	-24.9
3.7. – 10.7.	41.9	-10.1	31.8	-2.0	-10.7	0.0	-3.4	-15.7
11.7. – 17.7.	12.4	-3.5	8.9	-2.6	6.2	0.1	0.0	-12.6
18.7. – 5.8.	31.0	-12.8	18.2	-0.6	7.7	-0.2	0.0	-11.6
6.8. – 13.8.	3.0	-2.0	1.0	-0.3	1.3	0.6	0.0	-9.6
14.8. – 19.8.	73.6	-5.8	67.8	-0.7	13.4	0.6	0.0	-18.8
20.8. – 2.9.	21.5	-6.0	15.5	3.6	6.8	-0.5	0.0	-10.9
3.9. – 8.9.	35.0	-4.6	30.4	-8.0	-44.6	0.0	-4.5	-23.9
9.9. – 16.9.	8.8	-3.6	5.2	-7.3	14.6	-0.1	0.0	-9.5
17.9. – 25.9.	43.1	-11.0	32.1	-5.6	-7.0	-1.0	-0.8	-16.0
26.9. – 2.10.	27.1	-6.5	20.6	-5.4	4.0	4.2	0.0	-8.0
3.10. – 22.10.	39.2	-9.5	29.7	-40.7	-5.9	0.0	-18.0	-17.8
Sum	396.3	-90.8	305.5	-78.9	12.5	9.8	-30.7	-218.2
Percentage	100.0	-22.9	77.1	-19.9	3.1	2.5	-7.7	-55.1

Table 4. The hydrologic balance of the ERBP no. 1 "Nad cestou" in the "U Dvou louček" watershed

Note: E (t, s) transpiration of forest stand and evaporation from forest floor and ground vegetation [E(t, s) + Ei = ET]

Period	Precipitation P	Interception Ei	Net precipitation Pn	Capillary and semicapillary outflow Q(c, sc)	Soil water content variation Δ SW(a)	Gravitational water variation Δ SW(g)	Gravitational water outflow Q(g)	E (t, s)
Year 2008				mm				
28. 5. – 3.6.	2.5	-1.0	1.5	-4.0	10.8	1.2	0.0	-9.5
4.6. – 10.6.	9.9	-4.2	5.7	-3.2	12.0	2.3	0.0	-16.8
11.6. – 17.6.	4.2	-2.9	1.3	-0.5	2.9	0.0	0.0	-3.7
18.6. – 24.6.	5.4	-2.8	2.6	1.9	10.9	0.0	0.0	-15.4
25.6. – 2.7.	37.7	-4.5	33.2	-0.1	-15.3	0.0	-4.9	-12.9
3.7. – 10.7.	41.9	-10.1	31.8	-3.2	-8.3	-0.8	-4.1	-15.4
11.7. – 17.7.	12.4	-3.5	8.9	-3.5	11.5	0.8	0.0	-17.7
18.7. – 5.8.	31.0	-12.8	18.2	-7.3	15.7	0.0	0.0	-26.6
6.8. – 13.8.	3.0	-2.0	1.0	5.1	12.7	0.0	0.0	-18.8
14.8. – 19.8.	73.6	-5.8	67.8	0.0	-40.0	-1.5	-5.4	-20.9
20.8. – 2.9.	21.5	-6.0	15.5	-13.0	5.8	1.5	0.0	-9.8
3.9. – 8.9.	35.0	-4.6	30.4	-4.0	-6.1	-0.6	-0.9	-18.8
9.9. – 16.9.	8.8	-3.6	5.2	-3.4	7.8	0.0	0.0	-9.6
17.9. – 25.9.	43.1	-11.0	32.1	-6.4	-21.5	0.0	0.0	-4.2
26.9. – 2.10.	27.1	-6.5	20.6	-9.5	8.5	-6.4	-4.6	-8.6
3.10. – 22.10.	39.2	-9.5	29.7	-19.8	0.6	3.9	-10.8	-3.6
Sum	396.3	-90.8	305.5	-70.9	8.0	0.4	-30.7	-212.3
Percentage	100.0	-22.9	77.1	-17.9	2.0	0.1	-7.7	-53.6

Table 5. The hydrologic balance of the ERBP no. 2 "Pod cestou" in the "U Dvou louček" watershed
Note: E (t, s) transpiration of forest stand and evaporation from forest floor and ground vegetation [E(t, s) + Ei = ET]

The presented model of hydrological balance brings values of ET and Q consistent with experimental results from comparable conditions and with runoff data issued for streams in upper part of the Divoká Orlice River basin by Czech Hydrometeorological Institute (Horský 1970).
The results of investigation represent conditions of mountain land covered predominantly by coniferous forest in the Czech Republic.

3. Hydropedological and hydrological methods of ET assessment

3.1 ET obtained from continuous measuring the volumetric moisture of soil profile in a forest stand and calculation of E(t, s) from soil moisture differences

3.1.1 Outline of method procedure

A long-term observation of the water balance elements during the forest regeneration of a Norway spruce stand and that of European beech on the Deštenská hillside in the Orlické hory Mts serves also for comparing evapotranspiration (ET) of both stands. ET represents interception losses from the canopy (Ei), transpiration of forest stand (Et), and evaporation from forest floor and ground vegetation (Es). The first observations were made in mature spruce- and mature beach stands (1976–1981); a clearcut harvest with hole planting of spruce and beech transplants followed in 1982. Then, the research continued during the growth, development, and thinning of both stands from the stage of young plantation up to the small pole stage (1983–2006). The transpiration of the tree species was calculated as the only unknown of the water budget equation (cf. Kantor 1985, and the first method in this ET chapter). In 1998, we started a continuous measuring of soil moisture by volume in layers of the soil segments with the aim to come up with a new procedure of direct determination of ET. In 2005, a method of E(t, s) determination in the spruce and beech stands was devised on the basis of volumetric moisture changes in the soil profile - E(t, s) by Soil Water Content Variation, in abbreviation E(t, s)-SWCV and the results obtained in the growing season of 2005 were published (Šach et al. 2006). In the winter of 2005/06, the young spruce stand (25 years old) was completely damaged by crown and stem snowbreaks.
The extreme disturbance of the forest environment in the young experimental spruce stand after the snow breakage disaster in the winter of 2005/2006 became an impulse to carry out the next study. The investigation was based on two methodical procedures:

- assessing the E(t, s) of the forest stands based on continuous measuring of the water content in the root zone of the soil profile,
- intermittent measuring of the evaporation from the soil surface including the ground vegetation separately by dominant species (Es).

A comparative investigation was simultaneously done in the young experimental beech stand.
The aim of this part of the ET chapter is to present the method E(t, s)-SWCV, examples of its use, its validation, and some results obtained during recent growing seasons.

3.1.1.1 Description of forest stand development and present feature of the spruce and beech stands on experimental balance plots

1976–1981: observation of the water balance components run in the mature spruce and beech large-diameter stands (interception and transpiration of trees, soil surface evaporation, soil moisture changes, surface runoff, seepage of water, snow cover parameters, air temperature and humidity).
1982: forest regeneration by the clear felling method and hole planting of spruce and beech.

1983–2005: following the observation of the water balance components during the growth and progress, and tending both stands from plantation to small pole stage and pole stage stands (Kantor 1992, 1995) including foliage biomass.
2005–2006: 25-year-old spruce stand was severely damaged in winter by crown and stem snow breaks, the young beech stand was afflicted with snow breaks only minimally; 98% spruce trees were affected by snow breakage, the stand density decreased from 1550 to 950 trees per ha, the needle foliage of the stand was reduced to about 40%, and the stand canopy was markedly disturbed.
2006: following the observation of the water balance components in the remedying spruce and beech stands after the snow disaster.
2007: stand gaps began to get infested by forest weeds whose cover reached up to 80% in the summer and autumn.

3.1.2 E(t, s) of the young spruce and beech stands

During the growing season from May 1 to October 31 in 2005, 2006 and 2007, E(t, s) was determined by the calculation obtained through the continuous measurement of the volumetric moisture changes in the soil profile (Šach *et al.* 2006). An analogous procedure, e.g. Tesař et al. (1992) and Vilhar *et al.* (2005), was also used but with discrete data from discontinuous observations. By rooting through depth, we induced the thickness of the root zone equal to 500 mm for calculating E(t, s). The volumetric soil moisture was measured with the VIRRIB transducers belonging to sufficiently precise ones in the estimation of the changes in the volumetric soil moisture. The transducers were placed into the deductive root zone in depths of 50 mm, 200 mm, and 500 mm with 3 repetitions. The repeating followed the forest stand variability. In 3 depths with 3 repetitions the total of 9 transducers were placed into each forest stand.

3.1.3 Procedure of calculating E(t, s) of a forest stand

The calculation of E(t, s) in mm for a particular soil layer per month was done by using the formula:

$$E(t, s)_{LM} = \sum W_V \times D_{SL} \times (1 - S_{VP}) \ \text{(mm)} \tag{4}$$

where

$E(t, s)_{LM}$ E(t,s) for a soil layer (mm/month)
$\sum W_V$ sum of volumetric soil moisture decrements as decimal number
D_{SL} soil layer thickness (mm)
S_{VP} skeleton volumetric proportion as decimal number (an especially important entry)

The sum of $E(t, s)_{LM}$ for three observed soil layers of the root zone represents the E(t, s) of the forest stand in the respective month. Using the newly devised method, we can also calculate the daily values of E(t, s) (Šach et al. 2006).

3.1.3.1 Criteria for calculating E(t, s) of a forest stand

- We included the changes of the volumetric soil moisture into the calculation (the mean of 3 repetitions in the same depth) if the volumetric soil moisture in the subsequent record was lower than that in the preceding one.
- We did not usually include into the calculation small decreases in the volumetric soil moisture at night considering six-hour intervals (0, 6, 12, 18, 0, 6... hours of Central European Time – CET = UTC + 1).

- Decreases in the volumetric soil moisture during 12 hours after rain were considered to be a vertical flow and, especially at a low air temperature and a high air humidity (usually 100%), we did not take them into calculation (similarly Cheng 1987 under comparable conditions).
- During rain, when the volumetric soil moisture usually increases, we did not take E(t, s) into consideration.

3.1.4 Evaporation from soil surface including ground vegetation in the young spruce and beech stands – Es

Es was determined by intermittent accurate weighing sets of Popov's evaporimeters in the summer hydrologic half year of 2005, 2006, and 2007. The evaporimeter with evaporative circle cross-section equal to 160 square centimetres indicated the evaporation from the soil layer 0–20 cm. The evaporimeter set (more than 10, usually 16) represented the proportional soil cover in the spruce and beech stands, and newly for comparison also on the clearcut. The observations were realised during characteristic rainless periods on the beginning, in the middle, and at the end of the growing season. They provided the results about the morning, afternoon, and night evaporation. The determination of evaporation is based on an accurate weighing of evaporative vessels at regular time intervals 3 to 4 times per day by digital scales with accuracy of ± 0.1 g. Three comparable 5-day cycles were evaluated from the 14th to the 19th June in the respective years.

3.1.5 Validation and examples of the method use

The results of devised method are consistent with those represented by Kantor (cf. with the first method in this ET chapter). Average E(t, s) during growing season 1977-1982 equalled to 258 mm for the mature spruce stand and 234 mm for the undisturbed pole stage spruce stand in growing season 2005 found by the method of E(t, s)-SWCV (Šach et al. 2006). Similarly, average E(t, s) during growing season 1977-1982 equalled to 248 mm for the mature beech stand and 221 mm for the undisturbed small pole stage beech stand in growing season 2005 found by the method of E(t, s)-SWCV (Šach et al. 2006).

Methodical procedure of ET finding based on computing changes of recorded volumetric moisture in a soil profile and obtained results are consistent with procedures and data of further authors doing research in comparable mountain conditions; these results were discussed in papers by Šach et al. (2006) and Černohous, Šach (2008). However, accuracy of the devised method E(t, s)-SWCV is increased by continuous recording volumetric moisture and computing ET from its differences, and also taking volumetric stoniness into account in particular layers of a soil profile.

The method of E(t, s) computation in the young spruce- and the young beech stands based on the volumetric moisture changes in the soil profile (soil water content variation - SWCV) as applied on the Deštenská hillside in the Orlické hory Mts, comes from similar principals as the method based on tensiometric measuring of the suction pressure in the soil profile in mature spruce and beech stands on a NE slope in the experimental object Zdíkov-Liz in the Šumava Mts. (Mráz et al. 1990). Also the observed soil profile depths (100, 200 and 500 mm) corresponded practically to those on the Deštenská hillside including the features and course of drawing water for E(t, s). In the observed growing seasons of 1986–1989, the calculated ET in the mature spruce stand were equal to 274 mm on average. ET in the mature beech stand was calculated only in the growing season of 1989 and its value exceeded 300 mm.

The procedure of E(t, s) determination from differences of continuously measured soil moisture by volume presents sufficient accuracy, if E(t, s) values added into water budget equation completed hitherto the single unknown, cf. **Table 6** and **7** by Kantor et al. (2008).

Month	Precipitation P	Stemflow Sf	Through fall Tf	Net Precipitaton Pn	Interception Ei	Evapo transpiration E(t, s)	Overland flow OlF	Subsurface Lateral flow SslF	Vertical flow VF	Change in soil moisture ΔS
	mm	mm	mm	mm	mm	mm	mm	mm	mm	mm
May	196.0	0.2	144.9	145.1	50.9	51.3	0.4	0.0	92.5	+ 0.9
June	84.4	0.1	70.9	71.0	13.4	51.1	0.3	0.0	43.4	- 23.8
July	169.6	0.2	135.9	136.1	33.5	43.6	0.6	0.5	79.8	+ 11.6
August	97.6	0.1	84.1	84.2	13.4	46.0	0.3	0.2	47.6	- 9.9
September	69.8	0.2	53.1	53.3	16.5	23.3	0.3	0.1	23.6	+ 6.0
October	17.4	0.0	12.0	12.0	5.4	18.8	0.0	0.0	0.0	- 6.8
Sum	**634.8**	**0.8**	**500.9**	**501.7**	**133.1**	**234.1**	**1.9**	**0.8**	**286.9**	**- 22.0**
Percentage	**100**	**0.1**	**78.9**	**79.0**	**21.0**	**36.9**	**0.3**	**0.1**	**45.2**	**- 3.5**

Table 6. Water budget of Norway spruce pole stage stand (24 years old, stocking 10) in growing season 2005 (May – October)

Month	Precipitation P	Stemflow Sf	Through fall Tf	Net Precipitaton Pn	Interception Ei	Evapo transpiration E(t, s)	Overland flow OlF	Subsurface Lateral flow SslF	Vertical flow VF	Change in soil moisture ΔS
	mm	mm	mm	mm	mm	mm	mm	mm	mm	mm
May	196.0	13.5	155.0	168.5	27.5	36.5	0.7	0.1	127.0	+ 4.2
June	84.4	6.9	65.9	72.8	11.6	38.3	0.3	0.0	44.1	- 9.9
July	169.6	18.3	120.8	139.1	30.5	54.9	0.4	0.1	75.1	+ 8.6
August	97.6	9.1	76.0	85.1	12.5	52.0	0.2	0.0	43.0	- 10.1
September	69.8	9.4	47.5	56.9	12.9	22.8	0.3	0.0	29.4	+ 4.4
October	17.4	1.2	11.8	13.0	4.4	15.7	0.0	0.0	0.0	- 2.7
Sum	**634.8**	**58.4**	**477.0**	**535.4**	**99.4**	**220.2**	**1.9**	**0.2**	**318.6**	**- 5.5**
Percentage	**100.0**	**9.2**	**75.1**	**84.3**	**15.7**	**34.7**	**0.3**	**0.0**	**50.2**	**- 0.9**

Table 7. Water budget of European beech pole stage stand (24 years old, stocking 10) in growing season 2005 (May – October)

The E(t, s)-SWCV method may be made more precise if the results obtained from Es determination are used, especially at open canopy and weed infestation, e.g. after snow breakage of a forest stand (**Fig. 4**).
The method of E(t, s) estimation by SWCV may be employed also in computation of daily E(t, s) values (**Tab. 8**).

3.2 ET calculation from daily variation of baseflow between day and night from the small forest watershed

3.2.1 Outline of method procedure

Within the soil water regime observation of the partially waterlogged mountain catchment U Dvou louček (near the village Říčky in the Orlické hory Mts, Czech Republic), we recorded diurnal streamflow variation during the precipitation free period. The fluctuation emerged after doing the drainage in support of the forest stand regeneration and forest plantation growing out. The decline of the streamflow in the daytime against that in the night time, without influence of precipitation, is caused by fluctuation of baseflow from the catchment. We suppose the cause of the streamflow fluctuation in declining total evaporation in the night time. The principal component of total evaporation in rainless period is E(t, s), i.e. transpiration of forest stand and evaporation from forest floor and ground vegetation.
The changes of baseflow during day and night time in conditions of the catchment U Dvou louček were most likely caused by loss of soil water from saturated horizons of waterlogged and drained part of the catchment and from saturated horizons of natural watercourse surroundings connected directly hydraulically with streambed. The water loss was, in all likelihood, induced by total evaporation comprising transpiration of tree species and ground vegetation, evaporation from soil, interception evaporation from vegetation cover and evaporation from water surface of streams. The interception practically did not occur in precipitation free period and evaporation from water surface of streams was negligible in relation to total evaporation from the catchment area. To express water loss from the catchment, just only transpiration of trees and ground vegetation and evaporation from soil, i.e. evapotranspiration, were substantial. The deduction resulted from known increased soil water uptake by fine roots of trees on daytime transpiration, when concurrently under the drought period the evaporation from soil and transpiration of ground vegetation raised. We submitted the hypothesis on the basis of recording of outflow decrease at daytime against night time, when at night processes of total evaporation were not running so intensively, and on the basis of stream discharge analyses performed on the catchment U Dvou louček in the Orlické hory Mts.
Constantz et al. (1994, 1998) were interested in similarly causes of diurnal variation of streamflow. He substantiated the fluctuation by increasing of water temperature in a watercourse during daytime against night time and consequently greater infiltration into a streambed. Ronan et al. (1998) contemplated the relation between streamflow decrease in daytime and associated evapotranspiration of riparian vegetation, nevertheless, as the main factor of that streamflow decrease during daytime they considered temperature changes of water infiltrating into a streambed. We believe the theory can be accepted only in arid regions with conditions similar to those in the Middle West of the United States, where

Date	1	2	3	4	5
1.6.2005	10.0	76.9	0.4	1.9	1.0
2.6.2005	12.2	71.1	0.0	2.6	1.2
3.6.2005	18.0	65.9	0.0	2.9	1.7
4.6.2005	12.3	96.3	11.8	0.0	0.0
5.6.2005	10.2	100.0	11.2	0.7	1.0
6.6.2005	8.5	100.0	3.2	0.5	0.8
7.6.2005	4.4	99.6	6.2	0.3	0.9
8.6.2005	5.3	97.3	4.8	2.2	2.5
9.6.2005	7.9	77.6	0.0	1.6	2.1
10.6.2005	11.4	75.7	0.6	2.4	1.5
11.6.2005	7.3	100.0	3.0	0.3	0.7
12.6.2005	8.6	88.1	4.2	2.4	2.7
13.6.2005	15.2	86.9	3.6	1.7	1.2
14.6.2005	20.5	72.5	0.0	3.3	4.5
15.6.2005	16.7	94.1	3.4	0.6	0.8
16.6.2005	17.9	85.6	0.2	1.6	1.9
17.6.2005	17.5	87.4	0.0	1.6	1.2
18.6.2005	15.8	70.2	2.0	2.4	1.2
19.6.2005	16.1	73.8	0.0	2.1	1.6
20.6.2005	16.8	70.7	0.0	2.6	1.4
21.6.2005	19.4	63.5	0.0	2.9	1.7
22.6.2005	18.9	74.5	0.2	2.0	1.1
23.6.2005	17.2	67.7	0.0	2.3	1.4
24.6.2005	20.5	61.6	0.0	2.2	1.6
25.6.2005	22.1	71.8	4.8	2.3	1.5
26.6.2005	14.7	100.0	1.0	0.1	0.0
27.6.2005	18.5	72.5	0.0	1.8	1.3
28.6.2005	19.5	60.5	0.0	1.8	1.5
29.6.2005	17.6	68.5	0.0	1.3	1.2
30.6.2005	14.3	94.2	23.8	0.4	0.3

Table 8. Example of daily (24 hours') values of E(t, s) in Norway spruce and European beech pole-stage stand calculated from daily decrements of volumetric soil moisture
Notes:
1 - air temperature [°C], average for 6-20 CET, open area +200 cm above ground surface;
2 - air humidity [%], average for 6-20 CET, open area +200cm above ground surface;
3 - daily precipitation in mm, open area +50 cm above ground surface;
4 - daily E(t, s) in mm, spruce pole-stage stand; 5 - daily E(t, s) in mm, beech pole-stage stand;

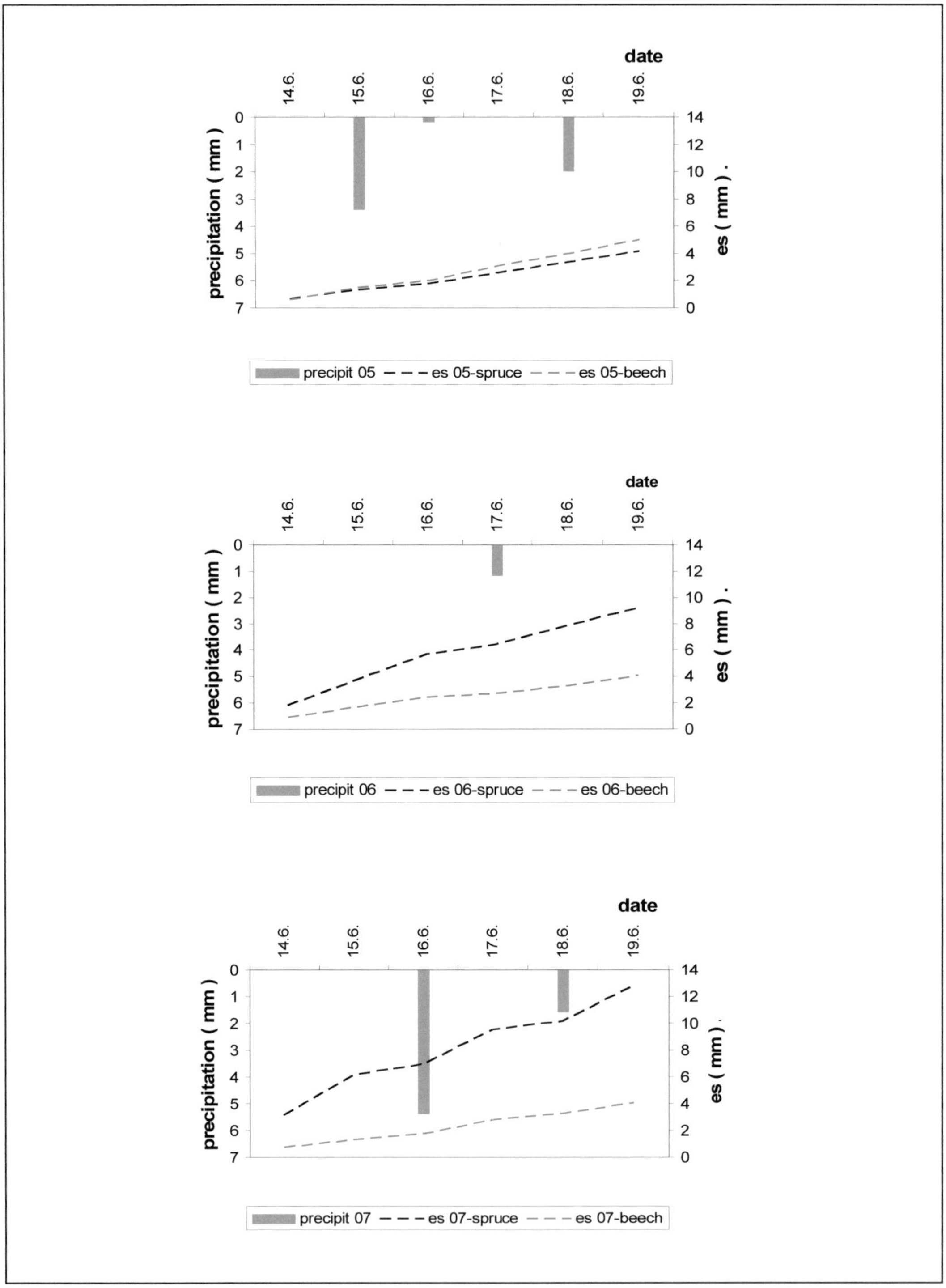

Fig. 4. Differing evaporation Es between young Norway spruce and European beech stands due to snowbreakage

Constantz et al. (1994, 1998) and Ronan et al. (1998) carried out their investigations. Under those conditions the evaporation from water surface and infiltration of water into streambeds represent the main loss factors. In humid regions, where streams predominantly drain landscape, the theory by Kobayashi et al. (1990, 1995) and Bren (1997) better corresponds with local natural conditions. They consider the evapotranspiration of riparian vegetation to be the main cause of diurnal streamflow fluctuation.
The new method of calculating evapotranspiration from decrease of day discharge in comparison with night one was developed. The procedure was demonstrated on computed example by **Fig. 5** and **Tab. 9** and it is a part of the next passage.

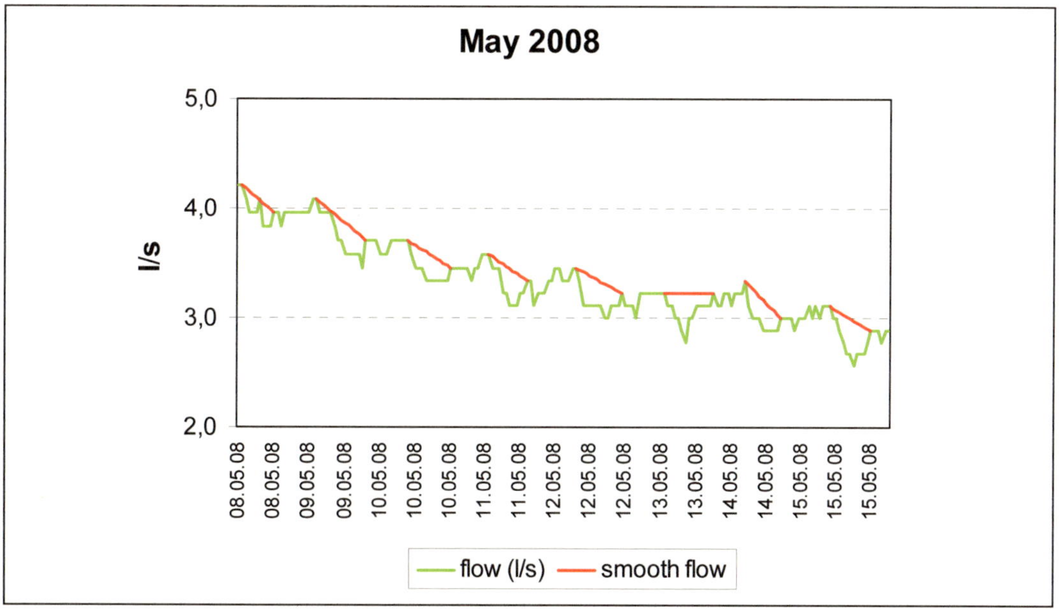

Fig. 5. Example of smoothing daily descend of streamflow

We examined the decrease of day discharge against night one from streamflows received by a manometric water-level recorder in the closing profile of the catchment. We analysed drought periods with streamflow fluctuation in revealing growing seasons 1997 - 1998, and the more recent year 2008.
In our special conditions the water loss per daytime ranged from 2,000 to 87,000 litres. By the theory of variable source areas (Hewlett & Hibbert, 1967), we determine 3 variable source areas (VSA). We derived VSA size from length of open ditches and natural streams (forming drainage network) and their possible lateral drainage reach. Hydraulic drainage reach was calculated using the Czech Government Standard No. ČSN 75 4200: "Treatment of water regime of agriculture soils by drainage". We carried out the enumeration by average weight portion of soil particles sized up to 0.01 mm and saturated hydraulic conductivity K (m per day) of the soils in our catchment. It is possible to suppose that use of agricultural procedures was not fully precise on computing values for non-homogeneous forest soils; nevertheless, its precision was sufficient for given purpose. From total length of natural streams and drainage ditches with hydraulic reach 8 m on each of both sides resulted in 3 VSA. The first VSA sized 8,000 square metres surrounds drainage open ditches

Date	Hour	Flow	Smooth flow	Difference	Litres per hour	Sum of litres	E(t, s) to 3rd VSA	E(t, s) to 2nd VSA	E(t, s) to 1st VSA
		l / s	l / s	l / s	l	l	mm	mm	mm
7.5.2008	9:00	4.63	4.63	0.000					
7.5.2008	10:00	4.35	4.59	0.244	877.22				
7.5.2008	11:00	4.35	4.56	0.212	763.84				
7.5.2008	12:00	4.35	4.53	0.181	650.45				
7.5.2008	13:00	4.35	4.50	0.149	537.07				
7.5.2008	14:00	4.22	4.47	0.252	907.08				
7.5.2008	15:00	4.08	4.44	0.353	1269.18				
7.5.2008	16:00	4.08	4.41	0.321	1155.79				
7.5.2008	17:00	4.22	4.37	0.157	566.93				
7.5.2008	18:00	3.96	4.34	0.388	1396.59				
7.5.2008	19:00	4.08	4.31	0.227	815.64				
7.5.2008	20:00	3.96	4.28	0.325	1169.82				
7.5.2008	21:00	3.96	4.25	0.293	1056.44	11,166	0.41	0.61	1.40
7.5.2008	22:00	4.22	4.22	0.000					
7.5.2008	23:00	4.08							

Table 9. Example of computing daily E(t, s)

of length equal to 500 m. The second VSA sized 18,240 square metres includes length of drainage ditches and main watercourse equal to 1,140 m in total. The third VSA sized 27,344 square metres involves strips passing along all natural and man-made watercourses on the watershed of the whole length equal to 1,709 m.

3.2.2 Results

If we recalculated the mean daytime loss of discharge equalled to 45,000 l on the probable area 27,344 square meters or 18,240 square meters influenced by drainage network, we obtained expression of runoff-depth loss in millimetres, i.e. 1.6 or 2.5 mm per day. The daily values of runoff-depth loss, computed for discharge loss interval 2,000 to 87,000 l from the VSA of 8,000 square meters, ranged between 0.25 to 2.5 mm per day. Decreases of 20,000 to 87,000 l related to the VSA of 18,240 or 27,344 square meters then response to 1.1 - 4.8 mm or 0.7 - 3.2 mm per day.

The range of values corresponded with transpiration and evapotranspiration data reported by Ladefoged (1963), Střelcová et al. (2004), Kantor (Krečmer et al., 2003), and most recently Šach et al. (2006).

Ladefoged (1963) stated maximum daily transpiration 1.9–4.9 mm per day for a mature Norway spruce stand and 2.2–4.8 mm per day for a mature European beech one under optimal climatic conditions. On the basis of his experiments in spruce stands he further specified interval of transpiration: 1.3–3.2 mm per day for the period of May to August (September). Střelcová et al. (2004) stated daily smoothed values of transpiration for a mature beech group in the Polana Mt. (the Slovak Republic) at an elevation of 850 m, calculated on the one hand from the measured sap flow and on the other hand using the SVAT model, for the period June–July 1996 in the range from 1.1 to 6.0 mm with mean value 1.6 mm per day.

Kantor (Krečmer et al., 2003) used his many years' investigation of water balance in Norway spruce stand and European beech one in the Orlické hory Mts for estimation of daily value of evapotranspiration. Young spruce or beech stand (from thicket to pole stand) gives daily evapotraspiration for cloudy spring, summer, and autumn day 1.1, 2.0, 1.0 mm respectively, for sunny spring, summer, and autumn day 2.6, 4.4, 2.5 mm respectively. Clear-cut with heavy weed infestation shows practically the same data: 1.1, 2.2, 1.0 mm respectively 2.6, 3.7, 2.4 mm. The interval of evapotraspiration 1.0–4.4 mm per day by Kantor is very close to interval 1.1–4.1 that we established for similar natural conditions on our experimental catchment U Dvou louček.

Šach et al. (2006) derived daily evapotranspiration from decrements continuously recorded volume soil moisture in young closed stands in the Orlické hory Mts. Daily evapotranspiration of spruce pole-stage stand and beech pole-stage one ranged in rainless days of June 2005 from 1.3 to 3.3 mm respectively from 1.2 to 4.5 mm. The evapotranspiration range from 1.2 to 4.5 mm again supported the range 0.7 - 4.8 mm from the examined catchment.

If we return the computing, i.e. from the known runoff loss and from E(t, s) per day, estimated by different procedure, it is possible to determine the concrete size of VSA for given streamflow per day.

3.2.3 Conclusion

The study resumes our hitherto knowledge on influence of evapotranspiration on discharge variations in conditions of the partially waterlogged catchment in growing seasons 1997, 1995, i. e. years of phenomenon revelation, and recent growing season in 2008. The daily discharge fluctuations of streamflow, which drained waterlogged pedon, were in accordance with the theory of Kobayashi et al. (1995) and Bren (1997) who also consider the evapotranspiration of riparian vegetation to be the main cause of that phenomenon in similar natural conditions. Our observations and computations in small forested watershed in the Orlické hory Mts. (Czech Republic, EU) correspond to this theory. Values of E(t, s) 0.25 to 4.8 mm per day found by us are consistent with those of other researchers.

From total length of natural streams and drainage ditches with hydraulic reach 8 m on each of both sides resulted in 3 variable source areas (VSA) different in size and area of drainage. If we return the computing procedure, i.e. from the known runoff loss and the E(t, s) per day, estimated by different procedure, it is possible to determine the concrete size of VSA for given streamflow per day.

4. General conclusion of the chapter

The evapotranspiration ET (Ei +Et + Es) plays an important role in ensuring of hydrologic and soil conservation services of forest and forestry.

Therefore knowledge of ET on sites in frame of forest stands and forest catchments is very important not only for a forest land itself but also for the landscape in lower elevations. The problem is to determine the ET promptly and accurately as much as necessary. Several methods were selected, made better or newly developed.

At the beginning, methods of ET calculation from water balance equation were used. ET was calculated from complex of measured water balance components in a forest stand with only single unknown Et. The method of ET assessment is performed on the Deštenská stráň Hillside experimental area. Investigations, from which the ET components for mature Norway spruce and European beech stands were obtained, we have done in a continuous sequence of five water years (from 1 Nov. 1976 to 31 Oct. 1981). Two components of vaporization process, i.e. interception losses (Ei) and evaporation from the litter and soil surface (Es) were ascertained directly; the tree canopy transpiration (Et) was estimated by calculation from water budget equation as the only unknown.

Measurement of water budget components during growing season, done in all days with measurable precipitation, i.e. usually 2 to 3 times in a week, resulted in very precise assessment of ET components of experimental pure Norway spruce and European beech mature stands during reproduction period. For this reason we might take it as a reference method for mountain slopes with various pure forest stands; it is corresponding to lots of research results mainly in EU.

The assessment ET values obtained from continuous measuring the volumetric moisture of soil profile in a forest stand and calculation of E(t, s) from soil moisture differences was developed as amendment and calibrated using the previous described method. It is simpler and if meeting all demands it gives the same precise results (moreover Et is directly enumerated and so the only unknown is eliminated). Of course, correct estimating volume stoniness and rooting layers is very important. Moreover, we supposed that using the method also helps to determine complicated and permanently unsolved problematic ET of mixed forest stands especially in dry periods and possible "hydraulic lift" in a soil profile (solum). The experiment was already established in the UDL experimental catchment.

The assessment of ET values obtained from complex of partially measured water balance components and partially derived from tensiometric measurements in a small watershed with only single unknown E(t, s) was performed in the UDL experimental catchment. The method taking into consideration water movement in the soil profile solves and removes the uncertainty of soil water outflow or inflow (discharge from soil mantle, eventually recharge). Up to now, the discharge from pedon (outflow from soil), measured with particular instruments (devices), greater than causal (effective) precipitation (including discharge from snowpack - so called delayed precipitation) was eliminated as outliers.

The presented model of hydrological balance brings values of ET and Q consistent with experimental results from comparable conditions, representing in the Czech Republic conditions of mountain lands covered predominantly by coniferous forest.

The method of ET calculation from daily variation of baseflow between day and night from the small forest watershed was designed. The daily discharge fluctuations of streamflow, which drained waterlogged pedon, were in accordance with the theory of Kobayashi et al. (1995) and Bren (1997) who also considered the evapotranspiration of riparian vegetation to be the main cause of that phenomenon in similar natural conditions. From total length of natural streams and drainage ditches with known hydraulic reach we determined number of variable source areas (VSA) differing in size. From concrete area of VSA and concrete runoff loss we calculated the daily ET. If we turn the computing procedure, i.e. from the known runoff loss and by other procedure estimated daily E(t, s), it is possible to determine the concrete size of VSA for given streamflow.

5. Acknowledgements

Supported by the Ministry of Agriculture of the Czech Republic, Projects No. MZE 0002070203 and QH92073.

6. References

Bren, L.J. (1997). Effects of slope vegetation removal on the diurnal variations of a small mountain stream. *Water Resources Research*, Vol. 33, No. 2, p. 321–331, ISSN 0043-1397.

Bruinzeel, L.A. (2004). Hydrology. Hydrological cycle, In: *Encyclopedia of Forest Sciences*, Burley, J.; Evans, J. & Youngquist, J.A. (Eds.), 340-397, Elsevier, ISBN 0-12-145160-7, Amsterdam.

Burley, J.; Evans, J. & Youngquist, J.A. (Eds.). (2004). *Encyclopedia of Forest Sciences*, Elsevier, ISBN 0-12-145160-7, Amsterdam.

Cheng, J.D. (1987) Root zone drainage from a humid forest soil in the west coast of Canada. *Proc. of the Vancouver Symposium on Forest Hydrology and Watershed Management*, August 1987, IAHS Publication No. 167, pp. 377-386, International Association of Hydrological Sciences, Wallingford.

Constantz, J.; Thomas, C.L. & Zellweger, G. (1994). Influence of diurnal variations in stream temperature on streamflow loss and groundwater recharge. *Water Resources Research*, Vol. 30, No. 12, p. 3253–3264, ISSN 0043-1397.

Constantz, J. (1998) Interaction between stream temperature, streamflow, and groundwater exchanges in alpine streams. *Water Resources Research*, Vol. 34, No. 7, p. 1609–1615, ISSN 0043-1397.

Černohous, V. & Šach, F. (2008). Daily baseflow variations and forest evapotranspiration. *Ekológia (Bratislava)*, Vol. 27, No. 2, p. 189-195, ISSN 1335-342X.

ČSN 75 4200, (1994). Treatment of water regime of agriculture soils by drainage (in Czech). Prague.

Hewlett, J.D. & Hibbert, A.R. (1967). Factors affecting the response of small watersheds to precipitation in humid areas. *Proc. of an International Symposium on Forest Hydrology*, Sopper, W.E. & Lull, H.W. (Eds.), pp. 275–290, Pennsylvania State University, Aug 29 - Sept 10, 1965, Pergamon Press, Oxford.

Hewlett, J.D. (1982). *Principles of Forest Hydrology*, University of Georgia Press, ISBN 0-8203-0608-8, Athens.

Horský, L. (1970). *Hydrologic conditions in the CSSR*, 3rd vol., Hydrometeorological Institute, Prague. (in Czech)

Kantor, P. (1985). Water balance of mature spruce stands in the middle-altitude mountains of the ČSR, In: *Water Balance of Spruce Stands (Picea abies (l.) Karst) in Different Grographical Regions*, Prax, A. & Raev, I. (Eds.), 33-54, University of Agriculture, Brno.

Kantor, P. (1992). Changes in a water balance of a spruce stand after its regeneration by clear cutting. Lesnictví-Forestry, Vol. 38, No. 9-10, p. 823–838, CS ISSN 0024-1105.

Kantor, P. (1995). Water relations in a beech stand before and after its regeneration by clear felling. Lesnictví-Forestry, Vol. 41, No. 1, p. 1–10, CS ISSN 0024-1105.

Kantor, P. & Šach, F. (2008). Water balance of young Norway spruce and European beech mountain stands in growing seasons 2005, 2006. *Folia Oecologica*, Vol. 35, No. 1, p. 6-14, ISSN 1336-5266.

Kobayashi, D.; Suzuki, K. & Nomura, M. (1990). Diurnal fluctuation in stream flow and in specific electric conductance during drought periods. *Journal of Hydrology*, Vol. 115, p. 105–114, ISSN 0022-1694.

Kobayashi, D.; Kodama, Y.; Ishii, Y.; Tanaka, Y. & Suzuki, K. (1995). Diurnal variations in streamflow and water quality during the summer dry season. *Hydrological Processes*, Vol. 9, No. 7, p. 833–841, ISSN 1099-1085.

Krečmer, V.; Kantor, P.; Šach, F.; Švihla, V. & Černohous, V. (2003). *Forests and floods* (in Czech), Ministry of Environment, ISBN 80-7212-255-X, Prague.

Ladefoged, K. (1963). Transpiration of forest trees in closed stands. *Physiologia Plantarum*, Vol. 16, No. 2, p. 378–414, ISSN 0031-9317.

Mráz, K. (1990). *Soil water regime, relation to wood increment and water outflow from different tree species stands*, Final report, Forest and Game Management Research Institute, Strnady, 50 p.

National Academy of Sciences. (2008). *Hydrologic effects of a Changing Forest Landscape*, National Academies Press, ISBN 978-0-309-12109-5, Washington, DC.

Novák, V. (1954). Water in soil - soil water regime. In: *Methods of study in phytocenology, ecology, climatology and pedology*, Klika, J.; Novák, V. & Gregor, A. (Eds.), 440-484, ČSAV, Prague. (in Czech)

Ronan, A.D.; Prudic, D.E.; Thodal, C.E. & Constanz, J. (1998). Field study and simulation of diurnal temperature effects on infiltration and variably saturated flow beneath an ephemeral stream. *Water Resources Research*, Vol. 34, No. 9, p. 2137–2153, ISSN 0043-1397.

Střelcová, K.; Matejka, F. & Kučera, J. (2004). Beech stand transpiration assessment - two methodical approaches. *Ekológia (Bratislava)*, Vol. 23, Suppl. No. 2, p. 147-162, ISSN 1335-342X.

Šach, F.; Kantor, P. & Černohous V. (2006). Determination of evapotranspiration of young Norway spruce stand and European beech one by method of continual measurement of volumetric moisture in soil profile. *Proc. of an International Conference on Stabilization of Forest Functions in Biotopes disturbed by Anthropogenic Activity*, Jurásek A.; Novák, J. & Slodičák, M. (Eds.), pp. 525-536, ISBN 80-86461-71-8, Forest Research Station, Opočno, Sept. 5 – 9, 2006, Forest and Game Management Research Institute, Strnady.

Tesař, M.; Šír, M.; Kubík, F.; Pražák, J. & Strnad, E. (1992). Forest transpiration in the vegetation season with a sufficient volume of soil water, *Lesnictví-Forestry*, Vol. 38, No. 11, p. 877-888, CS ISSN 0024-1105.

Thomas, D.L. & Beasley, D.B. (1984). *A Distributed, Management-oriented, Forest Hydrology Model*, Paper No. 84-2019, American Society of Agricultural Engineers., St. Joseph (Michigan).

Vilhar, U.; Starr, M.; Urbančič, M.; Smolej, I. & Simončič, P. (2005). Gap evapotranspiration and drainage fluxes in a managed and a virgin dinaric silver fir–beech forest in Slovenia: a modelling study. *European Journal of Forest Research*, Vol. 124, No. 3, p. 165–175, ISSN 1612-4677.

Weiler, M. & Mac Donell, J.J. (2004). Soil development and properties. Water storage and movement, In: *Encyclopedia of Forest Sciences*, Burley, J.; Evans, J. & Youngquist, J.A. (Eds.), 1253-1260, Elsevier, ISBN 0-12-145160-7, Amsterdam.

4

Analysis of Catchment Evapotranspiration at Different Scales Using Bottom-up and Top-down Approaches

Xiangyu Xu and Dawen Yang
State Key Laboratory Of Hydroscience and Engineering, Tsinghua University
China

1. Introduction

Evapotranspiration is a highly non-linear process in catchment hydrological cycle, and simulation of the evapotranspiration process is a major component of physically-based hydrological models. Most process-based hydrological models estimate actual evapotranspiration following the Penman's approach (Penman, 1948; Allen et al., 1998) by reducing potential evapotranspiration to actual evapotranspiration in response to water stress which is given by a function of soil water availability for a given vegetation condition. The soil water availability is governed by other hydrological processes such as infiltration and soil moisture redistribution (both vertically and laterally). Physically-based hydrological model predict the catchment water balance through detailed simulation of hydrological processes at small temporal and spatial scales. However, catchment annual evapotranspiration can be easily and simply predicted using lumped conceptual models such as Fu's model based on the Budyko hypothesis (Yang et al., 2007). Combination of these bottom-up and top-down approaches can help understand the dominant control on catchment evapotranspiration at different time scales.

The top-down approach in hydrology was first introduced by Klemes (1983) and extended further by Sivapalan and coworkers as outlined in Sivapalan et al. (2003). This approach was used to investigate the hydrological response at long temporal scale and large spatial scale (e.g. an annual time and watershed scale) and progressively narrowing down to processes at smaller scales (e.g. hourly time and hillslope scale). The major advantage of the top-down approach is that it can reduce data requirements and limit model complexity (Littlewood et al., 2003). The lumped conceptual model is a typical example. A simple lumped model can be easily used for hypothesis testing and as the starting point for developing a new generation of models. Many researchers carried out a series of studies using the top-down approach (Klemes, 1983; Jothityangkoon et al., 2001; Atkinson et al., 2002, 2003; Farmer et al., 2003; Eder et al., 2003; Son and Sivapalan, 2007; Zhang et al., 2008), and developed a collection of lumped conceptual models for simulating water balance responses across different temporal scales (annual, monthly and daily) in different watersheds located in different climatic regimes. Budyko (1974) proposed a semi-empirical relationship between the ratio of annual evapotranspiration to annual precipitation and the ratio of annual precipitation to annual net radiation. Considerable research has been performed on the

Budyko curve (Fu, 1981; Milly, 1994; Choudhury, 1999; Wolock and McCabe, 1999; Zhang et al., 2004, Yang et al., 2008a) through analyzing the interactions between climate, soils and vegetation in producing annual water balance. A simple water-energy balance equation (Yang et al., 2006 and 2007) based on the Budyko hypothesis (1974) has been used for predicting the long-term average evapotranspiration and the inter-annual variability of evapotranspiration. The top-down approach is usually used for explaining the observed responses at the watershed scale, in which one problem is how to generalize the catchment hydrological pattern and whether the hypothesis of conceptualization is reasonable. Another problem is how to explain the interactions between the response at watershed scale and the processes that contributed to it that occur at smaller scales. The top-down approach has not been widely used in hydrological prediction, but as Sivapalan (2003) mentioned this approach should become an important component of the toolkit for data analysis and hydrological model development.

The bottom-up approach in hydrology generally uses "process based" or "physically based" distributed hydrological models, which can predict overall catchment's water balance components based on process simulation at smaller spatial and temporal scales. In order to understand the hydrological processes at a local scale and to analyze the temporal and spatial variability of water resources at a watershed scale, a number of distributed hydrological models have been developed: SHE model (Beven et al., 1980; Abbott et al., 1986; Bathurst et al., 1995; Refsgaard and Storm, 1995), IHDM model (Morris, 1980), SLURP model (Kite, 1995), SWAT model (Arnold et al., 1998), GBHM model (Yang et al., 1998, 2002, 2004; Cong et al., 2009), WEP (Jia et al., 2001), VIC (Liang, 1994) etc. These and many other models have been developed and have been applied to interpret and predict the impacts of land-use change and climate variability. A bottom-up model requires vast amounts of input data for describing the variable rainfall and evapotranspiration fields, topography, vegetation, soil, and other land surface conditions, and also needs dozens of parameters for representation of the hydrological processes. Therefore, in the one hand, the costs would be prohibitively high. Beven (2001) discussed the problems of distributed model including the problem of nonlinearity, scale, equifinality, uniqueness and uncertainty. Because the physically-based distributed hydrological models can offer both detailed information of hydrological processes inside a catchment and general catchment hydrological pattern, this kind of models are usually used to analyze hydrological impacts of the changes of land use and climate.

Most recently there are discussions on the direction of hydrological research, which focus on either estimating predictive uncertainty or understanding catchment hydrological function (Beven, 2006; Sivakumar, 2008; Sivapalan, 2009). For better understanding of catchment hydrology, we need a change in the methodology of hydrological analysis. Though both the bottom-up and the top-down approaches are widely used in catchment hydrology, very little research has been carried out on the linkage between these two methods. Klemes (1983) first suggested the combination of these two approaches and mentioned that "the most promising route to significant new discoveries in hydrology is to combine the upward and downward research based on the existing facts and knowledge as well as on imagination and intuition, to form testable hypotheses—i.e. to apply the time-honored scientific method". Sivapalan et al. (2003) also anticipated that "in the near future, studies that address the same problem by both approaches at the same time will make a significant impact on a better understanding of hydrologic prediction issues". Based on the previous research, this chapter attempts to build the connection between the top-down and bottom-up approach and to understand the dominant factors on catchment evapotranspiration at different time scales using a distributed physically-based hydrological model and a simple water-energy balance model.

2. Study area and data availability

The Luan River basin is located in the region delimited by geographical coordinates 40.4°-42.6°N and 115.5°-118.9°E with the elevation ranging about 150-2000m above the sea level. The Panjiakou Reservoir is the largest reservoir on the Luan River in northern China, the upstream watershed of the Panjiakou Reservoir is selected as the study area in the present research (see Fig. 1).

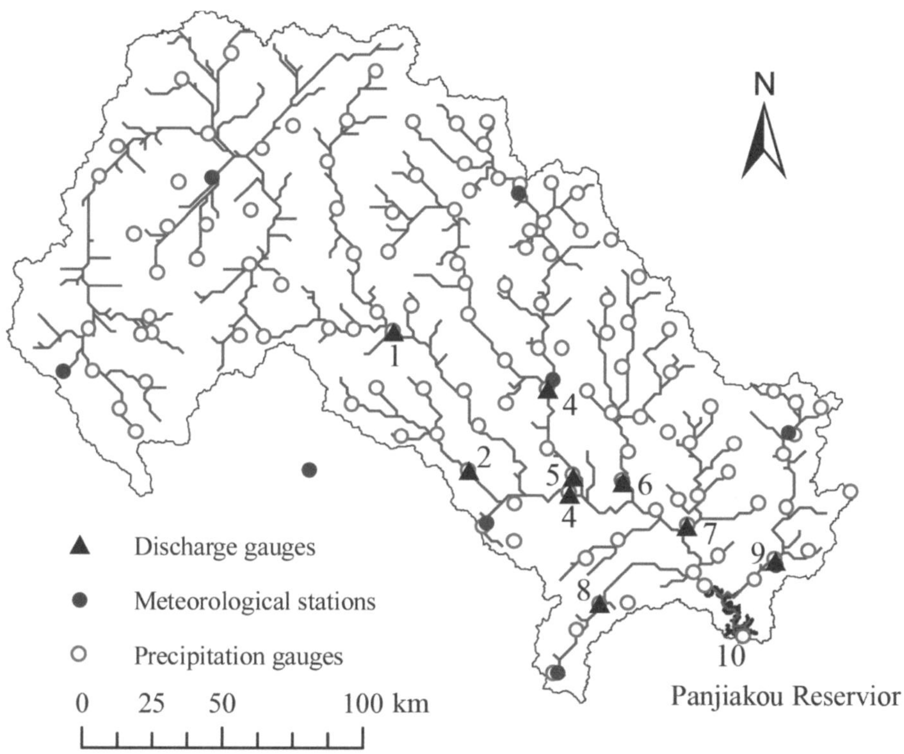

Fig. 1. The study area.

The catchment extent and topography are represented using the digital elevation model (DEM) of a 90m resolutionwhich is obtained from the global topography database (http://telascience.sdsc.edu/tela_data/SRTM/version2/SRTM3/). Land use map of 100m resolution was from Data Center for Resources and Environmental Sciences of Chinese Academy of Sciences. Based on the available data, the land use has been regrouped into 7 categories including water body, urban area, forest, irrigated cropland, upland, grassland and shrub for the study area. Seasonal change of vegetation is expressed by the monthly leaf area index (LAI) derived from monthly NDVI (NDVI data is obtained from the DAAC website of GSFC/NASA: http://daac.gsfc.nasa.gov/DAAC_DOCS/). The soil type used here is obtained from the FAO-UNESCO's 1:5 million scale Soil Map of the World (FAO,

2003). The soil properties used for the hydrological simulation including the porosity, the saturated hydraulic conductivity and the other soil water parameters corresponding to each soil type in this map are obtained from the Global Soil Data Task (IGBP-DIS, 2000) with an 8-km resolution.

Daily meteorological data are obtained from China Meteorological Administration and additional daily precipitation data are obtained from the Ministry of Water Resources. The daily meteorological data consists of precipitation, temperature, average wind speed, relative humidity, sunshine duration, etc. There are 10 meteorological stations and 121 precipitation gauges in this region (see Fig. 1). From the two data sets, a time period from 1980 to 1991 is selected according to the availability of records in these meteorological stations and precipitation gauges. The gridded meteorological data required as input to the distributed hydrological model is interpolated from the gauge data. Precipitation is interpolated using an angular-distance weighting method (New, et al., 2000). In the same way, the wind speed, relative humidity and sunshine hours are also interpolated into each grid. The temperatures (maximum, minimum and mean) are interpolated using an elevation-corrected angular-direction weighting method. The daily potential evapotranspiration is calculated using the wind speed, relative humidity, sunshine hours and temperature (Shuttleworth, 1993).

The discharge data are collected for the same period of 1980-1991 from the "Hydrological Year Book" published by the Hydrological Bureau of the Ministry of Water Resources. In this study, seven gauges located on the tributaries and two gauges located on the mainstream (see Fig. 1) are selected for analysis.

3. Methodology

The ratio of evapotranspiration to precipitation in the study area is nearly 70~80%, therefore, estimation of evapotranspiration is important for understanding the catchment water balance. The physically based hydrological model can estimate hourly evapotranspiration at hillslope scale, and through a bottom-up analysis the hourly evapotranspiration in each hillslope can be integrated into daily, monthly, annual and long-term mean annual evapotranspiration at catchment scale, which offers a reference for the top-down analysis. The long-term mean annual evapotranspiration of a catchment can be estimated using a water-energy balance model first, and through a top-down analysis the temporal scale is downward to year, month and day for identifying the dominant factors on catchment evapotranspiration at different time scales.

3.1 The distributed hydrological model—GBHM model

Since the GBHM model has been successfully applied in the Yellow River (Yang et al., 2004; Cong et al., 2009) and the Yangtze River (Xu et al., 2007, 2008), it is chosen as a bottom-up tool to simulate the hydrological processes and the evapotranspiration in the Luan River basin. As shown in Fig. 2, 2-km grid is used for discretization of the study catchment, in which a sub-grid parameterization scheme is employed to represent the hillslope topography (Yang et al., 2002). The procedures of model construction contain catchment subdivision, sub-grid parameterization, physically-based hydrological simulation on hillslope and kinematic wave flow routing.

A total of 97 sub-catchments were divided for the study catchment. For each sub-catchment, the geomorphologic property of stream-hillslope formation is used for representing the catchment topography and it is also assumed that a large grid is comprised of a set of hillslopes located along the streams. From a macro-scale viewpoint, the hillslopes located in a grid are viewed as being geometrically similar. A hillslope with a unit width is called a hillslope element, represented by a rectangular inclined plane (Fig. 2).

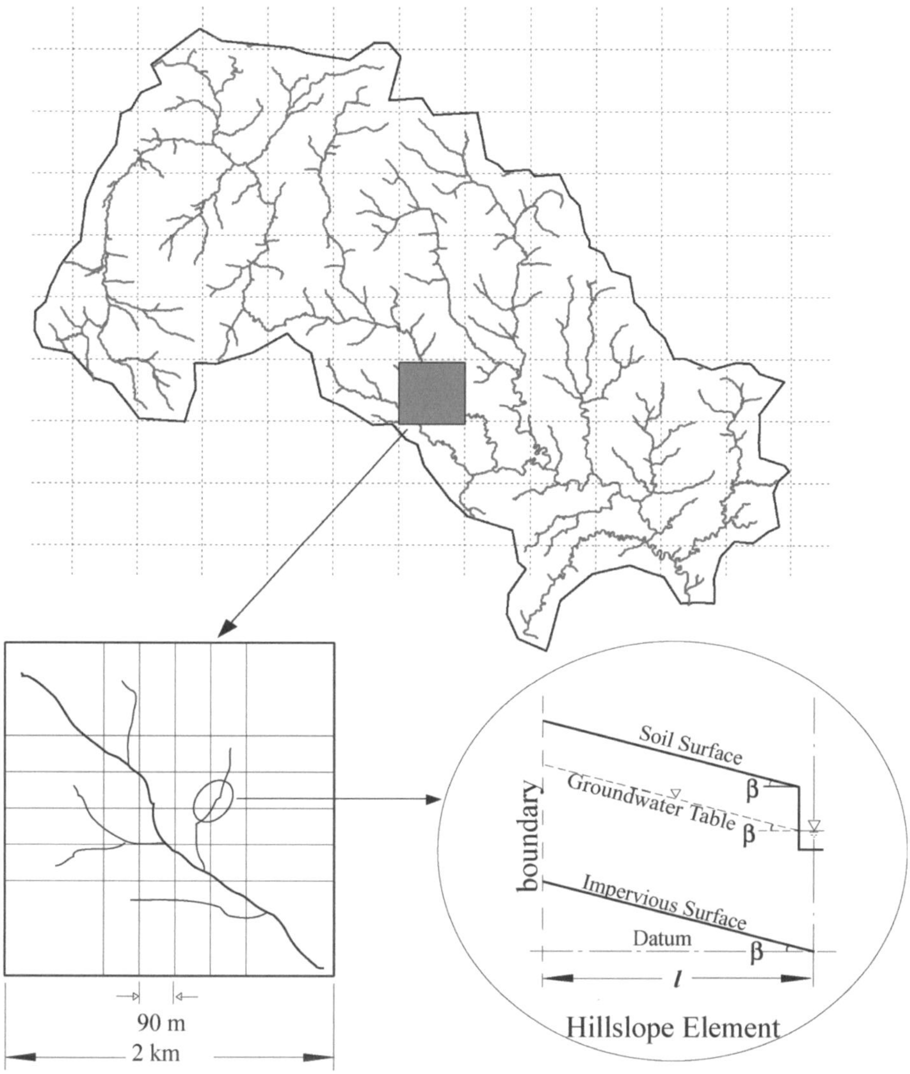

Fig. 2. Structure of the GBHM model.

Considering the land cover heterogeneity, hillslopes located in a 2-km grid are grouped. The soil within a grid is represented as a single dominant soil type. The runoff generated from one grid is the sum of surface and subsurface runoff. The vertical flux, the actual evapotranspiration, is the total evapotranspiration simulated from all of the hillslopes. The soil moisture content is taken as the area-averaged soil moisture of all the hillslopes.

The hydrological processes included in the GBHM model are snowmelt, canopy interception, evapotranspiration, infiltration, surface flow, subsurface flow and the exchange between the groundwater and the river (Yang et al., 1998, 2002 and 2004), which are simulated at 1-hour time step. Actual evapotranspiration is estimated from potential evapotranspiration, which is calculated by the Penman equation (Penman, 1984). Considering seasonal variation of LAI, root distribution and soil moisture availability, it is computed individually from the canopy water storage, root zone and soil surface. For each 1-hour time step, actual evaporation rate from the canopy storage is expressed as:

$$E_{\text{canopy}} = K_c E_p \tag{1}$$

where E_p is the potential evaporation rate and is the crop coefficient. The vegetation transpiration rate is estimated as the rate of water taken up from the root zone and follows the equation:

$$E_{\text{tr}}(z_j) = K_c E_p f_1(z_j) f_2(\theta_j) \frac{LAI}{LAI_0} \tag{2}$$

where $E_{\text{tr}}(z_j)$ is the transpiration rate from layer j of the root zone; $f_1(z_j)$ is the root distribution function that is given by a triangular distribution with its maximum near the surface; $f_2(\theta_j)$ is given as a linear function of soil moisture θ_j ; and LAI_0 is the maximum leaf area index of the vegetation in a year. For the bare soil, the evaporation rate is given by:

$$E_s = K_c E_p f_2(\theta) \tag{3}$$

where K_c is the evaporation rate from the soil surface. In ponding conditions, the value of K_c is 1. The actual transpiration from the root zone and evaporation from the soil surface are treated as sink terms in Richards' equation that is employed to model soil water movement in the unsaturated zone.

Infiltration and subsurface flow in the vertical direction and along the hillslope are described in a quasi-two-dimensional subsurface model. The vertical water flow in the topsoil is represented by Richards' equation. The soil water distribution along the hillslope is treated as uniform. When the soil water content in each layer is more than the field capacity, water moves to the stream along the hillslope by gravity.

The surface runoff, from the infiltration excess and saturation excess, is obtained by solving Richards' equation and flows to the river along the hillslope by a kinematic wave. The groundwater aquifer is treated as an individual storage corresponding to each grid. The exchange between the groundwater and the river water is considered as steady flow and is calculated by Darcy's law (Yang et al., 2002). The runoff generated from the grid is the lateral inflow into the river at the same flow interval. Flow routing in the river network is solved using the kinematic wave approach.

The parameters used in the model include vegetation parameters, land surface parameters, soil-water properties, river parameters, a snow-melting parameter and a groundwater parameter. Since most parameters have their physical meanings, they can be estimated through field tests. This study specifies the parameters by referring to the existing database and handbooks. A 6-year test run from 1980 to 1985 is carried out for calibrating the model parameters. One of the calibrated parameters in this model is the snowmelt factor in the temperature-based snowmelt equation. Another calibrated parameter, the hydraulic conductivity of the groundwater, is calibrated by checking the base flow in different sub-basins. Model validation is carried out from 1986 to 1991. In the calibration period, the

Nash-Sutcliffe coefficient for the simulated daily discharges at the outlet of the study area is nearly 0.80, and the absolute values of the relative error are less than 2%. For the major tributaries, the values of are larger than 0.65 and the absolute values of the relative error are less than 8% and the others are more than 10%. This showed that hydrological simulation by the GBHM model is reasonable.

3.2 The water-energy balance model

Evapotranspiration is the key component linking water balance with energy balance. The primary factors controlling the long-term mean evapotranspiration are the local interaction of water supply (precipitation) and demand (potential evapotranspiration). Budyko (1974) assumed that actual evapotranspiration is controlled by both water and energy availabilities and considered that the ratio of annual evapotranspiration and annual precipitation is the function of the ratio of annual precipitation and annual net radiation. Based on the Budyko hypothesis, and based on dimensional analysis and mathematical reasoning, Yang et al. (2008a) considered the important effect of the vegetation and derived an analytical equation of the coupled water–energy balance at an annual time scale, expressed as:

$$E = \frac{E_0 P}{(P^n + E_0^n)^{1/n}} \tag{4}$$

where the parameter n represents the effect of the catchment landscape characteristics, such as vegetation, soil properties and slope gradient. This theoretical equation has also been extended to a variety of time scales.

In order to consider the effect of the vegetation, Yang et al. (2009) changed the relative soil water storage $S_{\max}/E_0$ in the parameter ϖ to vegetation coverage M in the parameter n. The empirical formula of parameter n is given for the Haihe River basin as:

$$n = 2.721(\frac{K_s}{i_r})^{-0.393} M^{-0.301} \exp(4.351 \tan\beta) \tag{5}$$

where the monthly vegetation coverage was calculated from the NDVI data by using the method proposed by Gutman and Ignatov (1998):

$$M = (NDVI - NDVI_{\min}) / (NDVI_{\max} - NDVI_{\min}) \tag{6}$$

The actual evapotranspiration obtained from Eqs. (4) and (5) showed remarkable agreement with that derived from the long-term water balance at Haihe River Basin (Yang et al., 2009), so Eqs. (4) and (5) were used in this research to calculate the actual annual evapotranspiration. In order to validate the water-energy balance model given by Eqs. (4) and (5) at annual time scale, the actual annual evapotranspiration simulated for each sub-catchment were plotted against the actual annual evapotranspiration estimated from the observed precipitation subtracting the observed runoff by neglecting the change of water storage (see Fig. 3). Most values of predicted annual evapotranspiration are much close to the observed ones, which shows good simulation of annual evapotranspiration. The relative error between the long-term average evapotranspiration calculated by this model and the actual evapotranspiration estimated from the long-term water balance is less than 5%.

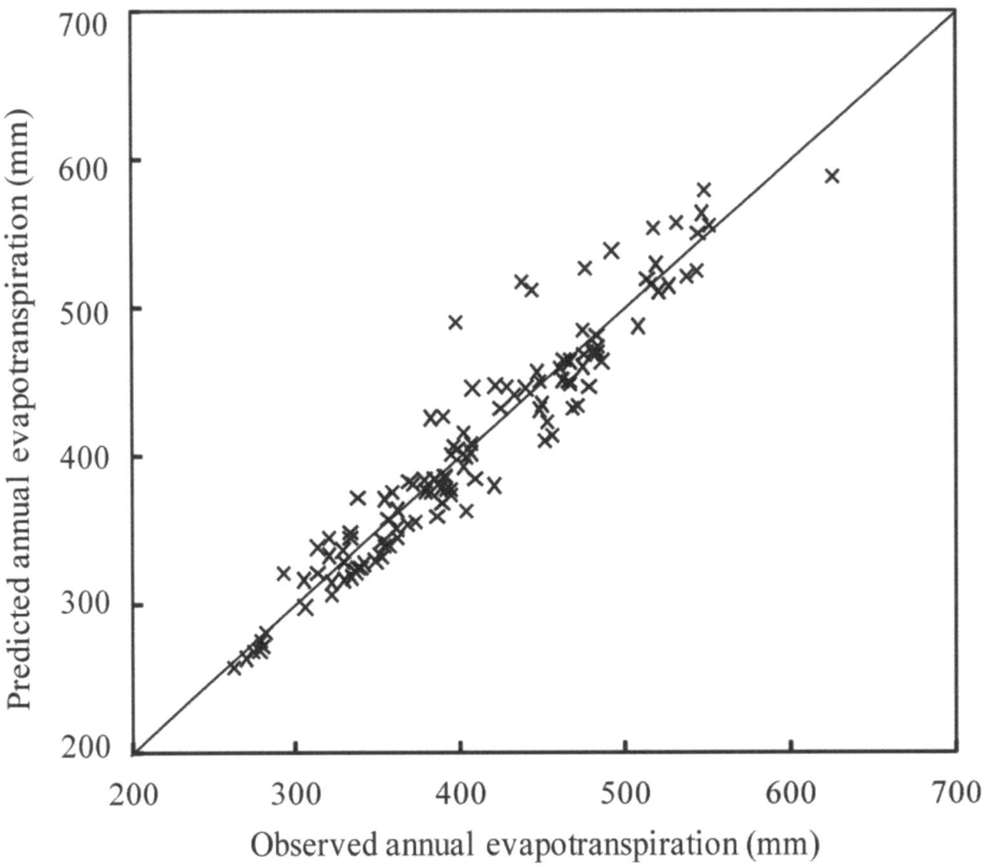

Fig. 3. Predicted value of annual evapotranspiration by the water-energy balance model plotting versus observed one (the 1:1 line is plotted for comparison).

As Eq. (4) doesn't consider inter-annual variability of soil water content, Yang et al. (2008b) derived another equation to estimate the evapotranspiration at arbitrary time scale which contains the variability of soil water, followed as:

$$E = E_0 (P + S) \Big/ \left((P + S)^n + E_0^n \right)^{1/n} \tag{7}$$

in which P is the precipitation, and S is the initial soil water of each simulation time period (i.e., annual, monthly, daily period).

To use the above equation for continuous simulation, we rewrite Eq. (7) as:

$$E_i = E_{0,i} (P_i + S_{i-1}) \Big/ \left((P_i + S_{i-1})^n + E_{0,i}^n \right)^{1/n} \tag{8}$$

in which i represents the ith simulation time period, S_{i-1} represents the initial soil water content at the ith period. And the soil water content at the end of the ith period can be derived from the water balance equation as:

$$S_i = P_i + G_i - R_i - E_i + S_{i-1} \tag{9}$$

in which G_i is the groundwater recharge. The value of G_i is ignored since its relatively small value in the research area.

The evaportranspiration at arbitrary time scales can be simulated using Eqs. (8) and (9) as long as the initial soil water content S_0 is given. The value of S_0 is derived from the continuous simulation at the first several simulation periods. The parameter n in Eq. (8) is calculated at different time scale according to Eq. (5), however, at the daily time scale the same parameter as the monthly scale is used due to the vegetation data availability.

4. Results and discussion

4.1 Bottom-up analysis

The GBHM model used a linear assumption at hourly temporal and hillslope spatial scale, that is, the actual evapotranspiration for a hydrological unit at hourly scale is estimated to be proportional to the potential evapotranspiration (Eqs. (1) - (3)). It is known that annual precipitation, potential and actual evapotranspiration are following the Budyko curve at catchment scale. Based the long-term evapotranspiration estimated by GBHM, Fig. 4 plots the annual precipitation, potential and actual evapotranspiration for the 9 sub-catchments and the whole watershed in two alternative forms of the Budyko curves, i.e. E/P vs. E_0/P and E/E_0 vs. P/E_0. The value of parameter ranges from 2.5 to 2.95 in the study area and all the points stand inside the two curves.

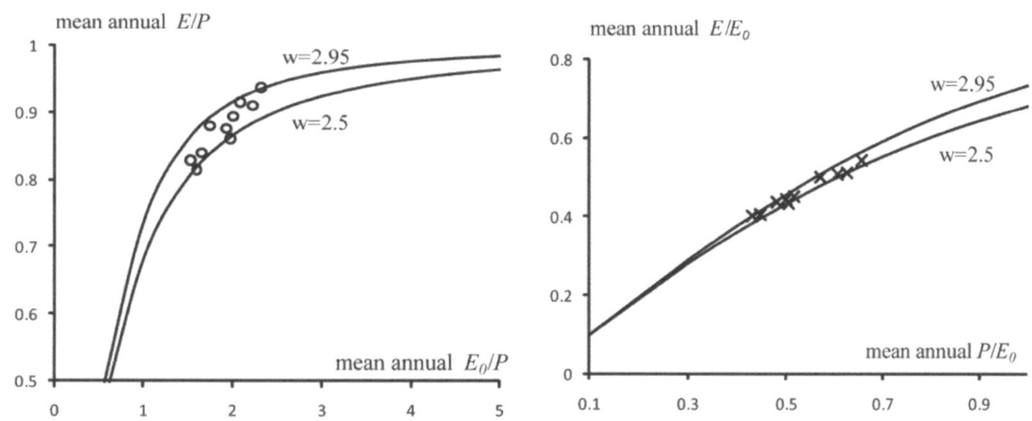

Fig. 4. Long-term mean values of annual actual evapotranspiration, precipitation, and potential evapotranspiration for the 10 sub-catchments are plotted in two different but equivalent Budyko–type forms (plotted in scattering points) together with the Budyko curves with the regional average values of parameter ϖ .

According to Yang et al.'s study (Yang et al., 2006), the actual evapotranspiration at annual and catchment scale is mainly controlled by precipitation rather than potential evapotranspiration in the non-humid regions of China, and the relationship between the actual and potential evapotranspiration should be complementary. The actual and potential evapotranspiration is usually plotted against the precipitation for validating the complementary relationship of evapotranspiration. Fig. 5 shows the complementary relationship between the actual and potential evapotranspiration, in which the actual

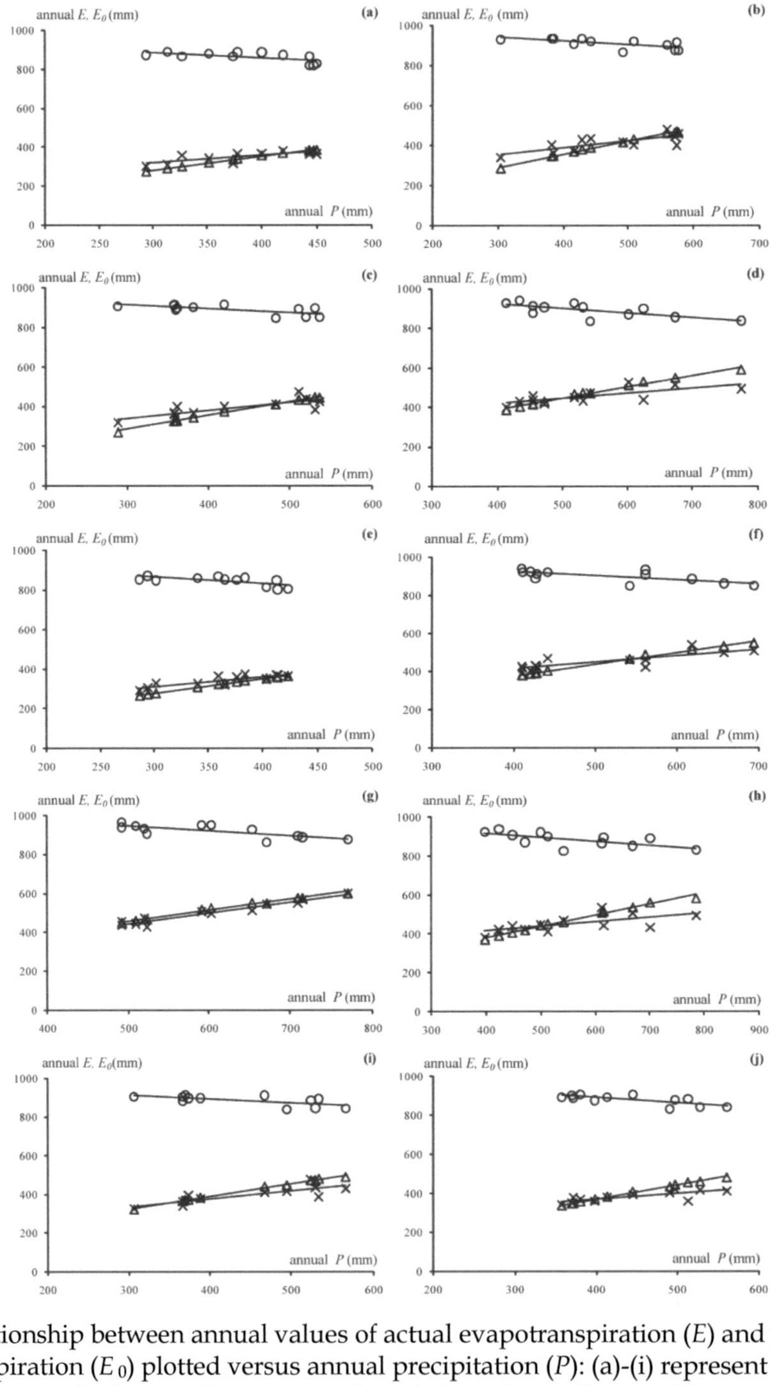

Fig. 5. Relationship between annual values of actual evapotranspiration (E) and potential evapotranspiration (E_0) plotted versus annual precipitation (P): (a)-(i) represent the 9 sub-catchment, respectively, and (j) represents the whole study catchment. (The circular dots represent potential evapotranspiration (E_0), the cross dots represent actual annual evapotranspiration (E) predicted by GBHM, and the triangular dots represent actual annual evapotranspiration (E) predicted by water-energy balance model)

evapotranspiration is simulated by the GBHM. From Fig. 4 and 5, it is known that annual catchment evapotranspiration simulated by the GBHM agreed with the Budyko curve and followed the complementary relationship with potential evapotranspiration, which showed that the GBHM could simulate interaction between the land surface and atmosphere.

For a general case, the actual evapotranspiration is determined mainly by the potential evapotranspiration, soil moisture and vegetation status, which can be expressed in a general form as $E = f(E_0, \theta, vegetation)$($\theta$ is soil moisture content). This is an unknown non-linear equation for the most cases in a real catchment. One analytical solution of this equation is given as Eq. (4) which shows a nonlinear relationship between the annual precipitation, potential and actual evaportranspiration. This equation can also be calculated as the summation of the actual evapotranspiration at each 1-hour time step over the same year. At hourly time scale over a uniform landscape condition, the soil moisture can be viewed as constant and expressed as a soil moisture-impact factor (that is expressed as a stage-linear function of soil moisture content) in the equation; and the leaf-area-index can also viewed as constant and expressed as a vegetation-impact factor. Therefore, the linear assumption (Eqs. (1)- (3)) for a hydrological unit with relative uniform landscape condition at a short time scale (1-hour in this model) can be adopted.

The complementary relationship between the actual and potential evapotranspiration results from the interaction between the land surface and atmosphere. For a given net radiation, a decrease in actual evapotranspiration will cause an increase in sensible heat flux, and then an increase in potential evapotranspiration due to the land surface-atmosphere interactions (Bouchet, 1963). Budyko (1974) assumed that actual evapotranspiration is controlled by both water and energy availabilities. At the annual time scale, the water availability is the amount of annual precipitation and the energy availability can be measured by the potential evapotranspiration. The water-energy balance given by Eq. (4) describes the non-linear relationship among annual precipitation, potential and actual evapotranspiration and the land surface-atmosphere interactions. In the GBHM, the potential evapotranspiration rate is estimated using Penman's equation based on the daily meteorological data, which is the synthesized result of net radiation and turbulent diffusion including the feedback from land surface (e.g. the vapor pressure and air temperature are affected by the land surface evapotranspiration). Therefore, the actual evapotranspiration simulated by the GBHM model also includes the interaction and feedback between land surface and atmosphere.

4.2 Top-down analysis

A lumped water-energy balance model is used in the top-down analysis for understanding the major controlling factors of the evapotranspiration at different temporal scales. The actual evapotranspiration simulated by the GBHM was used as a reference for this analysis. Analysis is from a long-term time scale down to an annual, monthly and daily time scale. At the long-term time scale, Eq. (4) is used to estimate the actual evapotranspiration; at the annual time scale, Eq. (4) is used to estimate the actual evapotranspiration neglecting of the inter-annual variability of soil moisture; at the monthly and daily time scale, Eq. (8) is used with consideration of the soil moisture availability.

1. Actual evapotranspiration at long-term time scale

The mean annual evapotranspiration is controlled by the local interaction of fluctuating water supply (precipitation) and demand (potential evapotranspiration) at long-term time scale. Fig. 6 and Table 1 show the comparison of the long-term mean values of actual

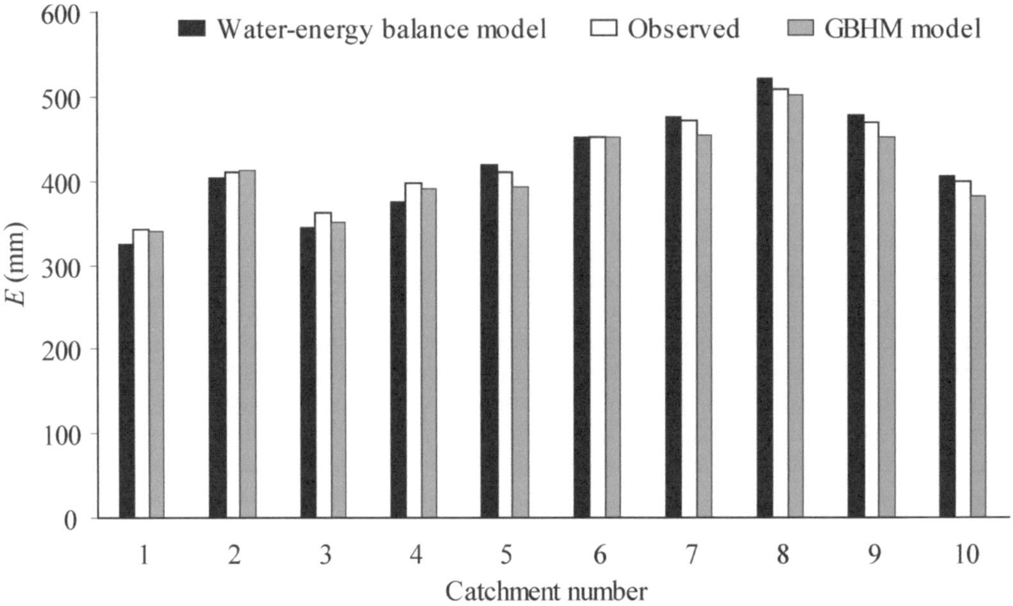

Fig. 6. Comparison of the long-term average evapotranspiration estimated by water-energy balance model, GBHM model and from the observed data. (number 1-9 represents the sub-catchment of the upstream of the Guojiatun station, Boluonuo station, Sandaohezi station, Xiahenan station, Hanjiaying station, Chengde station, Xiabancheng station, Liying station, Kuancheng station, respectively and number 10 represents the whole watershed).

No.	Guage	River	*P* (mm)	*R* (mm)	Simulated actual *E* (mm)			Relative error *RE* (%)	
					E_{wb}	E_{webm}	E_{gbhm}	RE_1	RE_2
1	Guojiatun	Luan River	356.8	21.7	335.1	319.3	340.7	-4.72	1.67
2	Boluonuo	Xingzhou	453.1	59.8	393.2	393.3	411.5	0.00	4.63
3	Sandaohezi	Luan	375.5	23.8	351.7	336.6	351.5	-4.29	-0.07
4	Xiahenan	Yimatu	420.7	29.8	390.9	372.5	389.7	-4.72	-0.32
5	Hanjiaying	Yixun	426.8	30.0	396.7	409.5	393.7	3.21	-0.77
6	Chengde	Wulie	503.4	62.1	441.2	444.2	452.3	0.69	2.51
7	Xiabancheng	Laoniu	524.2	71.1	453.2	463.4	454.9	2.27	0.37
8	Liying	Liu	595.3	96.6	498.6	516.0	501.0	3.47	0.47
9	Kuancheng	Pu	533.5	85.7	447.7	464.4	452.5	3.72	1.07
10	Panjiakou	Luan	430.0	45.2	384.8	395.1	381.8	2.68	-0.76

Table 1. Comparison of the long-term water balance simulated by the GBHM and water-energy balance model.

Notes: (1) E_{wb}, E_{fu}, E_{gbhm} represent mean values of actual evapotranspiration calculated by long-term water balance, the water-energy balance model and the GBHM model respectively. (2) RE_1 and RE_2 are the relative errors of the long-term mean values of actual evapotranspiration calculated using the water-energy balance model and the GBHM model, respectively, comparing with the value estimated by long-term water balance.

evapotranspiration estimated from the annual precipitation and runoff data by neglecting the change of water storage, simulated by the GBHM model and the water-energy balance model. It can be seen that the values of the actual evapotranspiration by different models are in a good agreement. Considering the value estimated from the long-term water balance by neglecting the change of water storage as the "true value" (which is denoted as the "observation"), the relative error for the simulated annual evaportranspiration are within by both GBHM and water-energy balance model.

2. Actual evapotranspiration at annual time scale

In order to understand inter-annual variability and the major controlling factors of catchment evapotranspiration at annual time scale, assuming vegetation condition is unchangeable or changeable and neglecting or consideing the variability of soil water four different conditions are used in the simulation for comparison: (1) using Eq. (4) with the same value of parameter n for every year in each sub-catchment (denoted as Sim-A1); (2) using Eq. (4) with the different value of parameter n estimated using Eq. (5) according to the vegetation coverage for every year in each sub-catchment(denoted as Sim-A2); (3) using Eq. (8) with the same value of parameter n for every year in each sub-catchment(denoted as Sim-A3); (4) using Eq. (8) with the different value of parameter n estimated using Eq. (5) according to the vegetation coverage for every year in each sub-catchment(denoted as Sim-A4). Taking average values of the annual evapotranspiration during the simulated period (1980-1991) for each sub-catchment simulated by the water-energy balance model and comparing with the observation, Table 2 shows the relative error. Results show that Sim-A3 and Sim-A4 have a better result than Sim-A1 and Sim-A2, which implies that it is essential to consider inter-annual variability of soil moisture for estimating annual catchment evapotranspiration. Table 2 also shows that catchment annual evapotranspiration is more

NO.	Gauges	Relative Error (*RE*, %)			
		Sim-A1	Sim-A2	Sim-A3	Sim-A4
1	Guojiatun	-5.17	-7.71	-1.32	-2.60
2	Boluonuo	-8.11	4.39	-4.02	4.20
3	Sandaohezi	-7.30	-7.62	-2.42	-2.64
4	Xiahenan	-5.28	-1.44	-2.11	-0.28
5	Hanjiaying	-7.66	-1.81	-3.88	-0.29
6	Chengde	-5.59	4.78	-2.04	4.42
7	Xiabancheng	-7.93	5.22	-4.57	4.80
8	Liying	2.32	2.77	2.76	3.05
9	Kuancheng	2.25	2.90	2.38	2.71
10	Panjiakou	-2.73	-1.11	-0.33	0.11

Table 2. The annual evapotranspiration simulated by the water-energy balance model with different consideration soil water storage and vegetation

Notes: (1) Sim-A1, Sim-A2, Sim-A3, Sim-A4 represent four conditions in simulation of the annual evaportranspiration, i.e., Sim-A1: neglecting the change of the vegetation and soil water; Sim-A2: neglecting the change of soil water; Sim-A3: neglecting the change of the vegetation; Sim-A4: considering the change of the vegetation and soil water. (2) *RE* is the relative errors of the mean values of annual actual evapotranspiration calculated using the water-energy balance model.

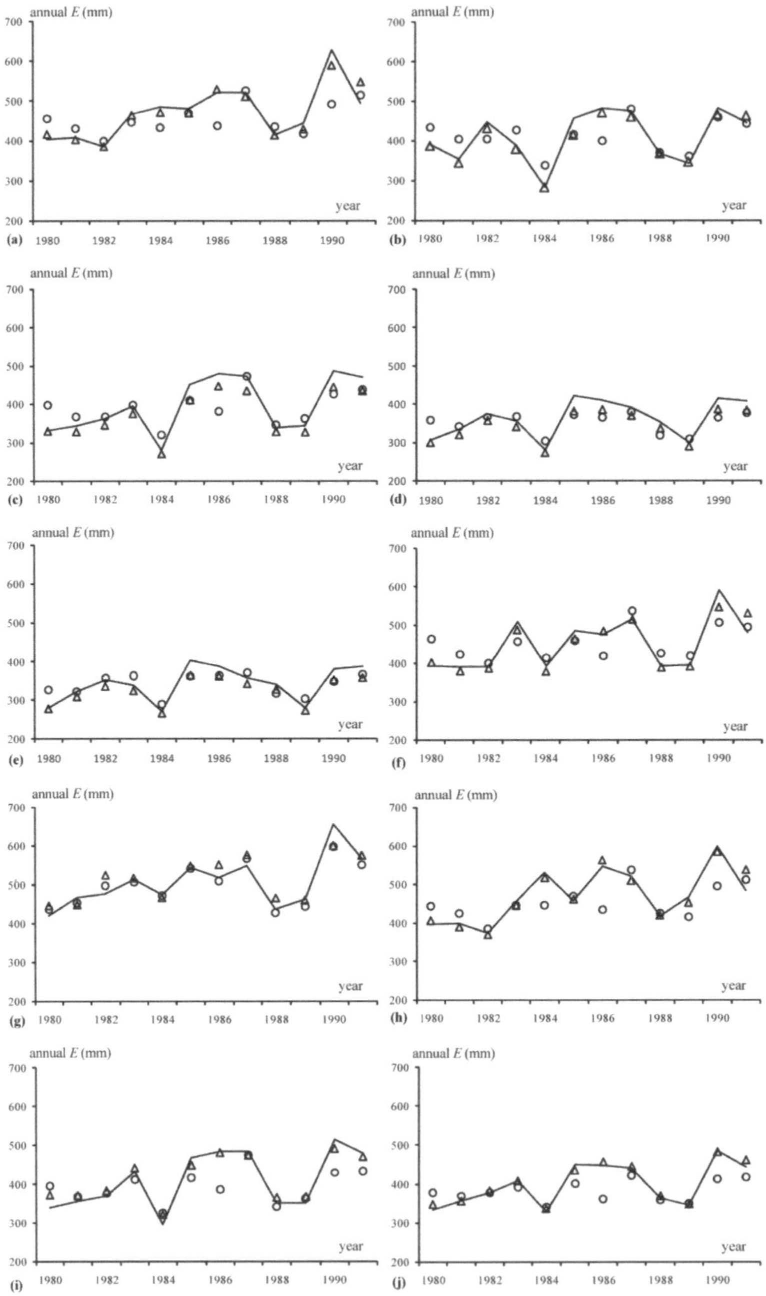

Fig. 7. Comparison of the annual evapotranspiration estimated by the water-energy balance model and GBHM model and the observed one. Figure (a)-(i) represent the 9 sub-catchments respectively. Figure (j) represents the whole watershed. (The real lines represent observed annual evapotranspiration (E), the circular dots represent actual annual evapotranspiration (*E*) predicted by GBHM, and the triangular dots represent actual annual evapotranspiration (*E*) predicted by water-energy balance model)

sensitive to the soil water than to the vegetation coverage. Fig. 7 shows the inter-annual variability of the actual evapotranspiration in each sub-catchment from 1980 to 1991 simulated by the water-energy balance model in Sim-A4 (considering inter-annual variability of both vegetation and soil water) and the GBHM model. It can be seen that both results have a very similar temporal pattern for the 10 sub-catchments. Results show that both the simple lumped water-energy balance model and the physically-based distributed hydrological model can predict catchment annual evapotranspiration and its inter-annual variability.

3. Actual evapotranspiration at monthly time scale

At monthly time scale, similar assumptions and similar four conditions as the above were used in the simulation of monthly evapotranspiration, i.e., (1) using Eq. (4) with a constant parameter n for each sub-catchment (denoted as Sim-M1); (2) using Eq. (4) with the different value of parameter n for every month in each sub-catchment (denoted as Sim-M2); (3) using Eq. (8) with a constant parameter n in each sub-catchment (denoted as Sim-M3); (4) using Eq. (8) with the different value of parameter n for every month in each sub-catchment (denoted as Sim-M4). Monthly evapotranspiration are integrated to the long-term annual evapotranspiration, which is used to compare with the observation. Table 3 shows the results of the relative error for each sub-catchment. It can be seen that the result of Sim-M4 has the least relative error, which implies that it is essential to include variability of both soil water and vegetation coverage for estimating catchment evapotranspiration at monthly time scale. The results also indicate that the soil water content is more important than the parameter n for estimating the monthly evapotranspiration. Taking the whole study area as an example, Fig. 8 shows the comparison of the monthly evapotranspiration estimated by Sim-M4 of the water-energy balance model and by the GBHM model. Both results have a

NO.	Gauges	Relative Error (*RE*, %)			
		Sim-M1	Sim-M2	Sim-M3	Sim-M4
1	Guojiatun	-19.88	-18.11	-2.39	-1.14
2	Boluonuo	-16.32	-13.24	-2.51	-0.26
3	Sandaohezi	-20.07	-17.72	-4.53	-2.47
4	Xiahenan	-16.31	-13.90	-5.03	-2.67
5	Hanjiaying	-17.30	-15.10	-7.71	-5.52
6	Chengde	-13.33	-11.22	-8.29	-5.89
7	Xiabancheng	-15.94	-13.83	-3.07	-2.01
8	Liying	-14.10	-8.22	-4.44	-1.82
9	Kuancheng	-12.61	-10.59	-5.29	-3.29
10	Panjiakou	-15.38	-12.77	-4.86	-2.68

Table 3. The monthly evapotranspiration simulated by the water-energy balance model with different consideration soil water storage and vegetation

Notes: (1) Sim-M1, Sim-M2, Sim-M3, Sim-M4 represent four conditions in simulation the monthly evaportranspiration, i.e., Sim-M1: neglecting the change of the vegetation and soil water; Sim-M2: neglecting the change of soil water; Sim-M3: neglecting the change of the vegetation; Sim-M4: considering the change of the vegetation and soil water. (2) *RE* is the relative errors of the mean values of annual actual evapotranspiration calculated using the water-energy balance model.

similar temporal pattern but the highest and lowest peaks estimated by the water-energy balance model are larger/smaller than the peaks calculated by the GBHM model. In Eq. (8), the soil is considered as a single storage with no interaction with the groundwater, and this simplification may be a reason for the error shown in Fig. 8.

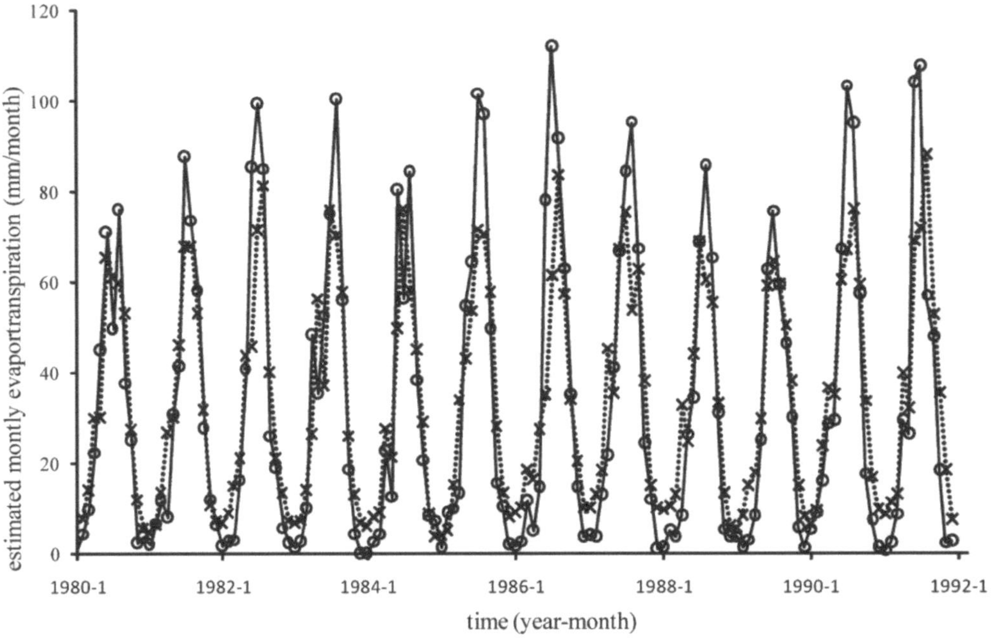

Fig. 8. Comparison of the monthly evapotranspiration estimated by the water-energy balance model and GBHM model for the whole watershed. (The circular real line represents the actual monthly evapotranspiration (*E*) predicted by water-energy balance model, and the cross dashed line represents the actual monthly evapotranspiration (*E*) predicted by GBHM)

4. Actual evapotranspiration at daily time scale

Since the monthly NDVI data was used in the study, the parameter n only can be obtained at monthly time scale and the daily parameter n is considered as the same in each month. At daily time scale, two different conditions are used: (1) neglecting the variability of soil water, which is using Eq. (4) (denoted as Sim-D1) with monthly parameter n; (2) considering the variability of soil water, which is using Eq. (8) (denoted as Sim-D2) with monthly parameter *n*. Daily evapotranspiration is integrated into the long-term mean value of annual evapotranspiration for comparing with the observation. Table 3 shows the relative error for each sub-catchment. It is clear that result of Sim-D2 is much better than the result of Sim-D1, which implies that the soil water content is a very important factor controlling the daily evapotranspiration. Comparing the results simulated at daily time scale by the water-energy balance model and by the GBHM model for the whole study area, it can be seen that the simple water-energy balance model is not possible to simulate the daily variability of evapotranspiration without consideration of the soil moisture storage. However, with consideration of the soil water storage, this simple model is also difficult to capture the daily variability appropriately. As known that soil water movement usually can not penetrate the whole soil layer (e.g. 1-m depth) within a day, therefore, it is important to describe the soil water dynamics at daily or shorter time scale rather than to consider a total storage of the soil water.

5. Conclusions

In this study a bottom-up and top-down analysis based on a distributed physically-based hydrological model (GBHM) and a simple water-energy balance model derived from Budyko hypothesis was carried out for understanding the dominant factors on catchment evapotranspiration at different scales in 9 sub-cachments and the whole basin of the Luan River in Northern China. At annual and long-term time scales, a consistent result has been achieved by the two models, which show the complementary relationship between the actual and potential evapotranspiration at annual time scale for the study areas. However, at a shorter time scale, the difference between the two simulations by GBHM and the water-energy balance model became larger. On the basis of the comprehensive inter-comparison in the framework of the top-down and bottom-up analysis through different temporal scales, it can be concluded as follows:

1. The catchment annual water balance pattern can be predicted successfully by both a simple lumped model and a detailed distributed model in the study areas;
2. The relationships between potential and actual evapotranspiration can be different at different temporal and spatial scales. It shows highly nonlinear relationship at annual and catchment scale, but can be simplified as a linear relationship at hourly and hillslope scale;
3. The soil water storage and the soil water dynamics become more important for simulation of evapotranspiration at a shorter temporal scale;
4. Consideration of vegetation coverage can improve the simulation of inter-annual and seasonal variability of catchment evapotranspiration in the lumped water-energy balance model;
5. Comparison of the top-down and the bottom-up methods can offer a new way for understanding the catchment hydrological behavior at different temporal and spatial scales, and is potentially helpful for developing new generation of hydrological model.

6. Acknowledgement

This paper was supported by the National Natural Science Foundation of China (Grant No. 50939004).

7. References

Abbott M B, Bathurst J C, Cunge J A, O'Connell P E, Rasmussen J(1986), An introduction to the European Hydrological System. SHE, 1: History and philosophy of a physically 2: based distributed modeling system. *J. Hydrol.* 87, 45 - 59.

Allen, R. G., L. S. Pereira, D. Raes, and M. Smith (Eds.) (1998), *Crop Evapotranspiration: Guidelines for Computing Crop Water Requirements.* FAO Irrig. and Drainage Paper 56, Rome.

Arnold, R. Srinivasan, R.S. Muttiah and J.R. Williams (1998), Large area hydrologic modelling and Assessment part I: model development, *Journal of the American Water Resources Association.* 34 (1), 73 - 89.

Atkinson S, Sivapalan M, Woods RA, Viney NR (2002). Dominant physical controls of hourly streamflow predictions and an examination of the role of spatial variability: Mahurangi catchment, New Zealand. *Adv. Water Resour.,* 26(2), 219 - 235.

Atkinson S, Sivapalan M, Viney NR, Woods RA (2003), Physical controls of space-time variability of hourly streamflows: Mahurangi catchment, New Zealand. *Hydrol. Process.,* 17, 2171 - 2193.

Bathurst, J.M. Wicks and P.E. O'Connell (1995), The SHE/SHESED basin scale water flow and sediment transport modeling system. In *Computer Models of Watershed Hydrology,* edited by V.P. Singh pp. 563–594. Water Resources Publications, Littleton, CO.

Beven, K. J., Warren, R., Zaoui, J. (1980), SHE: towards a methodology for physically-based, distributed forecasting in hydrology. *IAHS Publ. No. 129,* Wallingford, UK. 133 - 137.

Beven, K.J. (2001), How far can we go in distributed hydrological modelling? *Hydrology and Earth System Sciences,* 5(1), 1–12.

Beven K. (2006), On undermining the science? *Hydrological Processes,* 20, 3141 - 3146.

Budyko, M. I. (Ed.) (1974), *Climate and Life,* translated from Russian by Miller, D. H., Academic, San Diego, Calif..

Bouchet, R. (1963), Evapotranspiration reelle et potentielle, signification climatique, *Int. Assoc. Sci. Hydro. Pub.,* 62, 134-142.

Choudhury, B. J. (1999), Evaluation of an empirical equation for annual evaporation using field observations and results from a biophysical model, *J. Hydrol., 216,* 99 - 110.

Cong, Z., D. Yang, B. Gao, H. Yang, and H. Hu (2009), Hydrological trend analysis in the Yellow River basin using a distributed hydrological model, Water Resour. Res., 45, doi:10.1029/2008WR006852.

Eder G, Sivapalan M, Nachtnebel H.P. (2003), Modeling of water balances in Alpine catchment through exploitation of emergent properties over changing time scales. *Hydrol. Process.,* 17 (11): 2125-2149.

Farmer, D., M. Sivapalan, and C. Jothityangkoon (2003), Climate, soil, and vegetation controls upon the variability of water balance in temperate and semiarid landscapes: Downward approach to water balance analysis, *Water Resour. Res., 39*(2), 1035, doi:10.1029/2001WR000328.

FAO (2003), *FAO Map of World Soil Resources,* Rome.

Fu, B. P. (1981), On the calculation of the evaporation from land surface, *Scientia Atmospherica Sinica,* 5 (1), 23 - 31 (in Chinese).

Gutman, G., and A. Ignatov (1998). The derivation of the green vegetation fraction from NOAA/AVHRR data for use in numerical weather prediction models, *International Journal of Remote sensing,* 19, 1533 - 1543.

IGBP-DIS (Global Soil Data Task: Global Soil Data Products CD-ROM) (2000), International Geosphere-Biosphere Programme - Data and Information Services. Available online at (http://www.daac.ornl.gov/) from the ORNL Distributed Active Archive Center, Oak Ridge National Laboratory, Oak Ridge, Tennessee, U.S.A..

Jia, G. Ni, Y. Kawahara and T. Suetsugi (2001), Development of WEP model and its application to an urban watershed, *Hydrol. Process.* 15(11), 2175 - 2194.

Jothityangkoon C, Sivapalan M, Farmer D. (2001). Process controls of water balance variability in a large semi-arid catchment: downward approach to hydrological model development. *J. Hydrol.*, 254(1-4): 174 - 198.

Klemes, V. (1983). Conceptualization and scale in hydrology. *J. Hydrol.*, 65: 1-23.

Kite, G. W. (1995), The SLURP model, In *Computer Models of Watershed Hydrology*, edited by V. P. Singh. Water Resources Publications, Highlands, CO.

Liang, X. (1994), A two-layer variable infiltration capacity land surface representation for general circulation models. In: *Water Resour. Ser.* TR140, Univ. Washington, Seattle.

Littlewood, B.F.W. Croke, A.J. Jakeman and M. Sivapalan(2003), The role of 'top-down' modelling for Prediction in Ungauged Basins (PUB), *Hydrol. Process.*, 17, 1673 - 1679.

Milly, P. C. D. (1994), Climate, soil water storage, and the average annual water balance, *Water Resour. Res.*, *30*, 2143 - 2156.

Morris, E.M. (1980), Forecasting flood flows in grassy and forested basins using a deterministic distributed mathematical model. *IAHS Publ. No.129*, 247 - 265.

New, M., M. Hulme and P. Jones (2000), Representing twentieth-century space-time climate variability. Part II: Development of a 1961-96 monthly grids of terrestrial surface climate, *Journal of Climate*, 13, 2217-2238.

Penman, H. L. (1948), Natural evaporation from open water, bare and grass. *Proc. R. Soc. Lond., Ser.* A 193, 120 - 145.

Refsgaard J.C. and Storm B. (1995), MIKE SHE. In *Computer Models of Watershed Hydrology*, edited by V. P. Singh. pp. 809–846. Water Resources Publications, Highlands, CO.

Shuttleworth, W. J., (1993), Evaporation. In *Handbook of Hydrology*, edited by D. R. Maidment, pp.4.1 - 4.53. McGraw-Hill, New York.

Sivapalan M, Blöschl G, Zhang L, Vertessy R. (2003), Downward approach to hydrological prediction. *Hydrol. Process.*, 17, 2101-2111.

Sivapalan M. (2009), The secret to 'doing better hydrological science': change the question! *Hydrological Processes*, 23, 1391 - 1396.

Sivakumar, B. (2008), Undermining the science or undermining Nature? *Hydrologial Processes*, 22, 893 - 897.

Son, K. and Sivapalan, M. (2007), Improving model structure and reducing parameter uncertainty in conceptual water balance models through the use of auxiliary data. *Water Resour. Res.*, 43: W01415, doi: 10.1029/2006WR005032.

Sun, F.B., Yang D.W., Liu Z.Y., Cong Z.T., Lei Z.D. (2007), Validation of Coupled water-energy balance in the Haihe River basin and inland river basins, *Journal of China Hydrology*, 27(2),7-10 (in Chinese).

Wolock, D. M., and G. J. McCabe (1999), Explaining spatial variability in mean annual runoff in the conterminous United States, *Clim. Res.*, *11*, 149 - 159.

Xu, J., D. Yang, J. Ding, Z. Lei (2007), Development of space-nested watershed hydrological model and its application to floods forecasting of Three Gorges reservoir. *Journal of Hydraulic Engineering*, S1, 365 - 371 (in Chinese).

Xu, J., D. Yang*, Z. Lei, J. Chen and W. Yang (2008), Spatial and temporal variation of runoff in the Yangtze River basin during the past 40 years. Quaternary International, 186: 32–42, doi:10.1016/j.quaint. (SCI)

Yang, D., S. Herath, and K. Musiake (1998), Development of a geomorphology-based hydrological model for large catchments, Annual Journal of Hydraulic Engineering, JSCE, 42, 169-174.

Yang D, Oki T, Herath S, Musiake K. (2002), A geomorphology-based hydrological model and its applications. In *Mathematical Models of Small Watershed Hydrology and Applications,* edited by Singh VP, Frevert DK. Water Resources Publications: Littleton, CO; 259-300.

Yang, D., C. Li, H. Hu, Z. Lei, S. Yang, T. Kusuda, T. Koike, and K. Musiake, (2004), Analysis of water resources variability in the Yellow River of China during the last half century using historical data, *Water Resour. Res., 40,* W06502, doi:10.1029/2003WR002763.

Yang, D., F. Sun, Z. Liu, Z. Cong, and Z. Lei (2006), Interpreting the complementary relationship in non-humid environments based on the Budyko and Penman hypotheses. *Geophys. Res. Lett.*, 33, L18402, doi: 10.1029/2006GL027657.

Yang, D., F. Sun, Z. Liu, Z. Cong, G. Ni and Z. Lei (2007), Analyzing spatial and temporal variability of annual water-energy balance in non-humid regions of China using the Budyko hypothesis. *Water Resour. Res.,* 43, W04426, doi: 10.1029/2006WR005224.

Yang, D., W. Shao, P. J.-F. Yeh, H. Yang, S. Kanae, and T. Oki (2009), Impact of Vegetation Coverage on Regional Water Balance in the Non-Humid Regions of China, *Water Resour. Res.*, doi:10.1029/2008WR006948.

Yang, H., D. Yang, Z. Lei, and F. Sun (2008a), New analytical derivation of the mean annual water-energy balance equation, *Water Resour. Res.,* 44, W03410, doi: 10.1029/2007WR006135.

Yang, H., D. Yang, Z. Lei, and H. Lei (2008b), Derivation and verification of the coupled water-energy balance equation at arbitrary time scale, *Journal of Hydraulic Engineering*, 39 (5), 610 - 617 (in Chinese).

Zhang, L., K. Hickel, W. R. Dawes, F. H. S. Chiew, A. W. Western, and P. R. Briggs (2004), A rational function approach for estimating mean annual evapotranspiration, *Water Resour. Res., 40,* W02502, doi: 10.1029/2003WR002710.

Zhang, L., N. Potter, K. Hickel, Y.Q. Zhang, Q.X. Shao (2008), Water balance modeling over variable time scales based on the Budyko framework - Model development and testing, *Journal of Hydrology*, 360(1-4), 117-131.

5

Nonlinear Evapotranspiration Modeling Using MLP-NNM and SVM-NNM Approach

Sungwon Kim and Hung Soo Kim
Dongyang University and Inha University
Republic of Korea

1. Introduction

Evapotranspiration (ET) is the sum of volume of water used by vegetaion, evaporated from the soil, and intercepted precipitation (Singh, 1988). ET plays an important role in our environment at global, regional, and local scales. Water entering the evaporation phase of the hydrological cycle becomes unavailable and cannot be recovered for further use (Brutsaert, 1982). In many areas where water resources are scarce, the calculation of this loss becomes imperative in the planning and management of irrigation practices (Kisi, 2007). Evaporation and transpiration occur simultaneously and there is no easy way of distinguishing between the two processes (Allen et al., 1998). Transpiration consists of vaporization of liquid water contained in plant tissues and the vapor removal to the atmosphere. Evaporation occurs at the topsoil if the water is available. When the crop is small, water is predominantly lost by soil evaporation. Once the crop, however, is well developed and completely covers the soil, transpiration becomes the main process.

ET is one of the hydrologic cycle components and the accurate estimation of ET is very important for the researches such as water balance, irrigation design and management, crop yield modeling, and water resources planning and management (Kumar et al., 2002). ET is observed using a lysimeter directly or can be estimated using the water balance method or the climatic variables indirectly. Because the measurements of ET using a lysimeter directly, however, requires much unnecessary time and needs correct and careful experience, it is not always possible in field measurements. Thus, an empirical approach based on the climatic variables is generally used to estimate the ET (Penman, 1948; Allen et al., 1989). In the early 1970s, the Food and Agricultural Organization of the United Nations (FAO), Rome, developed practical procedures to estimate the crop water requirements (Doorenbos & Pruitt, 1977), which have become the widely accepted standard for irrigation studies. A common practice for estimating the ET from a well-watered agricultural crop is to estimate the reference crop ET such as the grass reference evapotranspiration (ETo) or the alfalfa reference evapotranspiration (ETr) from a standard surface and to apply an appropriate empirical crop coefficient, which accounts for the difference between the standard surface and the crop ET.

The emergence of neural networks model has provided many promising results in the field of hydrology and water resources modeling. Due to the ease of application and simple architecture, the neural networks model has become a promising research field with surprising potential. A comprehensive review of the application of neural networks model to hydrology can be found in ASCE (2000). The success using the neural networks model in many fields of science and engineering suggests that the neural networks model may prove be an effective and efficient way for the modeling of ET process. Recently, the outstanding results using the neural networks model in the fields of ET modeling have been obtained (Sudheer et al., 2003; Trajkovic et al., 2003; Trajkovic, 2005; Kisi, 2006; Kisi, 2007; Jain et al., 2008; Kim & Kim, 2008; Kumar et al., 2008; Landeres et al., 2008; Zanetti et al., 2008; Kumar et al., 2009). Kumar et al. (2002) developed the neural networks models to estimate the daily ETo. They used proper combinations of the observed climatic variables such as solar radiation, temperature, relative humidity, and wind speed for the neural networks models. Kisi & Ozturk (2007) used the neuro-fuzzy models to estimate the FAO-56 PM ETo using the observed climatic variables. They used proper combinations of the observed climatic variables such as air temperature, solar radiation, wind speed, and relative humidity for the neuro-fuzzy models. Kisi (2008) investigated the potential of different neural networks models in the ET modeling. He used proper combinations of the observed climatic variables such as solar radiation, mean temperature, mean relative humidity, and wind speed for the neural networks models. Traore et al. (2010) developed the neural network models to calculate the reference ET complex process in Sudano-Sahelian zone. They proper combinations of the observed climatic variables such as minimum temperature, maximum temperature, extraterrestrial radiation, relative humidity, and wind speed for the neural networks models.

This paper investigates the modeling of FAO-56 PM ETo using the neural networks models. The major objective of the study is to evaluate the potential of neural networks models for estimating the FAO-56 PM ETo using climatic data avaliable. A comparative evaluation of multiple linear regression model (MLRM) and neural networks models including multilayer perceptron neural netwprks model (MLP-NNM) and support vector machine neural networks model (SVM-NNM) are carried out. From this study, we evaluate the impact of MLP-NNM and SVM-NNM performances for the modeling of FAO-56 PM ETo. The optimal MLP-NNM and SVM-NNM can estimate the FAO-56 PM ETo with the least cost and endeavor. Finally, the FAO-56 PM ETo data can be constructed to provide the fundamental data for the drought analysis and irrigation networks systems, Republic of Korea.

2. Grass reference evapotranspiration model : FAO-56 PM ETo equation

Penman (1948) combination method links evaporation dynamics with the flux of net radiation and aerodynamic transport characteristics of the natural surface. Based on the observations that latent heat transfer in plant stem is influenced not only by these abiotic factors, Monteith (1965) introduced a surface conductance term that accounted for the response of leaf stomata to its hydrologic environment. This modified form of the Penman-Monteith (PM) ET model. Jensen et al. (1990) measured the ET using the lysimeters at 11 stations located in the different climatic zones of various regions around the world. They compared the results of the lysimeters with those of 20 different empirical equations and methodologies for the ET measurements. It was found that PM ET model showed the optimal results over all the climatic zones. If the observed/measured data for the ET does not exist, therefore, PM ET model can be considered as a standard methodology to estimate the ET. In 8 meteorological stations which were selected for this study, there are no observed data for the grass reference ET (ETo). The data calculated using PM ETo model can be assumed as the observed ETo, whose reliability was verified by many previous studies.

All calculation precedures as used in PM ETo model are based on the FAO guidelines as laid down in the publication No. 56 of the Irrigation and Drainage Series of FAO "Crop Evapotranspiration - Guidelines for Computing Crop Water Requirements" (1998). Therefore, FAO-56 PM ETo equation means the PM ETo equation suggested by the Irrigation and Drainage Paper No.56, FAO. FAO-56 PM ETo equation is given by Allen et al. (1998) and can be shown as the following equation (1).

$$\text{FAO-56 PM ETo} = \frac{0.408\Delta(R_n - G) + \gamma(900/(T+273))u_2(e_s - e_a)}{\Delta + \gamma(1+0.34u_2)} \quad (1)$$

where FAO-56 PM ETo = the grass reference evapotranspiration (mm/day); R_n = the net radiation at the crop surface (MJ/m^2·day); G = the soil heat flux density (MJ/m^2·day); T = the mean daily air temperature at 2m height (˚C); u_2 = the wind speed at 2m height (m/sec); e_s = the saturation vapor pressure (kPa); e_a = the actual vapor pressure (kPa); e_s - e_a = the saturation vapor pressure deficit (kPa); Δ = the slope vapor pressure curve (kPa/◦C); and γ = the psychometric constant (kPa/◦C). FAO CROPWAT 8.0 computer program has been used to calculate FAO-56 PM ETo and extraterrestrial radiation (R_a). FAO CROPWAT 8.0 computer program allows the user to enter the climatic data available including maximum temperature (T_{max}), minimum temperature (T_{min}), mean relative hymidity (RH_{mean}), mean wind speed (U_{mean}), and sunshine duration (SD) for calculating FAO-56 PM ETo. On the base of climatic data available, FAO CROPWAT 8.0 computer program estimates the solar radiation reaching soil surface. Fig. 1(a)-(b) show the calculation of FAO-56 PM ETo using FAO CROPWAT 8.0 computer program in Gunsan and Haenam stations, respectively.

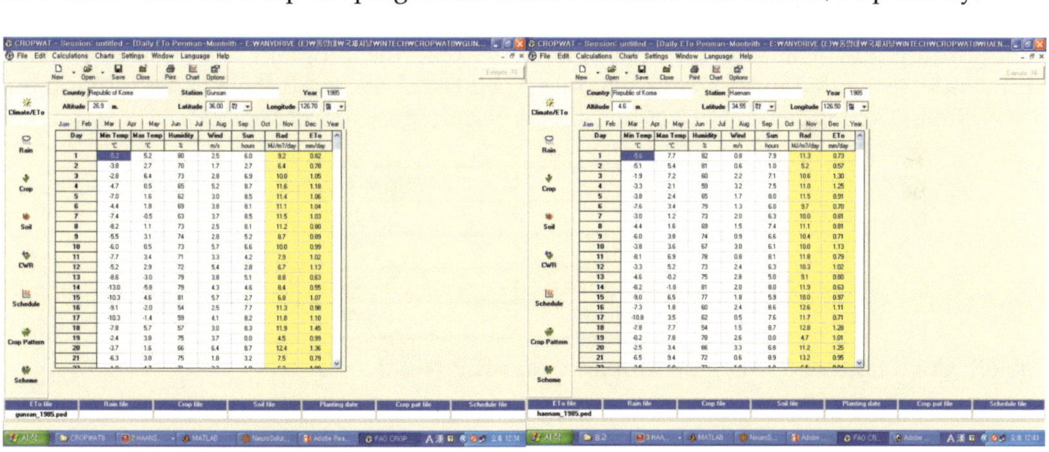

(a) Gunsan station (b) Haenam station

Fig. 1. Calculation of FAO-56 PM ETo using FAO CROPWAT 8.0

3. Neural networks models

3.1 MultiLayer perceptron neural networks model (MLP-NNM)

MLP-NNM has an input layer, an output layer, and one or more hidden layers between the input and output layers. Each of the nodes in a layer is connected to all the nodes of the next layer, and the nodes in one layer are connected only to the nodes of the immediate next layer. The strength of signal passing from one node to the other depends on the connection weights of the interconnections. The hidden layers enhance the network's ability to model complex functions. MLP-NNM is trained using the many kinds of backpropagation algorithms. Training performance is a process of adjusting the connection weights and biases so that its output can match the desired output best. Specifically, at each setting of the connection weights, it is possible to calculate the error committed by the networks simply by taking the difference between the desired and actual responses (Simpson, 1990; Specht, 1991; Gallant, 1993; Wasserman, 1933; Bishop, 1995; Tsoukalas & Uhrig, 1997; Haykin, 2009). In this study, MLP-NNM is trained with the Quickprop backpropagation algorithm (BPA). The QuickProp BPA is a training method that operates much faster in the batch mode than the conventional BPA. It has the additional advantage that it is not sensitive to the learning rate and the momentum. In MLP-NNM with five input nodes, the results of the output layer can be written as equation (2).

$$\text{FAO-56 PM ETo} = \Phi_2\left(\sum_{k=1}^{1} W_{kj} \cdot \Phi_1\left(\sum_{j=1}^{5} W_{ji} \cdot X(t) + B_1\right) + B_2\right) \qquad (2)$$

where i, j, k = the input layer, the hidden layer, and the output layer, respectively; FAO-56 PM ETo = the grass reference evapotranspiration (mm/day); $\Phi_1(\cdot)$ = the linear sigmoid transfer function of the hidden layer; $\Phi_2(\cdot)$ = the linear sigmoid transfer function of the output layer; W_{kj} = the connection weights between hidden and output layers; W_{ji} = the connection weights between input and hidden layers; X(t) = the time series data of input nodes including mean wind speed (U_{mean}), mean temperature (T_{mean}), sunshine duration (SD), mean relative humidity (RH_{mean}), and max temperature (T_{max}); B_1 = the bias in the hidden layer; and B_2 = the bias in the output layer. A number of MLP-NNM computer programs are now available. NeuroSolutions 5.0 computer program was used to develop MLP-NNM structure. Fig. 2 shows the developed structure of MLP-NNM with five input nodes. Table 1 shows the conditions of training performance for MLP-NNM.

Index	Assigned Value
Stepsize	1.0
Momentum	0.5
Maximum Iterations	50000
Training Threshold	0.001

Table 1. Conditions of training performance for MLP-NNM

3.2 Support vector machine neural networks model (SVM-NNM)

SVM-NNM has found wide application in several areas including pattern recognition, regression, multimedia, bio-informatics and artificial intelligence. Very recently, SVM-NNM is gaining recognition in hydrology (Dibike et al., 2001; Khadam & Kaluarachchi, 2004). SVM-NNM implements the structural risk minimization principle which attempts to minimize an upper bound on the generalization error by striking a right balance between the training performance error and the capacity of machine. The solution of traditional neural networks models including MLP-NNM may tend to fall into a local optimal solution, whereas global optimum solution is guaranteed for SVM-NNM (Haykin, 2009). SVM-NNM is a new kind of classifier that is motivated by two concepts. First, transforming data into a high-dimensional space can transform complex problems into simpler problems that can use

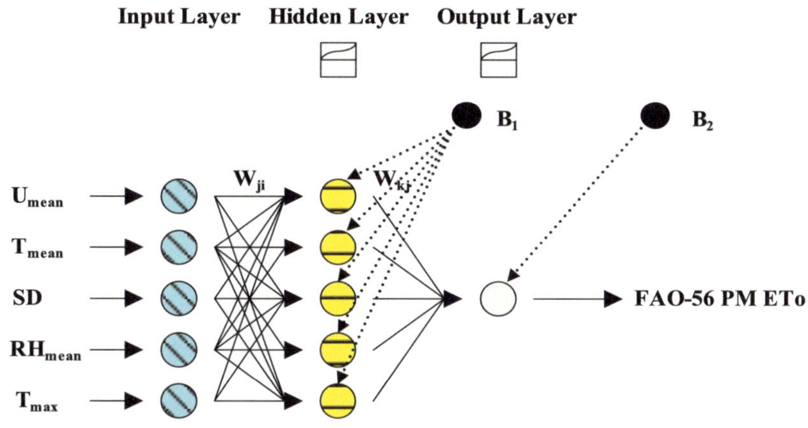

Fig. 2. The developed structure of MLP-NNM with five input nodes

linear discriminant functions. Second, SVM-NNM is motivated by the concept of training and using only those inputs that are near the decision surface since they provide the most information about the classification. The first step in SVM-NNM is transforming the data into a high-dimensional space. This is done using radial basis function (RBF) that places a Gaussian at each sample data. Thus, the feature space becomes as large as the number of sample data. RBF uses backpropagation to train a linear combination of the gaussians to produce the final result. SVM-NNM, however, uses the idea of large margin classifiers for training performance. This decouples the capacity of the classifier from the input space and at the same time provides good generalization. This is an ideal combination for classification (Vapnik, 1992, 2000; Principe et al., 2000; Tripathi et al., 2006).

In this sudy, ε SVM-NNM regression is used. The basic ideas of ε SVM-NNM regression are reviewed. Consider the finite training sample pattern (x_i, y_i), where $x_i \in \Re^n$ is a sample value of the input vector x considering of N training patterns and $y_i \in \Re^n$ is the corresponding value of the desired model output. A nonlinear transformation function $\varphi(\cdot)$ is defined to map the input space to a higher dimension feature space, $\Re^{n_h}$. According to Cover's theorem (Cover, 1965), a linear function, $f(\cdot)$, could be formulated in the high dimensional feature space to look for a nonlinear relationship between inputs and outputs in the original input space. It can be written as equation (3).

$$\bar{y} = f(x) = w^T \varphi(x) + b \tag{3}$$

where $\bar{y}$ = the actual model output. The coefficient w and b are adjustable model parameters. In the ε SVM-NNM regression, we aim at minimizing the emprical risk. It can be written as equation (4).

$$R_{emp} = \frac{1}{N}\sum_{i=1}^{N}\left|y_i - \bar{y}_i\right|_{\varepsilon} \tag{4}$$

where R_{emp} = the empirical risk; and $\left|y_i - \bar{y}_i\right|_{\varepsilon}$ = the Vapnik's ε-insensitive loss function. Following regularization theory (Haykin, 2009), the parameters w and b are estimated by minimizing the cost function. It can be written as equation (5).

$$\psi_{\varepsilon}(w,\xi,\xi^*) = \frac{1}{2}w^T w + C\sum_{i=1}^{N}(\xi_i + \xi_i^*) \tag{5}$$

subject to the constraints ① $y_i - \bar{y}_i \leq \varepsilon + \xi_i$ i = 1, 2, ... , N, ② $-y_i + \bar{y}_i \leq \varepsilon + \xi_i^*$ i = 1, 2, ... , N, and ③ $\xi_i, \xi_i^* \geq 0$ i = 1, 2, ... , N. where $\psi_{\varepsilon}(w,\xi,\xi^*)$ = the cost function; ξ_i, ξ_i^* = positive slack variables; and C = the cost constant. The first term of the cost function, which represents weight decay, is used to regularize weight sizes and to penalize large weights. This helps in improving generalization performance (Hush and Horne, 1993). The second term of the cost function, which represents penalty function, penalizes deviations of $\bar{y}$ from y larger than $\pm\varepsilon$ using Vapnik's ε-insensitive loss function. The cost constant C determines the amount up to which deviations from ε are tolerated. Deviations above ε are denoted by ξ_i, whereas deviations below - ε are denoted by ξ_i^*. The constrained quardratic optimization problem can be solved using the method of Lagrangian multipliers (Haykin, 2009). From this solution, the coefficient w can be written as equation (6).

$$w = \sum_{i=1}^{N}(\alpha_i - \alpha_i^*)\varphi(x_i) \tag{6}$$

where α_i, α_i^* = the Lagrange multipliers, which are positive real constants. The data points corresponding to non-zero values for $(\alpha_i - \alpha_i^*)$ are called support vectors.
In ε SVM-NNM regression to calculate FAO-56 PM ETo, there are several possibilities for the choice of kernel function, including linear, polynomial, sigmoid, splines and RBF. In this study, RBF is used to map the input data into higher dimensional feature space. RBF can be written as the equation (7).

$$k(x,x_j) = \Phi_1 = \exp(-B_1 R_j^2) = \exp\left(-\frac{\|x_i - x_j\|^2}{2\sigma^2}\right) \tag{7}$$

where i, j = the input layer and the hidden layer; $K(x,x_j) = \Phi_1$ = the inner product kernel function; $B_1 = \frac{1}{2\sigma^2}$, and has a constant value; and σ = the width/spread of RBF, which can be adjusted to control the expressivity of RBF. The function for the single node of the output layer which receives the calculated results of RBF can be written as the equation (8).

$$G_k = [\sum_{j=1}^{N}(\alpha_j - \alpha_j^*) \cdot K(x,x_j)] + B \tag{8}$$

where k = the output layer; G_k = the calculated value of the single output node; and B = the bias in the output layer. Finally, equation (8) takes the form of equation (9), which represents ε SVM-NNM regression for modeling of FAO-56 PM ETo.

$$\text{FAO-56 PM ETo} = \Phi_2(G_k) = \Phi_2 \cdot [[\sum_{j=1}^{N}(\alpha_j - \alpha_j^*) \cdot K(x,x_j)] + B] \tag{9}$$

where Φ_2 = the linear sigmoid transfer function. A number of SVM-NNM computer programs are now available. DTREG computer program was used to develop SVM-NNM structure. SVM-NNM in the DTREG computer program has been developed and modified using LIBSVM algorithm, a freeware program, developed by Chih-Chung Chang and Chih-Jen Lin (Chang & Lin, 2001). The basic algorithm of LIBSVM is a simplication of both SMO by Platt and SVM Light by Joachims. LIBSVM is capable of C SVM-NNM classification, one-class classification, ν SVM-NNM classification, ν SVM-NNM regression, and ε SVM-NNM regression, respectively. The accuracy of ε SVM-NNM regression is largely dependent on the selection of model parameters such as C, Gamma(γ), and P. DTREG computer program provides two methods, a grid search and a pattern search, for finding optimal parameters values. A grid search tries values of each parameter across the specified search range using geometric steps. A pattern search starts at the center of the search range and makes trial steps in each direction for each parameter. If the fit of model improves, the search center moves to the new point and the process is repeated. If no improvement is found, the step size is reduced and the search is tried again. The pattern search stops when the search step size is reduced to a specified tolerance. Fig. 3 shows the developed structure of SVM-NNM with five input nodes. Table 2 shows the conditions of training performance for SVM-NNM.

Index	Assigned Value
Type of SVM-NNM	ε-SVM regression
Kernel Function	RBF
Parameter Optimization	Grid search & Pattern search
Model Parameters	C, γ, P
Training Threshold	0.001

Table 2. Conditions of training performance for SVM-NNM

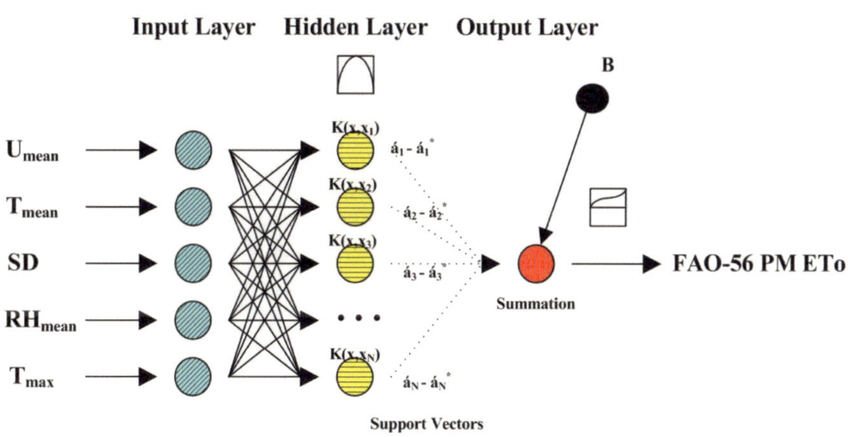

Fig. 3. The developed structure of SVM-NNM with five input nodes

4. Study scope and data

The meteorological stations were selected that could represent the entire lands of the Republic of Korea. They were selected from among the 71 meteorological stations including

Jeju-do under the control of the Korea meteorological administration (KMA). The selected meteorological stations should be distributed over the country, and represent each region/county. They should possess long-term climatic data dating back for at least 30 years. Thus, the meteorological stations, which are appropriate for these conditions, include a total of 8 meteorological stations. They are located in Gunsan, Daegu, Seoul, Seongsanpo, Ulsan, Jeonju, Tongyoung, and Haenam. Fig. 4 shows the locations of 8 meteorological stations. The climatic data, which was necessary for MLP-NNM and SVM-NNM application, were collected from the Internet homepage of water management information system (www.wamis.go.kr) and the Korea meteorological administration (www.kma.go.kr). The climatic data available including mean wind speed (U_{mean}), mean temperature (T_{mean}), sunshine duration (SD), mean relative humidity (RH_{mean}), and max temperature (T_{max}) were sufficient for MLP-NNM and SVM-NNM application. Furthermore, The climatic data available including maximum temperature (T_{max}), minimum temperature (T_{min}), mean relative hymidity (RH_{mean}), mean wind speed (U_{mean}), and sunshine duration (SD) were sufficient for estimating FAO-56 PM ETo using FAO CROPWAT 8.0 computer program. Therefore, the training and testing data were composed using the climatic data in daily units from 01/01/1985 to 12/31/1992.

Fig. 4. The locations of 8 meteorological stations, Republic of Korea

5. Application of MLP-NNM and SVM-NNM

5.1 Performance statistics

The performance of MLP-NNM and SVM-NNM to account for calculating the daily FAO-56 PM ETo was evaluated using a wide variety of standard statistics index. A total of three different standard statistics index were employed; the coefficient of correlation (CC), root mean square error (RMSE), and Nash-Sutcliffe coefficient (R^2) (Nash & Sutcliffe, 1970; ASCE, 1993). Table 3 shows summary of the statistics index in this study. where $\overline{y}_i(x)$ = the calculated FAO-56 PM ETo (mm/day); $y_i(x)$ = the observed FAO-56 PM ETo (mm/day); $\overline{u}_i$ = mean of the calculated FAO-56 PM ETo (mm/day); u_y = mean of the observed FAO-56

PM ETo (mm/day); and n = total number of the daily FAO-56 PM ETo considered. A model which is effective in the modeling of FAO-56 PM ETo accurately, and efficient in capturing the complex relationship among the various inputs and output variables involved in a particular problem, is considered the best. CC, RMSE, and R^2 statistics quantify the efficiency of MLP-NNM and SVM-NNM in capturing the extremely complex, dynamic, nonlinear, and fragmented rainfall-runoff relationships (Kim et al., 2009).

5.2 Input nodes determination

At the beggining of this study, the input nodes of MLP-NNM and SVM-NNM had to be determined. From the previous literatures on the ET modeling using the neural networks models, five kinds of climatic data available, which were used to cite frequently were determined. The climatic data available were mean wind speed (U_{mean}), mean temperature (T_{mean}), sunshine duration (SD), mean relative humidity (RH_{mean}), and max temperature (T_{max}). Therefore, MLP-NNM and SVM-NNM was prior fed with the mean wind speed (U_{mean}), which was the most frequently cited by the previous researchers. It was adopted as the minimum input combinations represented by MLP 1 and SVM 1 for determining the optimal input combinations. Then, the best network configuration determined was used to train and test the several other input combinations. MLP 2 and SVM 2 have two input nodes; mean wind speed (U_{mean}) and mean temperature (T_{mean}). MLP 3 and SVM 3 have three input nodes; mean wind speed (U_{mean}), mean temperature (T_{mean}), and sunshine duration (SD). MLP 4 and SVM 4 have four input nodes; mean wind speed (U_{mean}), mean temperature (T_{mean}), sunshine duration (SD), and mean relative humidity (RH_{mean}). Finally, MLP 5 and SVM 5 have five input nodes; mean wind speed (U_{mean}), mean temperature (T_{mean}), sunshine duration (SD), mean relative humidity (RH_{mean}), and max temperature (T_{max}). Table 4 shows the input combinations of MLP-NNM and SVM-NNM.

Statistics Index	Equation
CC	$\dfrac{\frac{1}{n}\sum_{i=1}^{n}[y_i(x)-u_y][\bar{y}_i(x)-\bar{u}_y]}{\sqrt{\frac{1}{n}\sum_{i=1}^{n}[y_i(x)-u_y]^2}\sqrt{\frac{1}{n}\sum_{i=1}^{n}[\bar{y}_i(x)-\bar{u}_y]^2}}$
RMSE	$\sqrt{\frac{1}{n}\sum_{i=1}^{n}[\bar{y}_i(x)-y_i(x)]^2}$
R^2	$1-\dfrac{\sum_{i=1}^{n}[y_i(x)-\bar{y}_i(x)]^2}{\sum_{i=1}^{n}[y_i(x)-u_y]^2}$

Table 3. Summary of statistics index

Neural Networks Model		Input Combinations
MLP-NNM	SVM-NNM	
MLP 1	SVM 1	U_{mean}
MLP 2	SVM 2	U_{mean}, T_{mean}
MLP 3	SVM 3	U_{mean}, T_{mean}, SD,
MLP 4	SVM 4	U_{mean}, T_{mean}, SD, RH_{mean}
MLP 5	SVM 5	U_{mean}, T_{mean}, SD, RH_{mean} , T_{max}

Table 4. Input combinations of MLP-NNM and SVM-NNM

5.3 Data normalization

The data used in this study including mean wind speed (U_{mean}), mean temperature (T_{mean}), sunshine duration (SD), mean relative humidity (RH_{mean}), and max temperature (T_{max}) were nomalized for preventing and overcomining problem associated with the extreme values. An important reason for the normalization of input nodes is that each of input nodes represents an observed value in a different unit. Such input nodes are normalized, and the input nodes in non-dimension unit are relocated. The similarity effect of input nodes is thus eliminated (Kim et al., 2009). According to Zanetti et al. (2007), by grouping the daily values into averages, ETo may be estimated due to their highest stabilization. For data normalization, the data of input and output nodes were scaled in the range of [0 1] using the equation (10).

$$Y_{norm} = \frac{Y_i - Y_{min}}{Y_{max} - Y_{min}} \tag{10}$$

where Y_{norm} = the normalized dimensionless data of the specific input node; Y_i = the observed data of the specific input node; Y_{min} = the minimum data of the specific input node; and Y_{max} = the maximum data of the specific input node.

5.4 Training performance

The method for estimating parameters is generally called the training performance in the neural networks model category. The training performance of neural networks model is iterated until the training error is reached to the training tolerance. Iteration means one completely pass through a set of inputs and target patterns or data. In general, it is assumed that the neural networks model does not have any prior knowledge about the example problem before it is trained. A difficult task with the neural networks model is to choose the number of hidden nodes. The network geometry is problem dependent. This study adopted one hidden layer for the construction of MLP-NNM and SVM-NNM since it is well known that one hidden layer is enough to represent the ETo nonlinear complex relationship (Kumar et al., 2002; Zanetti et al., 2007). The number of hidden nodes was determined as five for MLP-NNM with the various input combinations (MLP 1, MLP 2, MLP 3, MLP 4, and MLP 5). In SVM-NNM, however, the number of hidden nodes was determined by the training performance of SVM-NNM with the various input combinations (SVM 1, SVM 2, SVM 3, SVM 4, and SVM 5). Kisi (2007) varied the hidden nodes between 2 and 6 after trial and error method for Claremont, Ponoma, and Santa Monica stations, respectively. For three stations, the optimal number of hidden nodes was found at six based on minimum mean square error (MSE), minimum mean absolute error (MAE), and maximum determination coefficient (R^2). Khoob (2008a, b) trained the neural networks models with up to thirty hidden nodes using similar inputs set and found optimal results at six and nine hidden nodes in Safiabad and Khuzestan plain, respectively. For the training data of MLP-NNM and SVM-NNM, the six-year data from 01/01/1985 to 12/31/1990 in 8 meteorological stations were used. The total amount of data used for the training performance was composed of 2191 data for daily time series.

5.4.1 Results of MLP-NNM training performance

For the training performance of MLP-NNM, NeuroSolutions 5.0 computer program was used to carry out the training performance. Fig. 5 shows the building processes of MLP-NNM training performance using NueroSolution 5.0 computer program. Table 5 shows the summary of optimal MLP-NNM statistics results during the training performance for 8 meteorological stations. For 8 meteorological stations, the best statistics results were found at MLP 4 and MLP 5 on average. In Gunsan station, the performance statistics results of MLP 5 were 0.968, 0.365 (mm/day), and 0.936 for CC, RMSE, and R^2, respectively. In Daegu station, the performance statistics results of MLP 5 were 0.975, 0.364 (mm/day), and 0.950 for CC, RMSE, and R^2, respectively. In Seoul station, the performance statistics results of MLP 4 were 0.963, 0.413 (mm/day), and 0.927 for CC, RMSE, and R^2, respectively. In Seonsganpo station, the performance statistics results of MLP 4 were 0.842, 0.676 (mm/day), and 0.710 for CC, RMSE, and R^2, respectively. In Ulsan station, the performance statistics results of MLP 4 were 0.956, 0.412 (mm/day), and 0.914 for CC, RMSE, and R^2, respectively. In Jeonju station, the performance statistics results of MLP 5 were 0.966, 0.383 (mm/day), and 0.932 for CC, RMSE, and R^2, respectively. In Tongyoung station, the performance statistics results of MLP 5 were 0.945, 0.438 (mm/day), and 0.893 for CC, RMSE, and R^2, respectively. In Haenam station, the performance statistics results of MLP 4 were 0.959, 0.396 (mm/day), and 0.919 for CC, RMSE, and R^2, respectively. From the evaluation of MLP-NNM training performance, MLP 4 and MLP5 was found to show the better statistics results compared with MLP 1, MLP 2, and MLP 3.

Station	Model	CC	RMSE (mm/day)	R^2
Gunsan	MLP 5	0.968	0.365	0.936
Daegu	MLP 5	0.975	0.364	0.950
Seoul	MLP 4	0.963	0.413	0.927
Seongsanpo	MLP 4	0.842	0.676	0.710
Ulsan	MLP 4	0.956	0.412	0.914
Jeonju	MLP 5	0.966	0.383	0.932
Tongyoung	MLP 5	0.945	0.438	0.893
Haenam	MLP 4	0.959	0.396	0.919

Table 5. Summary of optimal MLP-NNM statistics results during the training performance

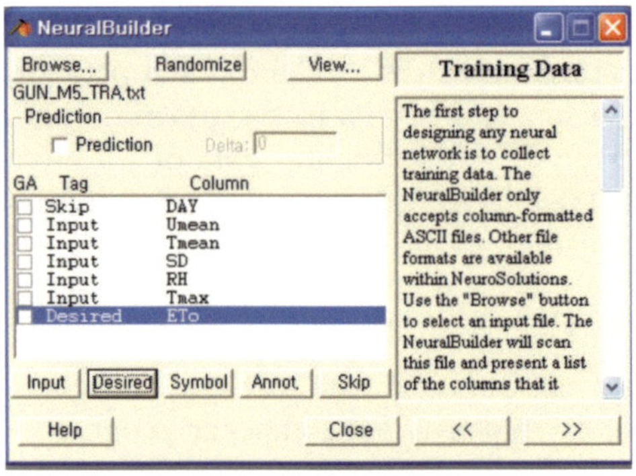

(a) Training Data

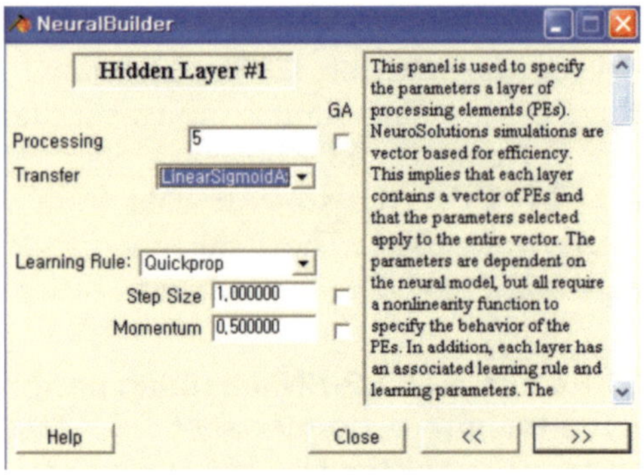

(b) Hidden Layer

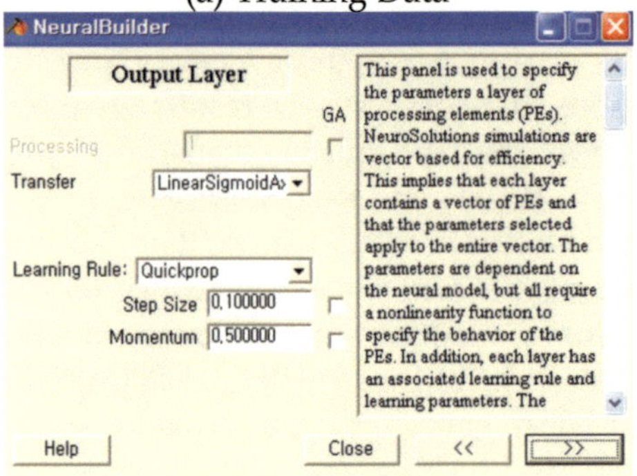

(c) Output Layer

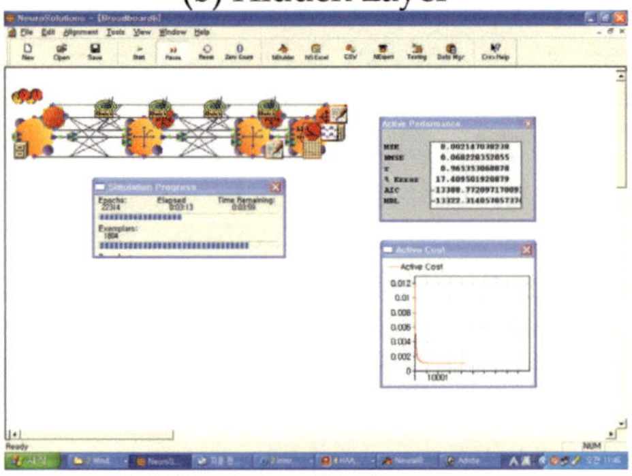

(d) Training Performance

Fig. 5. The building precesses of MLP-NNM training performance using NueroSolution 5.0

5.4.2 Results of SVM-NNM training performance

For the training performance of SVM-NNM, DTREG computer program was used to carry out the training performance. Fig. 6 shows the building processes of SVM-NNM training performance using DTREG computer program. Table 6 shows the summary of optimal SVM-NNM statistics results during the training performance for 8 meteorological stations.

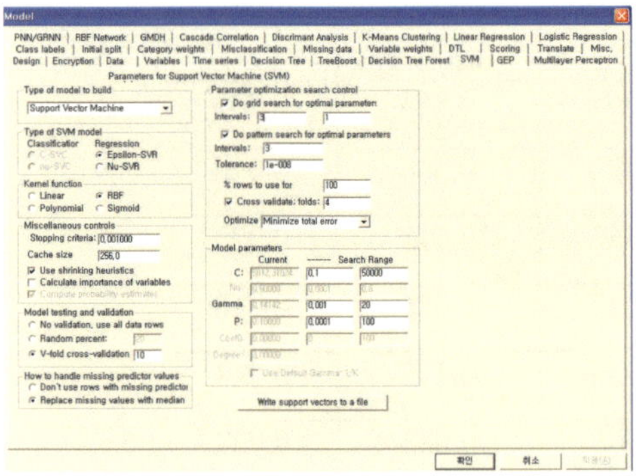

(a) Training Category

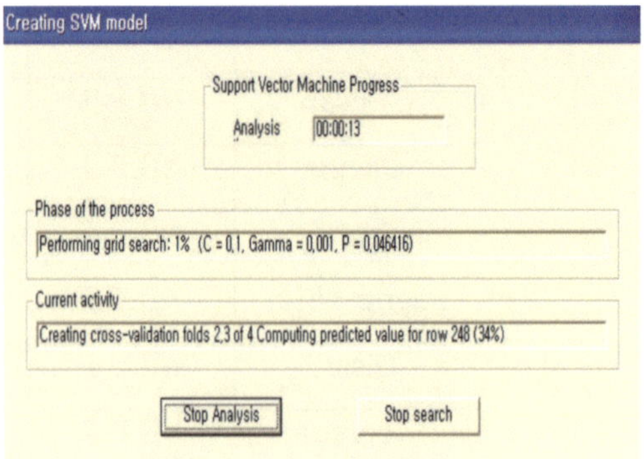

(b) Training Performance

Fig. 6. The building precesses of SVM-NNM training performance using DTREG

For 8 meteorological stations, the best statistics results were found at SVM 5. In Gunsan station, the performance statistics results of SVM 5 were 0.982, 0.274 (mm/day), and 0.964

for CC, RMSE, and R^2, respectively. In Daegu station, the performance statistics results of SVM 5 were 0.985, 0.278 (mm/day), and 0.971 for CC, RMSE, and R^2, respectively. In Seoul station, the performance statistics results of SVM 5 were 0.979, 0.315 (mm/day), and 0.957 for CC, RMSE, and R^2, respectively. In Seonsganpo station, the performance statistics results of SVM 5 were 0.857, 0.670 (mm/day), and 0.715 for CC, RMSE, and R^2, respectively. In Ulsan station, the performance statistics results of SVM 5 were 0.970, 0.336 (mm/day), and 0.940 for CC, RMSE, and R^2, respectively. In Jeonju station, the performance statistics results of SVM 5 were 0.979, 0.304 (mm/day), and 0.957 for CC, RMSE, and R^2, respectively. In Tongyoung station, the performance statistics results of SVM 5 were 0.963, 0.362 (mm/day), and 0.927 for CC, RMSE, and R^2, respectively. In Haenam station, the performance statistics results of SVM 5 were 0.971, 0.334 (mm/day), and 0.943 for CC, RMSE, and R^2, respectively. From the evaluation of SVM-NNM training performance, SVM 5 was found to show the better statistics results compared with SVM 1, SVM 2, SVM 3 and SVM 4. Furthermore, from the statistics results of training peformance for MLP-NNM and SVM-NNM, we could conclude that the statistics results of SVM-NNM were better than those of MLP-NNM.

Station	Model	CC	RMSE (mm/day)	R^2
Gunsan	SVM 5	0.982	0.274	0.964
Daegu	SVM 5	0.985	0.278	0.971
Seoul	SVM 5	0.979	0.315	0.957
Seongsanpo	SVM 5	0.857	0.670	0.715
Ulsan	SVM 5	0.970	0.336	0.940
Jeonju	SVM 5	0.979	0.304	0.957
Tongyoung	SVM 5	0.963	0.362	0.927
Haenam	SVM 5	0.971	0.334	0.943

Table 6. Summary of optimal SVM-NNM statistics results during training performance

5.5 Testing performance

Neural networks model is tested by determining whether the model meets the objectives of modeling within some preestablished criteria or not. Of course, the optimal parameters, which are determined during the training performance, are applied in the testing performance of neural networks model (Kim, 2004). For the testing data of MLP-NNM and SVM-NNM, the two-year data from 01/01/1991 to 12/31/1992 in 8 meteorological stations were used. The total amount of data used for the testing performance was composed of 731 data for daily time series. Generally, a maximum of 40% of the total training data are used as the testing data. The testing performance applied the cross-validation method in order to overcome the over-fitting problem of MLP-NNM and SVM-NNM. The cross-validation method is not to train all the training data until MLP-NNM and SVM-NNM reaches the minimum RMSE, but is to cross-validate with the testing data at the end of each training performance. If the over-fitting problem occurs, the convergence process over the mean square error of the testing data will not decrease but will increase as the training data are still trained (Bishop, 1994; Haykin, 2009). Furthermore, the statistics results of testing performance for MLP-NNM and SVM-NNM were compared with those of multiple linear regression model (MLRM).

5.5.1 Results of MLP-NNM testing performance

For the testing performance of MLP-NNM, NeuroSolutions 5.0 computer program was used to carry out the testing performance based on the statistics results of training performance. For 8 meteorological stations, the best statistics results were found at MLP 4 and MLP 5 on average. In Gunsan station, the performance statistics results of MLP 5 were 0.964, 0.369 (mm/day), and 0.928 for CC, RMSE, and R^2, respectively. In Daegu station, the performance statistics results of MLP 5 were 0.980, 0.349 (mm/day), and 0.961 for CC, RMSE, and R^2, respectively. In Seoul station, the performance statistics results of MLP 4 were 0.966, 0.402 (mm/day), and 0.933 for CC, RMSE, and R^2, respectively. In Seonsganpo station, the performance statistics results of MLP 4 were 0.955, 0.481 (mm/day), and 0.867 for CC, RMSE, and R^2, respectively. In Ulsan station, the performance statistics results of MLP 4 were 0.958, 0.398 (mm/day), and 0.918 for CC, RMSE, and R^2, respectively. In Jeonju station, the performance statistics results of MLP 5 were 0.959, 0.411 (mm/day), and 0.918 for CC, RMSE, and R^2, respectively. In Tongyoung station, the performance statistics results of MLP 5 were 0.949, 0.435 (mm/day), and 0.896 for CC, RMSE, and R^2, respectively. In Haenam station, the performance statistics results of MLP 4 were 0.956, 0.410 (mm/day), and 0.914 for CC, RMSE, and R^2, respectively. From the evaluation of MLP-NNM testing performance, MLP 4 and MLP5 was found to show the better statistics results compared with MLP 1, MLP 2, and MLP 3. The statistics results of testing performance were similar with those of training performance for MLP-NNM. In Gunsan, Jeonju, Haenam stations, the statistics results of training performance were better than those of testing performance. In Daegu, Seoul, Seongsanpo, Ulsan, and Tongyoung stations, vice versa. Table 7 shows the summary of optimal MLP-NNM statistics results during the testing performance for 8 meteorological stations. And, MLP 1 using only mean wind speed (U_{mean}) performed the worst results. However, adding mean temperature (T_{mean}) into the input combinations significantly increased the statistics results of testing performance. We can consider that adding the climatic variables into the input combinations increases the statistics results of testing performance for MLP-NNM. It can obviously be seen from CC, RMSE, and R^2 statistics of MLP-NNM. Table 8 shows the statistics results of each MLP-NNM for Daegu, Ulsan, Jeonju, and Tongyoung stations during the testing performance. Fig. 7 shows the comparison plots of observed and calculated FAO-56 PM ETo for optimal MLP-NNM. Fig. 8 shows the scatter plots between FAO-56 PM ETo and optimal MLP-NNM ETo.

Station	Model	CC	RMSE (mm/day)	R^2
Gunsan	MLP 5	0.964	0.369	0.928
Daegu	MLP 5	0.980	0.349	0.961
Seoul	MLP 4	0.966	0.402	0.933
Seongsanpo	MLP 4	0.955	0.481	0.867
Ulsan	MLP 4	0.958	0.398	0.918
Jeonju	MLP 5	0.959	0.411	0.918
Tongyoung	MLP 5	0.949	0.435	0.896
Haenam	MLP 4	0.956	0.410	0.914

Table 7. Summary of optimal MLP-NNM statistics results during the testing performance

Station	Statistics Index	Model Category				
		MLP 1	MLP 2	MLP 3	MLP 4	MLP 5
Daegu	CC	0.389	0.808	0.950	0.978	0.980
	RMSE (mm/day)	1.653	1.056	0.560	0.369	0.349
	R^2	0.129	0.644	0.900	0.957	0.961
Ulsan	CC	0.341	0.748	0.931	0.958	0.942
	RMSE (mm/day)	1.325	0.920	0.509	0.398	0.471
	R^2	0.086	0.559	0.865	0.918	0.884
Jeonju	CC	0.293	0.802	0.947	0.950	0.959
	RMSE (mm/day)	1.375	0.858	0.460	0.448	0.411
	R^2	0.077	0.641	0.897	0.902	0.918
Tongyoung	CC	0.368	0.739	0.933	0.932	0.949
	RMSE (mm/day)	1.288	0.913	0.534	0.497	0.435
	R^2	0.091	0.544	0.844	0.865	0.897

Table 8. Statistics results of each MLP-NNM during the testing performance

5.5.2 Results of SVM-NNM testing performance

For the testing performance of SVM-NNM, DTREG computer program was used to carry out the testing performance based on the statistics results of the training performance. For 8 meteorological stations, the best statistics results were found at SVM 5. In Gunsan station, the performance statistics results of SVM 5 were 0.983, 0.255 (mm/day), and 0.965 for CC, RMSE, and R^2, respectively. In Daegu station, the performance statistics results of SVM 5 were 0.990, 0.255 (mm/day), and 0.979 for CC, RMSE, and R^2, respectively. In Seoul station, the performance statistics results of SVM 5 were 0.985, 0.267 (mm/day), and 0.971 for CC, RMSE, and R^2, respectively. In Seonsganpo station, the performance statistics results of SVM 5 were 0.970, 0.332 (mm/day), and 0.937 for CC, RMSE, and R^2, respectively. In Ulsan station, the performance statistics results of SVM 5 were 0.971, 0.329 (mm/day), and 0.944 for CC, RMSE, and R^2, respectively. In Jeonju station, the performance statistics results of SVM 5 were 0.976, 0.311 (mm/day), and 0.953 for CC, RMSE, and R^2, respectively. In Tongyoung station, the performance statistics results of SVM 5 were 0.968, 0.353 (mm/day), and 0.932 for CC, RMSE, and R^2, respectively. In Haenam station, the performance statistics results of SVM 5 were 0.976, 0.307 (mm/day), and 0.951 for CC, RMSE, and R^2, respectively. From the evaluation of SVM-NNM testing performance, SVM 5 was found to show the better statistics results compared with SVM 1, SVM 2, SVM 3, and SVM 4. The statistics results of testing performance were similar with those of training performance for SVM-NNM. In every station except for Jeonju station, the statistics results of testing performance were better than those of training performance. In Jeonju station, vice versa. Table 9 shows the summary of optimal SVM-NNM statistics results during the testing performance for 8 meteorological stations. And, SVM 1 using only mean wind speed (U_{mean}) performed the worst results. However, adding mean temperature (T_{mean}) into the input combinations significantly increased the statistics results of testing performance. We can consider that adding the climatic variables into the input combinations increases the statistics results of testing performance for SVM-NNM. It can obviously be seen from CC, RMSE, and R^2 statistics of SVM-NNM. Table 10 shows the statistics results of each SVM-NNM for Daegu, Ulsan, Jeonju, and Tongyoung stations during the testing performance. Fig. 9 shows

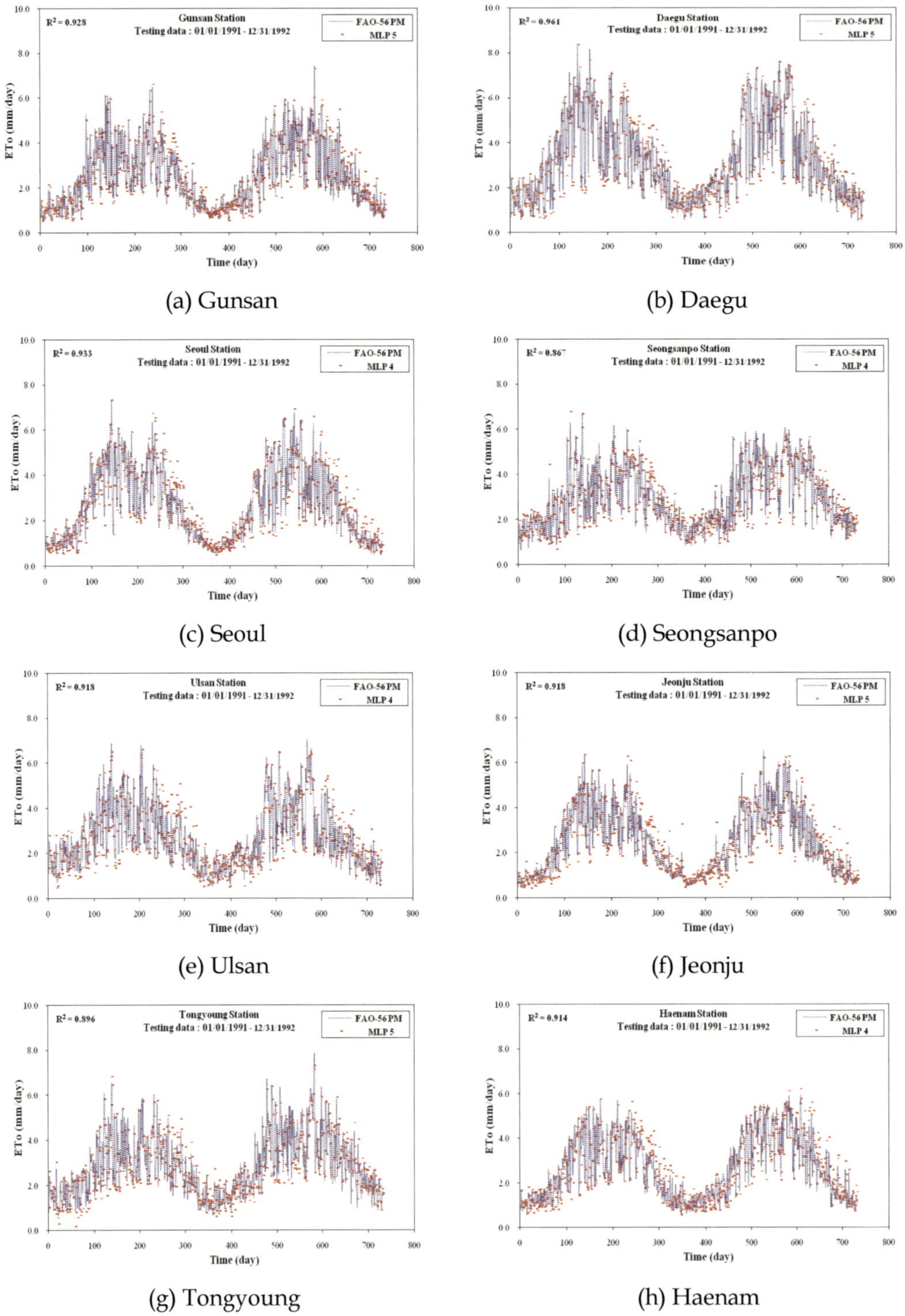

Fig. 7. Comparison plots of observed and calculated FAO-56 PM ETo

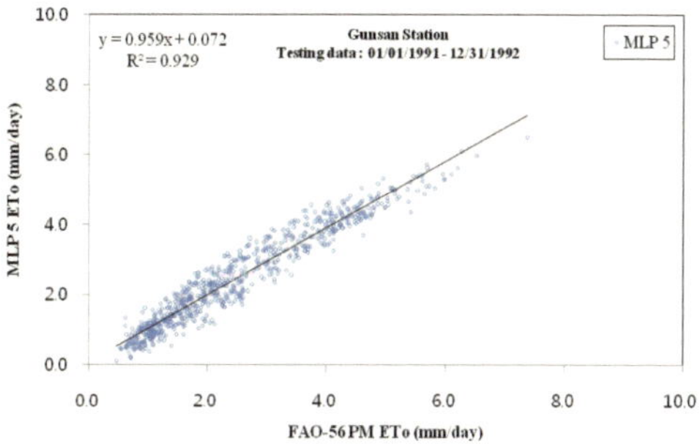

(a) Gunsan

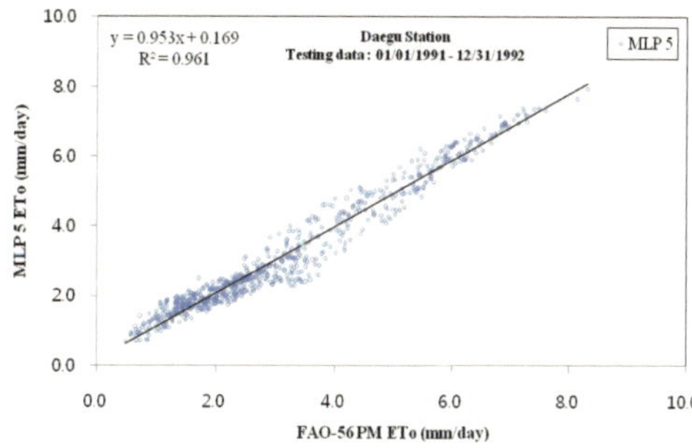

(b) Daegu

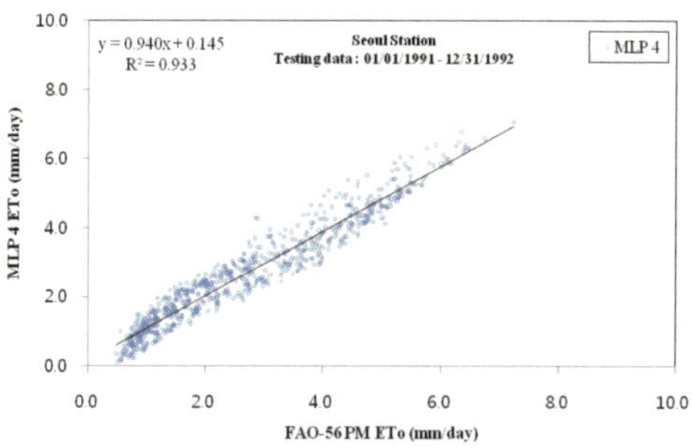

(c) Seoul

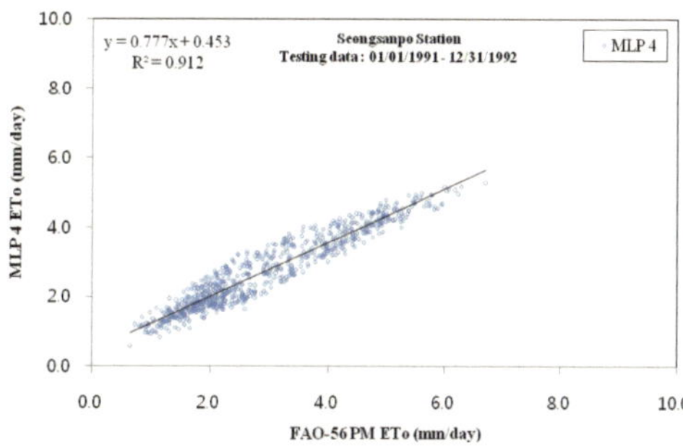

(d) Seongsanpo

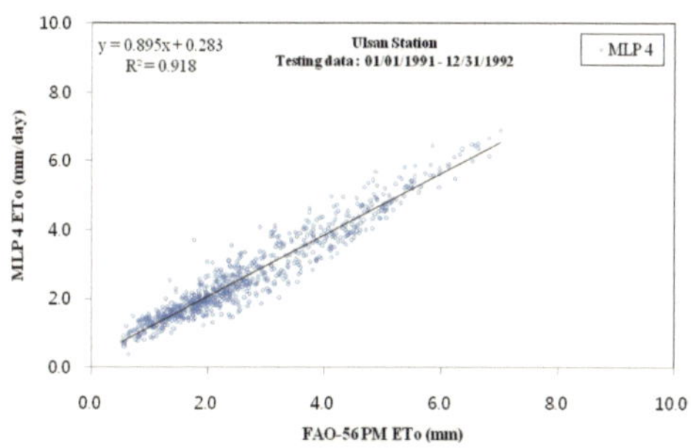

(e) Ulsan

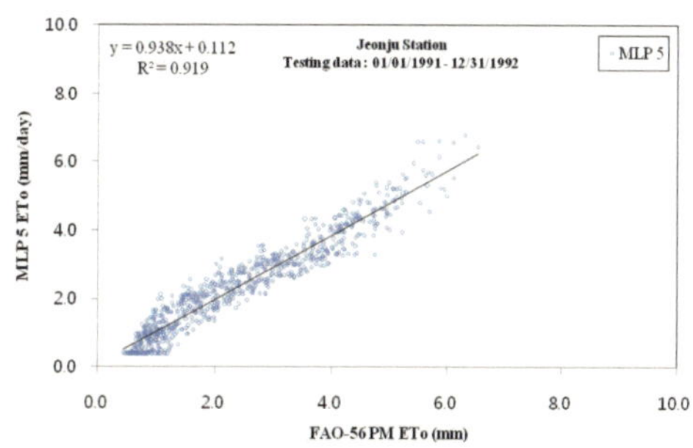

(f) Jeonju

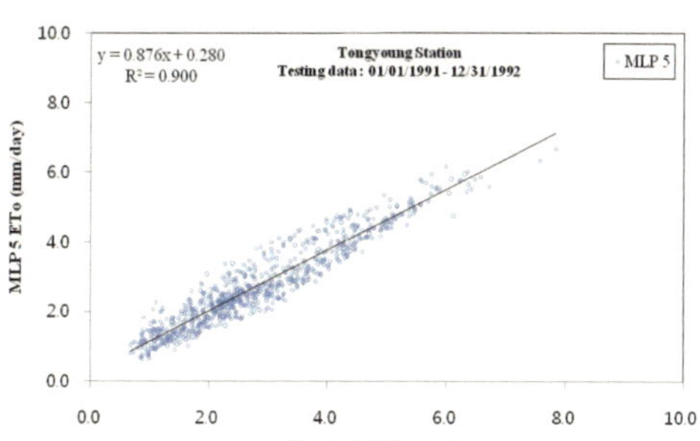

(g) Tongyoung

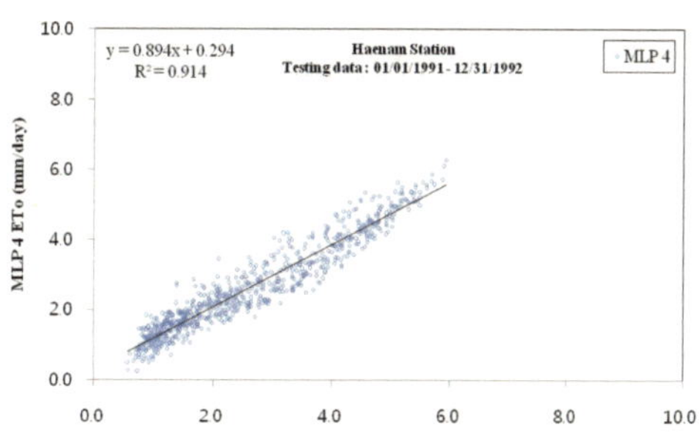

(h) Haenam

Fig. 8. Scatter plots between FAO-56 PM ETo and optimal MLP-NNM ETo the comparisons of observed and calculated FAO-56 PM ETo for the testing performance of optimal SVM-NNM. Fig. 10 shows the scatter plots between FAO-56 PM ETo and optimal SVM-NNM ETo. From the comparison of testing performance for MLP-NNM and SVM-NNM, the statistics results of SVM-NNM were better than those of MLP-NNM based on CC, RMSE, and R^2 statistics. We can consider that the performance of SVM-NNM is better than that of MLP-NNM for the modeling of nonlinear time series such as the FAO-56 PM ETo, which includes the natural uncertainty.

Station	Model	CC	RMSE (mm/day)	R^2
Gunsan	SVM 5	0.983	0.255	0.965
Daegu	SVM 5	0.990	0.255	0.979
Seoul	SVM 5	0.985	0.267	0.971
Seongsanpo	SVM 5	0.970	0.332	0.937
Ulsan	SVM 5	0.971	0.329	0.944
Jeonju	SVM 5	0.976	0.311	0.953
Tongyoung	SVM 5	0.968	0.351	0.932
Haenam	SVM 5	0.976	0.307	0.951

Table 9. Summary of optimal SVM-NNM statistical results during the testing performance

Station	Statistics Index	Model Category				
		SVM 1	SVM 2	SVM 3	SVM 4	SVM 5
Daegu	CC	0.345	0.829	0.963	0.984	0.990
	RMSE (mm/day)	1.706	1.004	0.484	0.313	0.255
	R^2	0.072	0.679	0.925	0.969	0.979
Ulsan	CC	0.363	0.767	0.953	0.970	0.971
	RMSE (mm/day)	1.318	0.894	0.424	0.336	0.329
	R^2	0.096	0.584	0.906	0.941	0.944
Jeonju	CC	0.332	0.823	0.962	0.966	0.976
	RMSE (mm/day)	1.394	0.828	0.399	0.372	0.311
	R^2	0.052	0.665	0.922	0.933	0.953
Tongyoung	CC	0.378	0.781	0.946	0.966	0.968
	RMSE (mm/day)	1.321	0.856	0.473	0.352	0.351
	R^2	0.044	0.599	0.877	0.932	0.932

Table 10. Statistics results of each SVM-NNM during the testing performance

5.5.3 Application and comparison of Multiple Linear Regression Model (MLRM)

The potential of MLP-NNM and SVM-NNM was tested for the application and comparison of multiple linear regression model (MLRM). The statistics result of testing performance for MLP-NNM and SVM-NNM were compared with those of MLRM. MLRM is important because the model enables more than one independent variable to be included in the structure. This can lead to significant increases in calculation accuracy and the ability to measure the effect of each X variable on Y. MLRM should provide more stable estimates of Y since calculations with the bivariate equation are subject to fluctuations due to extreme

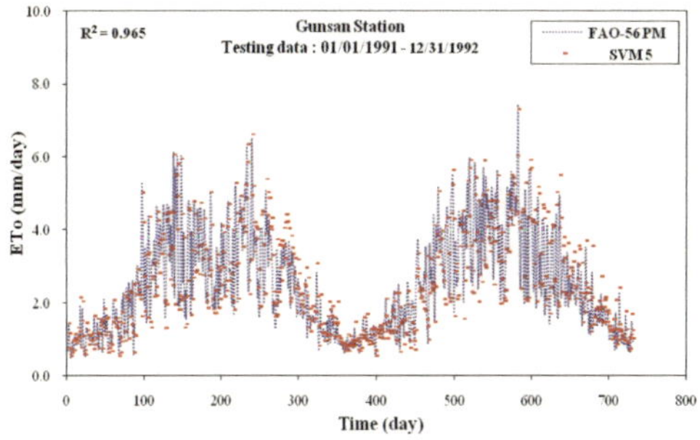

(a) Gunsan

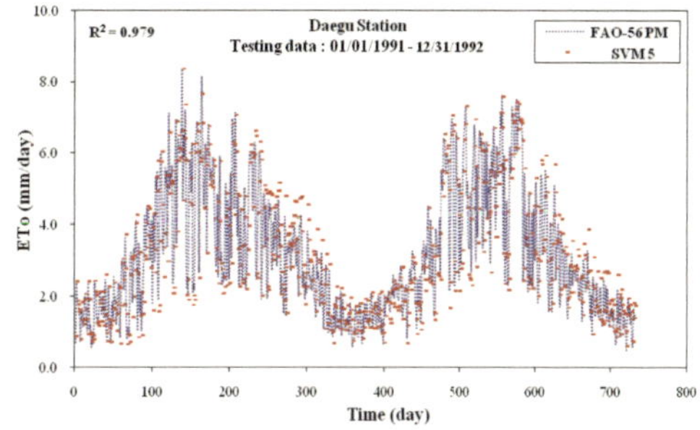

(b) Daegu

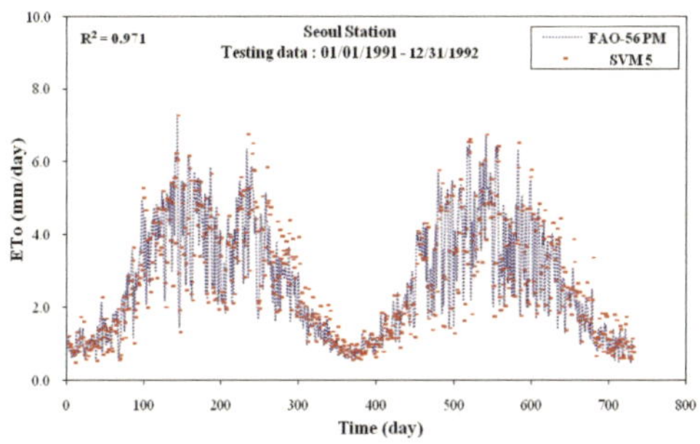

(c) Seoul

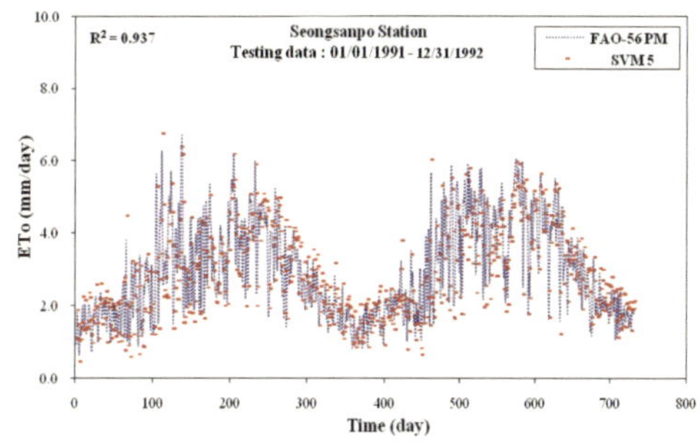

(d) Seongsanpo

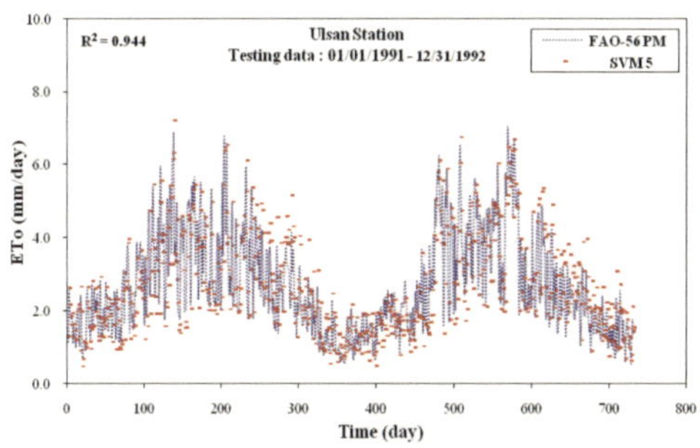

(e) Ulsan

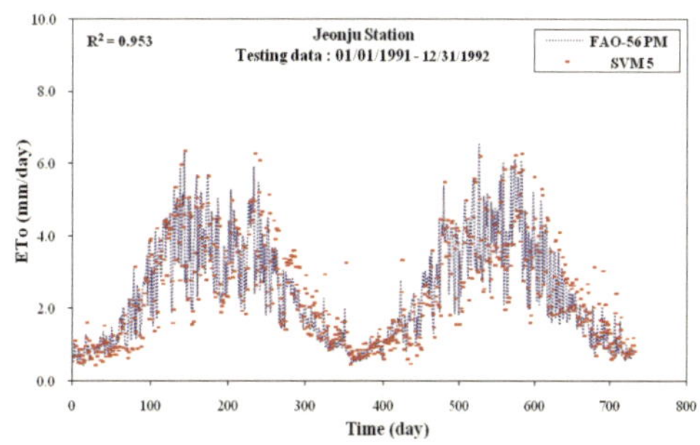

(f) Jeonju

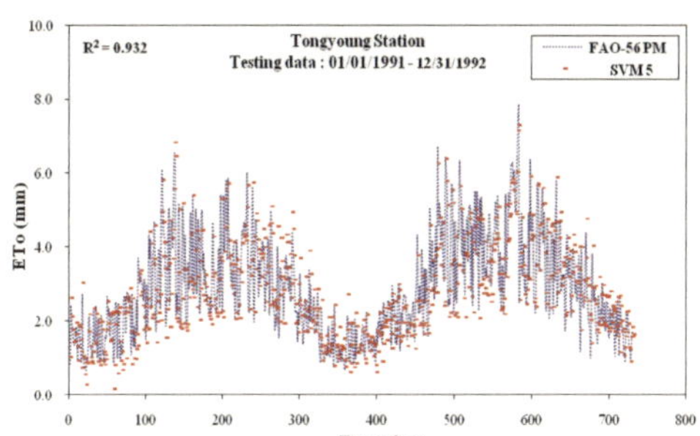

(g) Tongyoung

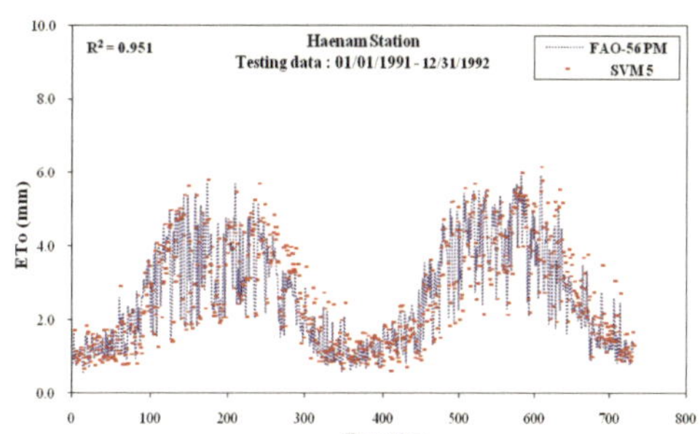

(h) Haenam

Fig. 9. Comparison plots of observed and calculated FAO-56 PM ETo

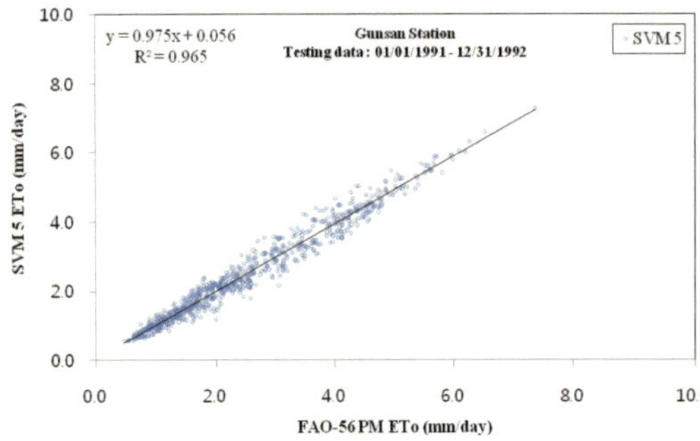

(a) Gunsan

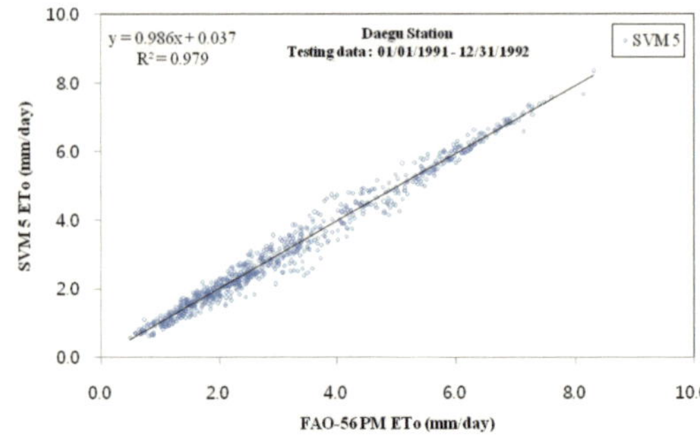

(b) Daegu

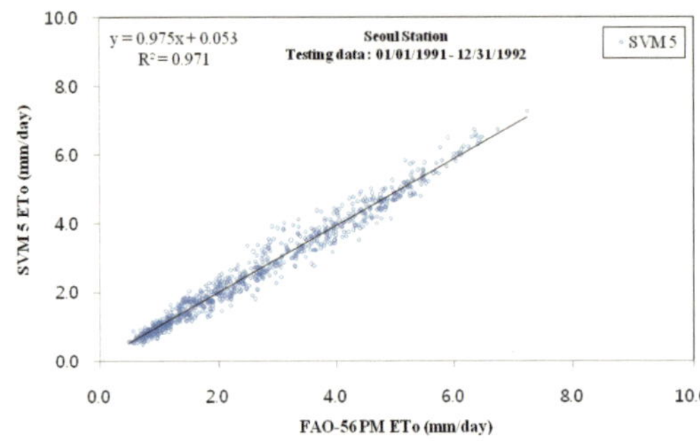

(c) Seoul

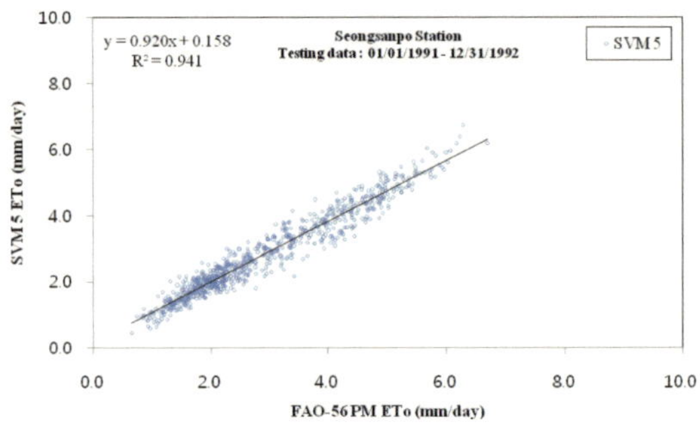

(d) Seongsanpo

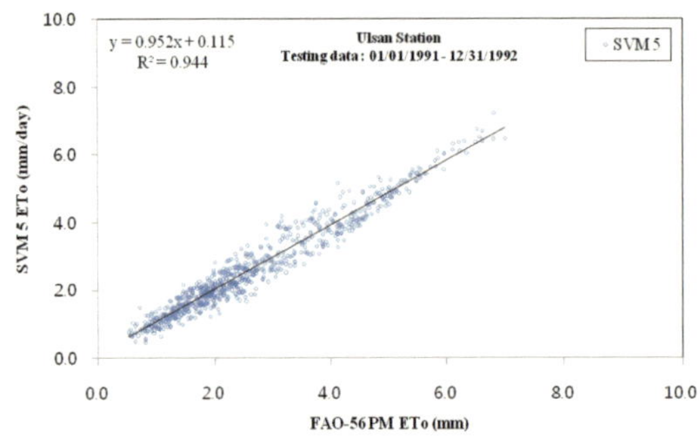

(e) Ulsan

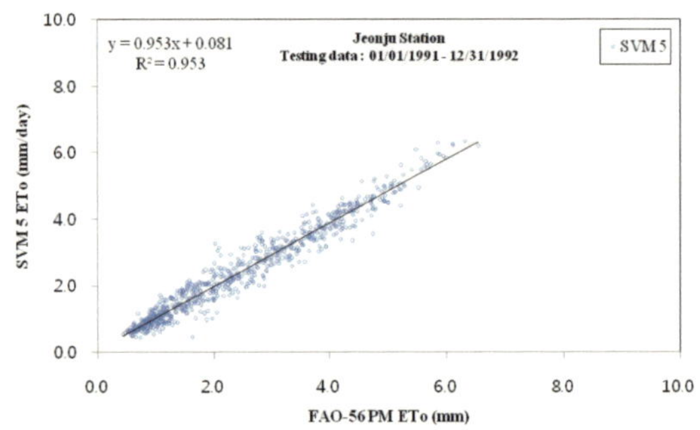

(f) Jeonju

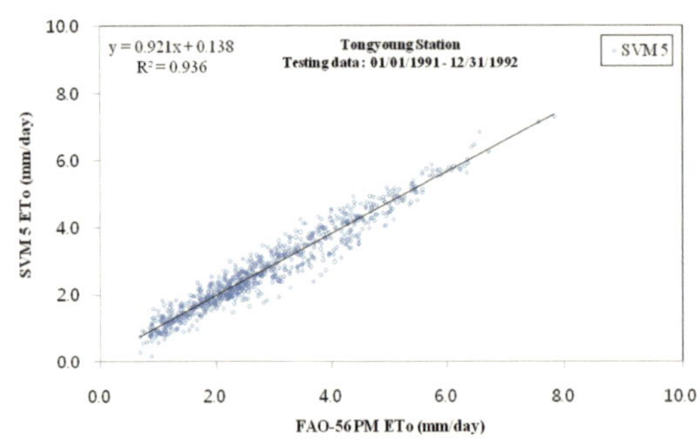

(g) Tongyoung

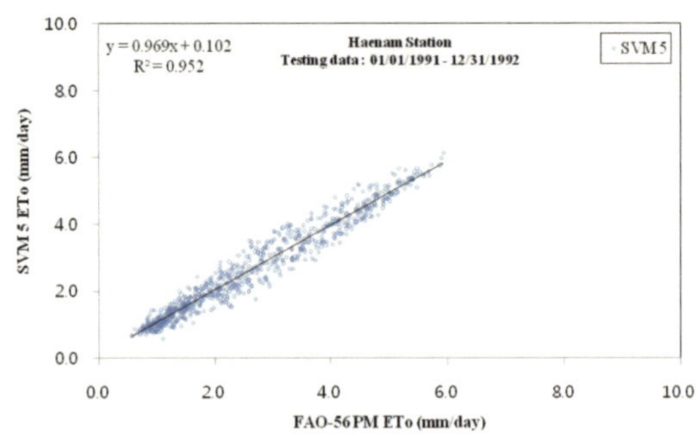

(h) Haenam

Fig. 10. Scatter plots between FAO-56 PM ETo and optimal SVM-NNM ETo variations in X. When the model includes more than one independent variable, extreme variation in one independent variable is less likely to cause extreme variation in the calculated values of Y (McCuen, 1993; Kottegoda, 1998; Salas et al., 2005). From the statistics results of testing performance for MLP-NNM and SVM-NNM, the best statistics results were found at MLP 4 and MLP 5 on average for MLP-NNM. The best statistics results, furthermore, were found at SVM 5 for SVM-NNM. So, two types of MLRM are adopted; MLRM 1 and MLRM 2. MLRM 1 has four independent variables including mean wind speed (U_{mean}), mean temperature (T_{mean}), sunshine duration (SD), and mean relative humidity (RH_{mean}). And, MLRM 2 has five independent variables including mean wind speed (U_{mean}), mean temperature (T_{mean}), sunshine duration (SD), mean relative humidity (RH_{mean}), and max temperature (T_{max}). That is, MLRM 1 corresponds to MLP 4 and SVM 4, and MLRM 2 corresponds to MLP 5 and SVM 5. MLRM 1 and MLRM 2 can be written as the equation (11) and (12).

$$\text{FAO-56 PM ETo} = b_0 + b_1 U_{mean} + b_2 T_{mean} + b_3 SD + b_4 RH_{mean} \quad (11)$$

$$\text{FAO-56 PM ETo} = b_0 + b_1 U_{mean} + b_2 T_{mean} + b_3 SD + b_4 RH_{mean} + b_5 T_{max} \quad (12)$$

where $b_i (i = 1,2,\ldots,p) =$ the slope coefficient, which is also known as the regression coefficient because it is claculated by the results of regression analysis, and b_0 = intercept. In this study, the slope coefficients of MLRM 1 and MLRM 2 were calculated using the training data, which were used for MLP-NNM and SVM-NNM. Table 11 shows equations of MLRM 1 and MLRM 2 calculated by the training data for Daegu, Ulsan, Jeonju, and Tongyoung stations, respectively.

Station	Model Type	Equation
Daegu	MLRM 1	FAO-56 PM ETo = $2.894+0.125U_{mean}+0.148T_{mean}+0.135SD-0.044RH_{mean}$
	MLRM 2	FAO-56 PM ETo = $2.099+0.192U_{mean}+0.041T_{mean}+0.100SD-0.041RH_{mean}+0.108T_{max}$
Ulsan	MLRM 1	FAO-56 PM ETo = $0.834+0.204U_{mean}+0.136T_{mean}+0.169SD-0.020RH_{mean}$
	MLRM 2	FAO-56 PM ETo =$0.552+0.237U_{mean}+0.094T_{mean}+0.156SD-0.019RH_{mean}+0.044\ T_{max}$
Jeonju	MLRM 1	FAO-56 PM ETo = $1.743+0.294U_{mean}+0.119T_{mean}+0.157SD-0.027RH_{mean}$
	MLRM 2	FAO-56 PM ETo = $1.188+0.332U_{mean}+0.048T_{mean}+0.137SD-0.024RH_{mean}+0.069T_{max}$
Tongyoung	MLRM 1	FAO-56 PM ETo = $1.356+0.094U_{mean}+0.134T_{mean}+0.160SD-0.024RH_{mean}$
	MLRM 2	FAO-56 PM ETo = $1.646+0.076U_{mean}+0.182T_{mean}+0.169SD-0.025RH_{mean}-0.049T_{max}$

Table 11. Equations of MLRM 1 and MLRM 2 calculated by the training data

It is worthwhile to compare the relative importance of the variables as indicated by the indendent variables correlations. Table 12 shows the correlation matrix of FAO-56 ETo data base for MLRM 2 of Daegu station. In Daegu station, the indendent variables correlations for MLRM 2 indicate that T_{max} is the most important (R=0.820), with T_{mean} and SD being less important (R=0.760 & R=0.522). RH_{mean} is more important U_{mean} (R=-0.169 vs 0.031). In this study, from the results of correlation matrix for MLRM 1 and MLRM 2 of 8 meteorological stations, T_{mean} (MLRM 1) and T_{max} (MLRM 2) are the most important independent variables, respectively.

Variable	X_1	X_2	X_3	X_4	X_5	Y
mean wind speed	1.000	-0.245	0.168	-0.327	-0.287	0.031
mean temperature		1.000	-0.024	0.426	0.982	0.760
sunshine duration			1.000	-0.667	0.105	0.522
mean relative humidity				1.000	0.330	-0.169
max temperature					1.000	0.820
FAO-56 ETo						1.000

Table 12. Correlation matrix of FAO-56 ETo data base for MLRM 2 of Daegu station

MLRM 1 and MLRM 2 were validated by the testing data of MLP-NNM and SVM-NNM. Table 13 shows statistics results of MLRM 1 and MLRM 2 calculated by the testing data for Daegu, Ulsan, Jeonju, and Tongyoung stations, respectively. We could consider that the performance of MLP-NNM and SVM-NNM was better than that of MLRM 1 and MLRM 2.

Station	Model	CC	RMSE (mm/day)	R^2
Daegu	MLRM 1	0.952	0.562	0.899
	MLRM 2	0.952	0.561	0.900
Ulsan	MLRM 1	0.932	0.503	0.869
	MLRM 2	0.930	0.510	0.865
Jeonju	MLRM 1	0.934	0.530	0.863
	MLRM 2	0.938	0.514	0.871
Tongyoung	MLRM 1	0.908	0.567	0.824
	MLRM 2	0.908	0.568	0.823

Table 13. Statistics results of MLRM 1 and MLRM 2 calculated by the testing data

6. Conclusions

Neural networks model provides a quick and flexible means for modeling of many hydrological processes and has showed better performance than the conventional methods. The hydrologic system under study may be nonlinear and multivariate, and the variables may have unknown interrelationships. Such problems can be efficiently explained by the neural networks model.

In this study, the potential of MLP-NNM and SVM-NNM for the modeling of FAO-56 PM ETo using climatic data has been illustrated. The study demonstrated that the modeling of FAO-56 PM ETo is possible through the use of MLP-NNM and SVM-NNM technique. For 8 meteorological stations which were selected for this study, there are no observed data for the ETo. The data calculated using FAO-56 PM ETo can be assumed as the observed ETo, whose reliability was verified by many previous studies. The following conclusions could be drawn from this study.

1. MLP 4, whose inputs are mean wind speed (U_{mean}), mean temperature (T_{mean}), sunshine duration (SD), and mean relative humidity (RH_{mean}) was found to perform the best among the input combinations for Seoul, Seongsanpo, Ulsan, and Hanam stations. And, MLP 5, whose input are mean wind speed (U_{mean}), mean temperature (T_{mean}), sunshine duration (SD), mean relative humidity (RH_{mean}), and max temperature (T_{max}) was found to perform the best among the input combinations for Gunsan, Daegu, Jeonju, and Tongyoung stations.
2. SVM 5, whose input are mean wind speed (U_{mean}), mean temperature (T_{mean}), sunshine duration (SD), mean relative humidity (RH_{mean}), and max temperature (T_{max}) was found to perform the best among the input combinations for 8 meteorological stations. This indicates that all these variables are needed for the better modeling of FAO-56 PM ETo using SVM-NNM.
3. The temperature, sunshine duration, and relative humidity were found to be more effective than the wind speed in the modeling of FAO-56 PM ETo.
4. The performance statistics results of SVM-NNM were better than those of MLP-NNM. It can obviously be seen from CC, RMSE, and R^2 statistics.
5. The potential of MLP-NNM and SVM-NNM were tested using MLRM 1 and MLRM 2. From the statistics results, the performance of MLP-NNM and SVM-NNM was better than that of MLRM 1 and MLRM 2.

MLP-NNM and SVM-NNM could be of use in water buget of watersheds and various other hydrological analysis where other models may be inappropriate. This study used only a single crop, grass reference crop, for a limited period and further studies using different crop such as alfalfa and rice reference crop may be required to strengthen these conclusions. FAO-56 PM ETo, furthermore, includes some errors in the estimation of many climatic variables. Because the ET are relatively important for the design of irrigation facilities and agricultural reservoirs, the spread of automatic measuring systems for the ET is important and urgent to ensure the reliable and accurate data from the measurements of ET, Republic of Korea.

7. References

Allen, R.G.; Jensen, M.E.; Wright, J.L. & Burman, R.D. (1989). Operational estimates of reference evapotranspiration. *Agronomy Journal*, Vol. 81, No. 4, pp. 650-662, ISSN: 0002-1962.

Allen, R.G.; Pereira, L.S.; Raes, D. & Smith, M. (1998). *Crop evapotranspiration. Guidelines for computing crop water requirement*, Irrigation and Drainage Paper No. 56, FAO, Rome, Italy.

ASCE Task Committee on Application of Neural Networks in Hydrology. (2000). Artificial Neural Networks in Hydrology. I : Preliminary concepts. *Journal of Hydrologic Engineering, ASCE*, Vol. 5, No. 2, pp. 115-123, ISSN: 1084-0699.

ASCE Task Committee on Definition of Criteria for Evaluation of Watershed Models. (1993). Criteria for evaluation of watershed models. *Journal of Irrigation and Drainage Engineering, ASCE*, Vol. 119, No. 3, pp. 429-442, ISSN: 0733-9437.

Bishop, C.M. (1994). Neural networks and their applications. *Review of Scientific Instruments,* Vol. 65, pp. 1803-1832, ISSN: 0034-6748.

Bishop, C.M. (1995). *Neural networks for pattern recognition,* Oxford University Press, ISBN: 0198538642, NY, USA.

Brutsaert, W.H. (1982). *Evaporation into the atmosphere,* Springer-Verlag, ISBN: 9027712476, NY, USA.

Chang, C.C & Lin, C. (2001). *LIBSVM : a library for support vector machines,* Software avaliable http://www.csie.ntu.edu.tw/cjlin/libsvm.

Cover, T.M. (1965). Geometrical and statistical properties of systems of linear inequalities with applications in pattern recognition. *IEEE Transactions on Electronic Computers EC-14,* pp. 326-334.

Dibike, Y.B.; Velickov, S.; Solomatine, D. & Abbott, M.B. (2001). Model induction with support vector machines: introductions and applications. *Journal of Computing in Civil Engineering, ASCE,* Vol. 15, No. 3, pp. 208-216, ISSN: 0887-3801.

Doorenbos, J. & Pruitt, W.O. (1977). *Guidelines for predicting crop water requirement,* Irrigation and Drainage Paper No. 24 2nd Edition, FAO, Rome, Italy.

Haykin, S. (2009). *Neural networks and learning machines 3rd Edition,* Prentice Hall, ISBN: 0131471392, NJ, USA.

Hush, D.R. & Horne, B.G. (1993). Progress in supervised neural network : What's new since Lippmann ?. *IEEE Signal Processing Magazine,* Vol. 10, pp. 8-39, ISSN: 1053-5888.

Jain, S.K.; Nayak, P.C. & Sudheer, K.P. (2008). Models for estimating evapotranspiration using artificial neural networks, and their physical interpretation. *Hydrological Processes,* Vol. 22, pp. 2225-2234, ISSN: 0885-6087.

Jensen, M.E.; Burman, R.D. & Allen, R.G. (1990). *Evapotranspiration and irrigation water requirements,* ASCE Manual and Report on Engineering Practice No. 70, ASCE, NY, USA.

Gallant, S.I. (1993). *Neural Network learning and expert system,* MIT Press, ISBN: 0262071452, MA, USA.

Khadam, I.M. & Kaluarachchi, J.J. (2004). Use of soft information to describe the relative uncertainty of calibration data in hydrologic models. *Water Resources Research,* Vol. 40, No. 11, W11505, ISSN: 0043-1397.

Khoob, A.R. (2008a). Comparative study of Hargreaves's and artificial neural network's methodologies in estimating reference evapotranspiration in a semiarid environment. *Irrigation Science,* Vol. 26, No. 3, pp. 253-259, ISSN: 0342-7188.

Khoob, A.R. (2008b). Artificial neural network estimation of reference evapotranspiration from pan evaporation in a semi-arid environment. *Irrigation Science,* Vol. 27, No. 1, pp. 35-39, ISSN: 0342-7188.

Kim, S. (2004). Neural Networks Model and Embedded Stochastic Processes for Hydrological Analysis in South Korea. *KSCE Journal of Civil Engineering,* Vol. 8, No. 1, pp. 141-148, ISSN: 1226-7988.

Kim, S.; Kim, J.H. & Park, K.B. (2009). Neural Networks Models for the Flood Forecasting and Disaster Prevention System in the Small Catchment. *Disaster Advances,* Vol. 2, No. 3, pp. 51-63, ISSN: 0974-262X.

Kim, S. & Kim, H.S. (2008). Neural networks and genetic algorithm approach for nonlinear evaporation and evapotranspiration modeling. *Journal of Hydrology,* Vol. 351, pp. 299-317, ISSN: 0022-1694.

Kish, O. (2006). Generalized regression neural networks for evapotranspiration modeling. *Hydrological Sciences Journal*, Vol. 51, No. 6, pp. 1092-1105, ISSN: 0262-6667.

Kish, O. (2007). Evapotranspiration modeling from climatic data using a neural computing technique. *Hydrological Processes*, Vol. 21, pp. 1925-1934, ISSN: 0885-6087.

Kish, O. (2008). The potential of different ANN techniques in evapotranspiration modeling. *Hydrological Processes*, Vol. 22 , pp. 2449-2460, ISSN: 0885-6087.

Kisi, O. & Ozturk, O. (2007). Adaptive neurofuzzy computing technique for evapotranspiration estimation. *Journal of Irrigation and Drainage Engineering, ASCE*, Vol. 133, No. 4, pp. 368-379, ISSN: 0733-9437.

Kottegoda, N.T. (1998). *Statistics, probability, and reliability for civil and environmental engineers*, McGraw-Hill, ISBN: 0070359652, Singapore.

Kumar, M.; Bandyopadhyay, A.; Raghuwanshi, N.S. & Singh, R. (2008). Comparative study of conventional and artificial neural network-based ETo estimation models. *Irrigation Science*, Vol. 26, No. 6, pp. 531-545, ISSN: 0342-7188.

Kumar, M.; Raghuwanshi, N.S. & Singh, R. (2009). Development and validation of GANN model for evapotranspiration estimation. *Journal of Hydrologic Engineering, ASCE*, Vol. 14, No. 2, pp. 131-140, ISSN: 1084-0699.

Kumar, M.; Raghuwanshi, N.S.; Singh, R.; Wallender, W.W. & Pruitt, W.O. (2002). Estimating evapotranspiration using artificial neural network. *Journal of Irrigation and Drainage Engineering, ASCE*, Vol. 128, No. 4, pp. 224-233, ISSN: 0733-9437.

Landeras, G.; Barredo, A.O. & Lopez, J.J. (2008). Comparison of artificial neural network models and empirical and semi-empirical equations for daily reference evapotranspiration estimation in the Basque country (Northern Spain). *Agricultural Water Management*, Vol. 95, No. 5, pp. 553-565, ISSN: 0378-3774.

McCuen, R.H. (1993). *Microcomputer applications in statistical hydrology*, Prentice Hall, ISBN: 0135852900, Englewood Cliffs, NJ, USA.

Monteith, J.L. (1965). The state and movement of water in living organism, *Proceedings of Evaporation and Environment*, pp. 205-234, Swansea, Cambridge University Press, NY, USA.

Nash, J.E. & Sutcliffe, J.V. (1970). River flow forecasting through conceptual models, Part 1 - A discussion of principles. *Journal of Hydrology*, Vol. 10, No. 3, pp. 282-290, ISSN: 0022-1694.

Penman, H.L. (1948). Natural evaporation from open water, bare soil and grass, *Proceedings of the Royal Society of London*, Vol. 193, pp. 120-146, London, England.

Principe, J.C.; Euliano, N.R. & Lefebvre, W.C. (2000). *Neural and adaptive systems: fundamentals through simulation*, Wiley, John & Sons, ISBN: 0471351679, NY, USA.

Salas, J.D.; Smith, R.A., Tabios III, G.O. & Heo, J.H. (2001). *Statistical computing techniques in water resources and environmental engineering*, Unpublished book in CE622, Colorado State University, Fort Collins, CO, USA.

Simpson, P.K. (1990). *Artificial neural systems: foundations, paradigms, applications and implementations*, Elsevier, ISBN: 0080378943, NY, USA.

Singh, V.P. (1988). *Hydrologic system rainfall-runoff modeling. Vol. 1*, Prentice Hall, ISBN: 0134480511, NJ, USA.

Specht, D.F. (1991). A general regression neural network. *IEEE Transactions on Neural Networks*, Vol. 2 No. 6, pp. 568-576, ISSN: 1045-9227.

Sudheer, K.P.; Gosain, A.K. & Ramasastri, K.S. (2003). Estimating actual evapotranspiration from limited climatic data using neural computing technique. *Journal of Irrigation and Drainage Engineering, ASCE,* Vol. 129, No. 3, pp. 214-218, ISSN: 0733-9437.

Traore, S.; Wang, Y.M. & Kerh, T. (2010). Artificial neural network for modeling reference evapotranspiration complex process in Sudano-Sahelian zone. *Agricultural Water Management,* Vol. 97, No. 5, pp. 707-714, ISSN: 0378-3774.

Trajkovic, S. (2005). Temperature-based approaches for estimating reference evapotranspiration. *Journal of Irrigation and Drainage Engineering, ASCE,* Vol. 131, No. 4, pp. 316-323, ISSN: 0733-9437.

Trajkovic, S; Todorovic, B. & Stankovic, M. (2003). Forecasting reference evapotranspiration by artificial neural networks. *Journal of Irrigation and Drainage Engineering, ASCE,* Vol. 129, No. 6, pp. 454-457, ISSN: 0733-9437.

Tripathi, S.; Srinivas, V.V. & Nanjundish, R.S. (2006). Downscaling of precipitation for climate change scenarios: A support vector machine approach. *Journal of Hydrology* , Vol. 330, pp. 621-640, ISSN: 0022-1694.

Tsoukalas, L.H. & Uhrig, R.E. (1997). *Fuzzy and Neural Approaches in Engineering,* Wiley, John & Sons, ISBN: 0471192473, NY, USA.

Vapnik, V.N. (1992). Principles of risk minimization for learning theory. In: *Advances in Neural Information Processing Systems Vol. 4,* Moody, Hanson & Lippmann, (Ed.), pp. 831-838, Elsevier, ISBN: 1558602224, NY, USA.

Vapnik, V.N. (2010). *The nature of statistical learning theory 2nd Edition,* Springer-Verlag, ISBN: 0387987800, NY, USA.

Wasserman, P.D. (1993). *Advanced methods in neural computing,* Wiley, John & Sons, ISBN: 0442004613, NY, USA.

Zanetti, S.S.; Sousa, E.F.; Oliveira, V.P.S.; Almeida, F.T. & Bernardo, S. (2007). Estimating evapotranspiration using artificial neural network and minimum climatological data. *Journal of Irrigation and Drainage Engineering, ASCE,* Vol. 133, No. 2, pp. 83-89, ISSN: 0733-9437.

Part 2

Remote Sensing

6

A Simple Remote Sensing EvapoTranspiration Model (Sim-ReSET) and its Application

Qinxue Wang[1], Zhigang Sun[2],
Bunkei Matsushita[3] and Masataka Watanabe[4]
[1]*National Institute for Environmental Studies*
[2]*Civil & Environmental Engineering, University of Connecticut*
[3]*Graduate School of Life and Environmental Sciences, University of Tsukuba, Tsukuba*
[4]*Faculty of Environmental Information, Keio University, Tokyo*
[1,3,4]*Japan*
[2]*USA*

1. Introduction

Evapotranspiration (ET) refers to the combination of two separate processes whereby water is lost on the one hand from the soil surface by evaporation and on the other hand from the crop by transpiration. Evaporation and transpiration occur simultaneously and there is no easy way of distinguishing between the two processes. Apart from the water availability in the topsoil, the evaporation from a cropped soil is mainly determined by the fraction of the solar radiation reaching the soil surface. This fraction decreases over the growing period as the crop develops and the crop canopy shades more and more of the ground area. When the crop is small, water is predominately lost by soil evaporation, but once the crop is well developed and completely covers the soil, transpiration becomes the main process. Weather parameters, crop characteristics, management and environmental aspects are factors affecting ET.

Evapotranspiration is not easy to measure. Specific devices and accurate measurements of various physical parameters or the soil water balance in lysimeters are required to determine evapotranspiration. The most direct method of measuring evapotranspiration is with the eddy covariance technique in which fluctuations of vertical wind speed are correlated with fluctuations in atmospheric water vapor density. This directly estimates the transfer of water vapor (evapotranspiration) from the land (or canopy) surface to the atmosphere. These methods are often expensive, demanding in terms of accuracy of measurement and can only be fully exploited by well-trained research personnel.

Monitoring evapotranspiration (ET) at large scales is important for assessing climate and anthropogenic effects on natural and agricultural ecosystems. Remote sensing is the only way that can efficiently and economically monitoring ET with regional and global coverage. Because the regional and global real-time ET data from satellite remote sensing play a vital role in researches on climate change and water resource management, many studies focus on this topic, which can be classified into three categories according to the contribution of remote sensing data to the ET algorithms. The first category is studies in which ET is estimated only using reanalyzed or simulated data from general circulation models (GCM).

For example, Thomas (2008) estimated monthly ET over China and adjacent areas using the Penman-Monteith (P-M) method and gridded meteorological data (0.25°) from 1951–1990. The second category is studies in which ET estimations depend partially on ground data (Bastiaanssen et al., 1998; Su, 2002; Leuning et al., 2008; Zhang et al., 2009). ET methods in this category, more or less, used ground-based or reanalyzed data. For example, Cleugh et al. (2007) and Mu et al. (2007) obtained regional and global ET rates using ground-based meteorological observations and MODerate Resolution Imaging Spectroradiometer (MODIS) data. The third category is studies in which ET is estimated entirely from satellite remote sensing data. ET can be partitioned from available energy by using an evaporation fraction (EF) that can be defined using the relationship of remotely sensed surface temperature (T_s) and vegetation index (VI) / albedo (Boegh et al., 1999; Fan et al., 2007; Gillies et al., 1997; Venturini et al., 2004; Verstraeten et al., 2005).

The algorithms in the first and second categories are often relatively accurate, but detailed input parameters entirely or partially depend on ground data or simulated data from GCM models. This will limit their application to areas where the necessary input data are unavailable. Even if these input data can be derived from reanalyzed datasets, the spatial resolution is usually poor. Furthermore, we cannot obtain real-time ET rates due to the time delay of acquiring input data. The algorithms in the third category can produce real-time ET rates entirely using latest remote sensing data, such as VI-T_s triangle method (Jiang and Islam, 1999) and S-SEBI method (Roerink et al., 2000), but these methods usually cannot consider the aerodynamic characters of the land surface, such as land surface roughness (Sun et al., 2009).

As remotely sensed data have the advantage of large spatial coverage with high spatial resolution, frequent updates, and consistent quality, the optimal way is to develop a robust algorithm whose input parameters can all be easily derived from remote sensing. We have developed such an algorithm so-called Sim-ReSET based on the energy balance of land surfaces (Sun et al., 2009). In this model, the calculation of aerodynamic resistance can be avoided by using a reference dry bare soil and the assumption that wind speed at the upper boundary of the atmospheric surface layer is laterally homogenous, but the aerodynamic characters of land surface are still considered using canopy height. And all inputs (net radiation, soil heat flux, canopy height, variables related to land surface temperature) can be potentially obtained from satellite remote sensing, which allows mapping routine real-time ET.

2. Model description

The Sim-ReSET model is dual-source model (Sun et al., 2009). ET from a pixel can be approximately considered as a combination of ETs from vegetation and bare soil within the pixel.

$$\mathrm{ET} = f_{\mathrm{veg}}\mathrm{ET}_{\mathrm{veg}} + (1 - f_{\mathrm{veg}})\mathrm{ET}_{\mathrm{soil}} \tag{1}$$

where f_{veg} is the vegetation cover fraction. The ET_{veg} and ET_{soil} are ETs of vegetation and soil components within the pixel, respectively. They are obtained under neutral or near-neutral conditions:

$$\mathrm{ET}_{\mathrm{veg}} = (R_{\mathrm{nveg}} - G_{\mathrm{veg}}) - (R_{\mathrm{nd}} - G_{\mathrm{d}})\frac{T_{\mathrm{veg}} - T_{\mathrm{a}}}{T_{\mathrm{sd}} - T_{\mathrm{a}}}\frac{\ln(\frac{z}{z_{\mathrm{0hd}}})\ln(\frac{A}{z_{\mathrm{0md}}})}{\ln(\frac{z-d_0}{z_{\mathrm{0h}}})\ln(\frac{A-d_0}{z_{\mathrm{0m}}})} \tag{2}$$

$$\mathrm{ET}_{\mathrm{soil}} = (R_{\mathrm{nsoil}} - G_{\mathrm{soil}}) - (R_{\mathrm{nd}} - G_{\mathrm{d}})\frac{T_{\mathrm{soil}} - T_{\mathrm{a}}}{T_{\mathrm{sd}} - T_{\mathrm{a}}} \quad (3)$$

where the subscripts of d, veg, and soil denote reference dry bare soil, vegetation, and soil, respectively. R_n and G are the net radiation and soil heat flux (W/m^2). T_{veg} and T_{soil} are the surface temperatures for vegetation and soil components within the pixel (°C), respectively. T_{sd} and T_a are the surface temperature of reference dry bare soil and air temperature (°C), respectively. The d_0, z_{0h} and z_{0m} are the zero plane displacement height and roughness lengths for heat and momentum transfers (m), respectively. The z and A are the reference height and height of the upper boundary of the atmospheric surface layer (m), respectively. In the model, z is defined as 2 m plus the height of the vegetation canopy, and A is given as 100 m (Brutsaert, 1998).

Three categories of model inputs are required for the implementation of the Sim-ReSET model: solar radiation-related fluxes, temperature-related parameters, and height-related parameters (Sun et al., 2009). At first, we adopted the simple scheme proposed by Bisht et al. (2005) to estimate instantaneous net radiation for cloud-free days only using remote sensing observations. The soil heat flux is calculated from R_n multiplied by the ratio of G/R_n. The roughness lengths are determined from prior literatures and from a look-up table (LUT) by means of canopy heights. The canopy heights of forest and shrub are determined according to the IGBP land cover types (Sun et al., 2009). Generally, the heights of forests and shrubs do not change significantly throughout the four seasons. However, grasses and crops change seasonally, so their canopy heights vary with time throughout their whole life cycles. Since the heights of grasses and crops have linear relationships with their leaf area indexes (LAIs) before their heights reach the maximums, the heights of crops and grasses can be approximately estimated by means of spectral vegetation indices or f_{veg} (Turner et al., 1999).

Finally, T_{sd} and T_a can be generally obtained from the dry (or warm) edge in a triangular VI-T_s diagram (e.g., Sandholt et al., 2002), and T_{soil} can also be simply obtained by a linear extrapolation in the triangular VI-T_s diagram under an assumption that T_{veg} approximates T_a (Nishida et al., 2003). However, the VI-T_s diagram cannot be well defined when the ranges of land surface moisture and VI are incomplete, such as in the rainy season or within a period or an area with a narrow VI range (Sun et al., 2008). This will result in more uncertainties in the determinations of T_{soil}, T_{veg}, T_{sd}, and T_a. Here, we proposed a strategy to decrease these uncertainties. If the spatial range of a sampling area for defining a VI-T_s diagram is large enough under the homogeneous condition of atmospheric forcing, the condition of $T_{sd} - T_a \leq 2$°C represents two cases: either thoroughly wet (($f_{veg} \geq 0.2$,e.g., humid areas, wetland, or areas after strong rainfall) or thoroughly dry ($f_{veg} < 0.2$, e.g., desert) within the sampling area; then, ET is considered approximately 0 for a dry case and $R_n - G$ for a wet case. If $T_{sd} - T_a > 2$°C, then ET is calculated by the model.

Considering the 1°C accuracy of MODIS Ts retrievals (Wan et al., 2004), we took 2°C as a threshold. For completely dry areas, f_{veg} is usually close to 0; for thoroughly wet areas, there is usually extensive vegetation coverage. Hence, we approximately took 0.2 of f_{veg} as a threshold to distinguish wet and dry conditions.

3. Data collection and observation

The MODIS satellite data were used as inputs to drive the Sim-ReSET model, and ground data were used to validate the outputs. Three MODIS/Terra land products were collected

(Tiles: h22-h29, v03-v07; Projection: sinusoidal; Period: Mar. 5 2000 - Mar. 6 2010): 8-day global 1 km land surface temperature / emissivity, yearly global 1 km land cover type (Figure 1), and 16-day global 1 km vegetation indices. These land products were used to generate ET maps from 2000 to 2010 based on our model (Figure 2).

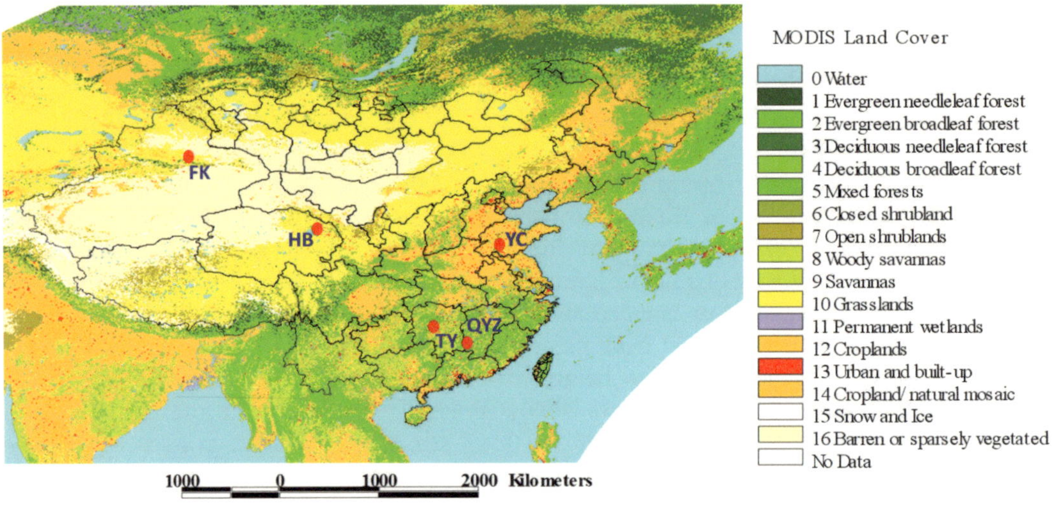

Fig. 1. Asian land-cover map and locations of ecological stations for ET validations in China (HB: Haibei, YC: Yucheng, TY: Taoyuan, QYZ: Qianyanzhou, FK: Fukang).

In order to validate remote sensing data products, a ground observation network was established in 2002 (Wang et al., 2004; Watanabe et al., 2005). Long-term micrometeorological, vegetation and soil measurements, and flux measurements of water vapor, energy, and CO_2 from a variety of ecosystems across China–grassland (HB: Haibei), irrigated cropland (YC: Yucheng), paddy cropland (TY: Taoyuan), forest (QYZ: Qianyanzhou), and desert (FK: Fukang) were implemented and integrated into a consistent, quality-assured, and documented dataset. These five stations represent different typical climates (from subtropical humid climate to temperate arid climate), terrains (from plain to plateau), and land covers. This dataset has been playing a vital role in validations of satellite remote sensing products and in related terrestrial studies (*e.g.*, Sun et al., 2007; Wang et al., 2005).

Compared against other ET methods, the Penman-Monteith (P-M) method has a better performance for ET estimations, and thus it is usually used as the standard for evaluating other methods (Jensen et al., 1990; Irmak et al., 2003). We also compared the eddy covariance observations with the estimation by the P-M method, and found that they were consistent. Due to many gaps of eddy covariance (EC) measurements at the Fukang, Taoyuan, and Yucheng stations and no EC data at the Haibei and Qianyanzhou stations, in this study, the P-M method together with intensive ground data was used to estimate ET at the five stations, and then these ETs at the Terra satellite's overpass time were selected to validate the ET estimations obtained from the MODIS-driven Sim-ReSET model.

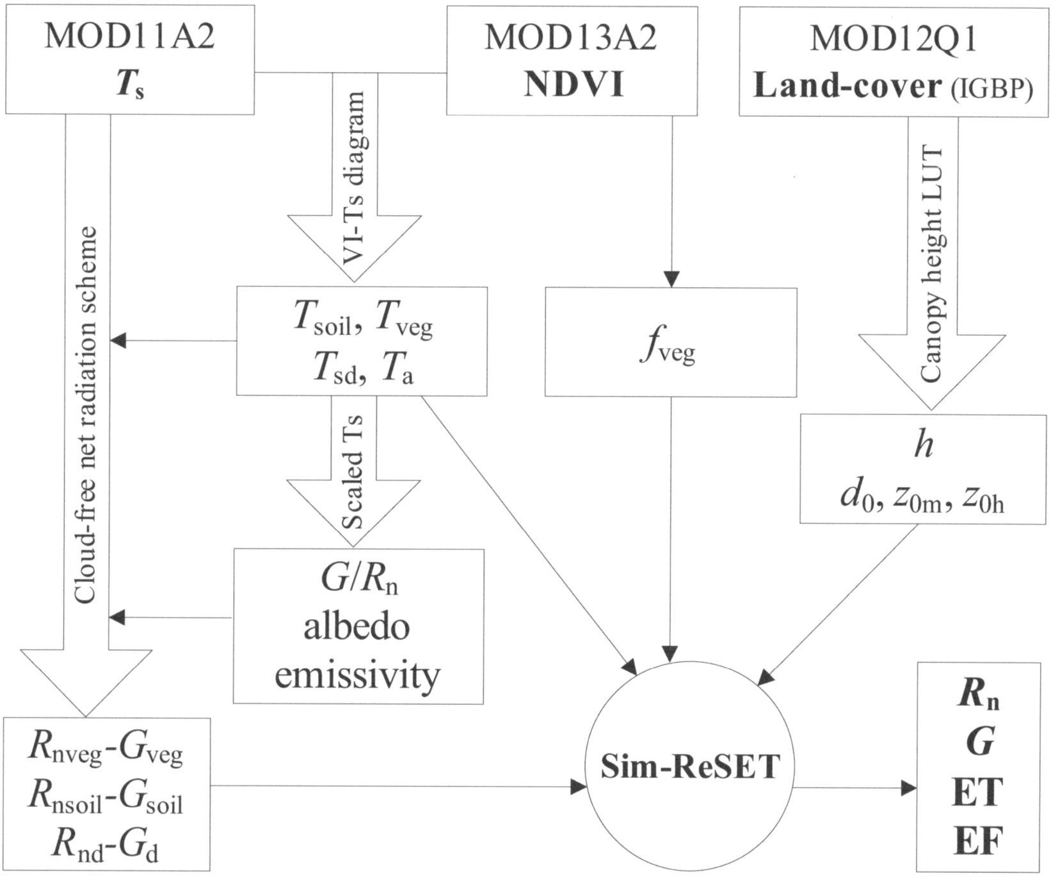

Fig. 2. Flowchart of ET estimations based on the Sim-ReSET model and MODIS land surface products.

4. Model estimation and validation

The 16-day 1 km Asian terrestrial R_n, G, ET, and EF maps from 05 Mar. 2000 to 06 Mar. 2010 were generated. Here we give an example of ET maps to represent the seasonal and spatial distribution of ET over Asia (Figure 3). The figure clearly shows that ET is relatively large in near-sea humid regions, such as Japan, the Korean Peninsula, the east and south of China, South Asia, and Southeastern Asia. In summer, ET over the high-latitude regions is also large because of the strong vapor exchange from a boreal forest ecosystem. ET is relatively low in arid and semi-arid areas, such as northwestern China, Mongolia, and Central Asia. These spatial and temporal distributions of Asian ET maps closely correspond to climates on the continental scale.

The ET and EF profiles extracted from 16-day Asian ET and EF time series maps were significantly different at the five stations (Figure 4). The ET and EF at the Haibei (grassland) and Fukang (desert) stations increase sharply in spring because ice and snow melt; however, the surface soil water supplied from ice, snow, and limited precipitation is soon completely evaporated, and then the ET and EF decrease sharply. Concentrated irrigation for crops and

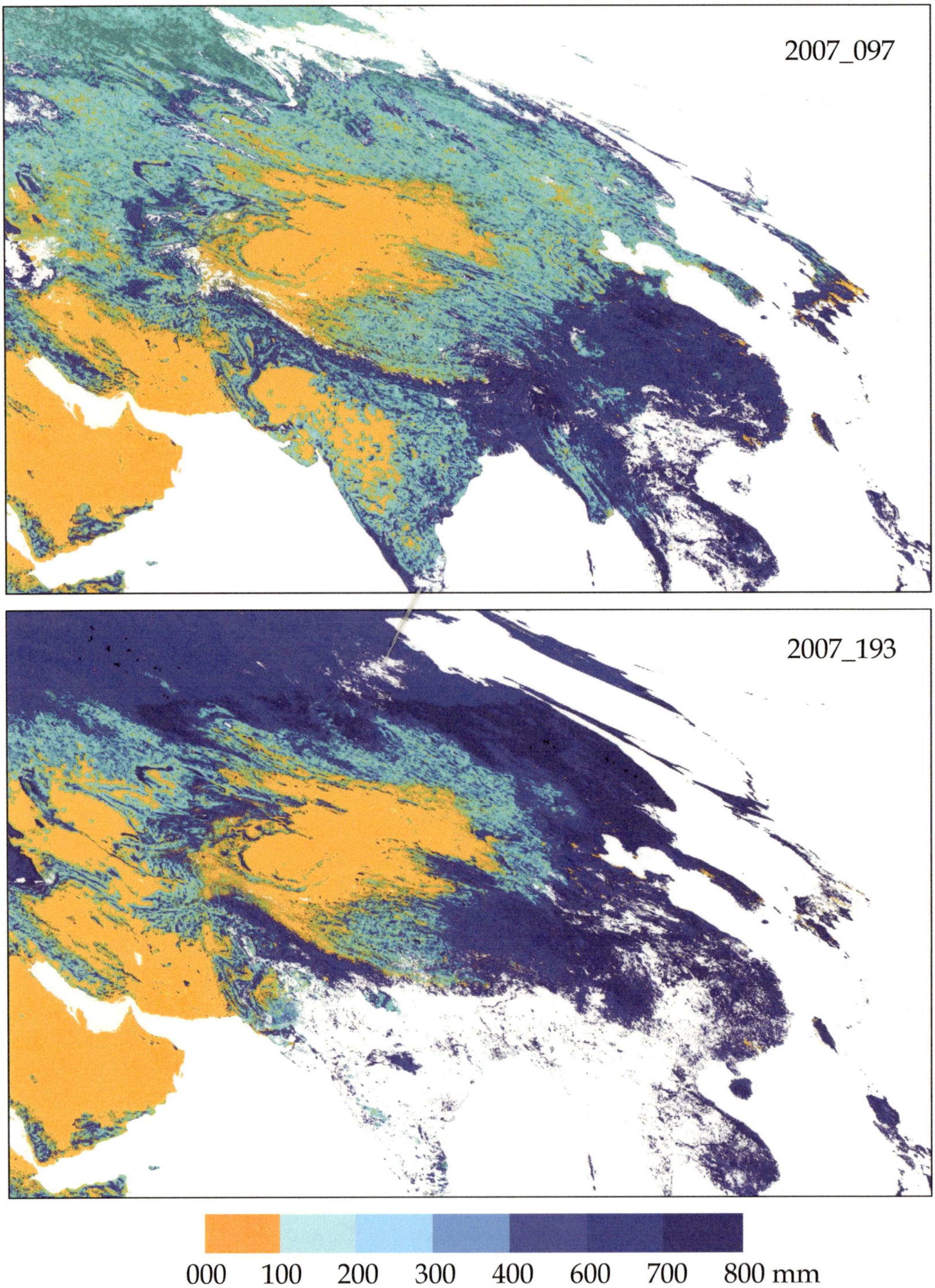

Fig. 3. An example of Asian ET maps (Apr. 7 and Jul. 12, 2007), in which white colour shows the areas no data due to clouds

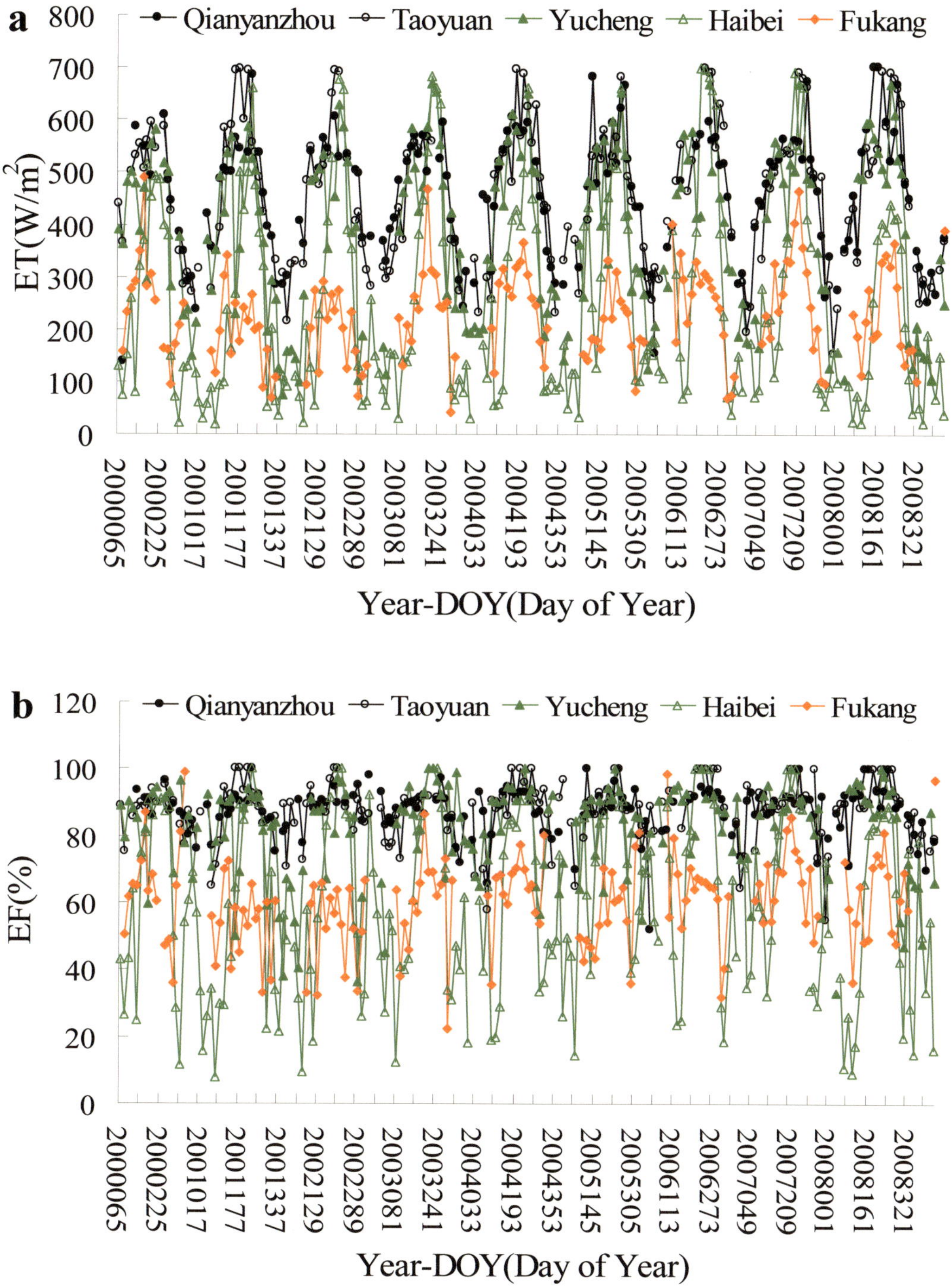

Fig. 4. Time series of MODIS-based ET (a) and EF (b) from 2000 to 2009 at 5 ecological stations

precipitation during growing seasons result in large ET and EF at the Yucheng (irrigated field) station. Due to plentiful precipitation, the EFs at the Qianyanzhou (forest) and Taoyuan (paddy field) stations are relatively large through the whole year, and the ET is close to the available energy.

The MODIS-based ETs were compared with the ETs calculated using the P-M method at the five experimental stations. Figure 5 shows that the MODIS-based ETs are in agreement with the ETs by the P-M method at the five stations. The slope, intercept, and R^2 are 0.84, 85.65, and 0.88, respectively. The MODIS-based ETs show smaller biases at the Qianyanzhou, Taoyuan, and Haibei stations, but larger biases at the Yucheng and Fukang stations. Error analysis shows that the mean absolute differences (MADs) of ET are 41.04, 50.38 and 52.36 W/m^2 at the Qianyanzhou, Taoyuan, and Haibei stations, respectively, and the MADs of ET are 68.60 and 66.95 W/m^2 at the Yucheng and Fukang stations, respectively. Because of the low evaporation and transpiration in the cold season, the ET biases are not significant in the cold season. Therefore, larger ET biases at the Yucheng and Fukang stations mainly occur in the warm season.

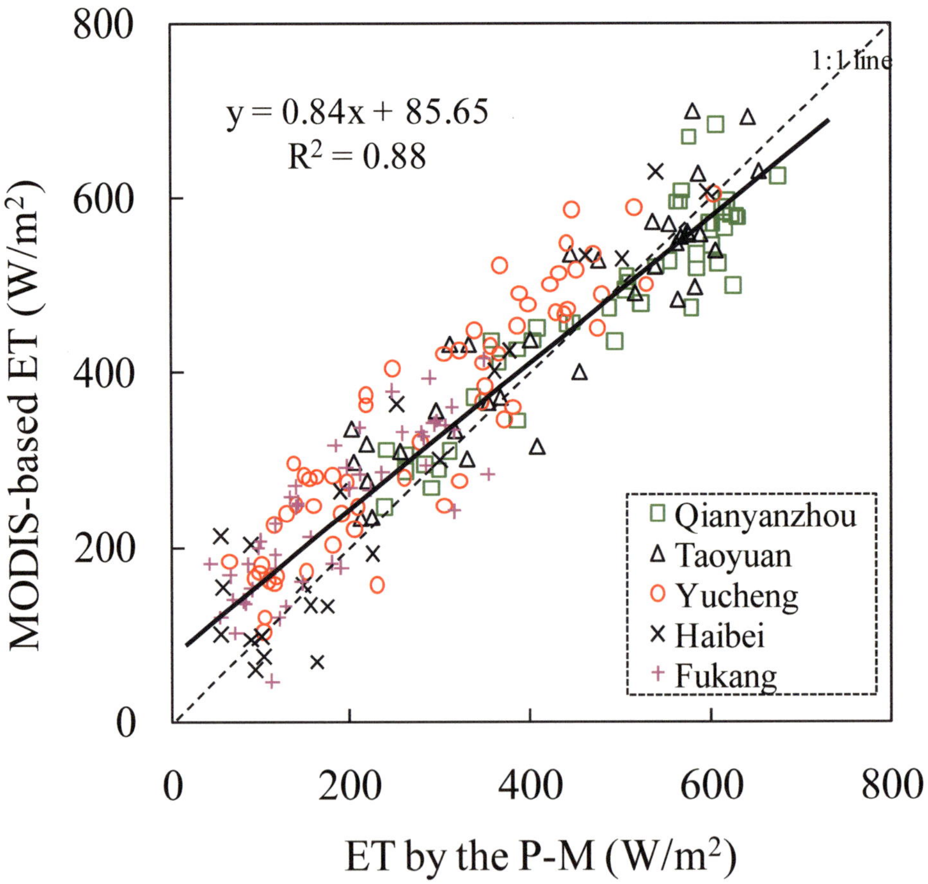

Fig. 5. Comparison of MODIS-based ET and ET by the P-M method at the five stations

5. Conclusions and discussions

A dual-source Simple Remote Sensing EvapoTranspiration model (Sim-ReSET) has been developed based solely on RS data. One merit of this model is that the calculation of aerodynamic resistance can be avoided by means of a reference dry bare soil and an assumption that wind speed at the upper boundary of atmospheric surface layer is laterally homogenous, but the aerodynamic characters are considered by means of canopy height. The other merit is that all inputs (net radiation, soil heat flux, canopy height, variables related to land surface temperature) can be potentially obtained from RS data, which allows obtaining regular RS-driven ET products.

The 16-day Asian MODIS-based ET maps from 2000 to 2010 were generated using the Sim-ReSET model. The ET maps are capable of showing the spatial and temporal variations of ET on a continental scale. Extensive ground data from a variety of ecosystems across China were used to validate the MODIS-based ET product. The MODIS-based ETs are in agreement with the ETs estimated by the P-M method for homogeneous forest, paddy, and grass lands, and the MADs are less than or equal to 52.36 W/m^2. Mainly because a 1km×1km heterogeneous MODIS pixel does not match the fetch of a micrometeorological / flux tower, the MODIS-based ETs show relatively large biases for heterogeneous desert and irrigated crop lands, and the MADs are larger than or equal to 66.95 W/m^2. These biases might come from assumptions in the model, mismatches of spatial scale in validations, and remote sensing input data. The ET accuracies vary between different stations and are mainly related to land covers and spatial scales. There are homogeneous forest, rice paddies, and grass from the observational sites to 1 km MODIS-pixel scales at the Qianyanzhou, Taoyuan, and Haibei stations, respectively, so ET estimations by the P-M method on the site scale can agree with the ET estimations by our model on the corresponding 1 km MODIS-pixel scale. However, within the range of 1 km around the observational site in the Yucheng station, there are not only cotton but also winter wheat, summer maize, and other crops. Near the observational site in the Fukang station, there is sparse shrub. Hence, ET estimations or observations on the site scale cannot well represent the ET on the 1 km MODIS-pixel scale at the Yucheng and Fukang stations. This might be the main reason for larger ET biases at the Yucheng and Fukang stations.

The ET accuracy depends on remote sensing input data. Compared with the ground-truth observations, remote sensing data still have uncertainties and errors due to their retrieving algorithms and the atmospheric effect on remote sensing observations. Of course, these uncertainties and errors in remote sensing inputs will be mitigated by the application of more new technologies for retrieving remote sensing data. Many of MODIS land products have been updated to new versions with more accuracy. These new data products will contribute to improving the ET accuracy.

The ET accuracy also depends on the derivative parameters from the remote sensing inputs, such as the surface temperature of reference dry bare soil (T_{sd}), air temperature (T_a), and canopy height (h). Our sensitivity analysis of the Sim-ReSET model showed that the model is sensitive to variables related to temperature, but insensitive to the heights of canopy and the atmospheric surface layer (Sun et al., 2009). Hence, the improvements of T_{sd} and T_a estimations will enhance the accuracy of ET product. We used a triangular VI-T_s diagram to obtain T_{sd} and T_a in this study. However, an ideal triangular VI-T_s diagram sometimes cannot be well constructed using 1 km MODIS data if the ranges of land surface moisture and VI are incomplete within a sampling window. This will result in more uncertainties and errors in the determinations of T_{sd} and T_a, and then in ET estimations

Although a daily ET is more useful in the applications of hydrology and water resources, only instantaneous ET products at the overpass time of the Terra satellite were generated using MODIS/Terra data in this study. Similar to the sinusoidal variation of solar radiation in the daytime on cloudless days, a daily ET can be estimated from an instantaneous ET at the satellite overpass time (Sun et al., 2009). This up-scaling method seems invalid on cloudy days. Li et al. (2009) tried to use the microwave difference vegetation index (EDVI) and ground data to estimate ET in a mid-latitude forest. The EDVI can be obtained under both clear and cloudy sky conditions, but the land surface temperature is still unavailable from remote sensing under cloudy sky conditions. Therefore, it is still an attractive challenge to obtain actual daily ET directly from remote sensing.

6. Acknowledgements

This research was supported by the project of "Establishment of Early Detection Network of the Global Warming Impacts", sponsored by the Ministry of the Environment, Japan, and the GCOM (Global Change Observation Mission) project, sponsored by the Japan Aerospace Exploration Agency (JAXA).

7. References

Bastiaanssen, W.G.M., Menenti, M., Feddes, R.A., Holtslag, A.A.M., 1998. A remote sensing surface energy balance algorithm for land (SEBAL): 1. Formulation. *Journal of hydrology*, 212-213, pp. 198-212.

Bisht, G., Venturini, V., Islam, S., Jiang L., 2005. Estimation of the net radiation using MODIS (Moderate Resolution Imaging Spectroradiometer) data for clear sky days. *Remote Sensing of Environment*, 97, pp. 52-67.

Boegh, E., Soegaard, H., Hanan, N., Kabat, P., Lesch, L., 1999. A remote sensing study of the NDVI-Ts relationship and the transpiration from sparse vegetation in the Sahel based on high-resolution satellite data. *Remote Sensing of Environment*, 69 (3), pp. 224-240.

Brutsaert, W., 1998. Land-surface water vapor and sensible heat flux: spatial variability, homogeneity, and measurement scales. *Water Resource Research*, 34 (10), pp. 2433-2442.

Cleugh, H.A., Leuning, R., Mu, Q., Running, S.W., 2007. Regional evaporation estimates from flux tow and MODIS satellite data. *Remote Sensing of Environment*, 106, pp. 285-304.

Fan, L., Liu, S., Bernhofer, C., Liu, H., Berger, F. H., 2007. Regional land surface energy fluxes by satellite remote sensing in the Upper Xilin River Watershed (Inner Mongolia, China). *Theoretical and Applied Climatology*, 88, pp. 231-245.

Gillies, R.R., Carlson, T.N., Cui, J., etc., 1997. A verification of the "triangle" method for obtaining surface soil water content and energy fluxes from remote measurements of the Normalized Difference Vegetation Index (NDVI) and surface radiant temperature. *International Journal of Remote Sensing*, 18(15), pp. 3145- 3166.

Irmak, S., Allen, R.G., Whitty, E.B., 2003. Daily grass and alfalfa-reference evapotranspiration estimates and alfalfa-to-grass evapotranspiration ratios in Florida. *Journal of Irrigation and Drainage Engineering-ASCE*, 129(5), pp. 360–370.

Jensen, M.E., Burman, R.D., Allen, R.G., 1990. Evaporation and Irrigation Water Requirement, ASCE Manual No. 70. American Society of Civil Engineers, New York, pp. 332.

Li, R., Min, Q., Lin, B., 2009. Estimation of evapotranspiration in a mid-latitude forest using the Microwave Emissivity Difference Vegetation Index (EDVI). *Remote Sensing of Environment*, 113, pp. 2011-2018.

Jiang, L., Islam, S., 1999. A methodology for estimation of surface evapotranspiration over large areas using remote sensing observations. *Geophysical Research Letters*, 26, pp. 2773-2776.

Leuning, R., Zhang, Y.Q., Rajaud, A., Cleugh, H., Tu, K., 2008. A simple surface conductance model to estimate regional evaporation using MODIS leaf area index and the Penman-Monteith equation, *Water Resources Research*, 44, W10419, doi:10.1029/2007WR006562.

Mu, Q., Heinsch, F.A., Zhao, M., Running, S.W., 2007. Development of a global evapotranspiration algorithm based on MODIS and global meteorology data. *Remote Sensing of Environment*, 111, pp. 519-536.

Nishida, K., Nemani, R.R., Running, S.W., 2003. An operational remote sensing algorithm of land surface evaporation. *Journal of Geophysical Research*, 108 (D9), 4270, doi: 10.1029/2002JD002062.

Roerink, G.J., Su, Z., Menenti, M., 2000. S-SEBI: a simple remote sensing algorithm to estimate the surface energy balance. *Physics and Chemistry of the Earth (B)*, 25, pp. 147-157.

Sandholt, I., Rasmussen, K., Andersen, J., 2002. A simple interpretation of the surface temperature/vegetation index space for assessment of surface moisture status. *Remote Sensing of Environment*, 79, pp. 213-224.

Su, Z., 2002. The Surface Energy Balance System (SEBS) for estimation of turbulent heat fluxes. *Hydrology and Earth System Sciences*, 6, pp. 85-99.

Sun, Z.G., Wang, Q.X., Ouyang Z., Watanabe, M., Matsushita, B., Fukushima, T., 2007. Evaluation of MOD16 algorithm using MODIS and ground observational data in winter wheat field in North China Plain. *Hydrological Processes*, 21, pp. 1196-1206.

Sun, Z.G., Wang, Q.X., Matsushita, B., Fukushima, T., Ouyang, Z., Waternabe, M., 2008. A new method to define the VI-Ts diagram using subpixel information: a case study over a semiarid agricultural region in the North China Plain. *Sensors*, 8, pp. 6260-6279.

Sun, Z.G., Wang, Q.X., Matsushita, B., Fukushima, T., Ouyang, Z., Waternabe, M., 2009. Development of a simple remote sensing evapotranspiration model (Sim-ReSET): algorithm and model test. *Journal of Hydrology*, 376, pp. 476-485.

Thomas, A., 2008. Development and properties of 0.25-degree gridded evapotranspiration data fields of China for hydrological studies. *Journal of Hydrology*, 358, pp. 145-158.

Turner, D.P., Cohen, W.B., Kennedy, R.E., Fassnacht, K.S., Briggs, J.M., 1999. Relationships between leaf area index and Landsat TM spectral vegetation indices across three temperate zone sites. *Remote Sensing of Environment*, 70, pp. 52-68.

Venturini, V., Bisht, G., Islam, S., Jiang, L., 2004. Comparison of evaporative fractions estimated from AVHRR and MODIS sensors over South Florida. *Remote Sensing of Environment*, 93, pp. 77-86.

Verstraeten, W.W., Veroustraete, F., Feyen, J., 2005. Estimating evapotranspiration of European forests from NOAA-imagery at satellite overpass time: Towards an operational processing chain for integrated optical and thermal sensor data products. *Remote Sensing of Environment*, 96, pp. 256-276.

Wan, Z., Zhang, Y., Zhang, Q., Li, Z.L., 2004. Quality assessment and validation of the MODIS global land surface temperature. *International Journal of Remote Sensing*, 25(1), pp. 261-274.

Wang, Q.X., Watanabe, M., Hayashi, S., 2004. Monitoring and simulation of water, heat and CO2 fluxes in various terrestrial ecosystems (in Chinese). *Acta Geographica Sinica*, 59 (1), pp. 13-24.

Wang, Q.X., Watanabe, M., Ouyang, Z., 2005. Simulation of water and carbon fluxes using BIOME-BGC model over crops in China. *Agricultural and Forest Meteorology*, 131(3-4), pp. 209-224.

Watanabe, M., Wang Q.X., Hayashi S., 2005. Monitoring and simulation of water, heat, and CO2 fluxes in terrestrial ecosystems based on the APEIS-FLUX system. *Journal of Geographical Sciences*, 15(2), pp. 131-141.

Zhang, K., Kimball, J.S., Mu, Q., Jones, L.A., Goetz, S.J., Running, S.W., 2009. Satellite based analysis of northern ET trends and associated changes in the regional water balance from 1983 to 2005. *Journal of Hydrology*, 379, pp. 92-110, doi:10.1016/j.jhydrol.2009.09.047.

7

Estimating Actual Evapotranspiration using ALARM and the Dimensionless Temperature

Ayman Suleiman and Jawad Al-Bakri
Land, Water and Environment Department, Faculty of Agriculture
University of Jordan, Amman,
Jordan

1. Introduction

Estimates of evapotranspiration, ET, are needed for many applications in diverse disciplines such as agriculture and hydrology. Many studies of long-term averages have shown that more than half of the net solar energy, and subsequently two thirds of precipitation, goes to ET (Brutsaert, 1982). ET is linked to the land surface energy budget as follows (e.g., Brutsaert, 1982):

$$R_n - G = H + E \tag{1}$$

where R_n (W m^{-2}) is the net incoming radiation, G is the heat flux into the ground (W m^{-2}), and H (W m^{-2}) and E (W m^{-2}) are the sensible and latent (evaporative) heat fluxes into the atmosphere. For the energy balance to close, any part of (R_n - G) that does not contribute to E must be converted into H. In order for that to happen, the surface has to have the temperature (T_s) that forces the energy balance to close. Estimation of H (or ET as a residual) over vegetated terrain is based on an aerodynamic temperature (T_i), which is the temperature that gives the correct value of H at a specified value (denoted $z_{0h,i}$) of the scalar roughness length, z_{oh}, based on Monin-Obukhov Similarity (MOS) theory in the surface sub-layer (Brutsaert, 1982; Stull, 1988). Specification of the value of z_{0h} to give the correct value of H for use with a radiometric surface temperature T_r is a difficult problem (e.g., Mahrt and Vickers, 2004); Crago and Suleiman (2005) outlined a method (discussed here in section 2.a) to specify $z_{0h,i}$ and to convert T_r to T_i. In the MOS theory, the flux is proportional to the difference between T_i and air temperature (T_a), with the ratio H / (T_i-T_a) depending on variables characterizing the atmospheric turbulence and the land surface. This relationship can be expressed as (e.g., Brutsaert, 1982):

$$H = \frac{(T_s - T_a) k u_* \rho c_p}{\left[\ln\left(\frac{z_a - d_0}{z_{0h}}\right) - \psi\left(\frac{z_a - d_0}{L}\right)\right]} \tag{2}$$

where T_s (C°) is the surface temperature, T_a (C°) is the air temperature at a height z_a (m) in the surface sublayer, k (where k=0.4) is von-Karman's constant, u_* (ms^{-1}) is the friction velocity, ρ (kg m^{-3}) is the density of the air, c_p (J kg^{-1} K^{-1}) is the specific heat at constant pressure, z_{0h} (m) is the scalar roughness length for sensible heat, and d_0 (m) is the displacement height. Atmospheric stability, which affects the efficiency of turbulent transport, is included by means of ψ, which is a function of the stability or buoyancy parameter $(z_a-d_0)/L$, where L (m) is the Obukhov length.

Once T_i is known it can be applied to calculate H (equation 2) and then actual ET can be obtained as a residual (equation 1). In one example, the accuracy of regional scale actual ET [obtained as a residual from (1) after finding H using (2), with T_a measured at a single point within the region] is approximately 70-80% (Wang et al., 2005). Models have been developed to improve accuracy through the use of radiometric surface temperature (Hatfield et al., 1983; Ben-Asher et al., 1992; Kustas et al., 2007; Anderson et al., 2007), leaf area index (LAI) (Consoli et al., 2005) and net radiation (Bandara, 2003) available from visible and infrared bands of satellite data. One example of ET modeling from remote sensing data is the Surface Energy Balance Algorithm for Land (SEBAL) which uses an empirical relationship between radiometric surface temperature and the difference between aerodynamic surface temperature and air temperature for each pixel (Bastiaanssen et al., 1998). Several studies (e.g. Menenti and Choudhury, 1993; Bastiaanssen and Chandrapala, 2003; Chandrapala and Wimalasuriya, 2003; Allen et al., 2005; Tasumi et al., 2005; Wang et al., 2005) used and modified SEBAL for spatial estimates of ET with remote sensing and weather data.

In order to implement SEBAL to estimate ET, it must be possible to identify a wet pixel and a dry pixel within an area of interest, and it must be reasonable to assume that atmospheric conditions aloft are horizontally uniform over that area. When one or both of these conditions cannot be met, the Surface Energy Balance Index (SEBI) can be used to calculate relative ET using a ratio of temperature differences (Roerrink and Menenti, 2000). The minimum surface temperature difference is obtained by solving the similarity equation for the minimum sensible heat flux that is found as a residual after determining the potential ET while the maximum surface temperature difference is determined by assuming that ET is zero. Since these bounds are pixel-dependant, potential ET has to be calculated for each pixel.

Suleiman and Crago (2004) developed a dimensionless temperature that does not require information from wet and dry pixels nor potential ET. For each pixel, the maximum surface temperature is determined by assuming that ET for the pixel is zero and the minimum temperature is determined by assuming that sensible heat flux is zero. This approach is advantageous in practice especially when a dry pixel is not available (Qiu et al., 2006). The dimensionless temperature Δ_T can be defined as $(T_i - T_a)/(T_{max} - T_a)$, where T_i is the aerodynamic surface temperature, T_a is the air temperature and T_{max} is the surface temperature that would occur if all the available energy (R_n - G) was converted to sensible heat flux (H) and no evaporation occurred. The dimensionless temperature procedure was used for mapping ET at a local scale with hydrological applications at riparian meadow restoration sites in California, USA (Loheide and Gorelick, 2005).

The Analytical Land Atmosphere Radiometer Model (ALARM) has been developed to convert the radiometric surface temperature T_r to the aerodynamic surface temperature T_i at any view angle (Crago, 1998; Suleiman and Crago, 2002a). ALARM converts radiometric surface temperature measured at any view angle to a well-defined aerodynamic surface temperature (T_i) by correcting for vegetation temperature profile and considering LAI, canopy height, fractional cover, leaf angle distribution, and sensor zenith view angle. ALARM worked well for varied canopy density when the zenith view angle was less than 20° and satisfactorily for view angles greater than 20° (Suleiman and Crago, 2002b; Zibognon et al., 2002). Other models such as Lhomme *et al.* (2000) and Massman (1999) also worked best at near-nadir view angles (Suleiman and Crago, 2002a).

2. Theoretical background

a. ALARM description

Within ALARM, the foliage is assumed to have an exponential vertical temperature profile (Brutsaert and Sugita, 1996) as follows:

$$T_f = T_{fg} + \left(T_{fh} - T_{fg}\right)e^{-b\zeta}, \tag{3}$$

where T_f is the temperature of the foliage at a height z above the soil surface, T_{fh} is the foliage temperature at the top of the canopy, T_{fg} is the asymptotic limit of the exponential foliage temperature profile, far below the bottom of the canopy, $\zeta=(h-z)/h$ is the dimensionless depth into the canopy, h is the canopy height, and b is a decay constant. Qualls and Yates (2001) observed an exponential vertical temperature profile within a grass canopy.

ALARM converts radiometric surface temperature measured at any view angle to a well-defined aerodynamic surface temperature (T_i) by correcting for vegetation temperature profile and considering LAI, canopy height, fractional cover, leaf angle distribution, and zenith sensor view angle as follows:

$$T_i = T_r + \left(T_{fg} - T_{fh}\right)\left(w - W\right). \tag{4}$$

where W is defined below in (7) and w can be derived (Crago, 1998) as:

$$w = \left(1 - f_{soil}e^{-b}\right)\left[\frac{\mu_r b}{g' LAI} + 1\right]^{-1} \tag{5}$$

In (5), $f_{soil}=\exp[-g'(LAI)/\mu_r]$ is the fraction of soil seen by the IRT (Friedl and Davis, 1994), μ_r is the cosine of the view zenith angle, and g' is taken as 0.5 which corresponds to a spherical leaf angle distribution and is representative of a wide range of vegetation types. When $T_{fh}=T_{fg}$, the canopy is isothermal. Under these conditions, Brutsaert and Sugita (1996) showed that the resulting scalar roughness length, $z_{0h,i}$, is given by:

$$z_{0h,i} = z_0 \exp\left[\frac{h}{(h-d_0)r_2} + \ln\left(\frac{h-d_0}{z_0}\right)\right]. \tag{6}$$

where d_0 is zero plane displacement height, z_o is momentum roughness length, and r_2 is defined below equation (7). The "aerodynamic" surface temperature T_i found with (4) is actually the "equivalent isothermal surface temperature" (Brutsaert and Sugita, 1996), or the value of T_s needed in (2) to estimate the correct H using the $z_{0h,i}$ for z_{0h}. Alternatively, T_i is the temperature the surface would require to give the correct sensible heat flux if the canopy was isothermal. The r_2 is given by $r_2= [a- (a^2+4C_2)^{1/2}]/2$ and In (4), W is:

$$W = -(r_2 + b)C_2 / \left[r_2\left(b^2 + ba - C_2\right)\right], \tag{7}$$

In turn, $C_2 =2(LAI)(Ct_f)\, h/[k\,(h-d_0)]$ and Ct_f is the transfer coefficient in the bulk transfer equation for the foliage elements, given by $Ct_f = C_L Re_*^{-m} Pr^{-n}$. The variable a is an exponential decay parameter of eddy diffusivity, Pr is the Prandtl number, and the Reynold's number

appropriate for transport through a leaf boundary layer is $Re_*=u_*L_f/\nu$, where L_f is the characteristic length scale of a leaf and ν is the kinematic viscosity.
In (5), w is a weighting coefficient, describing the importance of T_{fh} and T_{fg} in determining the radiometric surface temperature seen by a radiometer:

$$T_r = wT_{fh} + (1-w)T_{fg} \tag{8}$$

Similarly, W is a weighting coefficient describing the relative importance of T_{fh} and T_{fg} in producing sensible heat flux.
b. ALARM parameterization
The ALARM model has several variables (T_{fh}, *b*, *a*, and d_0/h) that need to be parameterized, all of which have real physical meanings independent of the means of measuring surface temperature. Crago and Suleiman (2005), on a study at different sites with varying LAI, found that the use of a generalized parameterization for these variables at the different sites gave sensible heat flux values comparable to those obtained using localized parameterization. Based on their findings, they recommended the following generalized parameterization for the four variables ((T_{fh}, *b*, *a*, and d_0/h):

$$T_{fh} = T_a \tag{9}$$

where T_a is the air temperature at the top of the canopy.
The parameter *b* controls the rate at which foliage temperature increases with depth into the canopy and was parameterized as a function of LAI:

$$b \;=\; 0.75 \quad for\, LAI \geq 1.87, \tag{10}$$

and

$$b \;=\; 3.7 - 1.58LAI \quad for\, LAI < 1.87. \tag{11}$$

Previous work with ALARM (Zibognon *et al*, 2002; Suleiman and Crago, 2002a and b) suggests that the parameters *a* and d_0/h can influence the estimates of ET. Specifically, larger values of *a* (near 5) effectively confine turbulence and turbulent transport to the top layers of the model canopy, while smaller values (near 0) allow turbulence. They were parameterized as follows:

$$a \;=\; 0.5LAI, \tag{12}$$

and

$$d_0 / h = 0.335a \tag{13}$$

c. Dimensionless temperature
Suleiman and Crago (2004) introduced a dimensionless temperature (Δ_T) as follows:

$$\Delta_T = \left(\frac{T_i - T_a}{T_{max} - T_a} \right) \tag{14}$$

A value of T_{max} can be obtained by solving for T_s in Eq. [2], assuming that H equals (R_n - G). Also T_{max} may be assumed equals to the surface temperature of a hot pixel for heterogonous surfaces. Such an assumption would reduce the number of weather data needed.

The relationship between H and Δ_T is approximately linear, and goes through the origin [assuming the denominator of (2) varies little as T_s goes from T_i to T_{max}]:

$$H = (R_n - G)\Delta_T \tag{15}$$

The relationship between E and Δ_T is:

$$E = (R_n - G)(1 - \Delta_T) \tag{16}$$

and the evaporative fraction EF is:

$$EF = \frac{E}{(R_n - G)} = 1 - \Delta_T = \frac{T_{max} - T_i}{T_{max} - T_a} \tag{17}$$

The dimensionless temperature Δ_T can be found from (14) using ALARM T_i, measured T_a, and T_{max} found as described above. Scaling T_i using the dimensionless procedure reduces the sensitivity of ET estimates to errors in T_a and T_r. The assumption of a constant evaporative fraction, EF = E / R_n, was implemented to extend instantaneous to daily ET because orbiting satellites usually provide coverage only once a day.

In all, there are three major assumptions in the integrated ALARM and dimensionless algorithms. Within ALARM, the foliage is assumed to have an exponential vertical temperature profile. Such an assumption should be generally valid during the satellite pass in the middle of the day because exponential foliage vertical temperature profiles are more evident in the middle of the day. The relationship between H and Δ_T is assumed to be linear, and goes through the origin. This assumption should not result in any serious errors in the estimation of ET especially that the variation of the dominator of (2) is little as T_s goes from T_i to T_{max}. The assumption of a constant evaporative fraction contributes to uncertainty in ET estimates but these uncertainties should be most of the time minimum in semiarid climatic conditions. The implementation of the constant evaporative fraction assumption usually yields accurate daily ET estimates for cloud-free days, as indicated by Zhang and Lemeur (1995). However, variations in cloudiness during the day may produce some uncertainties in ET estimates because clouds at times other than the time when the satellite passes may invalidate the constant-EF assumption. This effect, i.e., varying EF during mid day hours has been found from numerous observations to vary little (e.g., Shuttleworth *et al.*, 1989; Gurney and Hsu, 1990, Brutsaert and Chen, 1996; Crago and Brutsaert, 1996; Lhomme and Elguero, 1999) and it has been found that it depends on the site and time of year (Kustas et al., 1993). In semiarid areas under a wide range of conditions, Kustas et al. (1993) found that the correlation coefficient between midday and daytime EF was rather high (r = 0.92).

Figure1 shows the inputs and steps of actual hourly and daily ET calculations suing ALARM and the dimensionless temperature (ALARMD).

d. FAO-56 and ASCE evapotranspiration

Daily actual ET was obtained from the FAO-56 grass reference ET (ET_o) approach in order to compare it with the ALARM actual ET. Allen et al. (1998; 2005) emphasized that the FAO-56 Penman-Monteith (PM) reference ET (ET_o) would provide reasonable estimates of ET under various climatic conditions. The FAO-56 ET_o was developed for a hypothetical well-watered and actively growing uniform grass of 0.12 m height with a surface resistance of 70 s m^{-1} and an albedo of 0.23 (Allen et al., 1998). The equation for a grass reference crop according to Allen et al. (1998) is defined as follows:

$$ET_o\left(\text{mm d}^{-1}\right) = \frac{0.408\Delta(R_n - G) + \gamma \frac{900}{T+273} u_2\left(e_s - e_a\right)}{\Delta + \gamma\left(1 + 0.34u_2\right)} \tag{18}$$

where R_n is the net radiation (MJ m^2 d^{-1}), G the soil heat flux (MJ m^2 d^{-1}), T the mean daily air temp (°C), u_2 the mean daily wind speed at 2 m height (m s^{-1}), e_s-e_a the saturation vapor pressure deficit (kPa), Δ the slope of the vapor pressure-temperature curve (kPa °C^{-1}), and γ the psychometric constant (kPa °C^{-1}). The terms in the numerator on the right hand of the equation are available energy forcing and air dryness forcing, respectively (Kim and Entekhabi, 1997). Actual ET, ET_a is found from ET_0 as ET_a = $(K_s)(K_c)ET_0$, where K_c is the crop coefficient and K_s is a water stress coefficient (Allen et al., 1998). American Society of Civil Engineers (ACSE) (2005) alfalfa equation is similar to the FAO-56 with different resistance value.

3. Case studies

3.1 Jordan

3.1.1 Validation study

3.1.1.1 Data and methods

A validation study was undertaken using data from the Agricultural Research Station of the University of Jordan (ARSUJ) in the central Jordan Valley at 32° 10′ N latitude and 35° 37′E longitude at an altitude of -230 m (below mean sea level). The station has a warm climate in winter with a minimum temperature of 8.5 °C in January and a hot summer with a maximum temperature of 40.4 °C in July. The yearly average maximum and minimum temperatures are 30.9 and 18.5 °C, respectively, while the yearly mean temperature is 24.7 °C. The experiment site was selected in an alfalfa field where an automated weather station (Campbell Scientific, Logan, UT) was installed. The crop wass irrigated with a sprinkler irrigation system twice to three times a week and planted on a sandy loam soil with good internal drainage. Data collected by the weather station included hourly and daily net solar radiation measured by a NR-LITE-L net radiometer (Kipp & Zonen USA Inc., Bohemia, NY), hourly and daily wind speed at 2 m measured using a R.M. Young wind sentry 03101-5 system (Campbell Scientific, Logan, UT), and air temperature and humidity measured at a height of 2 m using a shielded and aspirated REBS THP. ALARM used MODIS LAI, albedo and T_r along with hourly solar radiation, air temperature, and wind speed and daily and solar radiation.

The soil water content was monitored with TRIME tube access probe (P3, IMKO Micromodultechnik GmbH, Ettlingen, Germany). Five access tubes of 1 m height were installed in the field. The measurements of the volumetric soil water content with the TRIME probe at depths of 0-20, 20-40, 40-60, and 60-80 cm were conducted manually once a day in the morning from March to October 2006. A water balance equation was used to calculate the measured ET using the soil moisture readings as:

$$ET_m = I - D - \Delta W \tag{19}$$

Where ET_m is measured ET (mm d^{-1}), I is irrigation (mm d^{-1}), D is vertical drainage (mm d^{-1}), and ΔW is the change in soil water (mm d^{-1}). Only ET_m for days of no irrigation (I = 0) and zero D (D= 0) were used in this study to minimize the errors of ET_m. Because of the limited number of usable satellite overpasses needed for use with ALARM, a total of twelve days of data were available for the validation study.

3.1.1.2 Results

Results from the validation study in the ARSUJ alfalfa field are shown in Figures 2, 3, and 4. Since the field was irrigated, ET rates ranged from about 6 to about 10 mm day^{-1}. For this range the Root Mean Square Error (RMSE) for ALARM (as compared to the TRIME probe reference values-referred to as "measured" values hereafter) was 0.87 mm day^{-1}, and coefficient of determination (r^2) was 0.36 while the RMSE for ASCE (2005) was 1.25 mm d^{-1} and r^2=0.06.
Errors in the ASCE (2005) and FAO-56 methods for moisture-stressed sites are likely to be dominated by errors in the water stress coefficient. Such errors are likely to be quite large when taken as a percent error of the daily ET, but relatively small in actual magnitude during very dry conditions. Assuming independent random errors equal to 1.25 mm day^{-1} in successive measurements, two daily FAO-56 measurements that differ by less than $(1.25^2+1.25^2)^{1/2}$ mm day^{-1} = 1.76 mm day^{-1} are not far enough apart to rule out random variability. Error magnitudes in the ALARM model are unlikely to vary greatly with the magnitude of the ET rate. Theoretically, EF varies with Δ_T, which is a ratio of two temperature differences . Absolute errors in both T_i and T_{max} are likely to be largest under conditions of high available energy, but under these conditions (T_i-T_a) and (T_{max}-T_a) are likely to be large, so the relative errors are likely to be similar for a wide range of conditions. Experimentally, Suleiman and Crago (2004) applied the dimensionless temperature approach (without the ALARM model) to grasslands under stressed and unstressed conditions, and found similar scatter of estimated to measured ET under high (up to about 5.5 mm day^{-1}) and low (down to about 1.5 mm day^{-2}) daily ET rates. Thus, assuming random and independent errors, it seems reasonable to assume that two daily ALARM measurements of ET that differ by less than $(0.87^2+0.87^2)^{1/2}$ mm day^{-1} = 1.2 mm day^{-1} are not far enough apart to rule out random variability. Finally, if ET_{aa} and ET_{fa} estimates are different by less than $(1.25^2+0.87^2)^{1/2}$ mm day^{-1}= 1.5 mm day^{-1}, they are not far enough apart to rule out random variability.

3.1.2 Comparison study

3.1.2.1 Data and methods

Six weather stations distributed within the different ecological zones in Jordan (Figure 5) were used in this study. The Rwaished site was located within the Saharo-Arabian region. The Safawi and Mafraq sites were located in the Irano-Turanian region. The Aqaba site was located within the Sudanian Penetration while the Irbed and Amman sites were located within the Mediterranean region. Monthly precipitation for the study sites is shown in Table 1. The high variation of monthly and total rainfall amounts from one site to another is apparent in Table 1.
Land use/cover of each site was derived from visual interpretation of high resolution Landsat ETM+ (Enhanced Thematic Mapper plus) images using an onscreen digitizing procedure for each of the 1 × 1 km pixels containing the study sites. Output from land use/cover mapping of each site was used to calculate the weighted average of water stress and crop coefficients (K_sK_c), shown in Table 2, which was required to convert ET_o to actual ET.

Eight-day 1-km leaf area index (LAI), albedo and instantaneous 4-km radiometric surface temperature (T_r) were obtained from the Moderate Resolution Imaging Spectroradiometer (MODIS) Terra instrument for April 2002 for the six sites. Linear interpolation was used to obtain daily LAI from the 8-day LAI. Although a functional relationship between LAI and plant canopy height (h) is vegetation dependent, h was estimated from LAI as h (m) = 0. 2 LAI when LAI was greater than or equal to 0.5 and h (m) = 0.1 when LAI was less than 0.5. The range of LAI and T_r and the number of days for which T_r was available in April for the different sites are shown in Table 3. The LAI was less than 0.5 for Rwaished, Safawi and Aqaba and more than 0.7 for Mafraq, Irbed and Amman. The MODIS T_r was lowest in Irbed and Amman and greatest in Rwaished and Safawi. The number of days for which MODIS T_r was available in April 2002 varied from one site to another. It ranged from 23 occurrences (77 %) at Aqaba to 13 in Mafraq (43 %).

The 2001/2002-season was wet, with few cold days (minimum air temperature of 4.8 C° at Mafraq) and some hot days (maximum air temperature of 34.5 C° at Aqaba and Rwaished) towards the end of April. According to the official climatic records (JMD, 2002; http://www.jmd.gov.jo), the 2001/2002 rainfall season was above average for all sites, except Aqaba. Irbed had the highest precipitation amounts in April, then Amman Airport, and then Mafraq while the other three stations received 0 or about 0 rainfall (Table 1). Wind speed (u) was measured at a height of 2 m four times a day at 12 am, 6 am, 12 pm and 6 pm local time. The *u* at 12 pm (noon) was used in this study since it was the closest to the time of Terra MODIS passes (10 am to 12 pm) over the six sites.

Hourly and daily solar radiation and air temperature at 2 m height were obtained from SoDa ((Solar Data) server (http://www.soda-is.com/eng/index.html) (Wald et al., 2002). The downward solar radiation data were found from the NCEP/NCAR daily values reanalysis. The server makes online query to the archive (1958-onwards) of NCEP/NCAR database for radiation parameters, air temperature and precipitation (National Centers for Environmental Prediction / National Center for Atmospheric research, USA). The server is mirrored by Institute Pierre-Simon Laplace, France. The hourly and daily net longwave irradiance was calculated using Allen et al. (1998) procedure. The heat flux into the ground (G) varies during the day and it depends on many factors such as leaf area index and soil moisture. However, it is strongly correlated with net radiation and it is often assumed to be 30% of the net radiation (e.g., Santanello and Friedl, 2003). Hence, in this study for simplicity the same assumption was employed. In open range areas, K_s was assumed to be 0.5 to account for the low soil moisture availability that resulted from limited rainfall amounts at the sites that have open range areas. For all other vegetation types, K_s was set equal to 1.0 throughout the study because green urban and irrigated olives areas were irrigated and because it was a wet year in Irbed. Following the recommendations and using the equations of Allen et al. (1998), the vapor pressure needed for ET_0 (equation 18) was found by assuming the dew point temperature for each day was equal to the minimum temperature for that day.

3.1.2.2 Results

Air and MODIS radiometric surface temperatures in April 2002 for the different sites are shown in Figure 6. The two highest radiometric surface temperatures and the highest differences between radiometric surface and air temperatures were seen in Rwaished and Safawi. The difference between radiometric surface and air temperature was lowest in Irbed where in some instances it was negative. Dimensionless temperature (Δ_T) was greater than 0 for all the days for all the locations except Irbed, which had 6 instances of negative Δ_T, two of which were < -0.50 (Figure 7). The influence of warm advection at Irbed will be discussed later. The positive Δ_T values ranged from 0.09 for Irbed to about 0.80 for Rwaished and Safawi.

The maximum ALARM actual ET, ET_{aa} ranged from 2.4 mm d^{-1} in Safawi to 6.5 mm d^{-1} in Irbed while the maximum FAO-56 actual ET, ET_{fa} ranged from 1.7 mm d^{-1} in Mafraq to 7.1 mm d^{-1} in Irbed (Table 4). The minimum ET_{aa} ranged from 0.7 mm d^{-1} in Safawi to 1.5 mm d^{-1} in Mafraq while the minimum ET_{fa} ranged from 0.7 mm d^{-1} in Aqaba to 3.7 mm d^{-1} in Irbed. The average of ET_{aa} and ET_{fa} were almost identical in Amman, within 0.7 mm d^{-1} in Rwaished, Safawi, Mafraq, and Aqaba and greater than 1 mm d^{-1} in Irbed (Table 4). These results indicate that generally there is a good agreement between minimum, maximum, and average ET_{aa} and ET_{fa} at all sites but Irbed for which the minimum and consequently the average ET_{aa} and ET_{fa} were different.

At three of the sites (Rwaished, Safawi and Amman), ET_{aa} and ET_{fa} were similar throughout April (Figure 8). At the Amman site, ET_{aa} appeared to fluctuate from one day to another throughout April. However, ET_{aa} and ET_{fa} from DOY 95- 104 both fluctuated over a wider range than they did from DOY 108- 120. In the case of ET_{fa}, the fluctuations were smaller than 1.76 mm day^{-1}, so random variations cannot be ruled out. The ET_{aa} fluctuations were on the order of 2 mm day^{-1}, which suggested they cannot be explained by random variations. One possible explanation is that soil moisture was reduced after DOY 104 and was unable to supply sufficient water to allow for high ET after this date. The drop in ET_{aa} near DOY 105 at Aqaba, which was also greater than 1.2 mm day-1, might have a similar explanation.

In Mafraq and Aqaba , ET_{aa} and ET_{fa} were close on most days. However, on some days there were larger differences. ET_{aa} was consistently a little higher than ET_{fa} in Mafraq and on most days in Aqaba. In Irbed, ET_{aa} and ET_{fa} were close on many days while ET_{aa} was lower than ET_{fa} on several days indicating that water stress may have taken place on these days. Due to its use of radiometric surface temperatures ET_{aa} is expected to respond automatically to water stress, while ET_{fa} can only respond by changing K_s. On the other hand, on four days ET_{aa} was greater than ET_{fa} indicating that warm advection was present. For five days, ET_{fa} was greater than 6 mm $d^{-1}$and for day 111 ET_{aa} was more than the reference ET (ET_o).

The ET_{aa} and ET_{fa} were much lower than ET_o for all the sites but Irbed. Generally, a good agreement was observed in the arid and semiarid sites which were utilized as open rangeland and in Amman where the irrigated area was 21%. On the other hand, little agreement was observed in Mafraq where 70% of the vegetation was denser due to the large fraction of protected rangeland and in Irbed where 75% of the site was cultivated, mainly with field crops (Table 2). For the sites of Safawi and Amman, the agreement between ET_{aa} and ET_{fa} was obvious and both curves nearly coincided after the day 105 (Figure 8). A similar trend with less agreement was observed in Rwaished and Aqaba.

In general, ET_{aa} had a well-defined linear relationship with Δ_T with some fluctuations due to the availability of energy (R_n-G) for ET (Figure 9). This relationship, which is mathematically described in equation (16), demonstrated that the dimensionless temperature and R_n-G are responsible for actual ET determination. Although R_n-G is the main factor that influences potential ET, Δ_T is a measure of the actual ET. Many points from the Irbed site deviated from the line because of the sensible heat advection, which resulted in higher actual ET_{aa} at very low (even negative) Δ_T.

Ideally, Δ_T ranges from 0 when ET is maximum ($ET=R_n-G$) to 1 when ET is minimum (ET=0). However, negative Δ_T can be found when warm advection takes place and the air temperature becomes higher than the aerodynamic surface temperature. The warm air in this case acts as another source of energy through downward (negative) sensible heat flux. A low value of Δ_T may suggest that a soil has sufficient moisture available to meet ET demands while a high value of Δ_T would indicate limited soil moisture availability. That is why Irbed, which received a relatively high rainfall amount, had the lowest Δ_T values and Rwaished and Safawi, which did not receive much rain, had the highest Δ_T values (Table 4). Continuous increase of Δ_T is a clear signal of decreasing water availability and increasing water stress. This coincided with the trend of herbaceous vegetation growth in Mafraq, where the green flush continued for a short period and quickly reduced the available soil moisture. Under such conditions, both soil water and vegetation type and density would likely affect the estimate of actual ET_{fa} from the FAO-56 equation (through the coefficients K_s and K_c, respectively). This was most clearly seen in Irbed and Mafraq where ET estimates from ALARM and FAO-56 were closest in periods following rainfall (Figure 7). Actual ET from both estimates was in relatively good agreement for the other sites that were less vegetated and were either urbanized or grazed rangelands.

Fluctuations of Δ_T from one day to another could also be due to the variability of the net solar radiation from one day to another. This was an expected behavior in all sites as the 1 km spatial resolution included a mixture of land use/covers. The presence of large increase and then decrease in Δ_T during drying periods (e.g., Mafraq around day 103) could be due to either a non-recorded small rain event that increased Δ_T in one day and then in the next Δ_T came back to where it was or that the high Δ_T was an outlier for some reasons. Nevertheless, a Δ_T value of about 0.35 to 0.40 suggests that potential ET was taking place while a Δ_T value of about 0.75 to 0.80 implies that soil evaporation is not contributing significantly to ET because of a dry soil surface and transpiration is not meeting the evaporative demand. Any Δ_T that is significantly less than about 0.30 is likely due to sensible heat advection because when Δ_T is about 0.30, ET seems to be close to potential ET.

The decreasing trends of actual ET may reflect drydown cycles of soil moisture due to ET and deep percolation. The contribution of deep percolation to the loss of soil water cannot be assessed without monitoring or simulating the change of the soil moisture profile due to drainage (Suleiman, 2008). The values of actual ET_{aa} and its range seemed reasonable although no site-specific calibration has been done. The trend of increasing ET_{aa} from DOY 100 to 104 in Aqaba could be attributed to the increase of air temperature during this period and the irrigation activities to the south of the site. The decrease of ET after that period was attributed to loss of soil water from the surrounding areas.

Because of the warm advection and the high wetness conditions the maximum and average actual ET were higher in Irbed than all the other sites (Table 4). The advection in Irbed is a well-known phenomenon that usually results in crop failure, particularly for lentil crops (NCARE, 2002a). Comparing the daily air temperature records (NCARE, 2002b) with the long-term averages (JMD, 2002) showed that the maximum air temperature was 7 C° above the average during this period. During the periods of advection the maximum air temperatures ranged from 29 to 31 C°, compared with an average value of less than 21°C. These results concurred with Rosenberg and Verma (1978) who found, in a study on irrigated alfalfa, that ET was greater than 10 mm d^{-1} on one-third of the days of studies. They attributed these high ET values to the contribution of sensible heat advection to ET. In this study, the effect of advection on ET estimates was observed by ALARM and MODIS data while potential ET equations such as Priestley-Taylor are not capable of predicting such effects because they are developed for minimally-advective conditions (e.g., Brutsaert, 1982). The interaction of soil water and climatic factors resulted in high ET values in Irbed. Suleiman and Ritchie (2003) demonstrated that soil evaporation could be more than 10 mm d^{-1} when the saturated soil layers are close to the surface. This was possible in Irbed as the area had heavy clayey soils (MoA, 1994) known as vertisols. These soils are well known for their high water holding capacity. Therefore, higher values of ET were obtained for this site, as water from heavy rainfall was stored in the soil. However, the other sites were characterized by arid soils that received low rainfall amounts and subsequently lower ET values were noticed at these sites. Therefore, the impact of soil water availability and advection on the estimation of actual ET would suggest that ALARM might be able to provide a more accurate estimate of ET than the FAO-56 approach which assumed no water stress (Allen et al., 1998)

3.1.3 California case

3.1.3.1 Data and methods

Atmospherically corrected 1-km leaf area index (MOD11A1), 8-day albedo (MCD43B3),and instantaneous radiometric surface temperature (T_r) ((MOD15A2), were obtained from the Moderate Resolution Imaging Spectroradiometer(MODIS) and atmospherically corrected 90-m T_r (AST08) and 15-m surface reflectance product (AST07) for July 5th, 2008. These images were for a fully-irrigated alfalfa field located in Blythe, California (33° 28' 8"N latitude and 114° 42' 25"W longitude). Hourly wind speed was obtained from a nearby automated weather station.

The observed ET_c was determined from the surface renewal meteorological method (Snyder et al., 1996). The surface renewal system consisted of a Q7.1 net radiometer, two soil heat flux plates (Radiation and Energy Balance Systems, Inc.), and one set of spatially averaging soil temperature probes to measure net radiation, soil heat flux and soil heat storage, respectively. Two fine wire thermocouples (0.076 mm diameter) (Campbell Scientific, Inc.) mounted 1.5 m above the ground surface to measure air temperature fluctuations. The sensors were connected to CR10X data logger (Campbell Scientific, Inc.). Data processing procedures to determine the sensible heat flux density can be found in Snyder et al. (1996). The latent heat flux was then determined from the energy balance equation. The latent heat flux was lower but close to the net radiation from 8:00 am to 2:00 pm after which and until the end of the day, the latent heat flux was somewhat higher than the net radiation due to advection of heat. The sensible heat flux was negative (downward) throughout the day while the ground heat flux was negative (upward) at nighttime and positive at daytime.

To obtain ALARM-D and SEBAL hourly and daily ET_c, the surface renewal net radiation and ground heat flux were used. For SEBAL, The wet and dry pixels were assigned to the lowest and highest radiometric surface temperatures pixels, respectively. The sensible heat flux was assumed 0 for the wet pixel and the latent heat flux was assumed 0 for the dry pixel. The assumption of a constant evaporative fraction, $EF = E / R_n$, was implemented to extend instantaneous to daily ET_c because orbiting satellites, i.e., Terra-MODIS provides a coverage once in the daytime which was about 11:30 am on July 5th, 2008. The evaporative fraction has been found from numerous observations to vary little during the daytime (Crago and Brutsaert, 1996). The EF on July 5th, 2008 was fairly constant at about 1.08 from 7:00 am in the morning to 2:00 pm after which it increased to about 1.35 by 7:00 pm (Figure 10).

3.1.3.2 Results

For the four ASTER 90-m pixels that have somewhat uniform vegetation, ALARM-D overestimated hourly ET_c in three pixels by 0.06 to 0.13 mm hr^{-1} and underestimated in one pixel, the pixel with maximum radiometric surface temperature, by 0.01 mm hr^{-1} while SEBAL underestimated hourly ET_c for all the pixels by 0.07 to 0.12 mm hr^{-1} (Table 5). On average ALARM-D overestimated hourly ET_c by 0.07 mm hr^{-1} whereas SEBAL underestimated hourly ET_c by 0.09 mm hr^{-1}. The relative error of hourly ET_c ranged from -1.1 to 14.9% for ALARM-D and -8 to -13.8% for SEBAL. Pixels that had relatively low absolute relative error for ALARM-D had relatively high absolute relative error for SEBAL and vice versa. Except for one pixel, the range of absolute relative error was comparable for ALARM-D and SEBAL.

For the ASTER 90-m pixels, ALARM-D overestimated daily ET_c in two pixels by 0.06 mm day^{-1} and underestimated in two pixels by 0.01 to 0.7 mm day^{-1} while SEBAL underestimated daily ET_c for all the pixels by 1.2 to 1.7 mm day^{-1} (Table 6). On average ALARM-D overestimated daily ET_c by 0.1 mm day^{-1} whereas SEBAL underestimated daily ET_c by 1.4 mm day^{-1}. The relative error of daily ET_c ranged from -8.1 to 7% for ALARM-D and -14 to –19.8% for SEBAL. The absolute relative error for ALARM-D was less than half of that for SEBAL for all the pixels. Having evaporative fraction higher towards the end of the day (after 2:00 pm) than the fairly constant evaporative fraction during the other hours (Figure 11) was the main reason why the ALARM-D daily ET_c overestimation was less than for the hourly ET_c.

For 1-km pixels at which the alfalfa field is spread, the difference between MODIS ALARM-D and SEBAL daily ET_c ranged from 0.2 to 1.2 mm day^{-1} (Table 7) while the difference between upscaled ASTER ALARM-D and SEBAL daily ET_c ranged from 0.7 to 2 mm day^{-1} (Table 8). That demonstrated that ALARM-D daily ET_c was higher than SEBAL daily ET_c for all the 1-km pixels. ASTER ALARM-D and SEBAL daily ET_c was greater than MODIS ALARM-D and SEBAL daily ET_c and closer to the surface renewal daily ET_c. Discrepancies between 1-km pixel and surface renewal daily ET_c are expected because of the variability in crop parameters such as leaf area index, soil moisture availability and microclimate.

4. Application of ALARM to monitor drought

4.1 Drought monitoring with earth observations

Drought is an extended period of deficiency in water supply results in a region due to insufficient precipitation, high evapotranspiration and overexploitation of water resources. Drought has adverse effects on the socioeconomic, agricultural, and environmental conditions. Jordan, located between 29° 11' N and 33° 22' N latitudes, and between 34° 19' E and 39 ° 18′ E longitudes in the eastern Mediterranean region, is among the world's ten most water-limited countries. Most of the country's land is arid with low rainfall and high potential evaporation (Figure 12). The occurrence of drought in the country is highly variable through time and space depending on the actual rainfall, which falls between December and March, with high variability in intensity and duration of the individual showers. The agricultural sector, which represents a source of income for more than 15 percent of the population, continues to grapple with challenges of scarce water and the frequent droughts. Therefore, monitoring of drought is one of the top priorities for rainfed agriculture.

Different methods of drought monitoring have been developed worldwide. The methods are based on drought-causative and drought-responsive parameters such as rainfall, soil moisture, potential evapotranspiration, and vegetation condition. The interest in the use of remote sensing data of different earth observation systems (EOS) started to increase in the last two decades. This could be attributed to the recent developments in these contemporary sources of information and to their availability in near real time. The use of remote sensing was also enhanced by the improved spatial and temporal resolutions, which could not be provided by the meteorological data. Among the most important sources of EOS data were those of MODIS land data (http://modis.gsfc.nasa.gov/), which could be downloaded from the web (https://wist.echo.nasa.gov/api/), free of charge. The free products of MODIS encouraged researchers and scientists to use the data for monitoring land resources of soil, vegetation and water. Recently, NASA launched the Land Atmosphere Near-Real-Time Capability for EOS (LANCE), which was designed to provide EOS data products within 3 hours of acquisition. The system was developed for near real-time applications and would be particularly relevant for agricultural monitoring. Access to LANCE data can be obtained at http://lance.nasa.gov/.

One of the most important products from EOS was the normalized difference vegetation index (NDVI), which was provided by the advanced very high resolution radiometer (AVHRR) in the 1980's and 1990's, followed by SPOT vegetation instrument in the 1990's and finally by MODIS in year 2000 and afterward. Two previous studies in Jordan (Al-Bakri and Taylor, 2003; Al-Bakri and Suleiman, 2004) showed that NDVI images were useful in tracing vegetation response to cumulative rainfall. Therefore, a drought monitoring unit at the National Center for Agricultural Research and Extension (NCARE) started to use MODIS-NDVI images for monitoring drought (Al-Naber et al., 2009). The implementation of NDVI images to monitor drought was based on comparing composite images of NDVI with the average of the time series for the particular period. This approach, however, was based on the assumption of normal distribution of NDVI which reduced the reliability of the outputs (Al-Naber et al., 2009). Alternatively, ALARMD was used to derive actual ET from MODIS and to calculate the Evapotranspiration Water Stress Index (EWSI) and to produce GIS maps that could be used to assess the severity of drought in Jordan. The EWSI was calculated as following:

$$EWSI = 1 - \left(\frac{ETa}{ETc} \right) \tag{20}$$

Where EWSI is evapotranpiration water stress indicator and ranges from 0 to 1 and ETa and ETc are daily actual and crop evapotranspiration.

4.2 Case study in Yarmouk basin in Jordan

The study area covers approximately 1400 km^2 and is characterized by an obvious rainfall gradient (Figure 12). The western parts of area are mainly rainfed where olive trees and wheat are planted, in addition to other crops of fruit trees and summer crops of vegetables. The eastern parts are extending in the low rainfall zone of the country where open grazing is practiced and rainfed barley is cultivated to support the grazing herds. Irrigation is taking place in different parts of the area using groundwater. The study area is extremely important to Jordan in that it is the main source of surface water at the present time that discharges through the Yarmouk River. Also, it forms an important part of the rainfed area in the north of Jordan.

For the computation of Evapotranpiration Water Stress Indicator (EWSI) the actual daily evapotranspiration (ETa) was determined using ALARMD model. Some input data for ALARMD were obtained from satellite images while the required weather data were acquired from weather stations. The MODIS products that were used in ALARMD to calculate ETa were: Land Surface Temperature (LST) from the Daily L3 Global 1 km Grid product (MOD11A1), the 16-day albedo product (MCD43B3) describing both directional hemispherical reflectance (black-sky albedo) at local solar noon and bihemispherical reflectance (white-sky albedo), The level-4 MODIS global Leaf Area Index (LAI) every 8 days (MOD15A2), and MODIS land cover product (MCD12Q1) derived from observations spanning a year's input of Terra and Aqua data with 17 land cover classes defined by the International Geosphere Biosphere Programme (IGBP).

Different image processing techniques and algorithms were used to prepare the data downloaded from MODIS website. Images were re-projected from Sinusoidal projection to geographic coordinates and resampled to a pixel size of 1 km. The output files were then clipped to the boundaries of the study area to reduce the size of files and to remove extraneous data. Daily weather data, on the other hand were used to calculate ETo (eq.18). MODIS land cover map was used to derive crop coefficients needed to calculate ETc.

Because of the difficulty of using the image processing modeler, images were exported from image processing software and were then imported and arranged in spreadsheets. Data of MODIS LST, passing time, view angle, LAI and Albedo along with solar radiation, air temperature, wind speed and plant height were inserted in spreadsheets. The model was applied on the data to calculate EWSI values for the corresponding data. Outputs were then exported, on the form of point data, to geographic information system (GIS) to generate maps of EWSI using bilinear interpolation method. An example will be discussed for the image of July 1st, 2009.

4.3 Initial results and future work

Output from the EWSI maps showed that the extent of drought was variable inside the study area. Within the study area, EWSI ranged from 0.56 to 0.84, which indicated that actual evapotranspiration was lower than the reference evapotranspiration. This could indicate the end of season for rainfed crops. Comparing the result from EWSI and the map of NDVI deviation, more details on drought could be obtained from the former. The trend of EWSI could be attributed to the land use/cover of the study area. For example, the high values were noticed in the middle parts of the study area, where soils were poor and land was sparsely vegetated and tended to be overgrazed by herds of sheep. The relatively lower values of EWSI observed in the remaining other parts of the study area could be attributed to three factors. The first factor was the presence of clay soils (Vertisols) in the west which resulted in more soil moisture storage that increased the ALARMD ETa. The second factor which could explain the EWSI values in the northern parts of the study area was irrigation, which was practiced in these areas. The third factor which explained EWSI was the relatively low temperatures in the southern parts of the study area which were characterized by high altitudes and lower air temperatures than the other parts.

Unlike EWSI, the NDVI maps were not able to detect patterns and spatial variation of drought in the study area (Figure 13). The NDVI map showed that most of the study area did not deviate from the 10-years mean of NDVI. The eastern parts of the study area showed higher NDVI values than the average. In terms of drought, the NDVI map was less-informative when compared with EWSI. This could be attributed to the fact that NDVI values were neither correlated with land use nor with vegetation parameters. Also, the range of NDVI deviation above which drought would occur was not identified. The EWSI, on the other hand, considered the most important vegetation parameters (LAI, crop type and Kc) in drought monitoring. These initial results could indicate the usefulness of ALARMD in modeling ETa and in mapping EWSI. Future work, however, is still needed to correlate EWSI to the severity and temporal distribution of drought. Also, the use of high spatial resolution data to improve ETc estimation and EWSI maps shall be investigated.

5. Acknowledgement

This publication and the Evapotranspiration Water Stress Index (EWSI) work were supported by the NATO's Science for Peace Program, project SfP-983368 (2009-2012) "Assessment and monitoring of desertification in Jordan using remote sensing and bioindicators". The Deanship of Academic Research at the University of Jordan supported the validation study in Jordan. The authors extend their thanks to Sari Shawash for processing MODIS data.

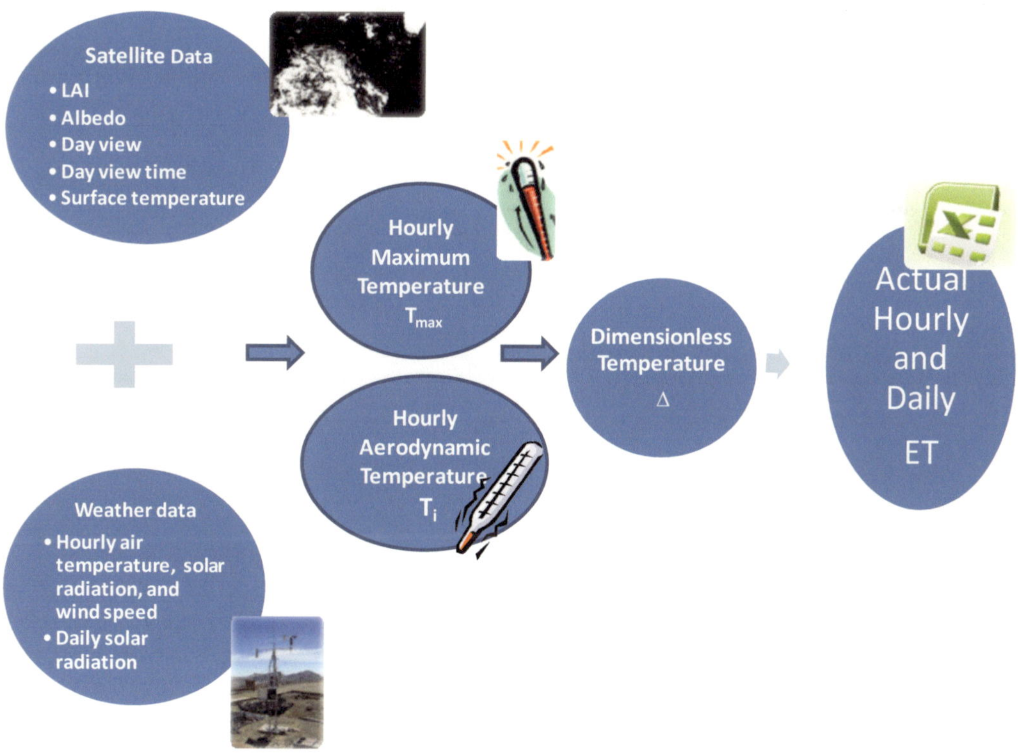

Fig. 1. ALARM and dimensionless temperature (ALARMD) calculations of actual hourly and daily evapotranspiration.

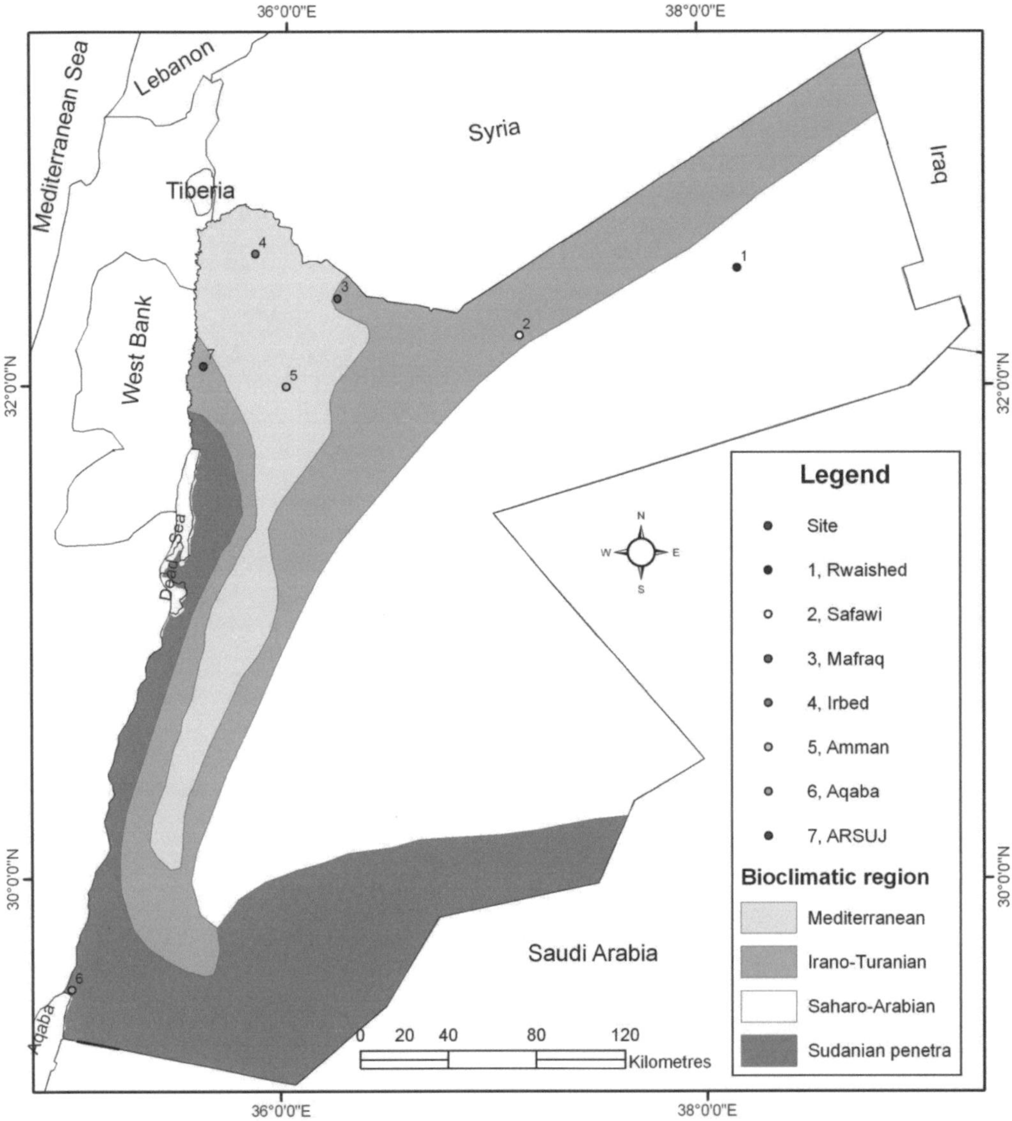

Fig. 2. Locations of the validation and comparison sites in Jordan.

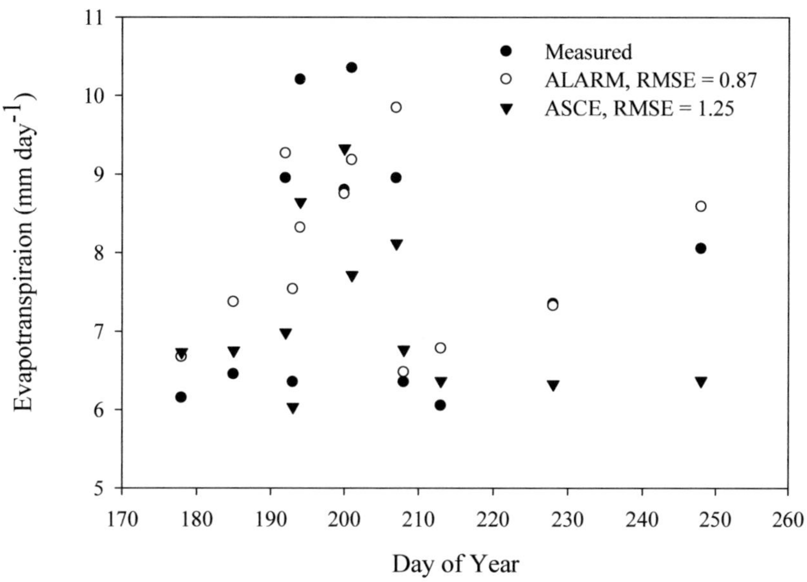

Fig. 3. Measured, ALARMD and ASCE evapotranspiration at the validation site.

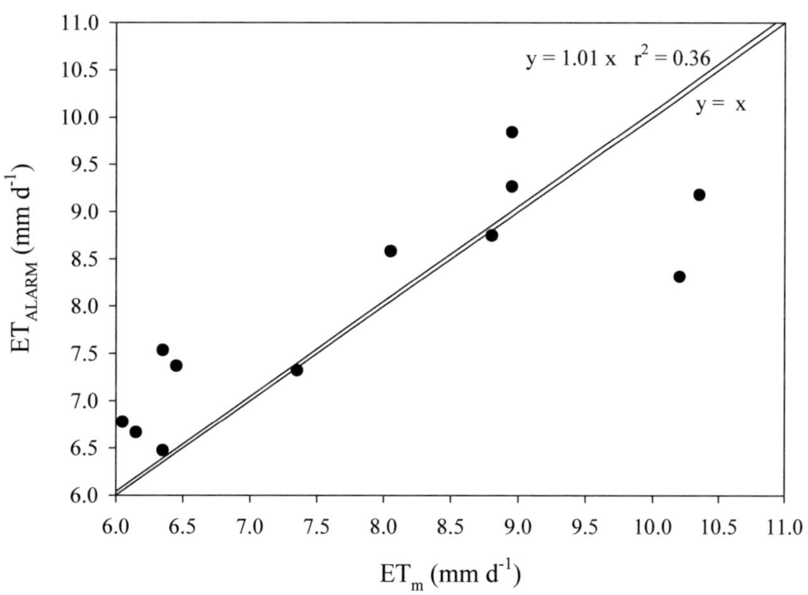

Fig. 4. ALARMD vs. measured evapotranspiration at the validation site

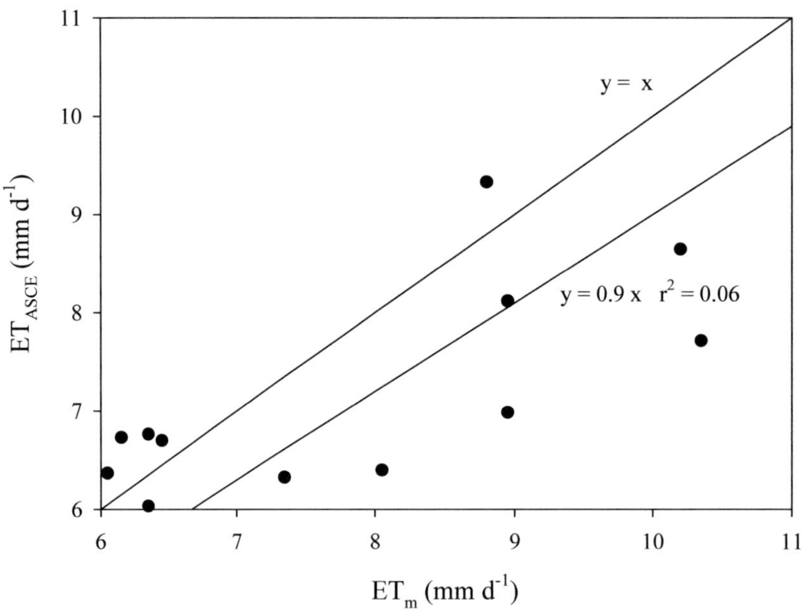

Fig. 5. ASCE vs. measured evapotranspiration at the validation site

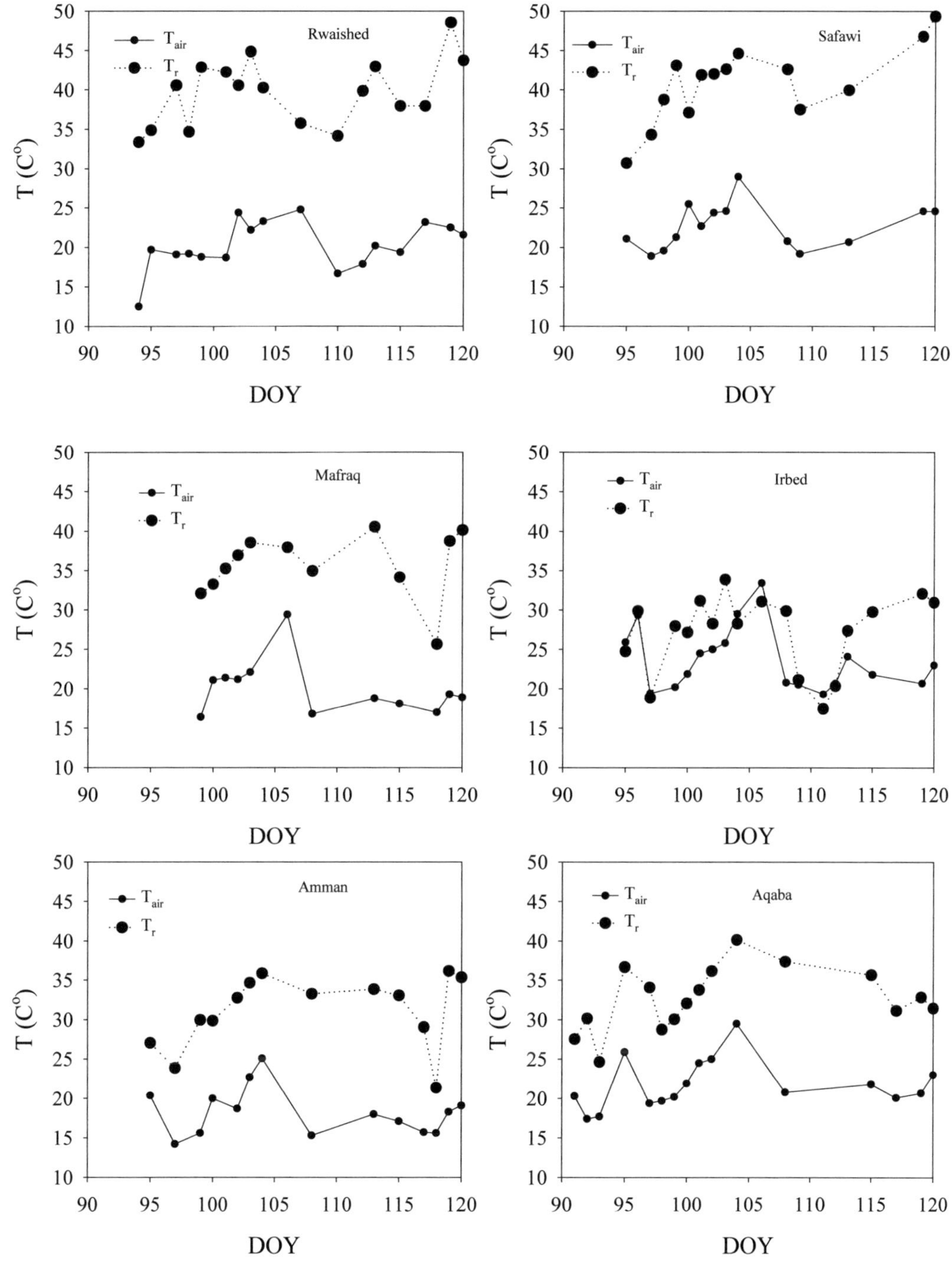

Fig. 6. Air (T_{air}) and radiometric surface temperature (T_r) on different days of year (DOY) during April 2005 for the six sites.

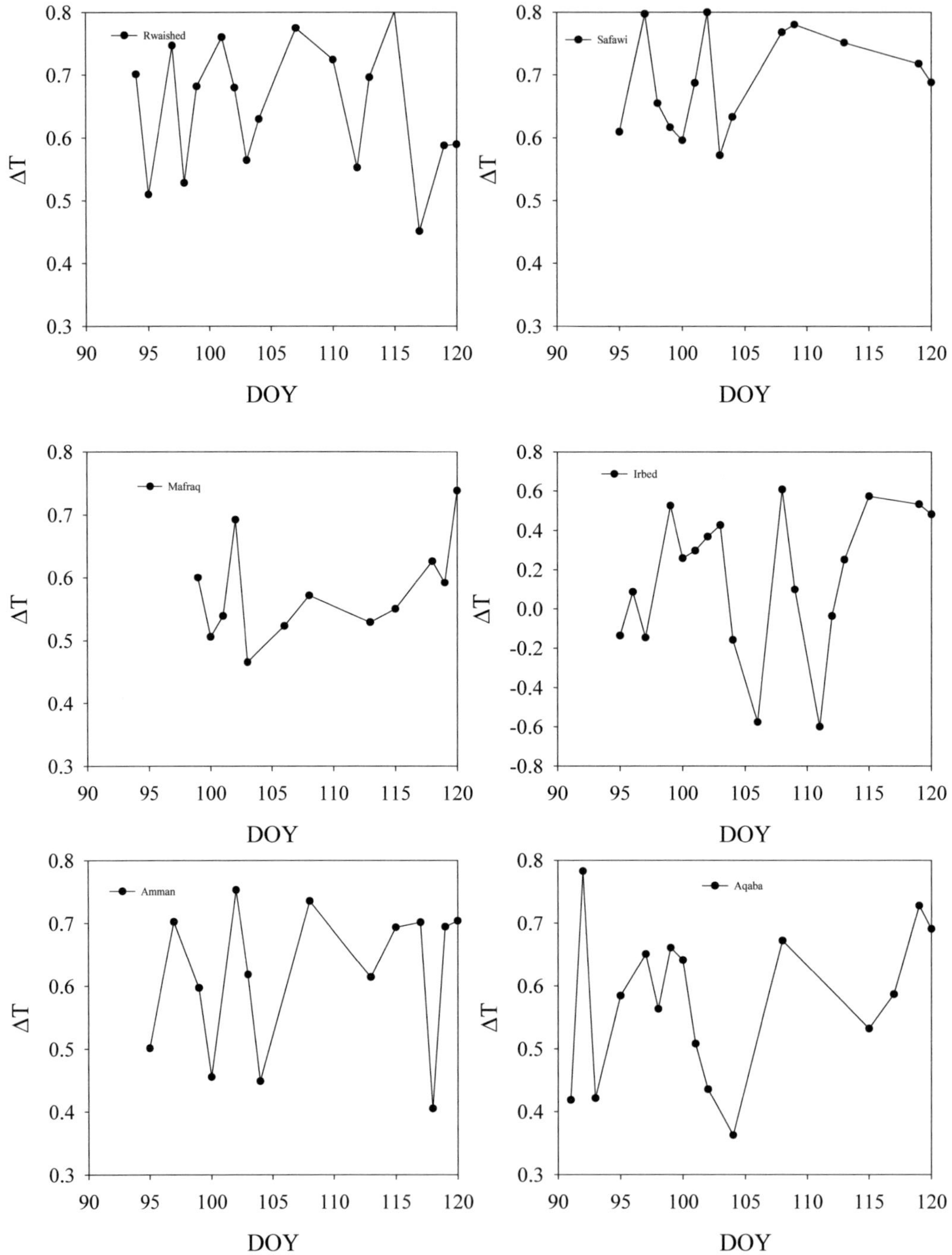

Fig. 7. Dimensionless temperature (Δ_T) on different days of year (DOY) during April 2005 for the six sites.

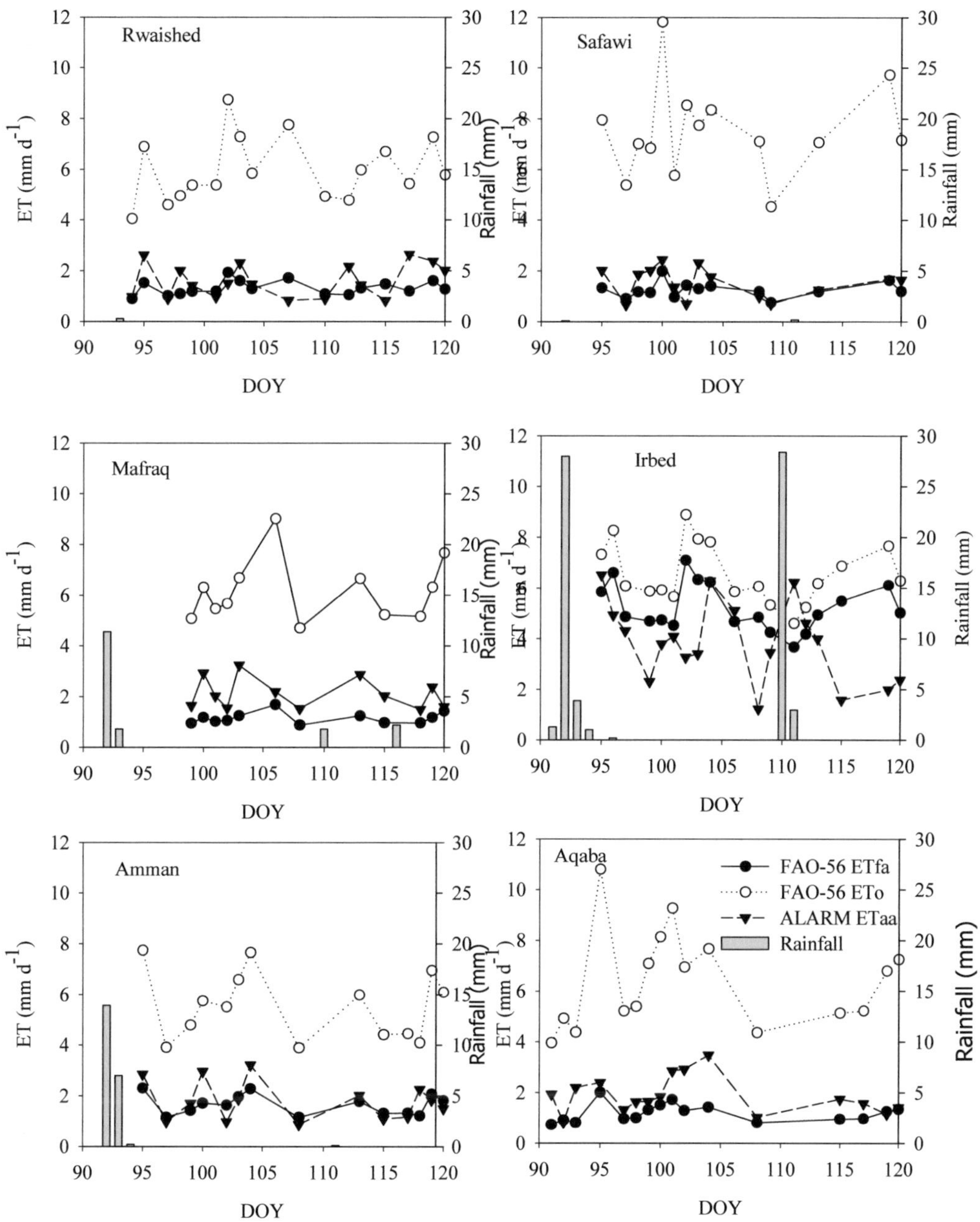

Fig. 8. Daily actual ALARMD (ET_{aa}), FAO-56 (ET_{fa}) and FAO-56 grass reference (ET_o) evapotranspiration and daily rainfall amounts on different days of year (DOY) April 2005 for the six sites.

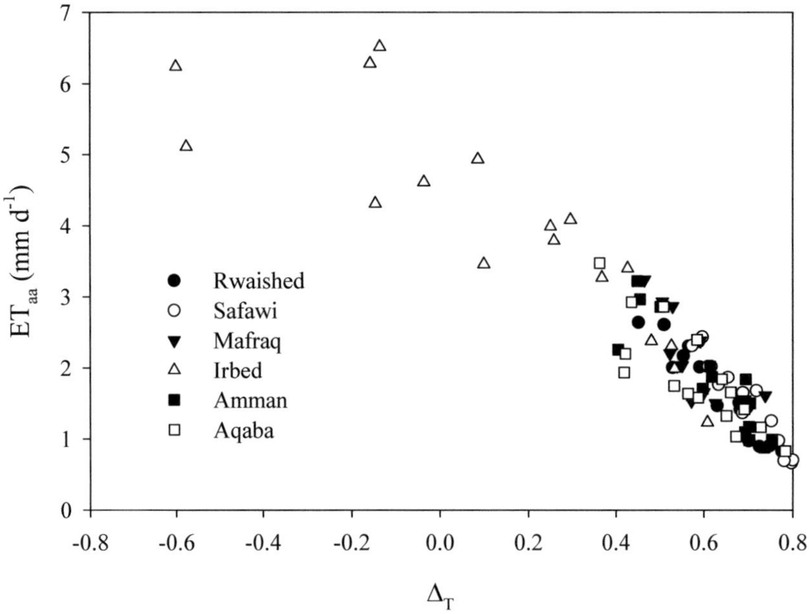

Fig. 9. Relationship between the ALARMD daily actual evapotranspiration (ET_{aa}) and the dimensionless temperature (Δ_T).

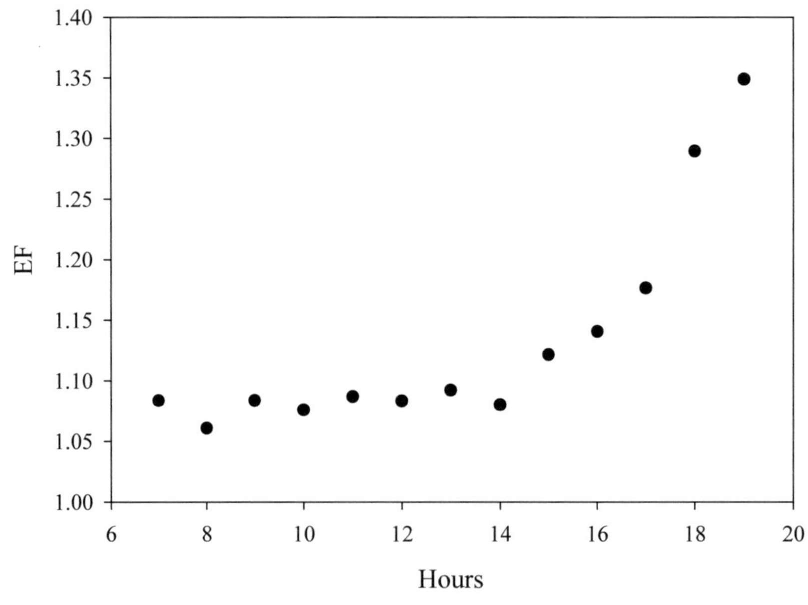

Fig. 10. Hourly evaporative fraction (EF) on July 5th, 2009 in Blythe, California.

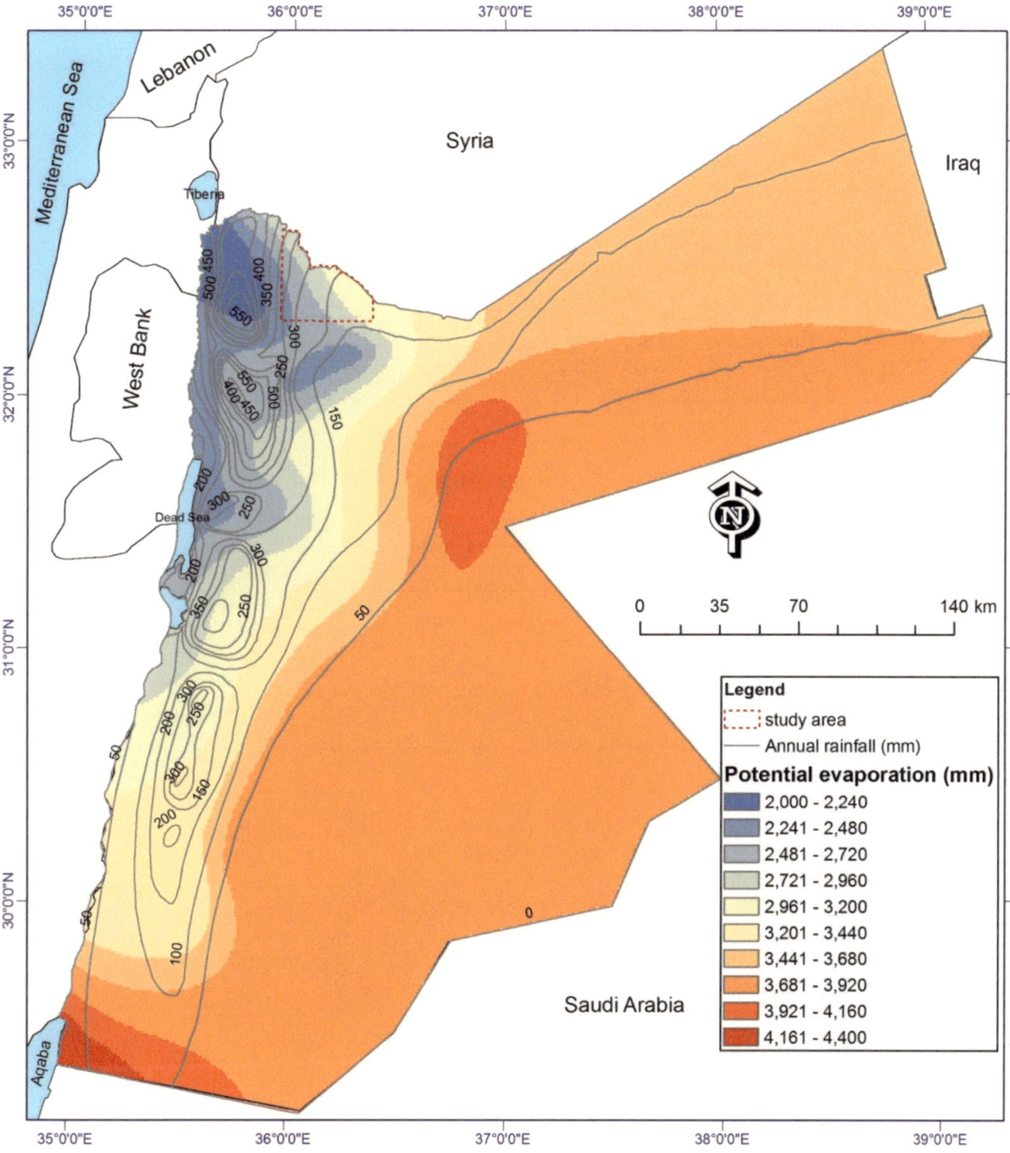

Fig. 11. Mean annual rainfall and potential evaporation in Jordan (1971-2005).

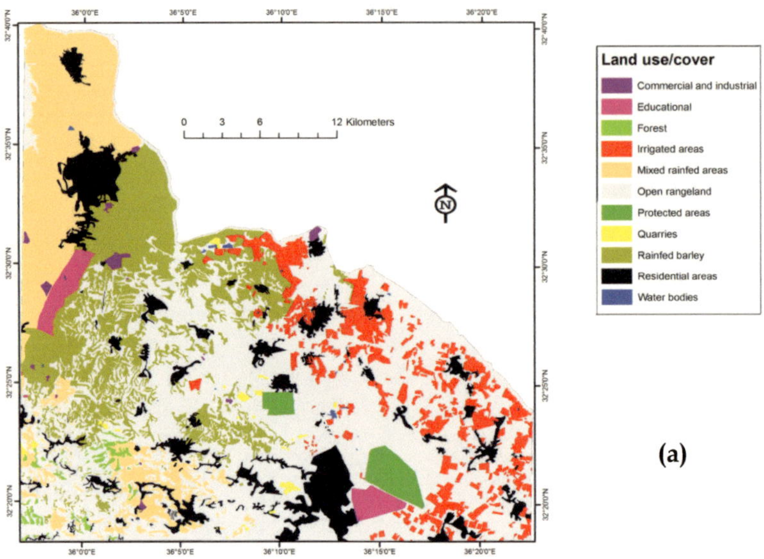
Land use/cover
Commercial and industrial
Educational
Forest
Irrigated areas
Mixed rainfed areas
Open rangeland
Protected areas
Quarries
Rainfed barley
Residential areas
Water bodies
0 3 6 12 Kilometers
N

(a)

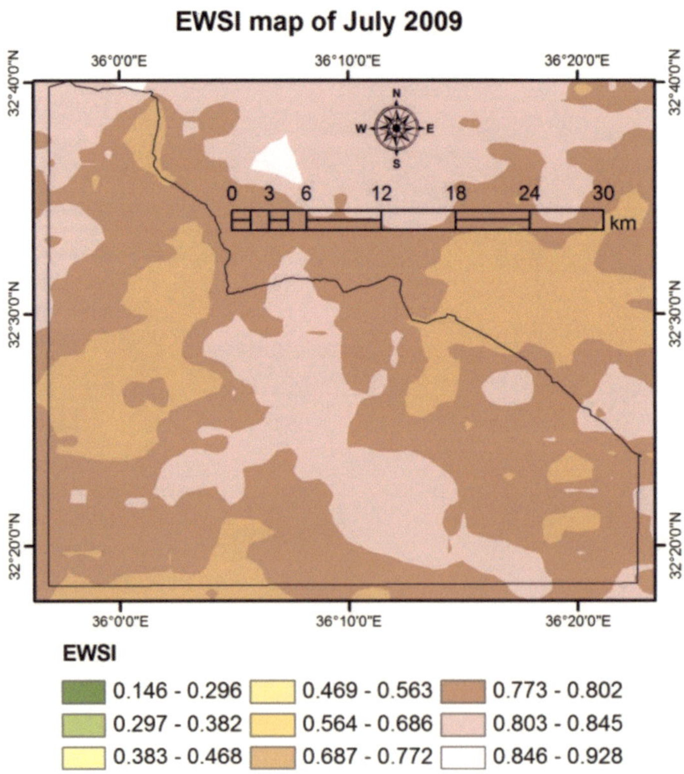
EWSI map of July 2009
36°0'0"E
36°10'0"E
36°20'0"E
32°40'0"N
32°30'0"N
32°20'0"N
N
W
E
S
0 3 6 12 18 24 30
km
EWSI
0.146 - 0.296
0.297 - 0.382
0.383 - 0.468
0.469 - 0.563
0.564 - 0.686
0.687 - 0.772
0.773 - 0.802
0.803 - 0.845
0.846 - 0.928

(b)

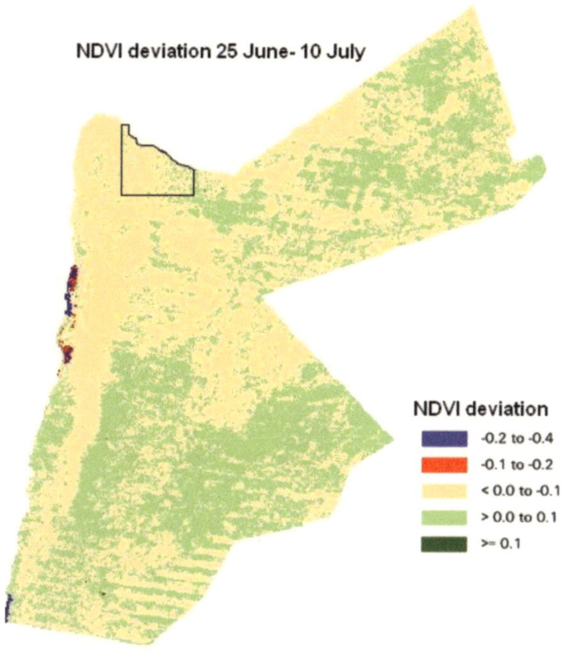

(c)

Fig. 12. Maps of land use/cover map of Yarmouk basin study area (a), EWSI of 1 July (b), and NDVI deviation map for the period 25 June-10 July for Jordan and the study area (c).

	Oct.-Nov. 2001	Dec. 2001	Jan. 2002	Feb. 2002	Mar. 2002	Apr. 2002	Total	% of Annual Average*
					(mm)			
Rwaished	32.1	16.2	28	5.9	3.2	0.3	85.7	106
Safawi	28.9	17.6	37.7	12.8	5.7	0.3	103	138
Mafraq	19	65.2	81.4	13.9	39	17.2	235.7	149
Irbed	51.2	205.1	133.3	40.4	117.7	65.8	617.5	136
Amman	20.3	93.8	109.8	29.3	38.4	21.2	314.2	115
Aqaba	9.3	1	3	4.4	0	0	17.7	58

*: Based on annual average from 1922/1923-1999/2000 (Jordan Meteorological Department, http://www.jmd.gov.jo)

Table 1. Monthly precipitation for the comparison study sites in Jordan (mm) in the 2001/2002 season.

	Continuous urban	Green urban	Rainfed crops	Irrigated olives	Rangeland*	Bare rock	Weighted K_sK_c
Rwaished	38.0	9.5	--	5.5	47.0	--	0.22
Safawi	29.6	7.4	--	1.7	51.1	10.2	0.17
Mafraq	3.6	0.9	--	--	95.5	--	0.19
Irbed	20.2	5.0	67.8	7.0	--	--	0.80
Amman	57.5	6.4	--	21.2	14.9	--	0.30
Aqaba	32.9	16.1	--	--	51.0	--	0.19
K_sK_c	0.00	1.00	1.00	1.00	0.15	0.00	

* Mafraq site includes 70.3 % of protected rangeland K_sK_c = 0.20) while Aqaba includes bare sandy soils with a very sparse vegetation (K_sK_c = 0.05).

Table 2. Land use/cover percent, water stress and crop coefficient (K_sK_c) and the weighted K_sK_c for each site.

Station	LAI	T_r	T_r Availability	T_r Availability
	($m^2\ m^{-2}$)	(C°)	(d)	(%)
Rwaished	0.10-0.40	33.4-48.6	18	60
Safawi	0.20-0.50	30.7-49.3	14	47
Mafraq	1.10-1.70	25.7-40.6	13	43
Irbed	0.85-0.98	17.5-33.9	18	60
Amman	0.71-1.10	21.4-36.2	15	50
Aqaba	0.20	24.6-40.2	23	77

Table 3. Leaf area index (LAI), MODIS radiometric surface temperature (T_r) and number of days for which MODIS T_r was available in April for the six sites.

	ET_{aa}			ET_{fa}			ΔT		
Stations	Min	**Max**	Ave	Min	**Max**	Ave	Min	**Max**	Ave
			($mm\ d^{-1}$)						
Rwaished	0.8	2.6	1.6	0.9	1.9	1.3	0.45	0.80	0.65
Safawi	0.7	2.4	1.5	0.8	2.0	1.3	0.52	0.80	0.65
Mafraq	1.5	3.2	2.1	0.9	1.7	1.6	0.47	0.74	0.58
Irbed	1.2	6.5	3.9	3.7	7.1	5.2	-0.6	0.61	0.16
Amman	0.9	3.2	1.8	1.2	2.3	1.7	0.40	0.75	0.62
Aqaba	0.8	3.5	1.9	0.7	2.0	1.2	0.36	0.78	0.58

Table 4. Minimum, maximum, and average FAO-56 (ETf_a) and ALARM (ET_{aa}) actual evapotranspiration and dimensionless temperature (ΔT) for the six sites.

	ET_c			Relative Error	
	ALARMD	SEBAL	Surface Renewal	ALARMD	SEBAL
		mm hr^{-1}		(%)	
Pixel 1	0.93	0.77	0.87	6.9	-11.5
Pixel 2	0.86	0.75	0.87	-1.1	-13.8
Pixel 3	1.0	0.8	0.87	14.9	-8.0
Pixel 4*	1.0	0.8	0.87	14.9	-8.0
Average	**0.94**	**0.78**	**0.87**	**8.0**	**-10.3**

*: The surface renewal station was located at Pixel 4.

Table 5. ALARMD and SEBAL ASTER 90-m pixel hourly Evapotranspiration.

	ET_c			Relative Error	
	ALARMD	SEBAL	Surface Renewal	ALARMD	SEBAL
		mm day^{-1}		(%)	
Pixel 1	8.5	7.1	8.6	-1.2	-17.4
Pixel 2	7.9	6.9	8.6	-8.1	-19.8
Pixel 3	9.2	7.4	8.6	7.0	-14.0
Pixel 4	9.2	7.4	8.6	7.0	-14.0
Average	**8.7**	**7.2**	**8.6**	**1.2**	**-16.3**

Table 6. ALARMD and SEBAL ASTER 90-m pixel daily Evapotranspiration.

	ALARMD	SEBAL
	ET_c (mm day^{-1})	
Pixel 1	6.6	6.4
Pixel 2	7.6	6.7
Pixel 3	6.3	6.1
Pixel 4	7.4	6.6
Pixel 5	6.6	6.1
Pixel 6	7.6	6.5
Pixel 7	7.4	6.4
Pixel 8	7.6	6.4
Average	**7.1**	**6.4**

Table 7. ALARMD and SEBAL MODIS 1-km pixel daily Evapotranspiration.

	ALARM-D	SEBAL
	ET_c (mm day^{-1})	
Pixel 2	8.1	6.9
Pixel 4	8.9	7.0
Pixel 6	9.0	7.0
Pixel 8	6.8	6.1
Average	**8.2**	**6.8**

Table 8. ALARM-D and SEBAL ASTER 1-km pixel daily Evapotranspiration.

6. References

Al-Bakri, J.T., Suleiman, A.A., 2004: NDVI response to rainfall in the different ecological zones in Jordan. *International J. Remote Sens.*, 25(19), 3897–3912.

Al-Bakri J. T., Taylor J. C., 2003: Application of NOAA-AVHRR for monitoring vegetation conditions and biomass in Jordan. *Journal of Arid Environments*, 54(3): 579-593.

Al-Eisawi, D.M., 1985: Vegetation of Jordan. In: A. Hadidi, (ed.), Studies in the History and Archaeology of Jordan. Vol I. (pp. 45-57). Dept. of Antiquities, Amman Jordan.

Al-Jitan, M.A., 2005: Evapotranspiration of major crops in the Jordan Valley using remote sensing techniques compared with estimated field measurement using eddy-correlation. PhD Thesis, University of Jordan, Amman, Jordan.

Al-Naber G., Al-Bakri J. , Saba M., 2009: Monitoring drought and desertification in Jordan with remote sensing. Proceedings of the remote sensing and GIS applications symposium, 20 April, 2009, Geography Department, Faculty of Arts, University of Jordan, Amman, Jordan.

Allen, R.G., Pereira, L.A., Raes, D., Smith, M., 1998: Crop evapotranspiration. FAO Irrigation and Drainage Paper 56. FAO, Rome, Italy, 293 p.

Allen, R.G., Tasumi, M., Morse, A.T., and Trezza R., 2005: A LANDSAT-based energy balance and evapotranspiration model in Western US water rights regulation and planning. *J. Irrig. and Drain. Systems.*, 19, 251-268.

Allen, R.G., Tasumi, and Trezza R., 2007: Satellite-Based Energy Balance for Mapping Evapotranspiration with Internalized Calibration (METRIC)—Model. *J. Irrig. and Drain. Engrg.*, 133(4), 380-394.

ASCE (American Society of Civil Engineers), 2005: The ASCE Standardized Reference Evapotranspiration Equation, edited by Allen, R.G., Walter, I.A., Elliott, R., Howell, T., Itenfisu, D., and Jensen, M., ASCE-EWRI Task Committee Report, January, 2005.

Anderson, M.C., Norman, J.M., Mecikalski, J.R., Otkin, J.A., and Kustas, W.P., 2007: A climatological study of evapotranspiration and moisture stress across the continental United States based on thermal remote sensing: 1. Model formulation. *J. Geophys. Res.*, 112(D10), D11112, doi:10.1029/2006JD007507.

Bandara, K.M.P.S., 2003: Monitoring irrigation performance in Sri Lanka with high-frequency satellite measurements during the dry season. *Agric. Water Manage.* 58(2), 159-170.

Bastiaanssen, W.G.M., Chandrapala, L., 2003: Water balance variability across Sri Lanka for assessing agricultural and environmental water use. *Agric. Water Manage.*, 58(2), 171-192.

Bastiaanssen, W.G.M., Menenti, M., Fwddws, R.A., Holtslag, A.A.M., 1998: A remote sensing surface energy balance algorithm for land (SEBAL). *J. Hydrol.*, 212/213, 198-212.

Ben-Asher, J., Phene, C.J., Kinarti, A., 1992: Canopy temperature to assess daily evapotranspiration and management of high frequency drip irrigation systems. *Agric. Water Manage.* 22(4), 379-390.

Brutsaert, W., 1982: Evaporation into the Atmosphere, D. Reidel, Boston, 299 pp.

Brutsaert, W., Chen, D., 1996: Diurnal variation of surface fluxes during thorough drying (or severe drought) of natural prairie. *Water Resour. Res.*, 32(7), 2013-2019.

Brutsaert, W., and Sugita, M., 1996: Sensible heat transfer parameterization for surfaces with anisothermal dense vegetation, *J. Atmos. Sci.*, 53(2), 209-216.

Consoli, S., D'Urso, G., Toscano, A., 2006: Remote sensing to estimate ET-fluxes and the performance of an irrigation district in southern Italy. *Agric. Water Manage.* 81 (3), 295-314.

Chandrapala, L., Wimalasuriya, M., 2003: Satellite measurements supplemented with meteorological data to operationally estimate evaporation in Sri Lanka. *Agric. Water Manage.* 58(2), 89-107.

Crago, R. D., 1998: Radiometric and equivalent isothermal surface temperatures. *Water Resour. Res.*, 34(11), 3017-3023.

Crago, R.D., Brutsaert, W., 1996: Daytime evaporation and the self-preservation of the evaporative fraction and the Bowen ratio. *J. Hydro.*, 178, 241-255.

Crago, R., Suleiman, A.A., 2005: Heat flux parameterization for sparse and dense grasslands with the Analytical Land-Atmosphere Radiometer Model (ALARM). *Boundary Layer Meteorol.*, 114(3), 557-572.

Friedl, M. A., Davis, F. W., 1994: Sources of variation in radiometric surface temperature over a tallgrass prairie. *Remote Sens. Environ.*, 48, 1-17.

Gurney, R.J., Hsu, A.Y., 1990: Relating evaporative fraction to remotely sensed data at the FIFE site. *In* F.G. Hall and P.J. Sellers (ed.). Symposium on FIFE, Anaheim, California, American Meterological Society, 112-116.

Hatfield, J.L., Perrier, A., and Jackson, R.D., 1983: Estimation of evapotranspiration at one time-of-day using remotely sensed surface temperatures. *Agric. Water Manage.*, 7(1-3), 341-350.

JMD (Jordan Meterological Department), 2002: Jordan agrometeorological Bulletin. Meteorological Department, Agrometeorological Directorate, Ministry of Transport, Amman, Jordan (http://www.jmd.gov.jo/).

Kim, C.P., Entekhabi, D., 1997: Examination of two methods for estimating regional evaporation using a coupled mixed layer and land surface model. *Water Resour. Res.*, 33(9), 2109-2116.

Kustas, W.P., Schmugge, T.J., Humes, K.S., Jackson, T.J., Parry, R., Weltz, M.A., Moran, M.S., 1993: Relationships between evaporative fraction and remotely-sensed vegetation index and microwave brightness temperature for semiarid rangelands. *J. of Appl. Meteorol.*, 32(12), 1781-1790.

Kustas, W.P., Anderson, M.C., Norman, J.M., and Li, F.Q., 2007: Utility of radiometric-aerodynamic temperature relations for heat flux estimation, *Boundary-Layer Meteorol.*, 122(1), 167-187

Liu, X.Y, Lin, E., 2005: Performance of the Priestley-Taylor equation in the semiarid climate of North China. *Agric. Water Manage*, 71(1), 1-17.

Lhomme, J. P., Chegbouni, A., Monteny, B., 2000: Sensible heat flux-radiometric surface temperature relationship over sparse vegetation: Parameterizing B^{-1}. *Boundary-Layer Meteorol.*, 97, 431-457.

Lhomme, J.P., Elguero, E., 1999: Examination of evaporative fraction diurnal behavior using a soil-vegetation model coupled with a mixed-layer model. *Hydrol. and Earth Sys. Sci.*, 3(2), 259-270.

Loheide II, S.P., and Gorelick, S.M. 2005: A local-scale, high resolution evapotranspiration mapping algorithm (ETMA) with hydrological applications at riparian meadow restoration sites. *Remote Sens. of Environ.*, 98, 182-200.

Mahrt, L., and Vickers, D. 2004: Bulk formulation of the surface heat flux. *Boundary-Layer Meteorol.*, 110, 357-379.

Massman, W. J., 1999: A model study of kB_H^{-1} for vegetated surfaces using 'localized near-field Lagrangian theory. *J. Hydrol.*, 223, 27-43.

Menenti, M., Choudhury, B.J., 1993: Parameterization of land surface evavotranspiration using a location dependent potential evapotranspiration and surface temperature range. In Exchange processes at the land surface for a range of space and time series, Bolle H.J., R.A. Feddes, and J.D. Kalma (Eds). *IAHS Publ*, 212, 561-568.

MoA (Ministry of Agriculture, Jordan), 1994: The Soils of Jordan: Semi-detailed Level (1:50 000). National Soil Map and Land Use Project, Ministry of Agriculture, Amman, Jordan (Maps available at http://alic.arid.arizona.edu/jordansoils/_html/about.html).

MoE (Ministry of Environment, Jordan), 2006: National Action Plan and Strategy to Combat Desertification. Amman, Jordan: Ministry of Environment (Document available at http://www.unccd.int/).

Nagler, P. L. , Cleverlyb, J., Glenna, E., Lampkinc, D., Huete, A., Wan, Z., 2005: Predicting riparian evapotranspiration from MODIS vegetation indices and meteorological data. *Remote Sens. of Environ.*, 94, 17-30.

NCARE, National Center for Agricultural Research and Extension. 2002a: Annual report of Rainfed Program: Maru Agricultural Station, Irbed. NCARE, Baqa'a, Jordan (Hardcopy can be supplied by authors).

NCARE, National Center for Agricultural Research and Extension. 2002b: Irrigation Management Information System (IMIS) (Data available under IMIS-NCARTT, http://www.ncare.gov.jo/).

Qiu, G.Y., Shi, P., and Wang, L., 2006: Theoretical analysis of a remotely measurable soil evaporation transfer coefficient. *Remote Sens. of Environ.*, 101, 390-398.

Qualls, R.J., Yates, D. N., 2001: Directional radiometric temperature profiles in a grass canopy. *Advances in Water Resour. Res.*, 24, 145-155.

Roerrink, G.L., Su, Z., Menenti, M., (2000). S-SEBI A simple remote sensing algorithm to estimate the surface energy balance. *Phys. Chim. Earth (B)*, 25(2), 147-157.

Rosenberg, N.J., Verma, S.B., 1978: Extreme Evapotranspiration by Irrigated Alfalfa: A Consequence of the 1976 Midwestern Drought. *J. Applied Meteorol.*, 17, 934-941.

Santanello Jr., J.A., Friedl M.A., 2003: Diurnal covariation in soil heat flux and net radiation. *J. Appl. Meteorol.*, 42, 851-862.

Senay, G.B., Budde, M., Verdin, J.P.), Melesse, A.M., 2007: A coupled remote sensing and simplified surface energy balance approach to estimate actual evapotranspiration from irrigated fields. *SENSORS*, 7(6), 979-1000.

Shuttleworth, W.J., Gurney, R.J., Hsu, A.Y., Ormsby, J.P., 1989: FIFE: The variation in energy partition at surface flux sites. *IAHS Publ.*, 186, 67-74.

Strahler, A.H., Muller, J.P. and MODIS Science Team Members, 1999: MODIS BRDF/Albedo Product: Algorithm Theoretical Basis Document. MODIS Version 5 Land Data Products (available at: http://modis.gsfc.nasa.gov/)

Stull, R.B., 1988: An Introduction to Boundary Layer Meteorology, Kluwer Academic, 666 pp.

Suleiman, A. A., Crago R.D., 2002a: Analytical land atmosphere radiometer model (ALARM) applied to a dense canopy. *Agric. Forest Meteorol.*, 112, 151-159.

Suleiman, A. A., Crago R.D., 2002b: Analytical land atmosphere radiometer model. *J. Appl. Meteorol.*, 41, 177-187.

Suleiman, A.A., Crago, R.D., 2004: Hourly and daytime evapotranspiration using radiometric surface temperatures. *Agronomy J.*, 96, 384-390.

Suleiman, A.A., Ritchie, J.T., 2003: Modeling soil water redistribution under second stage evaporation. *Soil Sci. Soc. Am. J.*, 67(2), 377-386.

Suleiman, A.A., Ritchie, J.T., 2004: Modifications to the DSSAT vertical drainage model for more accurate soil water dynamics estimation. *Soil Sci.*, 169(11), 745-757.

Tasumi, M., R. Trezza , R.G. Allen and J. L. Wright. 2005: Operational aspects of satellite-based energy balance models for irrigated crops in the semi-arid U.S *J. Irrig. and Drain. Systems.* 19, 355-376.

Venturini, V., Gautam Bisht, G., Islam, S., Jiang, L., 2004: Comparison of evaporative fractions estimated from AVHRR and MODIS sensors over South Florida. *Remote Sens. of Environ.*, 93, 77-86.

Wald, L., Albuisson, M., Best, C., Delamare, C., Dumortier, D., Gaboardi, E., Hammer, A., Heinemann, D., Kift, R., Kunz, S., Lefèvre, M., Leroy, S., Martinoli, M., Ménard, L., Page, J., Prager, T., Ratto, C., Reise, C., Remund, J., Rimoczi-Paal, A., Van der Goot, E., Vanroy, F., and Webb, A., 2002: SoDa: a project for the integration and exploitation of networked solar radiation databases. *In*: Environmental Communication in the Information Society, W. Pillmann, K. Tochtermann Eds, Part 2, pp. 713-720. Published by the International Society for Environmental Protection, Vienna, Austria.

Wan, Z., Zhang, Y., Zhang, Q., and Li, Z., 2002: Validation of the land-surface temperature products retrieved from Terra Moderate Resolution Imaging Spectroradiometer data. *Remote Sens. Environ.*, 83, 163-180.

Wang, J., Sammis, T.W., Meier, C.A., Simmons, L.J., Miller, D.R., and Samani, Z., 2005: A modified SEBAL model for spatially estimating pecan consumptive water use for Las Cruces, New Mexico. The 15th Conference on Applied Climatology/13th Symposium on Meteorological Observations and Instrumentation, Savannah, GA, June 2005.

Zhang, L., and Lemeur, R., 1995: Evaluation of daily evapotranspiration estimates from instantaneous measurements. *Agric. Forest Meteorol.*, 74, 139-154.

Zibognon, M., Crago, R.D., and Suleiman, A.A., 2002: Conversion of radiometric to aerodynamic surface temperature with an anisothermal canopy model. *Water Resour. Res.*, 38(6), 3-1 to 3-6.

8

Remotely Sensed Evapotranspiration Data Assimilation for Crop Growth Modeling

Baburao Kamble and Ayse Irmak
University of Nebraska-Lincoln
USA

1. Introduction

The agricultural sector will require more water in the near future to provide more food, fibre and fuels (Molden et al., 2007). As population increases and development calls for increased demand of food, a change in diet due to increased prosperity, and a recent focus on biofuels. This population growth - coupled with industrialization and urbanization - will result in an increasing demand for water and will have serious consequences on the conservation of water resources. Therefore, a rational approach to best water management practices is needed to balance water supply and demand. One approach to check if the supply is adequate to meet the demand is to account for the respective components in the water balance. Doing so provides an opportunity to search for possible ways to save water from one application and allocate it to another. Simulation models are strong in this regard; they can simulate the processes in the real system and predict the state variables at every stage in the simulation.

The role of simulation models in understanding the processes in the soil-plant-atmosphere system has increased significantly in recent years. This is attributed to increased computing capabilities available today (Ines et al., 2002). Such mechanistic ecophysiological models integrate knowledge from data collection by various methods (e.g. GPS, field sampling, satellite remote sensing, radar etc.) and laboratory research. Simulations from such models are widely used to predict and simulate crop growth, yield, water requirements and greenhouse gas emissions. For monitoring agricultural crop production, growth of crops is modeled, for example, by using simulation models. Estimates of crop growth often are inaccurate for practical field conditions. Therefore, model simulations must be improved by incorporating information on the actual growth and development of field crops, for example, by using remote sensing data.

Numerous researchers have also used remotely sensed data in conjunction with crop growth models via data assimilation for the purpose of improving soil moisture estimation (Entekhabi et al., 1994; Van Dams et al., 1997; Reichle et al. 2001; Ines et al., 2002; Kamble et al., 2008). The objective of data assimilation is to obtain the best estimate of the state of the system by combining observations with the forecast model's first guess. Genetic algorithms (GA) are designed to search, discover and emphasize good solutions by applying selection and crossover techniques, inspired by nature, to supply solutions (Holland, 1975; Goldberg, 1989). GA operates on pieces of information like nature does on genes in the course of evolution. Changes in the genes of individuals from a given population allow selection of certain groups of genes that are most important in fitting the environment pressures on the population. All individuals of one generation are evaluated by a fitness function. The strength of GA with respect to other local search algorithms is due to the fact that, in a GA framework, more strategies can be adopted together to find individuals to be added to the mating pool. Addition is made both in the initial population phase and in the dynamic generation phase. Thus, a more variable search space can be explored at each algorithm step. Based on the above biological evolution idea, a so-called "SWAP-GA" has been developed by Ines et al., (2002) to estimate input parameters of SWAP from remote sensing data. The SWAP-GA model was adopted and re-coded according to the objectives of this research. After recoding and recalibrating with local parameters simulation was then carried out for different generations and different populations.

Ines et al. (2002) developed an assimilation methodology for the Soil, Water, Atmosphere and Plant (SWAP) simulation model (Van Dams et al., 1997) with remote sensing data using a genetic algorithm (GA). Similar work was done by Chemin (2006), Kamble et al. (2006, 2008), Irmak et al. (2009) in which remotely sensed information was fed to SWAP-GA for optimization of soil hydraulic parameters. Proper evaluation of the water balance in the unsaturated zone depends strongly on the appropriate characterization of the soil hydraulic functions but direct measurement of soil hydraulic properties in the laboratory using soil core samples is the classic way to determine the soil hydraulic functions (van Genuchten et al., 1991). Unfortunately, direct measurement of these functions is impractical for most applications in research and management, especially for large-scale water management problems. The hydraulic parameters are mostly influenced by the water consumption by crop or evapotranpiration phenomenon which control the crop growth and water consumption and vice a versa. Therefore, it may be useful to have a method that assimilate evapotranpiration in a hydrological model and use as a function of flexible boundary conditions and can also give the optimized hydraulic property.

This chapter introduces ET data assimilation scheme was implemented with a SWAP model and genetic algorithm to optimize crop growth parameters. The goal of this system was to provide realistic description of hydrological balance in an analytically tractable way, as a basis for quantitative understanding of soil moisture response to different hydraulic parameter which controls ET. In addition, this chapter introduces few implementation results from two case studies performed in India and USA with different conditions.

2. Evapotranspiration by energy balance model

Evapotranspiration is one of the most critical parameters and has a considerable impact on water losses. ET is usually the largest hydrological flux through during the summer months in Great Plains. The ability is required to accurately estimate the magnitude of this flux will, therefore, go a long way towards being able to compute the water balance and plan the water resources and regimes. It is, however, the most difficult flux to quantify (Peacock and Hess, 2004). Furthermore, quantification of this flux on a watershed or a regional scale is much more difficult. ET is highly dynamic in space and time because of the complex interaction of soil, vegetation and climate. In the last few decades, analysis of this biophysical phenomenon has received much attention (Burman and Pochop, 1994). After FAO 56 by Food and Agriculture Organization of the United Nations, last decade has witnessed many investigations of up-scaling the point-scale evapotranspiration (ET) to regional scale (Allen, 2000) and of quantifying ET either directly from remotely sensed information (Bastiaanssen et al., 1998a, Bastiaanssen et al., 1998b, Schmugge et al., 2002, Su and Troch, 2003, Jia et al., 2003 Kamble et al., 2007 and Pan et al., 2008) or from simulation modeling.

These Land surface energy balance (EB)-based models convert satellite sensed radiances into land surface characteristics to estimate ET as a "residual" of the land surface energy balance equation. The Surface Energy Balance Algorithm for Land (SEBAL) was developed to quantify ET over large areas using remote sensing-based land surface energy fluxes (Bastiaanssen et al. 1998). It has been used to estimate riparian ET (Goodrich et al., 2000), basin wide ET (Bastiaanssen et al., 2002), mapping regional runoff and precipitation (Church et al., 1995), and developing crop coefficients (Singh and Irmak, 2009). Another satellite remote sensing model, the METRIC (Mapping Evapotranspiration at high Resolution using Internalized Calibration) was introduced by Allen et al. (2007a, b). The model originates from versions of SEBAL and is based on similar principles. Similar to SEBAL, METRIC models use near-surface temperature gradient (dT) estimated as an indexed function of radiometric surface temperature, thereby eliminating the need for absolutely accurate surface temperature or air temperature measurements to estimate sensible heat flux (H) in the computation of land surface energy balance.

Surface Energy Balance Algorithm for Land (SEBAL) was developed to quantify ET over large areas using satellite-based surface energy fluxes (Bastiaanssen et al. 1998). SEBAL is a one-source energy balance model that estimates the latent heat flux (evapotranspiration) as a residual of other energy balance components:

$$\lambda ET = R_n - G - H \tag{1}$$

where R_n is net radiation (W m^{-2}), G is the soil heat flux (W m^{-2}), H is the sensible heat flux (W m^{-2}), and λET is the latent heat flux (W m^{-2}). R_n is the difference between the incoming and outgoing fluxes, which is expressed as:

$$R_n = R_s \downarrow - R_s \uparrow + R_l \downarrow - R_l \uparrow - (1 - \varepsilon_s) R_l \downarrow \tag{2}$$

where $R_s\downarrow$ is the incoming shortwave radiation (W m^{-2}), $R_s\uparrow$ is the outgoing shortwave radiation (W m^{-2}), $R_l\downarrow$ is the incoming longwave radiation (W m^{-2}), $R_l\uparrow$ is the outgoing longwave radiation (W m^{-2}), and ε_s is the surface emissivity (unitless). Soil heat flux is mainly driven by a thermal gradient in the topsoil and this gradient is highly dynamic in space and time. The soil heat flux was estimated as a function of NDVI and R_n using the relationship developed by Singh et al. (2007) for south central Nebraska soil and crop management conditions:

$$G = \left[0.3811 \exp(-2.3187 NDVI)\right] R_n \tag{3}$$

$$H = \frac{\rho_a C_p dT}{r_{ah}} \tag{4}$$

where, ρ_a is the air density (kg m^{-3}), C_p is the specific heat of air (J kg^{-1} K^{-1}), dT is the near surface and air temperature difference (K), and r_{ah} is the aerodynamic resistance to heat transfer (s m^{-1}). Once the instantaneous R_n, G, and H are determined, the instantaneous evaporative fraction (Λ) was calculated as:

$$\Lambda = \frac{\lambda ET}{R_n - G} \tag{5}$$

Finally, the daily actual ET (ET_c) was estimated as:

$$ET_c = \frac{86400\Lambda(R_{n24} - G_{24})}{\rho_w \lambda} \tag{6}$$

where, ET_c is the daily crop ET (mm day^{-1}), R_{n24} is the daily net radiation calculated on a daily time step (W m^{-2}), G_{24} is the daily soil heat flux (W m^{-2}), λ is the latent heat of vaporization (J kg^{-1}), and ρ_w is the density of water (kg m^{-3}).
The output from the SEBAL model is an actual ET map calculated on a 30-m grid resolution basis using SEBAL algorithms (Eq. 1. through Eq. 6). Further descriptions of the SEBAL methodology are discussed in detail in Singh et al. (2008). We ran the model for each of the seven Landsat images to quantify the spatial distribution of ET_c. After obtaining the ET_r (via geostatistics) and ET_c maps (via SEBAL), the spatial distribution of K_{cr} was calculated by dividing the ET_c map values by the ET_r map values.

3. Crop growth modelling and SWAP

An intermediate version of the SWAP model (SWAP) (fig. 1) was used in this study (Ines et al., 2005). The SWAP model is physically based one-dimensional model to simulate vertical transport of water flow, solute transport, heat flow and crop growth at the field scale level (Van Dam et al., 1997). It requires inputs including management practices and environmental conditions to compute a daily soil water balance and crop growth. The major processes taken into account are phenological development, assimilation, respiration, and ET. The SWAP uses Richard's equation to simulate vertical soil water movement in variable saturated soils is given as follows:

$$\frac{\partial \theta}{\partial t} = \frac{\partial}{\partial z}\left[K(\psi)\left(\frac{\partial \psi}{\partial z} + 1\right)\right] \tag{7}$$

where K is the hydraulic conductivity (cm d-1), ψ is the pressure head (cm), z is the elevation above a vertical datum (cm), θ is the water content (cm3 cm-3), and t is time (d). The soil hydraulic functions in the model are defined by the Mualem-Van Genuchten (MVG) equations which describe the capacity of the soil to store, release and transmit water under different environmental and boundary conditions (Ines, 2002). Darcy's law is used to determine potential soil evaporation in wet soil conditions. Root water extraction at various depths in the root zone is calculated from potential transpiration, root length density and possible reductions due to wet, dry, or saline conditions (Eitzinger, 2000).
As SWAP simulates actual evaporation (Ea) and transpiration (Ta),ETa can be taken as the sum of Ea and Ta. The Penman–Monteith approach (Allen et al., 1994) was used to estimate the potential ET rate, ETp. The partitioning of ETp into potential soil evaporation (Ep) and potential transpiration (Tp) is established according to soil cover fraction. In the case of wet soil, Ea is determined by the atmospheric demand and equals to Ep. When the soil dries out, the soil hydraulic conductivity decreases, which reduces evaporation. SWAP calculates the maximum possible evaporation rate (Emax) according to Darcy's law and sets Ea equal to the minimum value of Ep and Emax. Hence, Emax is governed by the soil hydraulic

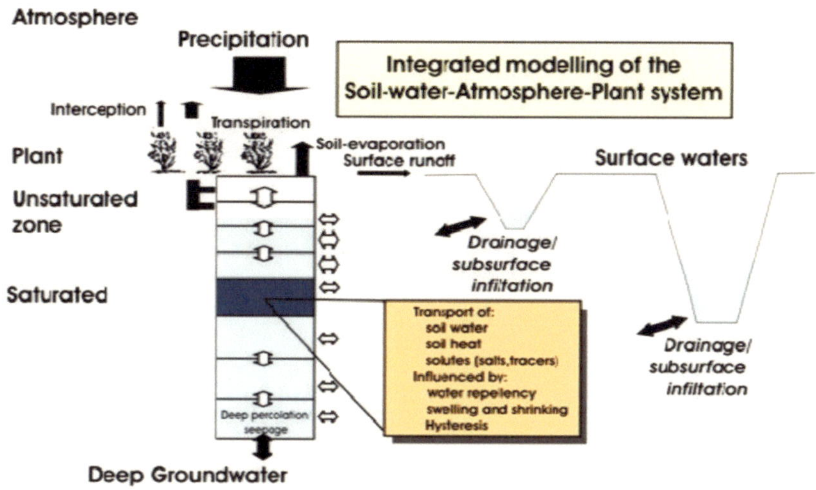

Fig. 1. Integrated modeling scheme of Soil Water Atmosphere Plant (SWAP) model

functions. The Ta is calculated using the root water uptake reduction due to water and/or salinity stress. Water requirements of a crop depend mainly on crop growth stage and environmental conditions. Different crops have different water-use requirements under the same weather conditions. For wheat, phenological stage was used along with the values of reference ET for computing the consumptive use at different growth stages by a water-balance approach.

4. Data assimilation with genetic algorithm

The objective of data assimilation is to obtain the best estimate of the state of the system by combining observations with the forecast model's first guess. Genetic algorithms (GA) are designed to search, discover and emphasize good solutions by applying selection and crossover techniques, inspired by nature, to supply solutions (Holland 1975; Goldberg 1989). GA operates on pieces of information like nature does on genes in the course of evolution. Changes in the genes of individuals from a given population allow selection of certain groups of genes that are most important in fitting the environment pressures on the population. All individuals of one generation are evaluated by a fitness function. The strength of GA with respect to other local search algorithms is due to the fact that, in a GA framework, more strategies can be adopted together to find individuals to add to the mating pool, both in the initial population phase and in the dynamic generation phase. Thus, a more variable search space can be explored at each algorithm step.

Based on the above biological evolution idea, a so called "SWAP-GA" has been developed by Ines and Honda (2005) to estimate input parameters of SWAP from remote-sensing data.

The proposed parameters were fed to SWAP by GA according to the evaluation of the difference processes between SWAP output ET and the target energy balance ET values. The GA searches for the optimum crop parameter set while SWAP tests the proposed parameters simultaneously by using them in forward simulations (Figure 2).

$$C = \sqrt{\left(\frac{ET_{METRIC} - ET_{SWAP}}{n}\right)^2} \qquad (8)$$

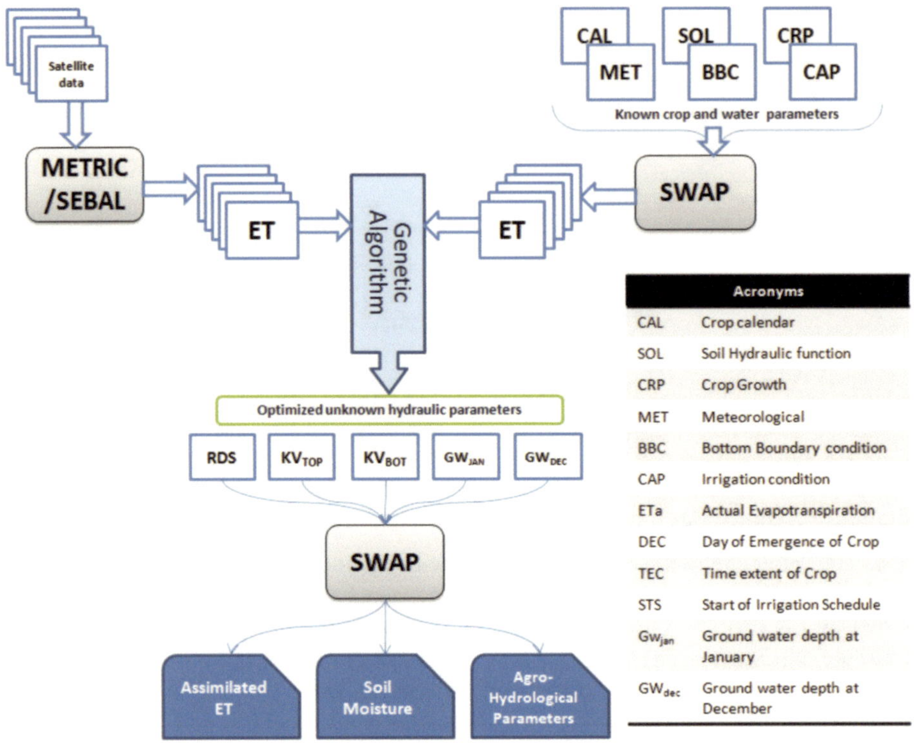

Fig. 2. SWAP-GA data assimilation framework

Consider C the cost function, and d is the satellite overpass date. where ET is estimated with energy balance model using remotely sensed data (cm) and it was treated as observations. ET_{SWAP} is an estimated ET with SWAP (cm), n is the time domain and C_{xy} is the objective function (root mean square error: RMSE) for the pixel at x, y location (cm). When a minimum-difference defined threshold was reached, SWAP parameters were stored for reconstruction of ET for any required day in the cropping season. The fitness of an individual having x,y pixel location characteristics is the inverse of the cost function times the constraints aimed at minimizing the RMSE between ET_{SWAP} and target ET_{METRIC}.

$$F_{xy} = \frac{1}{\left(C_{xy} * (1.0 + \text{Constraint})\right)} \tag{9}$$

In Equation 13, the constraint is a function of the emergence date of the second crop (DEC) and the date of harvest:

$$\text{Constraint} = \text{Date of emergence} - \text{Harvest date} \tag{10}$$

Subject to the possible range of sowing dates:

$$b_{\min j} \leq sd_j \leq b_{\max j} \quad (j=1,\ldots,6)$$

where b_{minj} is the earliest possible sowing date, b_{maxj} is the latest possible sowing date, and sd_j is the actual sowing date. The units are day-of-year (ordinal day).

5. Research case studies

5.1 On-demand irrigations scheduling in Sirsa Irrigation Circle, India

The proposed approach was tested using a dataset on irrigated cotton field in the Sirasa Irrigation Circle for On-demand irrigation scheduling. Data used in this study was previously collected as part of comprehensive research conducted by the Wageningen Agricultural University, The Netherlands during 2002 for calibrating the SWAP model (Van Dam et al., 2003).

Figure 3 compares temporal distribution of SEBAL ET with predicted SWAP ET with SWAP-GA. Both SEBAL ET and SWAP-GA showed similar patterns of under and over estimations of actual ET. The SWAP-GA slightly overestimated ET early in the season when the soil surface was dry and underestimated late in the season when the soil surface was wet and covered by the crop which influences efficiency of water use, high water productivity and efficient farming activities.

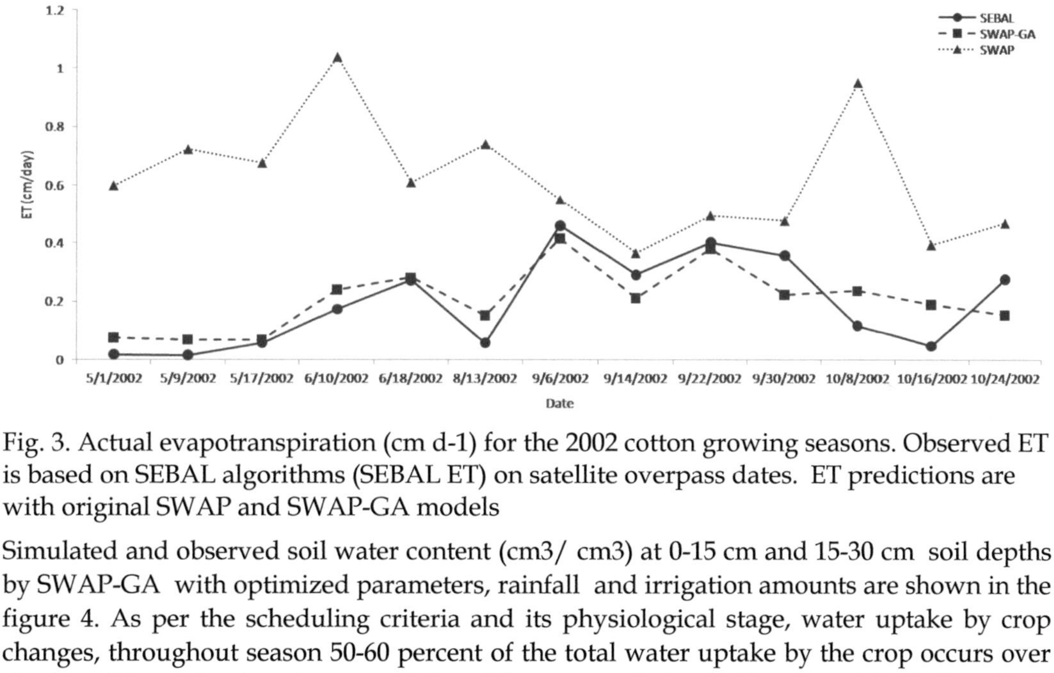

Fig. 3. Actual evapotranspiration (cm d-1) for the 2002 cotton growing seasons. Observed ET is based on SEBAL algorithms (SEBAL ET) on satellite overpass dates. ET predictions are with original SWAP and SWAP-GA models

Simulated and observed soil water content (cm3/ cm3) at 0-15 cm and 15-30 cm soil depths by SWAP-GA with optimized parameters, rainfall and irrigation amounts are shown in the figure 4. As per the scheduling criteria and its physiological stage, water uptake by crop changes, throughout season 50-60 percent of the total water uptake by the crop occurs over the first 90 cm depth, where more than 90 percent of the total root weight is found. It reveals that the top layer of the soil (0-15 cm and 15-30 cm) has greater fluctuations. It is because the top layer forms the sphere of life which receives moisture in pulses of rainfall and irrigation also the same water is eliminated through evaporation and transpiration by plants. The simulated and observed soil moisture levels show the increasing value from June to September, and then decreasing from September to November, which corresponds to the variation of irrigation and precipitation. But the dramatic difference between the simulated and observed soil moisture was found in July. The simulation has the lowest value in August due to the large relative contribution of ET (Fig. 3), whereas this phenomenon does not occur in the observation.

5.2 Optimization of hydraulic parameters and sensitivity analysis in Clay Center, Nebraska-USA

This study was conducted at the University of Nebraska-Lincoln South Central Agricultural Laboratory (SCAL) near Clay Center, Nebraska, USA (Latitude: 40° 34' N; Longitude: 98° 08' W; elevation: 552 m above MSL). In this research, the METRICTM was used to compute complete radiation and energy balances along with the resistances for momentum, heat, and water vapor transport for each pixel in the experimental area.

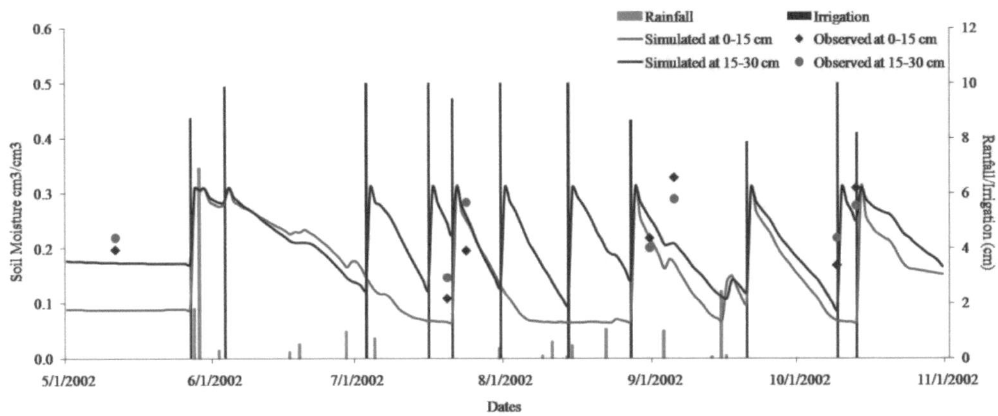

Fig. 4. Comparison of observed (SEBAL ET) and predicted evapotranspiration for the 2002 cotton growing seasons. The potential (dashed line) and actual (solid line) evapotranspiration were estimated by SWAP-GA with optimized parameters

Fourteen satellite images were selected from June through October in 2006 because these images were with a cloud cover less than 10%. The relative ET in this study was expressed as the reference ET fraction (ETrF) and was computed using the procedures outlined by Allen et al. (2007a and b). Figure 4a shows the ET for the MODIS satellite overpass on August 14, 2006. As expected, the ET was highly variable in south-central Nebraska due to variation in cropping practices, irrigation, and vegetation development. Furthermore, the ET values were usually lower for agricultural lands than rangeland/natural vegetation in August due to less green vegetation fraction because of crop maturity and harvest stages at this time. On the other hand, grazed rangeland/natural vegetation has green vegetation in August as evidenced by high NDVI on the scene. Higher ET values from grazed rangeland/natural vegetation pixels indicated that most of the available energy was used for transpiration. The spatial distribution of daily ET predictions on August 14, 2006 was between the range of 0.9-1.15 for agricultural lands and was as high as 0.13 for the natural vegetation/rangeland across the image (Fig. 4a). Figure 4b shows seasonal ET map corresponding to the 2006 season for the entire Clay, York, Hamilton Adams and Fillmore counties (Figure 1). Seasonal ET varied from 400 mm for bare soil to 950 mm for irrigated crops. Rain fed areas surrounding the Fillmore County (in the south east) had ET values around 400 mm which depicted the bare fields and fallow lands. The ET over Adams County showed the ET in between 400 mm to 650 mm, while ET values are for the SCAL fields in York and Hamilton Counties due to shallow water table, lateral seepage from the SCAL fields and an open network of irrigation canals. The ET map further showed a spatial gradient of increasing ET from the southern parts towards the northern parts of the irrigation system except low ET in the Howard County due to settlements. All of these ET values are important for the agro-hydrological balance of the area as well as ground water modeling. According to table 1, average daily ET (ET) was 0.426 cm.day^{-1} with a mode and maximum values of 0.75 cm. day^{-1} and 0.71 cm.day^{-1}, respectively, for the study field. It is evident from the numerical figures in table that some crops are still developing on May and others are transpiring at higher rates. On June23, all the crops in the area are established. This indicates the variability of sowing dates and water management practices as influenced by water availability.

The ET assimilation is carried out to obtain the best estimate of the state of the hydraulic system by combining observations with the forecast model at first guess. Figure 5 reveals the actual ET for the 2006 corn growing seasons. ET predictions are with original SWAP and SWAP-GA models. The result shows excellent fitness between the observed ET and simulated ET. There is a bias condition due to the comparison of point observation with model explicitly taken into account to prevent unnecessary forcing towards the biased observations.

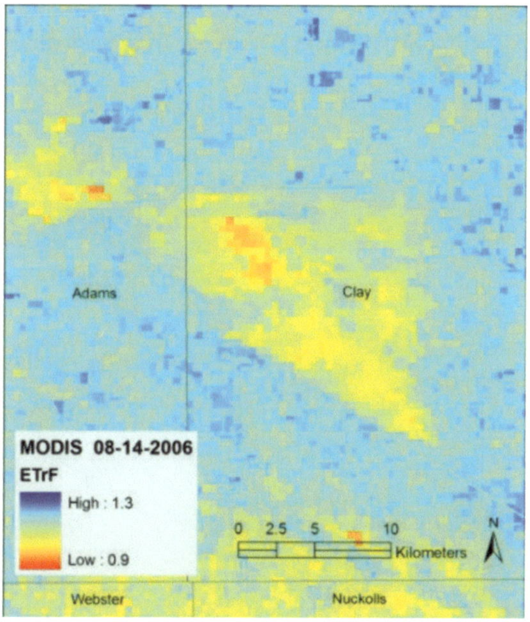

Fig. 6. MODIS derived spatial distribution of evapotranspiration (ET) (August 14, 2006) during the growing season in the study area

Figure 7 revised the relationship between the observed and predicted ET simulated by SWAP-GA (with data assimilation) and SWAP (without data assimilation) model for 2006 season. The independent regression analysis for two dataset shows the fitness of the SWAP-GA data with observed data and compares it with fitness of the SWAP data without data assimilation. This regression coefficient of SWAPGA (with data assimilation) provides good estimates than previously SWAP (without data assimilation). The relations obtained were statistically significant for ET data assimilation method. This fact indicates the strong need of assimilating observed data in crop growth model to minimize errors between simulation and reallife crop growth.

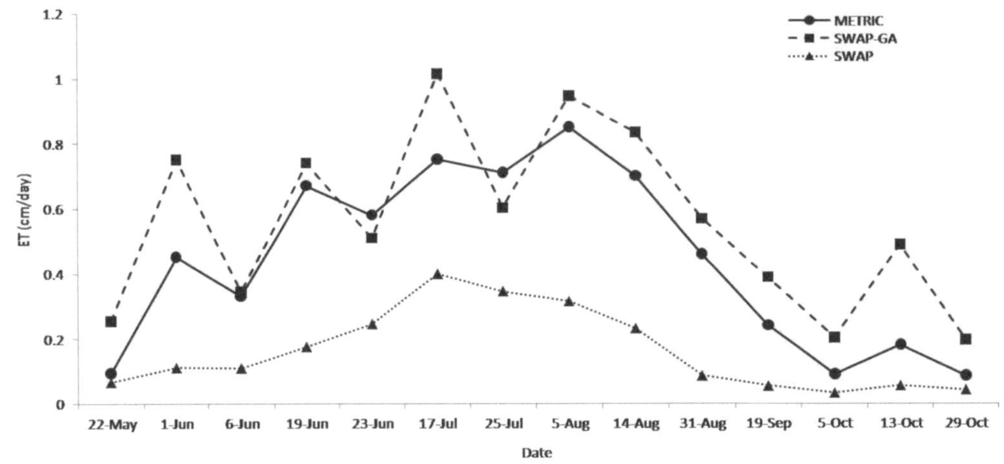

Fig. 7. Actual evapotranspiration (cm d-1) for the 2006 corn growing seasons. Observed ET is based on METRIC algorithms (METRIC) on satellite overpass dates. ET predictions are with original SWAP and SWAP

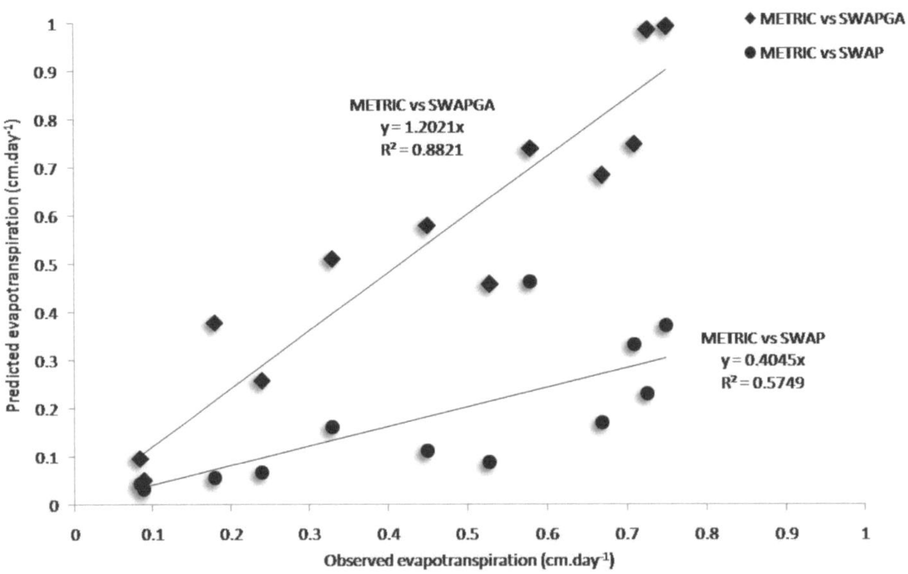

Fig. 8. Relationship between ET by remote sensing data and (observed) and Simulated ET from SWAPGA (with data assimilation) and SWAP (without data assimilation) model. The solid line corresponds to the 1:1 relation to the regression equation for the 2006 growing seasons

Optimization of crop growth parameters and sensitivity analysis

The optimized parameters were determined by minimizing the RMSE between SWAP-ET and the target METRIC-ET values. Generally, the actual remote sensing data contains errors due to atmospheric conditions, cloud cover, and errors in the remote sensing-based models/algorithms used to estimate ET. Therefore, we tested the procedure assuming that some degree of error in remote sensing observations (METRIC ET) was present in the dataset. We compared the results from GA for different populations and different generations. Best results were obtained by applying the algorithm which was configured for 10 populations and 10 generations with up to nine variable parameters (three crop and six hydraulic parameters). The data inventory during the 2006 corn growing season was used to verify the optimized parameters from SWAP-GA simulation. Table 1 shows the values of optimized parameters as well as data from the experimental field. Best results are obtained by applying the algorithm which configured for 100 Population and 100 generation with up to nine variable parameters, which are selected according to the hydraulic sensitivity to water management problem and fitness function is based on crop parameter sensitivity. In simulations, hydraulic properties were based on measured values where possible; some values were altered slightly by optimizing the model to the local conditions until good agreement with measured ET was attained (fig 9).

Optimized parameters	Definition	Unit	Minimum value	Maximum value	Optimized value
GWjan	Groundwater at start of season	cm	100	160	120
GWdec	Groundwater at end of season	cm	100	160	130
BASEGW	Level of impervious layer	cm	170	230	185
KHBOT	horizontal hydraulic conductivity bottom layer	cm	15	30	25
KVTOP	Vertical hydraulic conductivity top layer	cm	5	20	18
RDS	maximum rooting depth allowed by the soil	cm	120	240	180

Table 1. Definition, unit, minimum, and maximum values of optimized parameters in SWAP-GA

The implementation of this sensitivity analysis after the estimation process aims at determining if the parameters estimated previously are well identified. Therefore, to examine the SWAP model response to changes of specific input data, i.e., to have an indication of the required accuracy at which each hydraulic parameter should be available; a sensitivity analysis of the model was performed (Figure 8). From the original dataset, obtained from field measurements and literature as explained above, the base simulation was established. From this, the sensitivity analysis was performed, assessing the effect produced by a given variation of the input data range on the SWAP output.

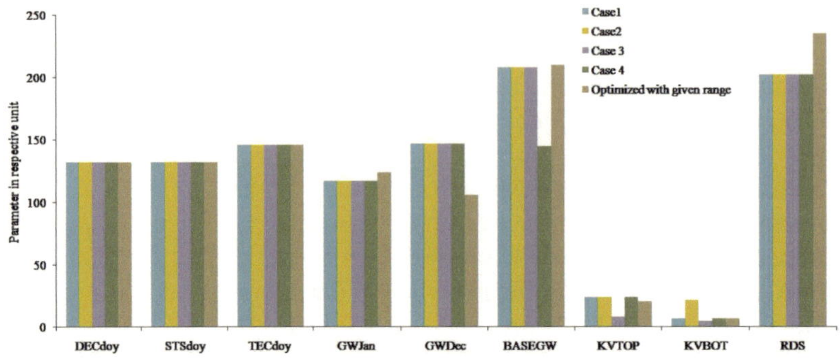

Fig. 8. Hydraulic parameter sensitivity analysis for 4 different cases according to change (+/- 20) in GW, BASEGW, KVTOP, KVBOT and RDS

6. Conclusion

We used remote sensing-based SEBAL and METRIC ET data to characterize our model via a stochastic data assimilation approach (GA), and the derived information was then used as inputs to SWAP. This methodology was evaluated in India and North American climatic conditions with different objectives. The methodology developed in this research to estimate hydraulic parameters and application to on-demand irrigation from calibrated crop model parameters gave good results. Parameter estimations were successful, and the ability of the model to produce similar ET values to the observed values (SEBAL ET) was promising, although, in general, the performance of SWAP-GA for on-demand irrigation can be described as reasonable. GA-based optimization retains the advantageous features of forward modeling, while reducing the number of required function evaluations to a level that is often much more computationally manageable. These conclusions suggest that it is indeed necessary to couple a remotely sensed ET with a pixel-based hydrological model in order to study and explore the water management options.

7. References

Allen, R.G., Tasumi, M. and Trezza, R. (2007a). Satellite-based energy balance for mapping evapotranspiration with internalized calibration (METRIC) -Model. *Journal Irrigation and Drainage Engineering*, 133 (4), 380-394.

Allen, R.G., Tasumi, M., Morse, A.T., Trezza, R., Wright, J.L., Bastiaanssen, W., Kramber, W., Lorite, I. and Robison, C.W. (2007b). Satellite-based energy balance for mapping evapotranspiration with internalized calibration (METRIC)-Applications. *Journal Irrigation and Drainage Engineering*, 133 (4), 395-406.

Bastiaanssen, W. G. M., Menenti, M., Feddes, R. A., and Holtslag, A. A. M. (1998). "A remote sensing surface energy balance algorithm for land (SEBAL): 1. Formulation." *Journal of Hydrology.*, 212–213, 198–213.

Bastiaanssen, W. G. M. (2000). SEBAL based sensible and latent heat fluxes in the irrigated Gediz Basin, Turkey. *Journal of Hydrology*, 229, 87-100.

Bastiaanssen, W. G. M., Ahmad, M. D., and Chemin, Y. (2002). Satellite surveillance of evaporative depletion across the Indus Basin. *Water Resources Research*, 38 (12), 1273.

Bastiaanssen, W. G. M., Noordman, E. J. M, Pelgrum, H., Thoreson, B. P., and Allen, R. G. (2005). SEBAL model with remotely sensed data to improve water-resources management under actual field conditions. *Journal Irrigation and Drainage Engineering*, 131 (1), 85-93.

Chemin Y., Honda, K., 2006. Spatiotemporal fusion of rice actual evapotranspiration with genetic algorithms and an agrohydrological model. *IEEE Transactions on Geosciences and Remote Sensing* 44: 3462-3469.

D'Urso, G. 2001. Simulation and management of on-demand irrigation systems: A combined agrohydrological and remote sensing approach, *Ph.D. Waginengen, The Netherlands.dissertation*,Wageningen University,

Eagleson, 1970 P.S. Eagleson, Dynamic hydrology, McGraw-Hill, New York (1970).

Eitzinger, J., Z. Žalud, C.A.M. Van Diepen, M. Trnka, D. Semerádová, M. Dubrovský and M. Oberforster, 2000: Calibration and evaluation of the WOFOST model for winter wheat. – *In : Proc. 8th International Poster day ' Transport of Water, Chemicals and Energy in the System Soil-CropCanopy-Atmosphere'* 16.11.200, Bratislava.

Entekhabi, D., H. Nakamura, and E.G. Njoku. 1994. Solving the inverse-problem for soil moisture and temperature profiles by sequential assimilation of multifrequency remotely sensed observations. *IEEE Transactions on Geosciences and Remote Sensing*., 32:438–448.

Goldberg, 1989 D.E. Goldberg, Genetic algorithms for search, Optimization and Machine Learning, *Addison-Wesley Publishing Co., Reading, Mass (*1989).

Holland, J.H. 1975. Adaptation in natural and artificial systems. *Univ. of Michigan Press*, Ann Arbor.

Ines, A. V. M., (2002), Improved Crop Production Integration GIS and Genetic Algorithms. Doctoral Thesis. AIT. Bangkok, Thailand. *AIT Diss no.* WM-02-01.

Ines, A.V.M. and K. Honda. 2005. On quantifying agricultural and water management practices from low spatial resolution RS data using genetic algorithms: a numerical study for mixed pixel environment. *Advances in Water Resources*. 28: 856-870.

Irmak, A and B. Kamble. 2009. Evapotranspiration Data Assimilation with Genetic Algorithms and SWAP Model for On-demand Irrigation. *Irrigation Science*. 28:101-112. DOI 10.1007/s00271-009-0193-9

Kamble, B.,(2006) Evapotranspiration Data Assimilation with Genetic Algorithms and SWAP Model for On-Demand Irrigation, *M.Engg Thesis, Asian Institute of Technology, Thailand,* 2006

Kamble, B., and Irmak A (2008), Combining Remote Sensing Measurements and Model Estimates through Data Assimilation, *IGARSS 2008*

Kustas, W. P., and Norman, J. M. (1996). Use of remote sensing for evapotranspiration monitoring over land surfaces. *Hydrological Sciences Journal, 41(4), 495-516.*

Malik, R.S., Kumar, R., Dabas, D.S., Dhindwal, A.S., Singh, S., Singh, U., Singh, D., Mal, J., Singh, R., Bessembinder, J.J.E., 2003. Measurement program and description database. In: Van Dam, J.C. and Malik, R.S. (Eds.), 2003. Water productivity of irrigated crops in Sirsa district, India. Integration of remote sensing, crop and soil models and geographical information systems. *WATPRO Final Report, Including CD-ROM.* ISBN 90-6464-864-6, pp. 29–39.

Meneti, M., and Choudhary, B. J. (1993). Parameterization of land surface evapotranspiration using a location dependent potential evapotranspiration and surface temperature range. Exchange Processes at the land Surface for a Range of Space and time series, Bolle, H. J., Feddes, R. A., and Kalma, J. D. (Eds), *International Association of Hydrological Sciences* Publication, 212, 561-568.

Molden, D., T. Oweis, P.Steduto, J.W. Kijne, M.A. Hanjra, P.S. Bindraban. 2007. Pathways for increasing agricultural water productivity. In Molden, D. (ed.). Water for Food, *Water for Life: A Comprehensive Assessment of Water management in Agriculture*. Earthscan, London, and International Water Management Institute, Colombo.

Norman, J. M., Kustas, W. P., and Humes, K. S. (1995). A two-source approach for estimating soil and vegetation energy fluxes in observations of directional radiometric surface temperature. Agricultural *and Forest Meteorology*, 77, 263-293.

Reichle R. H., D. Entekhabi, and D. B. McLaughlin, 2001: Downscaling of radiobrightness measurements for soil moisture estimation: A four-dimensional variational data assimilation approach. *Water Resources Research.*, 37, 2353–2364.

Singh, R., J.C. van Dam, R.A. Feddes, Water productivity analysis of irrigated crops in Sirsa district, India, *Agricultural Water Management,* Volume 82, Issue 3, 24 April 2006, Pages 253-278, ISSN 0378-3774, DOI: 10.1016/j.agwat.2005.07.027.

Singh, R., A. Irmak, S. Irmak, and D.L. Martin., 2007, "Application of SEBAL for mapping evapotranspiration and estimating surface energy fluxes in south central Nebraska.," *Journal of Irrigation and Drainage Engineering,* ASCE (accepted).

Singh et al., 2001 K.B. Singh, P.R. Gajri and V.K. Arora, Modelling the effect of soil and water management practices on the water balance and performance of rice, *Agricultural Water Management* 49 (2001), pp. 77–95

Su, Z. (2002). The surface energy balance system (SEBS) for estimation of turbulent heat fluxes. *Hydrology and Earth System Sciences,* 6(1), 85-89.

Van Dam J.C., & Malik, R.S. (Eds.), 2003. Water productivity of irrigated crops in Sirsa district, India. Integration of remote sensing, crop and soil models and geographical information systems. *WATPRO final report, including CD-ROM.* ISBN 90-6464-864-6. 29-39.

Van Dam, J.C., J. Huygen, J.G. Wesseling, R.A. Feddes, P. Kabat, P.E.V. Van Waslum, P. Groenendjik, and C.A. Van Diepen. 1997. Theory of SWAP version 2.0: Simulation of water flow and plant growth in the soil–water–atmosphere–plant environment. *Tech. Doc. 45.* Wageningen Agric. Univ. and DLO Win and Staring Centre, Wageningen, The Netherlands.

Z wart, S.J. and W.G.M. Bastiaanssen, 2003. Review of measured crop water productivity values for irrigated wheat, rice, cotton and maize. *Agricultural Water Management,* 69(2): 115-133.

9

The Effect of Temperature on Actual Evapotranspiration based on Landsat 5 TM Satellite Imagery

Preeyaphorn Kosa
School of Civil Engineering, Institute of Engineering, Suranaree University of Technology Thailand

1. Introduction

In the water cycle, evapotranspiration is one of the most important components, but it is one of the most difficult to measure and monitor. Evapotranspiration relates to the exchange of energy in the atmosphere, ground surface, and root zone. Some elements of calculated evapotranspiration can be measured by weather stations, while others are estimated from empirical equations. Then, the calculated evapotranspiration has some inaccuracy. To improve upon this problem, the combination of meteorological data and remote sensing observersions are an alternative evapotranspiration approach. (Hongjie et al., 2002; Kalluri et al., 2003) On the other hand, temperature is normally measured in a number of weather stations. Since temperature relates to many weather data, temperature can imply the characteristic of other weather data. For example, low temperature is included high humidity but low evaporation can be occurred in the condition of low temperature. Moreover, the variable spatial resolution is the characteristic of both actual evapotranspiration and temperature. At present, satellite image are used for studying the Earth's surface and it bolsters spatial resolution. Then, the purpose of this study is to determine the effect of temperature on actual evapotranspiration using satellite image.

Evapotranspiration occurs from evaporation and transpiration, and it can be obtained from weather data and satellite images. Evaporation is the primary process of water transfer in the hydrological cycle. The water is transformed into vapour and transported into the sky. Evaporation can be classified into potential evaporation and actual evaporation. The potential evaporation is defined as the amount of evaporation that would occur if a sufficient water source were available. On the other hand, the actual evaporation is the amount of water which is evaporated a normal day. The potential evaporation is the maximum value of the actual evaporation. Transpiration is included by the vaporization of liquid water contained in plant tissues and the vapor removal to the atmosphere. (Hongjie et al., 2002; Kalluri et al., 2003) Evapotranspiration is normally computed from the Penman-Monteith FAO 56 equation using weather data. This equation is affected by principal weather parameters such as radiation, air temperature, humidity and wind speed. These parameters can be measured by weather station and computed by the equation of FAO irrigation and Drainage Paper No. 56 (Allen et al., 1998). In addition to the Penman-Monteith FAO 56 equation, evapotranspiration can be estimated from the concept of energy balance. The main components of the energy balance equation are sensible heat flux, latent heat flux, soil heat flux, and net radiation. These elements relate with incoming and outgoing radiation in the atmosphere, ground surface, and root zone. They are estimated from remote sensing data and weather data. (Wei and Sado, 1994; Amin *et al.*, 1997; Xihua *et al.*, 1997; Andrew *et al.*, 2002)

Remotely sensed data are used for studying the Earth's surface. Current technology allows continuous acquisition of data, regular revisit capabilities (resulting in up-to-date information), broad regional coverage, good spectral resolution (including infra-red bands), good spatial resolution, ability to manipulate/enhance digital data, ability to combine satellite digital data with other digital data, cost-effective data, map-accurate data, possibility of stereo viewing, and large archives of historical data. Remote sensing helps to record data from remote locations. Satellite data provides timely and detailed information about the Earth's surface, especially in relation to the management of our renewable and non-renewable resources. The advantages of satellite data for many fields include for example; assessment and monitoring of vegetation types and their status, soil surveys, mineral exploration, map making and revision, production of thematic maps, water resources planning and monitoring, urban planning, agricultural property planning and management, crop yield assessment, and natural disaster assessment.

The Surface Energy Balance Algorithm for Land or SEBAL (Bastiaanssen *et al.*, 1998a and Bastiaanssen *et al.*, 1998b) is an image processing model to calculate evapotranspiration by using satellite imagery and weather data with the concept of energy balance at the land surface, so evapotranspiration for each pixel is calculated. Satellite images that can be used in SEBAL are LANDSAT, NOAA-AVHRR, MODIS, and ASTER. The advantages of SEBAL are easily applicable because of minimal collateral data needed, applicable to various climates because of physical concepts, no need for land use classification, no need to involve data demanding hydrology, and is a suitable method for all visible, near-infrared and thermal radiometers. However, the disadvantages of SEBAL include cloud-free conditions being required and that surface roughness is poorly described, so that SEBAL is only suitable for flat terrain. (Timmermans *et al.*, 2001; Allen *et al.*, 2001; Lal *et al.*, 2001; Ines, 2002; Abou and Tanton, 2003; Jacob *et al.*, 2003)

2. Landsat 5 TM

Landsat 5 TM has a unique and necessary role in the realm of earth observing satellite orbits, because no other satellite of the earth observing system matches with the synoptic coverage, high spatial resolution, spectral range and radiometric calibration of Landsat's system. Landsat 5 TM satellite imagery Path/Row:127/48 was mainly input data. The spatial resolution of Landsat 5 TM satellite imagery is 30 m and the swath width covered by the image is 185 km. The details of Landsat 5 TM characteristic are shown in Table 1 and Table 2. (Farr, 1999) At the ground stations, the Landsat ground system consists of a spacecraft control center, ground stations for uplinking commands and receiving data, a data handling facility and a data archive, which were developed by the Goddard Space Flight Center, Greenbelt, MD., in conjunction with the U.S. Geological Survey (USGS) EROS Data Center (EDC), Sioux Falls, SD. These facilities will receive, process, archive, and distribute ETM+ data to users. Raw ETM+ data can be transmitted to the EROS Data Center by the ground system within 24 hours of its reception. (Farr, 1999; Liu, 2000; and Liang et al., 2001)

Characteristic	Specification
Swath width:	185 km
Repeat coverage interval:	16 days (233 orbits)
Altitude:	705 km
Quantization:	Best 8 of 9 bits
On-board data storage:	~ 375 Gb (solid state)
Inclination:	Sun-synchronous, 98.2 degrees
Equatorial crossing:	Descending node; 10:00 am +/- 15 min
Launch vehicle:	Delta ll
Launch date:	April 1999

Table 1. Landsat 5 TM mission specification

Channel	Spectral Range (microns)	Ground Resolution (m)
1 (Visible and near infrared: VNIR)	0.450 - 0.515	30
2 (Visible and near infrared: VNIR)	0.525 - 0.605	30
3 (Visible and near infrared: VNIR)	0.630 - 0.690	30
4 (Visible and near infrared: VNIR)	0.750 - 0.900	30
5 (Short wavelength infrared: SWIR)	1.550 - 1.750	30
6 (Thermal long wavelength infrared: LWIR)	10.40 - 12.50	60
7 (Short wavelength infrared: SWIR)	2.090 - 2.350	30
Pan (Visible and near infrared: VNIR)	0.520 - 0.900	15

Table 2. Landsat 5 TM characterist

This Landsat 5 TM image covers Sri Songkhram sub-river basin where locates in the lower part of Mekhong river basin. Mekhong river basin is in the northeast of Thailand. The time period of this study is from November 2006 to January 2007. They were selected because they are in the period of the dry season. Also, They are in the condition of clear sky. Moreover, there is the cultivation during this time so there are both evaporation and transpiration that is evapotranspiration.

3. Radiation

Extraterrestrial radiation (R_a) is the solar radiation received at the top of the earth's atmosphere on a horizontal surface. The values of extraterrestrial radiation depend on seasons change, the position of the sun, and the length of the day. Therefore, the extraterrestrial radiation is a function of latitude, and the date and time of day. The solar constant is the radiation striking a surface perpendicular to the sun's rays at the top of the earth's atmosphere and it is some 0.082 MJ m^{-2} min^{-1}. If the position of the sun is directly overhead, the incidence angle of extraterrestrial radiation is zero. In this case, extraterrestrial radiation is some 0.082 MJ m^{-2} min^{-1}.

Solar or shortwave radiation (R_s) is the amount of radiation penetrating from the atmosphere to a horizontal plane. The sun emits energy by electromagnetic waves that include short wavelengths so solar radiation is referred to as shortwave radiation. In the atmosphere, radiation is absorbed, scattered, or reflected by gases, clouds, and dust. For a cloudless day, the solar radiation is about 75% of the extraterrestrial radiation, while it is about 25% of the extraterrestrial radiation on a cloudy day. The solar radiation, which is known as global radiation, is a summation of direct shortwave radiation from the sun and a diffuse sky radiation from all upward angles.

Relative shortwave radiation (R_s / R_{so}) is a relationship between shortwave radiation (R_s) and clear-sky solar radiation (R_{so}). The shortwave radiation is solar radiation that actually reaches to the earth's surface in a given time, while clear-sky solar radiation is solar radiation that reaches to the same area with a clear-sky condition. The relative shortwave radiation is affected by the cloudiness of the atmosphere. On a cloudy day, the ratio is smaller than on a cloudless day. The range of this ratio is between 0.33 (cloudy condition) to 1.00 (cloudless condition).

Relative sunshine duration (n/N) shows the cloudiness in the atmosphere. It is the relationship between the actual duration of sunshine (n) and the maximum possible duration of sunshine, or daylight hours (N). For the cloudless condition, n is equal to N, while n and n/N are nearly zero for the cloudy condition. The maximum possible duration of sunshine, or daylight hours (N), depends on the position of the sun, so it is a function of latitude and date. The daily values of N throughout a year differ with latitude.

Albedo (α) is a relationship between reflected radiation and total incoming radiation. It varies with both the characteristics of surface and the angle of incidence, or the slope of ground surface. Albedo can be more than 0.95 for freshly fallen snow, and it is smaller than 0.05 for wet bare soil. The range of albedo for green vegetation is about 0.20 - 0.25 and albedo for the green grass reference crop is 0.23.

Net solar radiation (R_{ns}) is the fraction of the solar radiation that is reflected from the ground surface. It can be calculated by Equation (1):

$$R_{ns} = (1-\alpha)R_s \tag{1}$$

Net longwave radiation (R_{nl}) is the difference in value between outgoing and incoming longwave radiation. The longwave radiation is solar radiation absorbed by the earth and turned to heat energy. Since the temperature of the earth is less than the sun, so the earth emits longer wavelengths. Terrestrial radiation is referred to as longwave radiation. The emitted longwave radiation ($R_{l,up}$) is absorbed by the atmosphere or lost into space. The longwave radiation received by the atmosphere ($R_{l,down}$) increases its temperature. Therefore, the earth's surface both emits and receives longwave radiation. The value of outgoing longwave radiation is normally more than incoming longwave radiation, so the net longwave radiation is used to present the energy loss.

Net radiation (R_n) is the difference in value between incoming and outgoing radiation of both short and long wavelengths. It is the balance among energy absorbed, reflected, and emitted by the earth's surface. The net radiation is also the difference in value between the incoming net shortwave (R_{ns}) and the net outgoing longwave (R_{nl}) radiation. It is a positive value during daytime, while it is a negative value during nighttime. For the total daily value, it is a positive value except for the condition of high latitude.

Soil heat flux (G) is energy that is used in heating the soil. It is a positive value under the condition of warming soil and negative value under the condition of cooling soil. The soil heat flux is very small when compares with net radiation but it cannot be ignored.

4. Temperature calculation

Normally, Landsat 5 TM satellite image data is in the form of Digital Number (DN) so it's necessary to convert from Digital Number to Radiances for all bands in Landsat 5 TM image. An equation for converting is shown in follow equation.

$$L_{\lambda} = \frac{(LMAX_{\lambda} - LMIN_{\lambda})}{(QCALMAX - QCALMIN)} \times (QCAL - QCALMIN) + LMIN_{\lambda} \tag{2}$$

where L_{λ} is spectral radiance at the sensor aperture in watts/(meter squared *ster* μ m), QCAL is the quantized calibrated pixel value in DN, $LMIN_{\lambda}$ is the spectral radiance that is scaled to QCALMIN in watts/(meter squared *ster* μ m), $LMAX_{\lambda}$ is the spectral radiance that is scaled to QCALMAX in watts/(meter squared *ster* μ m), QCALMIN is the minimum quantized calibrated pixel value (corresponding to $LMIN_{\lambda}$) in DN and QCALMAX is the maximum quantized calibrated pixel value (corresponding to $LMAX_{\lambda}$) in DN.

$LMAX_{\lambda}$ and $LMIN_{\lambda}$ are the spectral radiances for each band at digital number 1 and 255 (i.e QCALMIN, QCALMAX), respectively. QCAL or DN, $LMAX_{\lambda}$ and $LMIN_{\lambda}$ are input data. These elements are values in header file information.

Thereafter a thermal band or band 6 imagery is converted to the effective at satellite temperature (T_{bb}) calculated by follow equation.

$$T_{bb} = \frac{K_2}{\ln\left(\frac{K_1}{L_6} + 1\right)} \tag{3}$$

For thermal band, calibration constants, K_1 and K_2, are 666.09 watts/(meter squared *ster* μm) and 1282.71 Kelvin, respectively. L_6 is the spectral radiance for band 6 in watts/(meter squared *ster* μm). Then, surface temperature (T_s) is computed by follow equation.

$$T_s = \frac{T_{bb}}{\varepsilon_o^{0.25}} \tag{4}$$

where ε_o is surface emissivity.

5. Actual evapotranspiration calculation

SEBAL is a tool to estimate actual evapotranspiration for flat areas with the most accuracy and confidence. Satellite image and weather data are used in the SEBAL model to calculate actual evapotranspiration by using a surface energy balance at the land surface. SEBAL evaluates an instantaneous actual evapotranspiration flux for the image time, because the satellite image provides information for the overpass time only. The actual evapotranspiration flux can be calculated for each pixel of the image as a residual of the surface energy budget equation. SEBAL needs both shortwave and thermal bands. The required ground-based data is wind speed. The SEBAL energy balance calculates actual evapotranspiration for each pixel for the time of the satellite image, so the results are instantaneous actual evapotranspiration. To obtain actual evapotranspiration using the conception of SEBAL, following equations are applied (Bastiaanssen et al., 1998a; Bastiaanssen et al., 1998b; Bastiaanssen, 2000; Chemin and Ahmad, 2000; Bastiaanssen *et al.*, 2002).

Firstly, data in the format of radiance is converted to reflectance for all bands. For this converting, thermal band (band 6) is not considered. In practice, band 6 is converted, but it is a dummy in this file. The equation used to convert radiance to reflectance is presented as following.

$$\rho_\lambda = \frac{\pi \times L_\lambda}{ESUN_\lambda \times \cos\theta \times d_r} \tag{5}$$

where ρ_λ is unitless planetary reflectance, L_λ calculates from equation (1), $ESUN_\lambda$ is mean solar exoatmospheric irradiances from Table 3, θ is solar zenith angle in degrees, and d_r is the Earth-Sun distance in astronomical that can be obtained from follow equation

$$d_r = 1 + 0.033\cos\left(DOY\frac{2\pi}{365}\right) \tag{6}$$

where DOY (or J) is number of day in one year for example DOY of January 1 is 1 while DOY of December 31 is 365.

$$\cos\theta = \cos(90 - \beta) \tag{7}$$

where β is sun elevation angle in degree and $\cos\theta$ is in degree.

$$[\text{Radians}] = \frac{\pi}{180} \times [\text{decimal deg rees}] \tag{8}$$

Albedo for the top of atmosphere (α_{toa}) can be considered from follow equation

$$\alpha_{toa} = \Sigma(\omega_\lambda \times \rho_\lambda) \tag{9}$$

$$\omega_\lambda = \frac{ESUN_\lambda}{\Sigma ESUN_\lambda} \tag{10}$$

where ω_λ is weighting coefficient, which is constant value.

Band	Solar Spectral Irradiances in watts/(meter squared *ster* µm)
1	from 1969 to 1957
2	from 1840 to 1829
3	from 1551 to 1557
4	from 1044 to 1047
5	from 225.7 to 219.3
7	from 82.07 to 74.52
Pan	from 1969 to 1957

Table 3. Solar Spectral Irradiances

Surface albedo equation (α) can be considered from follow equation

$$\alpha = \frac{\alpha_{toa} - \alpha_{path\_radiance}}{\tau_{sw}^2} \tag{11}$$

where α_{toa} is calculated from previous step. $\alpha_{path\_radiance} \approx 0.03$.

$$\tau_{sw} = 0.75 + 2 \times 10^{-5} \times z \tag{12}$$

where z is an elevation of area where is defined from Digital Elevation Map (DEM).
Incoming solar radiation ($R_{s\downarrow}$) is estimated in spreadsheet using follow equation.

$$R_{s\downarrow} = G_{sc} \times \cos\theta \times d_r \times \tau_{sw} \tag{13}$$

where G_{sc} is solar constant value, 1367 W/m²
Vegetation indices can be considered from follow equations.

$$NDVI = \frac{\rho_4 - \rho_3}{\rho_4 + \rho_3} \tag{14}$$

$$SAVI = \frac{(1 + L)(\rho_4 - \rho_3)}{L + \rho_4 + \rho_3} \tag{15}$$

$$LAI = -\frac{\ln\left(\frac{0.69 - SAVI}{0.59}\right)}{0.91} \tag{16}$$

where ρ_3 and ρ_4 are reflectance value in red and near-infrared bans (band 3 and 4), respectively. L is constant for SAVI (L= 0.5, when an area have no information for L). L = 0.5 is suitable for this practice.
Surface emissivity (ε_o) can be considered from follow equation

$$\varepsilon_o = 1.009 + 0.047 \times \ln(NDVI) \tag{17}$$

Outgoing longwave radiation ($R_{L\uparrow}$) can be calculated by following equation

$$R_{L\uparrow} = \varepsilon_o \sigma T_s^4 \tag{18}$$

where σ = 5.67 x 10^{-8} W/(m^2-K^4).
For the selection of "anchor pixel", SEBAL process utilizes two "anchor" pixels to fix boundary condition for the energy balance. (a) "Cold" pixel: a wet, well-irrigated crop surface with full cover ($T_s \cong T_{air}$). In cold pixel, sensible heat flux (H) is usually zero so cold pixel should be selected from water area. (b) "Hot" pixel should be located in a dry and bare agricultural field where one can assume there is no evapotranspiration taking place, and should have a surface albedo similar to other dry and bare field in the area of interest. It should have a LAI in the range of 0 to 0.4. After the temperatures of both cold and hot pixel are defined, these values are used for calculation in the next step.

Incoming longwave radiation ($R_{L\downarrow}$) is computed in spreadsheet using following equation.

$$R_{L\downarrow} = \varepsilon_a \times \sigma \times T_{cold}^4 \tag{19}$$

$$\varepsilon_a = 0.85 \times (-\ln \tau_{sw})^{0.09} \tag{20}$$

where ε_a is an atmospheric emissivity.
Net surface radiation flux (R_n) can be computed by following equation.

$$R_n = R_{s\downarrow} - \alpha R_{s\downarrow} + R_{L\downarrow} - R_{L\uparrow} - (1 - \varepsilon_o) R_{L\downarrow} \tag{21}$$

The soil heat flux (G) can be calculated from following equation.

$$\frac{G}{R_n} = \frac{T_s}{\alpha}\left(0.0038\alpha + 0.0074\alpha^2\right)\left(1 - 0.98 NDVI^4\right) \tag{22}$$

After the above equations are computed, the main equation for sensible heat flux, latent heat flux, evaporative fraction, and 24-hour actual evaporatranspiration are presented as following.
Sensible heat flux (H) is the flux of heat from the earth's surface to the atmosphere. It is not associated with phase changes of water.

$$H = \frac{\rho \times c_p \times dT}{r_{ah}} \tag{23}$$

where ρ is air density (kg/m^3), c_p is air specific heat (1004 J/kg/K), dT (K) is the temperature difference (T_1-T_2) between two heights (z_1 and z_2), and r_{ah} is the aerodynamic resistance to heat transport (s/m).
The estimation of sensible heat flux is the largest drawback, because the temperature difference and aerodynamic resistance to heat transport are unknown for the sensible heat flux calculation. To find these unknowns, SEBAL first calculates the sensible heat flux at extreme dry and wet locations. They are manually identified by the user on the image. The aerodynamic resistance to heat transport is computed from the lower integration constant for r_{ah} (z_1 = 0.1 m.) and the upper integration constant for r_{ah} (z_2 = 2 m.).
For a dry pixel or hot pixel, it should be located in a dry and bare agricultural field where one can be assumed that there is no evapotranspiration taking place. The wet pixel or cold pixel will include a surface temperature equal to air temperature. The sensible heat flux for the cold pixel is usually zero. The linear relationship between the temperature difference (dT) and the surface temperature (T_s), $dT = a + bT_s$, is created, and the coefficients a and b are defined from the two (dT, T_s) pairs applicable to the hot and cold pixels. Then sensible heat flux can be computed for every pixel that has the condition of free convection. Next, the values of friction velocity (u^*) are estimated from the wind speed at the blending height, a value of 200 m will be used. Thereafter, the condition of mixed convection is applied, and the pixel-dependent aerodynamic resistance to heat transfer, r_{ah}, is calculated by using the Monin-Obukhov hypothesis. The new temperature difference is calculated. Finally the processes from the calculation of sensible heat flux to the temperature difference

are repeated until the aerodynamic resistance to heat transfer and temperature difference are stable values. To compute sensible heat flux, following processess are considered.

1. Friction velocity ($u*$) can be computed as follow equation.

$$u* = \frac{ku_x}{\ln\left(\frac{z_x}{z_{om}}\right)} \tag{24}$$

   The calculation of the friction velocity requires a wind speed measurement (u_x) at a known height (z_x) in the time of the satellite image. k is a constant (0.41). Then, u_x and z_x are know, but z_{om} is unknown. z_{om} can be calculated in many ways: from z_{om} = 0.12 x height of vegetation (h) for agricultural area, from a land-use map, or from a function of NDVI and surface albedo. At weather station, u_x , z_x , z_{om} , and $u*$ can be determined.

2. Wind speed at a height 200 m above the weather station can be computed as follow equation.

$$u_{200} = \frac{u*\ln\left(\frac{200}{z_{om}}\right)}{k} \tag{25}$$

3. The friction velocity for each pixel is calculated using the wind speed at a height 200 m (u_{200}) that is assumed to be constant for all pixels of the image because it is defined as occurring at a "blending height" unaffected by surface features. From equation (26), z_{om} is unknown to calculate the friction velocity, so z_{om} need to be fined.

$$u* = \frac{ku_{200}}{\ln\left(\frac{200}{z_{om}}\right)} \tag{26}$$

   where z_{om} is the particular momentum roughness length of each pixel, z_{om} for each pixel can be computed by two methods: using a land-use map or using NDVI and surface albedo data (z_{om} is calculated in spreadsheet).
   For this pattern, a land-use map is not available, and then NDVI and surface albedo data are used. In the method used NDVI and surface albedo, z_{om} is computed from the following equation:

$$z_{om} = \exp\left[\left(a \times \frac{NDVI}{\alpha}\right) + b\right] \tag{27}$$

   where a and b are correlation constants derived from a plot of ln(z_{om}) vs $\frac{NDVI}{\alpha}$ for two or more sample pixels representing specific vegetation types. To determine a and b, a series of sample pixels representing vegetation types and conditions of interest are selected and the associated values for NDVI and surface albedo are obtained.

Typical surface albedo values for rice field and deciduous forest are 0.17 – 0.22 and 0.15 – 0.20, respectively.

4. Aerodynamic resistance to heat transport (r_{ah}) is computed as follow equation.

$$r_{ah} = \frac{\ln\left(\frac{z_2}{z_1}\right)}{u* \times k} \tag{28}$$

where z_1 and z_2 are 0.1 m and 2 m, respectively.

5. Near surface temperature difference (dT) for each pixel is calculated using equation (29) and the assumption of a linear relationship between T_s and dT.

$$dT = T_s - T_a \tag{29}$$

where T_a is unknown.

$$dT = b + aT_s \tag{30}$$

where a and b are the correlation coefficients.
To define these coefficients, SEBAL uses the "anchor" pixel where a value for sensible heat flux (H) can be reliably estimated.

a. At the "Cold" pixel

$$H_{cold} = R_n - G - LE_{cold} \tag{31}$$

$$dT_{cold} = \frac{H_{cold} \times r_{ah}}{\rho \times c_p} \tag{32}$$

If "Cold" pixel is chose from a body of water, $H_{cold} = 0$

$$LE_{cold} = R_n - G \tag{33}$$

$$dT_{cold} = 0 \tag{34}$$

b. At the "Hot" pixel

$$H_{hot} = R_n - G - LE_{hot} \tag{35}$$

In the case of H_{hot}, $LE_{hot} = 0$

$$H_{hot} = R_n - G \tag{36}$$

$$dT_{hot} = \frac{H_{hot} \times r_{ah}}{\rho \times c_p} \tag{37}$$

Then, $dT_{cold} = 0$ (T_{cold}, from selection in above step) and $dT_{hot} = \frac{H_{hot} \times r_{ah}}{\rho \times c_p}$ (T_{hot}, from selection in above step)

6. The sensible heat flux is calculated in this step, called initial sensible heat flux.

7. Monin-Obukhov theory in an iterative process is applied in SEBAL to account for the buoyancy effects, which are generated by surface heating. The Monin-Obukhov length (L) is used to define the stability conditions of the atmosphere in the iterative process (this is not the same "L" as used in the SAVI computation). It is a function of the heat and momentum fluxes and is computed as follow:

$$L = -\frac{\rho c_p u*^3 T_s}{kgH} \tag{38}$$

or

$$L = -\frac{1 \times 1004 \times u*^3 T_s}{0.41 \times 9.81H} \tag{39}$$

8. The values of the stability corrections for momentum and heat transport (ψ_m and ψ_h) are computed as follow. These values depend on the condition of atmosphere.
If L < 0; unstable condition:

$$\psi_{h(200m)} = 2\ln\left(\frac{1 + x_{(200m)}}{2}\right) + \ln\left(\frac{1 + x_{(200m)}^2}{2}\right) - 2ARCTAN\left(x_{(200m)}\right) + 0.5\pi \tag{40}$$

$$\psi_{h(2m)} = 2\ln\left(\frac{1 + x_{(2m)}^2}{2}\right) \tag{41}$$

$$\psi_{h(0.1m)} = 2\ln\left(\frac{1 + x_{(0.1m)}^2}{2}\right) \tag{42}$$

where

$$x_{(200m)} = \left(1 - 16\frac{200}{L}\right)^{0.25} \tag{43}$$

$$x_{(2m)} = \left(1 - 16\frac{2}{L}\right)^{0.25} \tag{44}$$

$$x_{(0.1m)} = \left(1 - 16\frac{0.1}{L}\right)^{0.25} \tag{45}$$

If L > 0; stable condition:

$$\psi_{h(200)} = -5\left(\frac{2}{L}\right) \tag{46}$$

$$\psi_{h(2m)} = -5\left(\frac{2}{L}\right) \tag{47}$$

$$\psi_{h(0.1m)} = -5\left(\frac{0.1}{L}\right) \quad (48)$$

If L = 0; neutral condition: ψ_m and $\psi_h = 0$

9. The friction velocity ($u*$), which is a corrected values, is now computed for each successive iteration as:

$$u* = \frac{u_{200}k}{\ln\left(\frac{200}{z_{om}}\right) - \psi_{m(200m)}} \quad (49)$$

where u_{200} is in m/s and k is 0.41.

10. The aerodynamic resistance to heat transport (r_{ah}), which is a corrected value, is now computed during each iteration as:

$$r_{ah} = \frac{\ln\left(\frac{z_2}{z_1}\right) - \psi_{h(z_2)} + \psi_{h(z_1)}}{u* \times k} \quad (50)$$

11. Repeat the step of calculation from step 5) to 10) until the successive values for dT_{hot} and r_{ah} at the hot pixel have stabilized.

The latent energy of evaporation (LE) was computed using following equation.

$$LE = R_n - G - H \quad (51)$$

After the latent energy of evaporation is computed, the evaporative fraction (Λ) is the next value that is obtained using Equation 52. The evaporative fraction at each pixel of a satellite image can be estimated using the 24-hour evaporatranspiration for the day of the image. The evaporative fraction is assumed to be a constant value over the full 24-hour period.

$$\Lambda = \frac{LE}{R_n - G} = \frac{LE}{LE + H} \quad (52)$$

To estimate 24-hour actual evaporatranspiration, the following equation is utilized.

$$ET_{24} = \frac{86400\Lambda(R_{n24} - G_{24})}{\lambda} \quad (53)$$

where R_{n24} is daily net radiation, G_{24} is daily soil heat flux, 86,400 is the number of seconds in a 24-hour period, and λ is the latent heat of vaporization (J/kg). The 24-hour actual evaporatranspiration, ET_{24}, can be expressed in mm/day.

Since energy, on average, is stored in the soil during the daytime and released into the air at night, G_{24} is very small for the combined vegetative and soil surface, so it can be assumed as zero at the soil surface (Morse et al., 2000). Then, Equation 53 can be rewritten as:

$$ET_{24} = \frac{86400\Lambda R_{n24}}{\lambda} \quad (54)$$

As above equations and calculation process, it can be also presented using Figure 1. Thereafter, to determine the relationship between the temperature and actual evapotranspiration, the maximum and minimum values of actual evapotranspiration in each temperature were ignored. The selected actual evapotranspiration in each temperature was averaged. Finally, the relationship between temperature and actual evapotranspiration was determined using a polynomial equation.

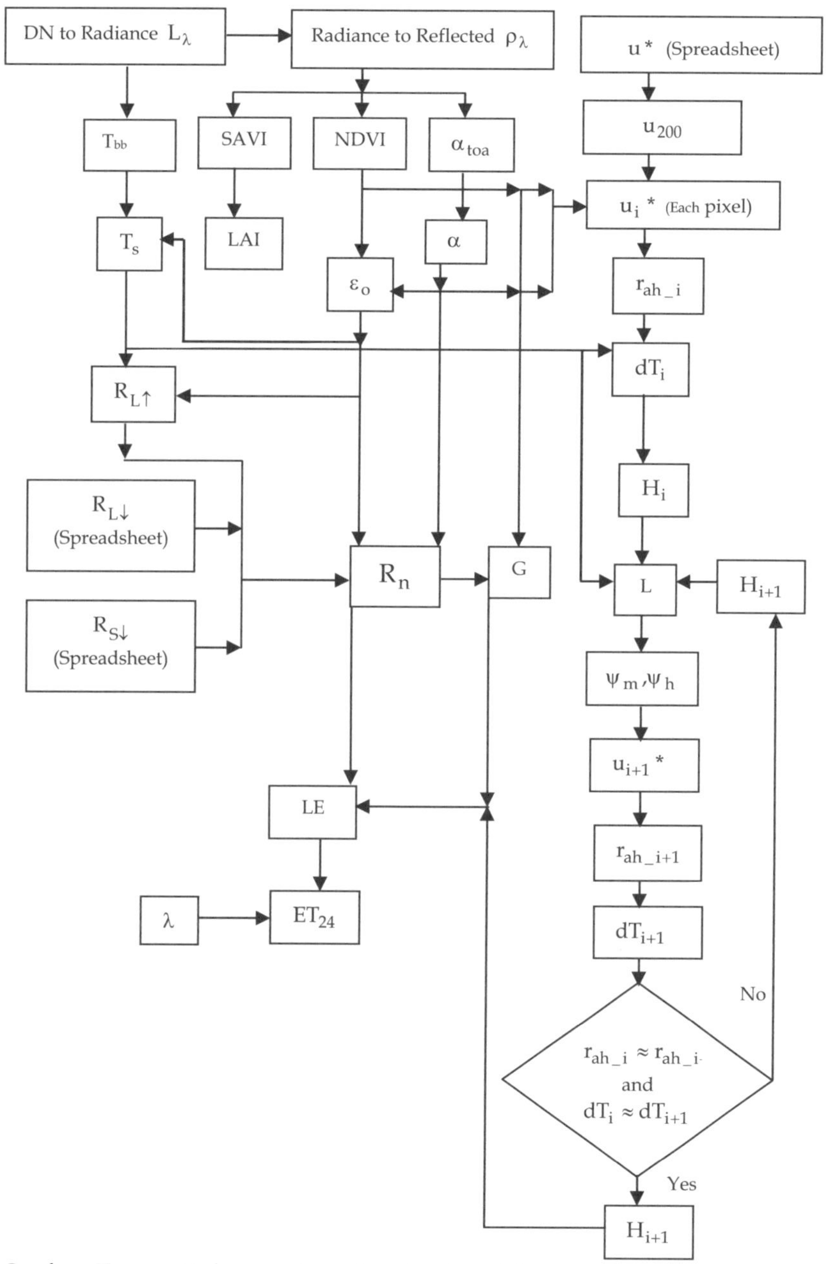

Fig. 1. The Surface Energy Balance Algorithm for Land Process

6. Validation

For the validation of temperature computed by using Landsat satellite image, it was compared with recorded temperature. On the other hand, to validate actual evapotranspiration calculated by SEBAL, recorded pan evaporation was used. Pan evaporation is the amount of water evaporated during a period (mm/day) with an unlimited supply of water (potential evaporation). It is a function of surface and air temperatures, insolation, and wind, all of which affect water-vapor concentrations immediately above the evaporating surface (Chong-yu et al., 2006). On the other hand, actual evapotranspiration is a function of temperature, wind, humidity and net radiation. It can be concluded that there is the relationship between the pan evaporation and actual evapotranspiration. (Humphreys et al., 1994; Grismer et al., 2002; Marco, 2002) Chong-yu et al. (2006) presents that the decreasing trend detected in the pan evaporation and actual evapotranspiration can be attributed to the significant decreasing trends in the net radiation and in the wind speed. Also, it can be attributed to the significant increasing trend in the air temperature.

7. Result

Since there are three main parts for the calculation of this study, there are three main parts of the result that are the spatial temperature, the spatial actual evapotranspiration, and the relation between temperature and actual evapotranspiration. The results for each part are presented as following.

7.1 Spatial temperature

The spatial distributions of temperature calculated by using Landsat 5 TM satellite images are presented from Figure 2 to Figure 4 (Kosa, 2009). These figures can be presented that the mean temperatures from Figure 2 to Figure 4 are 297.34, 295.74 and 296.25 ^{o}K, respectively. On the one hand, the minimum temperatures from these three figures are 283.21, 278.93 and 284.02^{o}K, respectively while the maximum temperatures from these three figures are 308.62, 313.63 and 310.65^{o}K, respectively.

7.2 Spatial actual evapotranspiration

The spatial distributions of actual evapotranspiration calculated by using Landsat 5 TM satellite images and SEBAL are presented from Figure 5 to Figure 7 (Kosa, 2009). These figure cab be presented that the mean actual evapotranspiration from these three figures are 3.67, 4.50 and 4.26mm, respectively.

7.3 Relation between temperature and actual evapotranspiration

After the spatial temperature and spatial actual evapotranspiration were calculated as show in Figure 2 to Figure 7, the temperature and actual evapotranspiration were plotted as present in Figure 8 and 9 (Kosa, 2009). These figures present that equations explained the relation between the temperature and actual evapotranspiration are consisted of $y = -0.028x^2 + 17.069x - 2593.2$ ($R^2 = 0.987$) and $y = -0.028x^2 + 1.7608x - 22.932$ ($R^2 = 0.987$) where y is the actual evapotranspiration (mm/day) and x is the temperature in the unit of ^{0}K and ^{0}C, respectively.

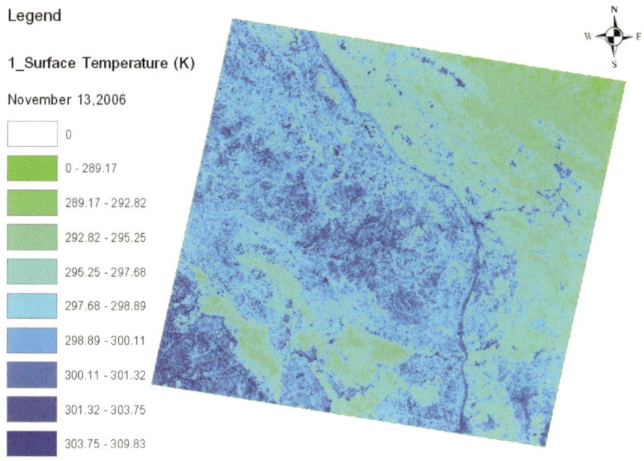

Fig. 2. The surface temperature on November 13, 2006

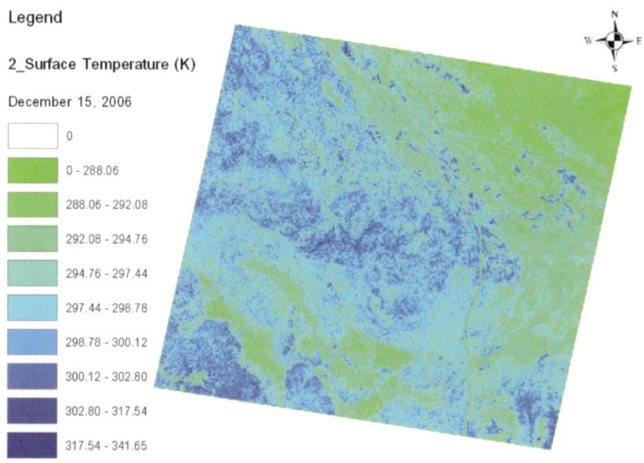

Fig. 3. The surface temperature on December 15, 2006

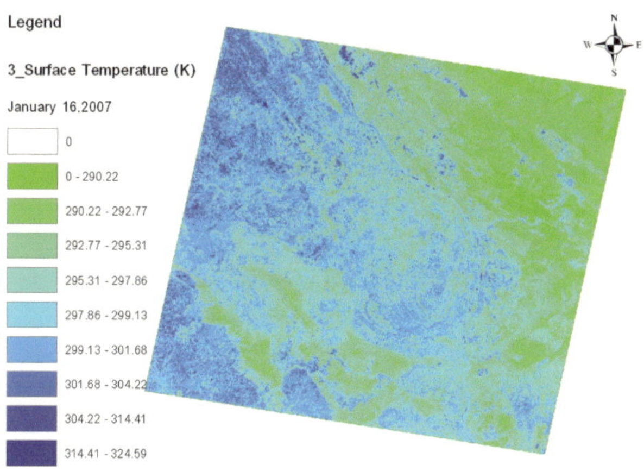

Fig. 4. The surface temperature on January 16, 2007

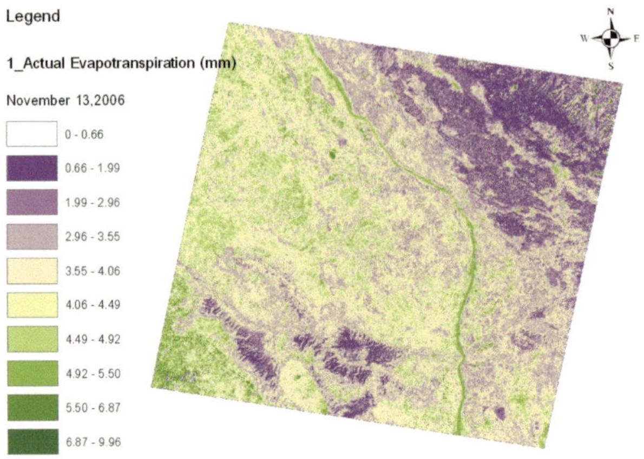

Fig. 5. The actual evapotranspiration on November 13 2006

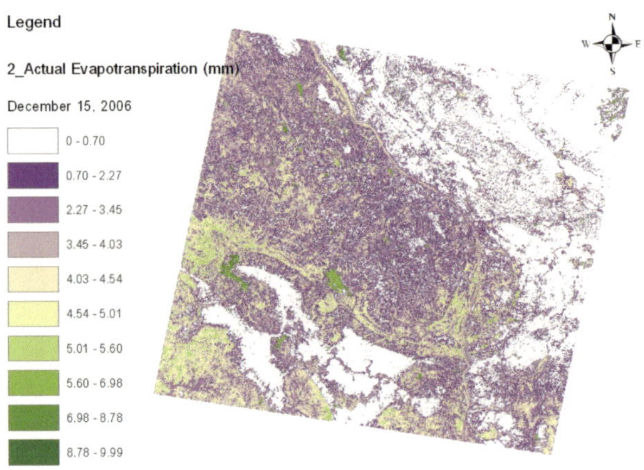

Fig. 6. The actual evapotranspiration on December 15, 2006

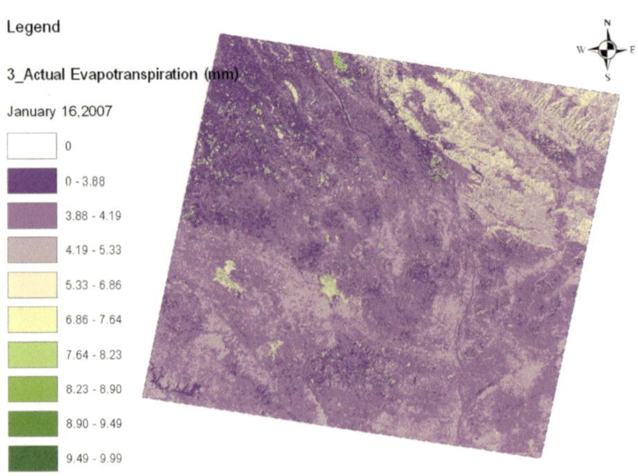

Fig. 7. The actual evapotranspiration on January 16, 2007

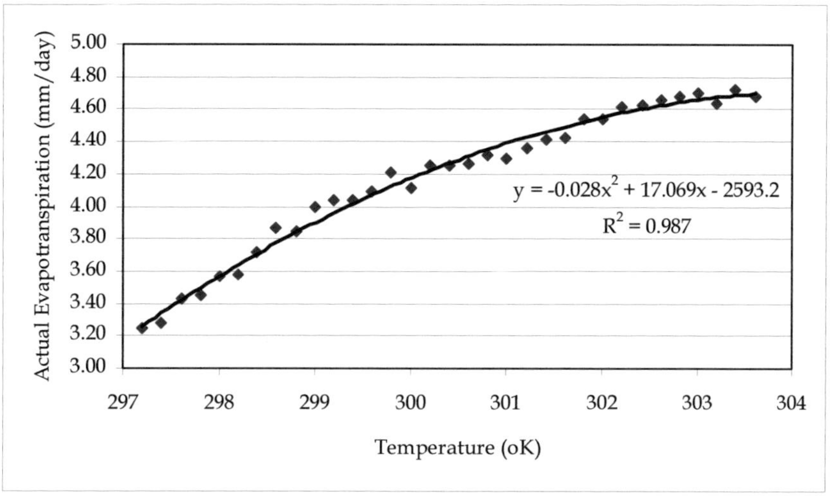

Fig. 8. The relation between the temperature (°K) and actual evapotranspiration (mm/day)

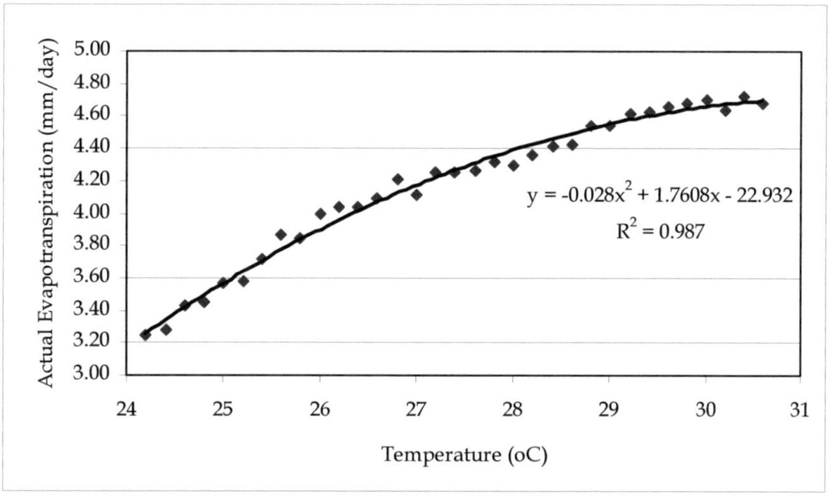

Fig. 9. The relation between the temperature (°C) and actual evapotranspiration (mm/day)

8. Conclusion and recommendation

As above result, it can be concluded that the mean temperature and the mean actual evapotranspiration are 296.44 °K (23.44 °C) and 4.14 mm, respectively. The relationship between the temperature (°K and °C) and actual evapotranspiration (mm/day) is in the format of the polynomial equation. For the temperature in Kelvin, an equation is $y = -0.028x^2 + 17.069x - 2593.2$ and for the temperature in Celsius, an equation is $y = -0.028x^2 + 1.7608x - 22.932$.

The spatial temperature and spatial actual evapotranspiration during November 2006 to January 2007 that are the result are in the condition of both dry season and clear sky. Also, the relationship between temperature and actual evapotranspiration is a result. These results are useful for irrigation project and water management. For spatial temperature, it presents temperature for each area. Since temperature relates to many weather data, temperature can imply the characteristic of other weather data. For example, high temperature is included low humidity but high evaporation can be occurred in the condition of high temperature. For spatial actual evapotranspiration, it can be used to present daily actual evapotranspiration. This daily actual evapotranspiration leads to the planning of water management in each area. Then, it is easy to manage water by irrigation project. For the relationship between temperature and actual evapotranspiration, it can be used to estimate actual evapotranspiration when temperature is not unknown.

9. Acknowledgements

The authors are grateful to the Suranaree University of Technology for funding, facilities and equipment provided. Satellite images from the Geo-informatics and Space Technology Development Agency (Public Organization) are appreciated.

10. References

Abou E. and T. Tanton. (2003). Real time crop coefficient from SEBAL method for estimating the evapotranspiration, *Proceedings of SPIE*, Vol. 5232.

Allen R.G., Pereira L.S., Dirk R. and Smith M. (1998). Crop evapotranspiration-Guidelines for computing crop water requirements, *FAO Irrigation and drainage paper 56*.

Allen, R. G., A. Morse, M. Tasumi, W. Bastiaanssen, W. Kramber and H. Anderson. (2001). Landsat Thematic Mapper for evapotranspiration via the SEBAL process for water rights management and hydrology water balances, SEBAL, Available Source: http://www.agu.org/meetings/sm01/sm01_pdf/sm01_B42C.pdf, August 8, 2003.

Amin, M. S. M., A. Nabi and S. Mansor. (1997). Use of satellite data to estimate areal evapotranspiration from a tropical watershed, *Asian Conference on Remote Sensing*, 20-24 October 1997, GIS Development Pvt. Ltd., Malaysia.

Andrew, N. F., T. J. Schmugge and W. P. Kustas. (2002). Estimating evapotranspiration over El Reno, Oklahoma with ASTER imagery. *EDP Sciences*, Vol. 22, pp. 105-106.

Bastiaanssen, W. G.M., M. Meneti, R. A. Feddes and A. A. M. Holtslag. (1998a). A remote sensing surface energy balance algorithm for land (SEBAL) 1 Formulation. *Journal of Hydrology*, Vol. 212-213, pp. 198-212.

Bastiaanssen, W. G.M., M. Meneti, R. A. Feddes and A. A. M. Holtslag. (1998b). A remote sensing surface energy balance algorithm for land (SEBAL) 1 Validation. *Journal of Hydrology*, Vol. 212-213, pp. 213-229.

Bastiaanssen, W. G.M. (2000). SEBAL-based sensible and latent heat fluxes in the irrigated Gediz Basin, Turkey. *Journal of Hydrology*, Vol. 229, pp. 87-100.

Bastiaanssen, W. G.M., M. Ahmad and Y. Chemin. (2002). Satellite surveillance of evaporative depletion across the Indus Basin. *Water Resource Research*, Vol. 38, No. 12.

Chemin, Y. and M. Ahmad. (2000). Estimating evaporation using the surface energy balance algorithm for land (SEBAL). *A Manual for NOAA-AVHRR in Pakistan. International Water Management Institute (IWMI)*. Report No. R-102.

Chong-yu XU, Lebing G., Jlang T., and Deliang C. (2006) Decreasing Reference Evapotranspiration in a Warming Climate-A Case of Changjiang (Yangtze) River Catchment During 1970-2000. *Advances Atmospheric Sciences* Vol. 23, pp. 115-131.

Farr, R. (1999). Landsat 7 Project, Policy, and History, Landsat, Available Source: http://landsat.gsfc.nasa.gov/main/project.html, October 21, 2003.

Grismer, M. E., M. Orang, R. Snyder, and R. Matyac (2002). Pan Evaporation to Reference Evapotranspiration Conversion Methods. *Journal of Irrigation and Drainage Engineering*, Vol. 128, No. 3, pp. 180-184.

Hongjie X, Jan H., Shirley K. and Eric S. (2002). Comparison of evapotranspiration estimates from the surface energy balance algorithm (SEBAL) and flux tower data, middle Rio Grande Basin, Evapotranspiration, Available http://www.nmt.edu/~hjxie/sebal-agu.htm, November 29, 2002.

Humphreys, E., W. S. Meyer, S. A. Prathapar and D. J. Smith. (1994). Estimation of evapotranspiration from rice in southern New South Wales: a review. Australian. *Journal of Experimental Agriculture,* Vol. 34, No.7, pp. 1013 - 1020.

Ines A. V. M. (2002). Improved crop production integration GIS and Genetic Algorithm. *Thesis of Doctor Degree*, Asian Institute of Technology, Thailand.

Jacob, F., A. Olioso, X. F. Gu, J. F. Hanocq, O. Hautecoeur and M. Leroy. (2003). Mapping surface fluxes using Visible-Near Infrared and Thermal Infrared data with the SEBAL Algorithm. *Physics and Chemistry of the Earth.*

Kalluri S., Gilruth P., Bergman P. and Plante R. (2003) Impacts of NASA's remote sensing data on policy and decision making at state and local agencies in the United State. Evapotranspiration. Available http://earth-outlook.east.hitc.com:1500/05_07_11.00_kallyri.pdf, August 5, 2003.

Kosa P. (2009) Air Temperature and Actual Evapotranspiration Correlation Using Landsat 5 TM Satellite Imagery. *Kasetsart Journal (Natural Sciences)*, Vol. 43, pp. 605-611.

Lal M., Y. Chemin and L. Chandrapala. (2001). Variability of soil moisture in the Walawe River Basin: A case study in Sri Lanka using low-resolution satellite data, *Asian Comference on Remote Sensing,* 5-9 November 2001. GIS Development Pvt. Ltd., Singapore.

Liang, S., H. Fang and M. Chen. (2001). Atmospheric correction of Landsat ETM+ Land Surface Imagery-Part I: Methods. *IEEE Transactions on Geoscience and Remote Sensing,* Vol. 39, No. 11, pp. 2490-2498.

Liu, J. G. (2000). Evaluation of Landsat-7 ETM+ panchromatic band for image fusion with multispectral bands. *Natural Resources Research,* Vol. 9, No. 4, pp. 269-276.

Marco, A. F. C. (2002). Reference Evapotranspiration Based on Class A Pan Evaporation. *Scientia Agricola Journal,* Vol. 59, No. 3, pp. 417-420.

Morse, A., M. Tasumi, R. G. Allen and J. W. Kramber. (2000). Application of the SEBAL methodology for estimating consumptive use of water and stremflow depletion in the BEAR River Basin of IDAHO through remote sensing. Final Report submit to The Raytheon System Company Earth Observation System Data and Information System Project.

Timmermans, W. J., A. M. J. Meijerink and M. W. Lubczynski. (2001). Satellite derived actual evapotranspiration and groundwater modeling, Botswana, *Proceedings of a symposium held at Santa Fe, New Mexico, USA*, April 2000. IAHS Publ, No. 267.

Wei, Y. and Sado. K. (1994). Estimation of Areal Evapotranspiration Using Landsat TM data Alone, *Asian Conference on Remote Sensing*, 17-23 November 1994, GIS Development Pvt. Ltd., Bangalore, India.

Xihua, Y., Q. Zhou and M. Melville. (1997). Estimating local sugarcane evapotranspiration using Landsat TM image and a VITT concept. *International Journal of Remote Sensing*, Vol. 18, No. 2, pp. 453-459.

10

Use of Visible Geostationary Operational Meteorological Satellite Imagery in Mapping Reference and Potential Evapotranspiration over Florida

John R. Mecikalski[1], David M. Sumner[2], Jennifer M. Jacobs[3], Chandra S. Pathak[4], Simon J. Paech[5], and Ellen M. Douglas[6]

[1]*University of Alabama in Huntsville, Huntsville, Alabama*
[2]*U. S. Geological Survey (USGS), Orlando, Florida*
[3]*University of New Hampshire, Durham, New Hampshire*
[4]*South Florida Water Management District, West Palm Beach, Florida*
[5]*University of Alabama in Huntsville, Huntsville, Alabama*
[6]*University of Massachusetts, Boston, Massachusetts*
USA

1. Introduction

In Florida, the fraction of mean, annual precipitation returned to the atmosphere as evaporation and transpiration ranges from ~50% in settings of relatively deep water table, shallow-rooted vegetation, and sandy soils (Sumner, 2001) to almost 110% from lakes (Swancar et al., 2000). The prominence of evapotranspiration (ET) in Florida water budgets necessitates its quantification for reliable water-resources management. Water-resources planning often requires use of hydrologic models to assess the impact (e.g., reduction in streamflows, wetland dehydration, or salt-water intrusion into an aquifer) of possible stresses to a hydrologic system. Hydrologic models require spatial and temporal quantification of fluxes into and out of the hydrologic system. The prominence of the outgoing ET flux within the water budget dictates that this flux must be included in many hydrologic models. As actual ET (AET) data are often lacking, ET is usually implemented within hydrologic models with a specified potential ET (PET), defined as the maximum ET at a given location in the absence of water stress, and a conceptualization of the relation between AET and PET. Water allocation and crop use requirements are often estimated using reference ET (RET) and crop coefficients. RET is defined as the ET from a standard reference crop of either grass or alfalfa. In Florida, an assumption of a hypothetical 12-cm grass reference is commonly used; the grass is assumed to be dense and actively growing. RET is typically computed at specific locations based on weather station data.

Estimates of incoming solar radiation (insolation, or R_s) have been made from geostationary satellite data over a 14-year period (1 June 1995 to 31 December 2009, continuing) for use by State of Florida Water Management Districts in ET estimation (Paech et al., 2009). Clearly, geostationary satellites provide spatially and temporally continuous data across all regions in their view (between ±55° latitude), a major advantage over ground-based instrumentation. The most desirable ET datasets are spatially continuous, rather than point values derived from traditional weather station networks. Thus, Statewide mapping of ET is greatly facilitated by satellite-derived estimates of the spatial distribution of insolation. Use of a satellite-based insolation algorithm also ensures that a consistent algorithm is applied across an entire region. Insolation is the primary determinant for temporal variation in PET and RET and is also a large determinant of spatial variation in these values, particularly in areas with heterogeneous cloud cover. RET is valuable for irrigation scheduling and water management, and PET can be used as input into surface and groundwater hydrological models, whereas the insolation data themselves may be used as data input in certain ecosystem models. The five Florida Water Management Districts have not had access to consistent, spatially continuous methods of computing RET and PET. A robust insolation calibration framework coupled to a satellite-based insolation model is described, toward providing a key radiative dataset for the formulation of an initial 9.5-year long (June 1995 - December 2004) ET climatology (which has subsequently been extended through 2009, with annual updates). These insolation datasets are used in conjunction with other information, such as net radiation (R_N), air temperature, relative humidity, wind speed, and land cover information in the formulation of daily, 2-km estimates of PET and RET throughout Florida.

2. GOES insolation data

Satellite visible data have been used for estimating insolation for a number of years, with methods ranging from statistical-empirical relations such as Tarpley (1979), to physical models of varying complexity (see Gautier et al., 1980; Diak & Gautier, 1983; Gautier et al. 1984; Möser & Raschke, 1984; Pinker & Ewing, 1985; Dedieu et al., 1987; Darnell et al., 1988; Frouin & Chertock, 1992; Pinker & Laszlo, 1992; Weymouth & Le Marshall, 1999). Studies such as Schmetz (1989) and Pinker et al. (1995) have proven the utility of satellite-estimated insolation methods, showing that such models produce fairly accurate results – with hourly insolation estimates within 5-10% of pyranometer data during clear-sky conditions (15-30% for all sky conditions) and daily all-sky estimates within 10-15%. Additional works such as those of Stewart et al. (1999) and Otkin et al. (2005) have further bolstered the utility of this technique. Advantages of using satellite-estimated insolation data over those collected by pyranometer networks include large spatial coverage, high spatial resolution, the availability of data in remote, inaccessible, or potentially hazardous regions, over oceans and large water bodies (e.g., Frouin et al., 1988), and in countries that may not have the means to install a ground-based pyranometer network. Cosgrove et al. (2003a,b) describes the use of the Global Energy and Water Cycle Experiment (GEWEX) Surface Radiation Budget (SRB) downward solar flux algorithm [Pinker & Laszlo, 1992] as demonstrated within the North American Land Data Assimilation (NLDAS) project. The error statistics for the SRB product are comparable to those shown by Paech et al. (2009), while the resolution of the SRB solar flux data are 0.5° (Meng et al., 2003).

Approximately 182,000 individual GOES images from the National Oceanic and Atmospheric Administration (NOAA) Geostationary Operational Environmental Satellite (GOES) "East" series of satellites were processed and used for this effort as of 31 December 2009. These data were processed using the method of Gautier et al. (1980) to produce half-hourly and daily-integrated insolation, and 2-week running noontime minimum surface albedo data throughout the State of Florida at 2-km horizontal spatial resolution. This 2 km resolution is chosen to provide insolation observations between cumulus clouds, which make up a significant component of Florida's cloud climatology.

Following the data collection and processing, an extensive calibration activity for the insolation product was undertaken by comparing satellite-derived insolation estimates to ground-based pyranometer measurements and clear-sky radiation models. This comparison allowed for bias corrections to be applied to the daily-integrated insolation dataset for local environmental conditions. Bias corrections were achieved using a three-step process: (1) comparison with ground-based pyranometer measurements on clear (non-cloudy) "reference" days, (2) correcting for bias related to cloudiness, and (3) deriving a monthly bias correction factor. This resulted in a significant reduction in bias errors and henceforth the development of a robust ET product (for RET and PET).

Described in this chapter are the production and calibration of the 2 km GOES-based insolation product, along with the methodology for obtaining PET and RET.

2.1 The insolation model

The insolation estimation method developed by Gautier et al. (1980), with modifications by Diak & Gautier (1983) and updated application methods by Diak et al. (1996), is referred to as the Gautier-Diak-Masse (GDM) method, and employs a simple model representing cloud and atmosphere radiative processes. The GDM method has been shown to perform as well as, or even better than, more complex methods over a variety of land-surface and climatic conditions (Gautier et al., 1984; Frouin et al., 1988; Raphael & Hay, 1984; Diak et al., 1996; Jacobs et al., 2002, 2004; Otkin et al., 2005). Compared to pyranometer data, these studies reported root mean square errors in hourly and daily insolation estimates (as a percentage of the mean pyranometer observed value) from 17-28% and 9-10%, respectively. The high ends of these errors (~28% and ~10%, respectively) were reported in the study by Jacobs et al. (2002), which took place over north and central Florida and was characterized by significant convective-cloud activity. The GDM method has also been proven in operational use, producing near-real-time insolation estimates for regional- and continental-scale land-surface carbon, energy and water flux assessments (Mecikalski et al., 1999; Anderson et al., 2003, 2004), subsurface hydrologic modeling, and the generation of agricultural forecasting products (Diak et al., 1998; Anderson et al., 2001). A full description of the GDM is given by Gautier et al. (1980), Diak & Gautier (1983) and Diak et al. (1996) – a basic overview is given here.

The GDM method is based on conservation of radiant energy in the Earth-atmosphere column. The method has two modes for determining insolation received at the Earth's surface: one for clear and one for cloudy conditions, based on satellite-derived surface albedo (α) data. A running 2 week minimum of this α data, reassessed at solar noon daily, is stored for each GOES satellite visible data pixel, yielding a reference α grid representative of clear-sky conditions and capturing temporal changes in land-surface characteristics. This approach represents the true land-surface α more accurately than using the daily estimated value because the latter can be corrupted by high α values when clouds are present during the course of a day. This minimum α is wavelength-specific, unique to the GOES visible sensor (which does in fact include contributions from the near infrared), and therefore does not represent a true surface α.

Within the GDM, for a given GOES image, the digital brightness at each image pixel is compared to that of the stored clear-sky reference α data for that pixel. If the brightness exceeds a given threshold [a function of the 2-week running minimum noontime α; Diak & Gautier, (1983)], the pixel is deemed cloudy, otherwise it is varying degrees of partly cloudy to clear. Based on this determination, either a clear or cloudy model of atmospheric radiation processes is used to calculate insolation received at the surface for each pixel. Both clear and cloudy models incorporate parameterizations for ozone absorption, Raleigh scattering, and water vapor absorption within the atmospheric column using simple bulk relations - the use of fixed ozone and aerosol contents being sufficient given that these produce secondary sources of error. The cloudy-sky component GDM method estimates a cloud-top α, and accounts for atmospheric effects above and below the cloud separately.

For the water vapor absorption parameterization, a fixed, approximate annual median value of 3.0 cm was used to estimate atmospheric column-integrated precipitable water (PW) during the initial processing. PW is the amount of water that would precipitate out of a vertical column of the atmosphere if all the water vapor were condensed into liquid. PW data are needed to calculate the slantwise path, and, subsequently, the absorption coefficients in the GDM method. Post-processing adjustments were then made to account for day-to-day variations of precipitable water (i.e. PW values greater or less than the 3.0 cm median value), given the logistical difficulty of including these data within the modeling stage. These adjustments were made by deriving adjustment factors based on daily representative PW values over Florida from numerical weather prediction model data. In many instances, daily PW values over Florida were well above 3.0 cm, especially during summer, while wintertime values were often much lower. No accounting was made for intra-day variations in PW considering the relatively small amount of variability that typically occurs over Florida, especially during summer, and because this would have required a reliance on forecasts from numerical weather prediction models, which are often incorrect. We also did not account for meso-γ scale (2-25 km) variations in PW given that such variations are most often quite small (<5%).

2.2 GOES data processing and quality control

The GOES East series of satellites (the most recent additions being GOES–8, –12 and –13) have been placed in geostationary orbit above the Earth's equator at longitude 75° W since 1975. These satellites provide continuous observations in visible and infrared radiation bands of much of the western hemisphere at high spatial (≥1 km) and temporal (≥15 min) resolution, making data collected by them ideal for high-resolution estimates of insolation. The GOES–8 satellite was launched in April 1994, became operational in late May 1995, and was replaced by GOES–12 on 1 April 2003. Subsequently, GOES–13 became operational on 15 April 2010. The Florida insolation processing began on 1 June 1995.

Although the GOES visible sensors have a nadir (the point directly below the satellite) spatial resolution of 1 km, this resolution decreases the farther from nadir the instrument scans. For the State of Florida, the highest resolution attainable is about 1.5-2.0 km, which was the input and output resolution of the GDM in this analysis. Half-hourly insolation values were calculated using GOES data from 15 and 45 min past the hour, and daily values were calculated by integrating the half-hourly values over the period of daylight using the trapezoidal integration method. A simple method for computing sunrise and sunset times per pixel across the domain was used. The running 2-week minimum α was calculated using data at solar noon. The original insolation and α fields, in satellite projection, were then translated to a 2 x 2 km grid that has been used for the statewide NEXRAD radar-derived rainfall estimation product (Hoblit et al., 2003).

Potential GOES data quality issues include sensor degradation with time and sun glint effects. The effects of the latter are small. Sensor degradation is addressed and corrected through the calibration of the product, detailed in following sections. In general, GOES satellite data have high reliability. Under specific conditions, however, the instruments are shut down (for example, when sunlight shines directly into the sensors), and other issues such as receiving-station problems can result in the occasional loss or corruption of an image or series of images. For this reason, if more than 5 half-hourly satellite images are missing on a given day, the daily insolation value for that day was flagged as "unusable." Days with 3-5 missing images are designated "usable," and those with zero to two missing images are designated as "good quality" data. Gaps in the usable data were filled by linear interpolation.

3. Pyranometer data

Pyranometer data used to calibrate the GOES satellite isolation product, and subsequently assess calibration performance, were obtained from five weather station networks across Florida, each maintained by a different agency. The State of Florida is divided into five regional Water Management Districts (WMDs): Northwest Florida (NWF), South Florida (SF), St. Johns River (SJR), Suwannee River (SR), and Southwest Florida (SWF). Historical pyranometer data were provided by three of the WMDs (SF, SJR, and SWF), and the remaining data were obtained from the University of Florida (UF) Institute of Food and Agricultural Sciences (IFAS) Florida Automated Weather Network (FAWN; fawn.ifas.ufl.edu), and from a single long-term station operated by the U. S. Geological Survey (USGS). Data from 57 stations were used; with stations divided into three groups, "Group 1," "Group 2," and "Group 3". Group 1 stations were used for GDM performance assessment (validation), and Group 2 and 3 data were used for calibration. For performance assessment (Group 1), we used nine stations - two within each WMD (except NWF, where only one quality station was available) so that each part of the State would be represented. Group 1 stations had good data quality over the longest periods of record. For Groups 2 and 3, high-quality data also were needed, but many different stations were used over the initial 10-year period (1995-2004) of analysis. Figure 1 shows the locations of all stations within groups over Florida.

Each weather station network used the LI-200 pyranometer produced by LI-COR, Inc. (Lincoln, NE), with recalibration performed every 1-2 years and quoted accuracy of <5%. Pyranometer locations varied from open fields to water bodies (lakes), but saltwater settings were avoided whenever possible in this calibration effort to minimize issues such as salt deposit contamination of the sensors. Temporal resolution of the pyranometer data ranged from 15 min to 1 hr averages, and daily-integrated insolation values were calculated using the midpoint integration method.

A practical issue for the calibration of satellite-based estimates of insolation is the availability of good quality pyranometer data. Most of the data were provided with quality assurance and quality control (QA/QC) flags, but these flags were not adequate for our purposes. Hence, an additional method for evaluating the data, developed by the American Society of Civil Engineers (ASCE), was employed (see Allen et al., 2005). This method involved comparing daily-integrated insolation data (half-hourly insolation data integrated

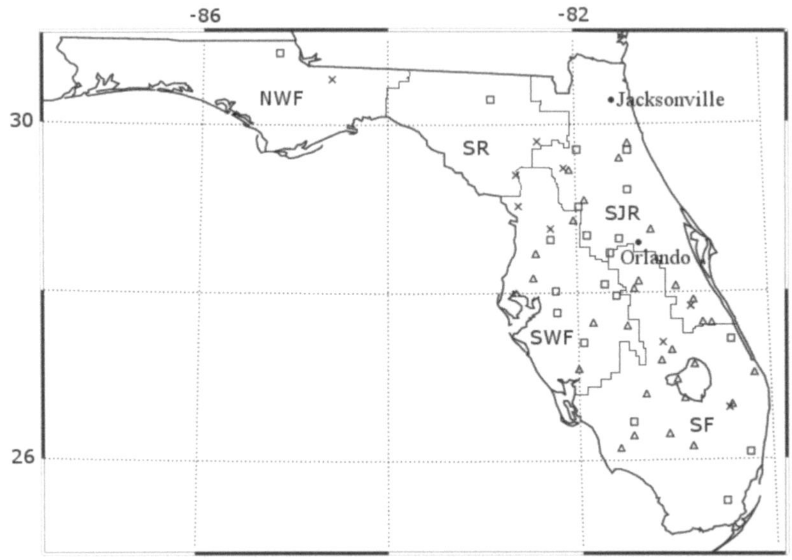

Fig. 1. Locations of pyranometer stations used in GDM calibration. Group 1, 2 and 3 datasets are denoted by crosses, triangles, and squares, respectively. State boundaries and WMD region boundaries are thick and thin black lines, respectively. WMD acronyms are shown. Latitudes are given on the left side and longitudes at the top. Lake Okeechobee can be seen in the southeast part of the state.

into one daily total) with estimated clear-sky radiation, R_{so} (MJm^{-2}day^{-1}), estimated as a function of station elevation (z) and extraterrestrial radiation (R_a: short-wave solar radiation in the absence of an atmosphere) over a 24-hour period by

$$R_{so} = (0.75 + 2 \times 10^{-5} z) R_a , \tag{1}$$

where R_a is a function of day of year, solar constant, solar declination, and latitude, given by

$$R_a = \frac{24}{\pi} G_{sc} d_r [\omega_s \sin(\varphi)\sin(\delta) + \cos(\varphi)\cos(\delta)\sin(\omega_s)] . \tag{2}$$

Here, G_{sc} is the solar constant (4.92 MJm^{-2}h^{-1}), d_r is the inverse relative distance factor for the Earth-Sun (unitless), w_s is the sunset hour angle (radians), φ is latitude (radians), and δ is the solar declination (radians) (Duffie & Beckman, 1980; Allen, 1996; Allen et al., 2005).

The assumption is that measured daily insolation should be close to estimated clear-sky values on at least some days during the year – those days being considered "cloud-free." When examining annual plots of both measured and cloud-free insolation, it was possible to identify when a station had significant data quality issues not indicated by QA/QC flags; specifically, under complete sunshine, quality pyranometer measurements should be near the R_{so} values. For pyranometer data not provided with any QA/QC information, the above method was employed as an initial filter, following that, data greater than 105% of the estimated clear-sky value were removed (Allen et al., 2005). Periods of the record, or even entire station records, that had measured values either substantially above or below the cloud-free "envelope" (by ≥5%) were eliminated from analysis.

4. GDM method calibration for insolation data

The GDM method performs well over a variety of land-surface and climatic conditions, as well as over a range of spatial and temporal resolutions. However, daily-integrated GOES-estimated insolation has biases that can be reduced by adjustments for clear sky conditions, cloudy sky conditions, and seasonality. GOES insolation estimates were fine-tuned through a cumulative three-step process by calibration to ground-based pyranometer data. In this section, we will discuss each of these bias corrections in detail. Briefly, the initial insolation data that are estimated from the GDM method (referred to as the "DAILY_A" dataset) were compared with pyranometer observations on clear (non-cloudy) reference days, resulting in a set of initial calibration coefficients, the application of which produced the second "DAILY_B" dataset. Then, a "cloudiness" bias correction was determined and applied to the DAILY_B data, resulting in the "DAILY_C" dataset. Lastly, a monthly correction factor was applied to the DAILY_C data, yielding the final dataset "DAILY_D". At each calibration step, GOES-estimated and pyranometer-measured insolation data were matched spatially by choosing the satellite data pixel that corresponded to the location of the pyranometer station.

To assess the performance of each calibration step, the A, B, C, and D datasets were compared to data from the "Group 1" stations, which were nine independent stations not used in the calibration process. The results of each step are shown for the entire data period and each of the nine stations in Fig. 2. Statistics used for this comparison included on each plot are the Root Mean Square Error (RMSE, also expressed as a percentage of the mean pyranometer observed value), the Mean Bias Error (MBE), and the coefficient of determination (r^2). Figure 3a shows station-averaged statistics and seasonal-station averaged GDM MBE. The number of stations in the averaged statistics (Fig. 3b) varied from 2 to 7 depending on the length of record for each station.

4.1 Clear day comparison

Disparities between satellite-estimated and pyranometer-measured insolation on clear-sky days should be at a minimum, because without clouds, insolation received at the surface will be spatially homogeneous, providing reference conditions for the comparison of the two datasets. These "clear day comparisons" were made every 6 month on days as free of cloud as possible over Florida - one day in each summer and winter season. Clear days over the entire State of Florida are rare, and therefore many times the comparison was limited to available cloud-free regions.

For each clear day, the half-hourly GDM data were compared with pyranometer data for up to three "Group 2" stations within each of the SF, SJR and SWF WMD regions (no data were available for NWF or SR for this analysis). Pyranometer time-stamps were adjusted to the middle of their data-averaging periods, and paired with unmodified GOES data times on the top of the hour and half-hour. For each WMD region that was analyzed, the GDM and corresponding pyranometer datasets were averaged across the selected stations, resulting in two diurnal curves (one each for the GDM and pyranometer). These two averaged datasets were then compared over the same 24-hour periods, and the satellite data calibration coefficient for each WMD was determined by multiplying the averaged GDM satellite data by a factor necessary for its diurnal insolation curve to closely align with the averaged pyranometer curve. This factor was manually determined as a means of correcting for the satellite–pyranometer differences in clear sky radiation.

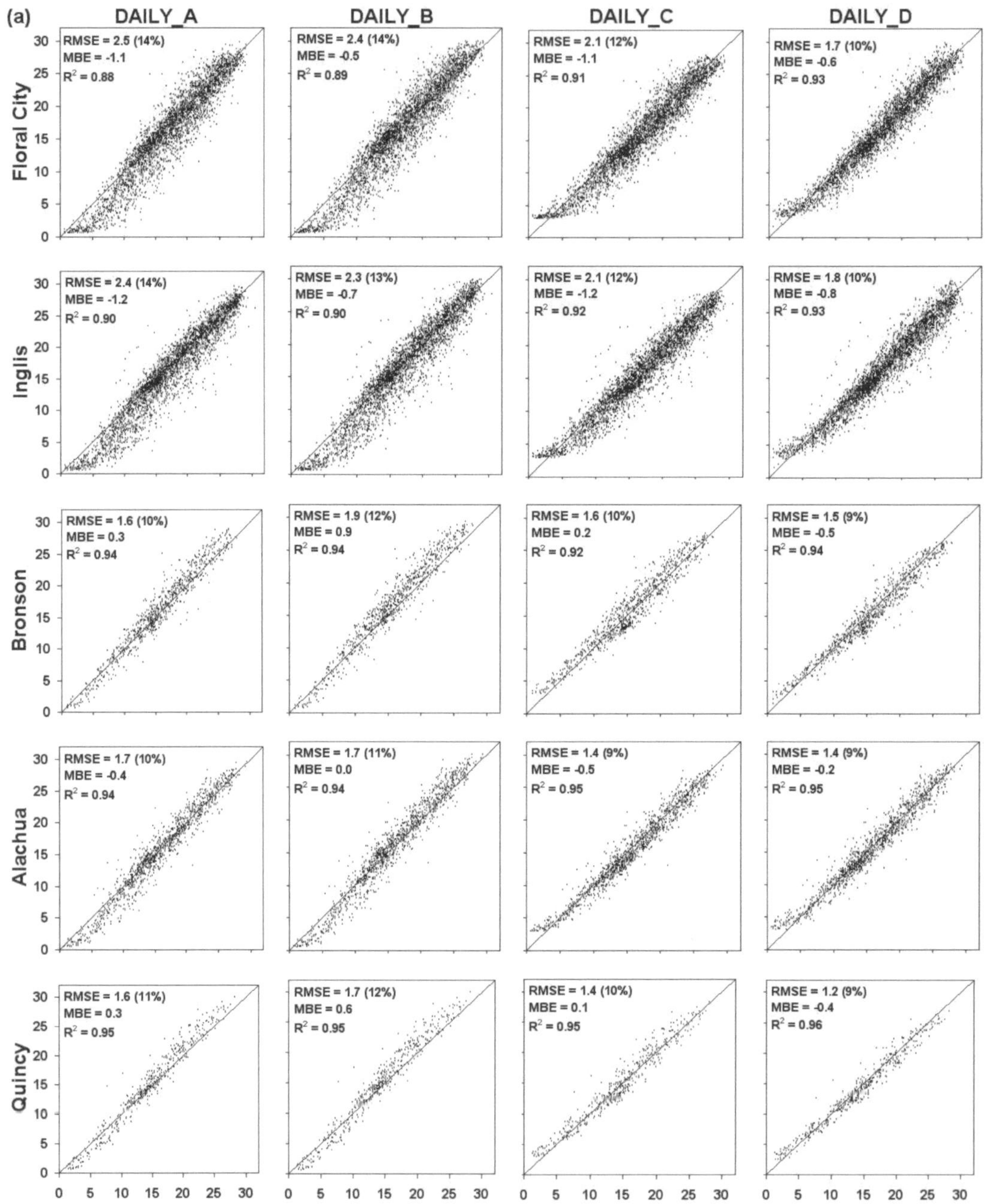

Fig. 2. Comparison of satellite-estimated (ordinate) and pyranometer-measured (abscissa) daily integrated insolation [$MJm^{-2}day^{-1}$] for the nine model calibration locations. Station names are given along the left-side, and comparison satellite-estimated dataset names are given at the top. RMSE values [$MJm^{-2}day^{-1}$], RMSE as a percentage of the mean observed value (in parentheses), MBE [$MJm^{-2}day^{-1}$], and coefficient of determination (R^2 as in figures) for each station and calibration step combination also are shown.

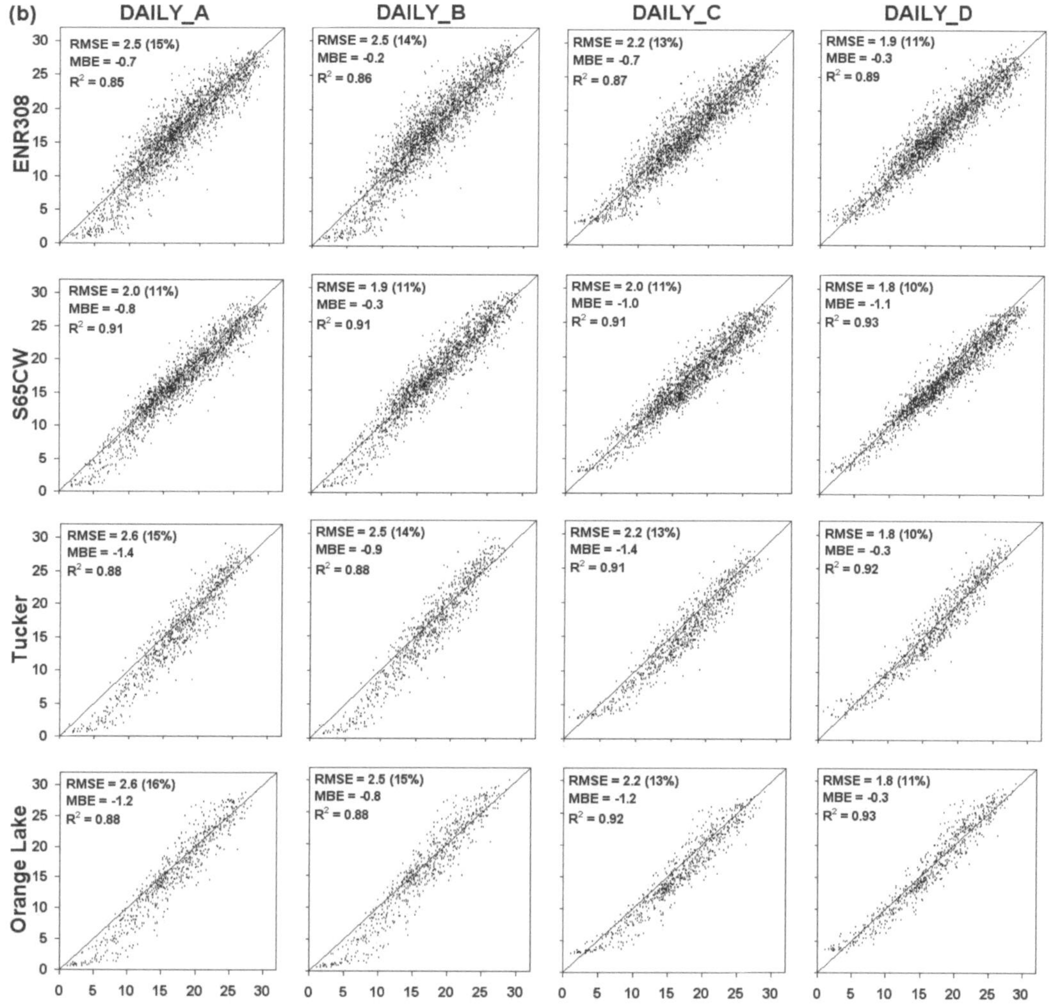

Fig. 2. Continued.

Subsequently, the average of all available WMD correction factors was taken to obtain a calibration coefficient for that particular day to apply over the entire State of Florida. This process was carried out for the entire observation period, resulting in a set of 20 approximately bi-annual clear sky calibration coefficients spanning the data record. The individual coefficients, obtained only on days when pyranometer data were available, were then interpolated in time to obtain a calibration coefficient set that could be applied to each day across the data record.

For calibration of the GDM, the clear sky coefficients were applied to the initial (DAILY_A) data, yielding the DAILY_B dataset. The results of this calibration are shown in Fig. 2 and indicate that the 9-station averaged RMSE remained the same as that of the DAILY_A dataset, the coefficient of determination increased to 0.91, and the average MBE decreased in magnitude from –0.7 to –0.2 $MJm^{-2}day^{-1}$. Figure 3a shows that after the correction for clear sky bias, the MBE generally has been reduced, but a temporal trend of MBE (positive shift with time) is present, and the station and seasonal averages have a similar range (beginning to end) as the initial DAILY_A dataset. A seasonal MBE oscillation is also evident, as is a cloudiness-related bias (Fig. 2).

4.2 Cloudiness bias correction

In an effort to correct for a known cloudiness-related bias in GDM data, a "cloudiness" bias correction was developed using the DAILY_B satellite-estimated insolation dataset (the result of the clear-day calibration). These data were compared with pyranometer data from the "Group 3" dataset with GDM bias values calculated for each individual station and day, and plotted versus "cloudiness index". The cloudiness index is defined here as the ratio of the DAILY_B satellite-estimated insolation to estimated daily clear-sky insolation, R_{so}. The model bias due to cloudiness [$MJm^{-2}day^{-1}$] is approximately linearly related to cloudiness by

$$Cloudiness\_bias = 4.44[Cloudiness\_index] - 2.55 \quad (3)$$

The cloudiness bias given by Eq. (3) was calculated for each Group 3 station for all days and subtracted from the DAILY_B data, resulting in the DAILY_C dataset. Examination of the DAILY_C dataset revealed that the bias related to cloudiness was almost negligible, leading to an increase in the coefficient of determination from 0.91 to 0.92 and a decrease in the average RMSE from 2.2 to 1.9 $MJm^{-2}day^{-1}$. This small improvement is also evident in Fig. 2; although the low end of the model data has been "raised" somewhat, this affects only a small percentage of the data. For the ultimate purpose of this dataset, which is the estimation of ET in Florida, correcting for the cloudiness bias has a small effect seen primarily on very cloudy days, when ET is typically low. With the cloudiness bias correction, the station and time-averaged MBE becomes more negative and increases in magnitude from -0.2 to -0.8 $MJm^{-2}day^{-1}$ (Fig. 3).

4.3 Monthly bias correction and final insolation product

The final calibration step was the development of a monthly bias correction. DAILY_C model and "Group 3" pyranometer data were averaged over all calibration stations for each month. The pyranometer data were then subtracted from the DAILY_C data, resulting in a set of monthly bias correction coefficients spanning the data period. Due to data availability and time constraints, June 1996 through June 1997 coefficients were used for the June 1995 through June 1996 period. This was deemed acceptable as the most important bias features (for example the seasonal oscillation) were captured by this surrogate set of coefficients. These bias corrections were then subtracted from the DAILY_C data, giving us the final dataset, DAILY_D (as shown in Figs. 2 and 3).

The result of all bias corrections was that the station-averaged statistics all improved. The average RMSE and MBE values decreased in magnitude to 1.7 (10% of the mean observed value) and -0.5 $MJm^{-2}day^{-1}$, respectively, and the coefficient of determination increased to 0.93. Compared to the initial dataset (DAILY_A), the RMSE and MBE decreased in magnitude by 0.5 and 0.2 $MJm^{-2}day^{-1}$, respectively, and the coefficient of determination increased by 0.03. Although the final average MBE is still negative and of greater magnitude than the result of the DAILY_B calibration, Fig. 3 shows that the effect of the monthly bias correction led to the removal of both the seasonal oscillation and the positive shift of MBE with time, with the final station average ranging between about -1 to 0 $MJm^{-2}day^{-1}$ across the data record period compared to DAILY_A. Lower GOES insolation values (from 0 to 3

$MJm^{-2}day^{-1}$) were completely removed during the calibration process, as evident in all DAILY_C and DAILY_D datasets.

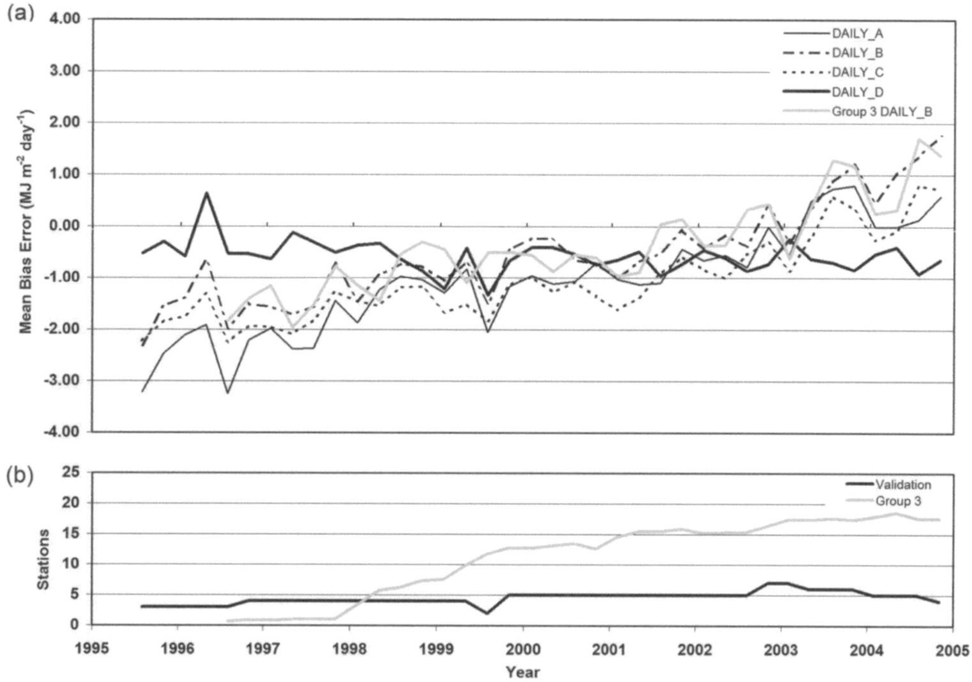

Fig. 3. (a) Season and station averaged daily integrated insolation MBE, and (b) number of stations in the average at any given time. The "Group 3" dataset is included in both (a) and (b) for comparison. Season months are December to February (winter), March to May (spring), etc.

5. Estimating potential ET

5.1 Utility of potential ET methods

Numerous methods exist to estimate PET using atmospheric parameters including wind speed, net radiation (R_N), temperature, and relative humidity (Vörösmarty et al., 1998; Oudin et al., 2005). The study by Douglas et al. (2009) used observed daily ET (DET) from 18 sites in Florida having measured DET and ancillary climate data to compare the performance of three common methods for estimating PET: the Turc, the Priestley-Taylor, and the Penman-Monteith methods. The sites were distributed throughout the State of Florida and represent a variety of land cover types: open water, marshland, grassland/pasture, citrus, and forest. The performance of the three methods when applied to conditions close to PET was used to judge relative merit. Under such PET conditions, the annually aggregated Priestley-Taylor and Turc methods performed comparably across land covers, and outperformed the Penman-Monteith method, possibly due to the use of generic literature values for coefficients in the Penman-Monteith method (Douglas et al., 2009).

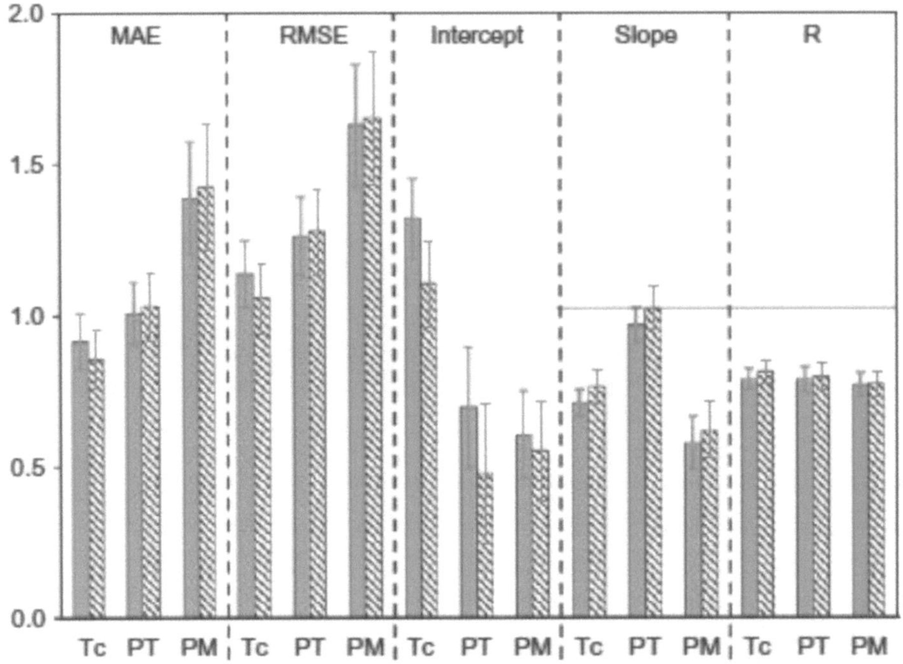

Fig. 4. From Douglas et al. (2009). Comparison of aggregate error statistics (in mm day^{-1}) mean absolute error (MAE), RMSE, Intercept, Slope, and regression coefficients (R), where Tc = Turc method; PT = Priestley–Taylor method; PM = Penman–Monteith method. Filled bars represent statistics computed on all "good" days, hashed bars represent statistic computed for days when the Bowen ratio (β) < 1. Mean values are represented by the heights of the bars. Error bars show the standard error of the mean. The horizontal line on Slope and R represents a value of 1.

The relative ranking of PET methods apparent in this study is (from best to worst): Priestley-Taylor, Turc, and Penman-Monteith, as shown in Fig. 4 (from Douglas et al. 2009). Performance at a daily time scale is indicated by the values of the regression intercept and slope and the correlation coefficient, R (or r). From Fig. 4, at a daily scale, the Turc intercept is much higher and statistically different than either the Priestley-Taylor or Penman-Monteith statistics. The Priestley–Taylor intercept statistics are closest to zero and slope and R statistics closest to 1. At a daily scale, the performance of all three methods does improve when applied to conditions close to PET (β ≤ 1). However, probably due to the lower sample size, the improved statistics are not significantly different from those computed for all "good" data. Interestingly, R values for all models are nearly identical. In aggregate, the Turc and Priestley–Taylor methods perform comparably and both outperform the Penman-Monteith method. But at a daily scale, the Priestley–Taylor performance appears to be superior to the other two methods. In fact, the slope and intercept show that the Turc method significantly overestimates low DET values and underestimates high values. Hence, the Priestley–Taylor method appears to be the best model for estimating PET in Florida.

The Priestley-Taylor method used in this study to estimate PET requires direct measurements of R_N, or R_N derived from incoming solar radiation. As the ground-based network of R_N instrumentation in Florida is sparse, an alternate method must be used for obtaining solar radiation to obtain statewide spatially distributed estimates of R_N and, therefore, PET. The GOES satellites are able to provide hourly estimates of solar radiation that are critical to PET calculations, and that have a spatial resolution that is significantly better than what is available from ground-based pyranometer networks.

5.2 Priestley-Taylor method

The Priestley-Taylor method estimates PET using the concept of the theoretical lower limit of evaporation from a wet surface as the "equilibrium" evaporation

$$PET_o = \alpha_{PT} \frac{\Delta}{\Delta + \gamma}(R_N - G), \tag{4}$$

where PET_o is the Priestley-Taylor-estimated PET (MJ m^{-2} d^{-1}), α_{PT} is the Priestley-Taylor coefficient, Δ is the slope of the saturation vapor pressure curve, γ is the psychrometric constant, R_N is the net radiation (MJ m^{-2} d^{-1}), and G is the soil heat flux (MJ m^{-2} d^{-1}). "Equilibrium" is defined here as evaporation from a wet surface under conditions of minimum advection that result in the actual vapor pressure of the air approaching the saturation vapor pressure. Priestley & Taylor (1972) showed that for conditions of minimum advection with no edge effects, $\alpha_{PT} = 1.26$. For this project, the Priestley-Taylor equation was used with $\alpha_{PT} = 1.26$ and, because daily PET computations were made, G was assumed to be negligible. Dividing PET in MJ m^{-2} d^{-1} by $\lambda\rho_w$ [λ is the latent heat of vaporization (MJ kg^{-1}), ρ_w is the density of water (= 1,000 kg m^{-3})] converts PET to a daily depth of water (mm).
The parameters Δ (kPa°C^{-1}) and γ (kPa°C^{-1}) were computed using

$$\Delta = \frac{4098 e_s}{(237.3 + T_{\min})^2} \tag{5}$$

$$\gamma = \frac{c_p P}{\varepsilon \lambda} \times 10^{-3} = 0.0016286 \frac{P}{\lambda}, \tag{6}$$

where e_s is the saturated vapor pressure (in kPa), c_p is the specific heat of moist air (1.013 kJ kg^{-1} °C^{-1}), P is atmospheric pressure (set equal to 101.3 kPa) and T_{min} is the minimum daily temperature (°C). Saturated vapor pressure was computed as

$$e_s = 0.6108 \exp\left(\frac{17.27 T_{\min}}{237.3 + T_{\min}} \right) \tag{7}$$

The latent heat of vaporization, MJkg^{-1}, is determined as

$$\lambda = 1000(2.501 - 0.002361T). \tag{8}$$

In addition, climate variables, solar radiation (R_s; MJm^{-2}day^{-1}), maximum (T_{max}) and T_{min} daily temperatures (°C), and maximum (RH_{max}) and minimum (RH_{min}) daily relative humidities (%) are used directly and indirectly in the PET methodology, Eqs (4)-(8). The average temperature (T) is computed as the average of T_{max} and T_{min}.

5.3 Estimating net radiation for PET computation

Jacobs et al. (2008) compared multiple methods of calculating R_N, the difference between downwelling and upwelling radiation of both short and long wavelengths, to identify the most robust method for Florida. The recommended method uses a four-component approach where incoming solar radiation, surface albedo, and upwelling and downwelling longwave (LW) radiation must each be measured or estimated. In the recommended method, R_N is estimated using

$$R_N = R_s(1-\alpha) + 0.0864\varepsilon_s R_{LWD} - 0.0864 R_{LWU}, \tag{9}$$

where R_s is the daily solar radiation ($MJm^{-2}day^{-1}$), α is the surface albedo, ε_s is the surface emissivity, R_{LWD} is the downwelling L radiation (Wm^{-2}), R_{LWU} is the upwelling LW radiation (Wm^{-2}), and 0.0864 is the conversion between Wm^{-2} and $MJm^{-2}day^{-1}$. R_N is determined from measured or estimated solar radiation, estimates of α, and modeled LW radiation values from ancillary meteorological data. For this project, the GOES R_s product is used to estimate R_s. Field measurements of downwelling LW, upwelling LW radiation, and reflected insolation were made by SJRWMD using Kipp & Zonen CNR1 sensors at 11 sites in SJRWMD during 2004-2005. The site land covers included water, wetland, urban, rangeland, forest, and agriculture. These radiation data were used to identify suitable methods to estimate the two LW radiation terms and the surface α (ratio of reflected insolation to incoming insolation) throughout Florida.

5.4 Estimating longwave radiation for PET computation

5.4.1 Downwelling longwave for PET computation

Estimation of downwelling LW radiation (R_{LWD}; Wm^{-2}) for PET computation requires two steps: 1) estimation of the clear sky radiation (R_{LWDc}) and 2) implementation of a correction for cloud cover. Choi et al. (2008) determined that R_{LWDc} is best estimated using the Sellers (1965) method (Eq. 10) with Florida-specific parameterization.

$$R_{LWDc} = (a_1 + a_2 e_a^{1/2})\,\sigma T^4, \tag{10}$$

where e_a is the (mean actual) atmospheric vapor pressure (mb), σ is the Stefan-Boltzman constant ($Wm^{-2}K^{-4}$), and T is the average air temperature (K). For this analysis, the Florida-specific coefficients a_1 and a_2 (0.575 and 0.054, respectively) are quite similar to Sellers' original values (0.605 and 0.048).

The vapor pressure, e_a (kPa), requires first calculating the saturation vapor pressure e_s (kPa), using Eq. (7). Then, the daily mean actual vapor pressure, e_a, is calculated from daily extreme temperature and relative humidity values using

$$e_a = \frac{e_s(T_{max})\frac{RH_{min}}{100} + e_s(T_{min})\frac{RH_{max}}{100}}{2}. \tag{11}$$

and converted from kPa to mb using a multiplier of 10. Choi et al. (2008) determined that the (Crawford and Duchon, 1999) method provides the best approach to obtain an equation that calculates daily downwelling LW under both clear and cloudy conditions. Specifically,

$$R_{LWD} = R_{LWDc}(1 - c) + c\sigma T^4, \qquad (12)$$

where c is fractional cloud cover estimated from the incoming solar radiation (Crawford & Duchon, 1999) as

$$c = 1 - R_s/R_{so}, \qquad (13)$$

where R_s is in this study the GOES-estimated incoming solar radiation at the surface, and R_{so} is from Eqs. (1) and (2).

5.4.2 Upwelling longwave for PET computation

Upwelling LW radiation is calculated using surface measurements of temperature

$$R_{LWU} = \varepsilon_s\sigma\, T_s^4 \qquad (14)$$

where ε_s is the surface emissivity, σ is the Stefan-Boltzmann constant, and T_s is the surface temperature. Average daily air temperature (T) is used in Eq. (14) instead of T_s when estimates of daily temperature are needed (Brutsaert, 1982). For typical surfaces, ε_s is approximately 0.97 and this value was assumed in the present study.

5.5 Estimating surface albedo for PET computation

An important step in the four-component calculation of R_N is to estimate surface α accurately. Ultimately, each 2 km cell within the Statewide ET grid will require an α estimate. Three approaches were considered; using measured α values to provide estimates, estimating α using literature values by land use, and using MODerate Resolution Imaging Spectroradiometer (MODIS) or GOES remotely sensed data. A review of site-specific daily α measurements (from the 11 SJRWMD radiation sites) provides insight into typical values across Florida land uses. For all sites, values were mainly between 0.10 and 0.20. A much lower value, 0.062 was observed for an open water site. The highest and lowest average α values were found for a rangeland site and the water sites, respectively. Albedo values were not consistent within land uses. The α data showed differences in magnitude and temporal variability, even within the same land use type. These differences may reflect either the heterogeneous nature of some land uses, or the relatively small number of sites with four-component radiometer data within each land use type.

Most sites where four-component radiometer data were collected exhibited an annual cycle of α, with highest α in December and lowest in July. Differences between the low and high values were on the order of 0.05. The annual cycle was less pronounced for the citrus and forest sites. Apart from the annual cycle, day-to-day values are fairly consistent. Exceptions include a wetland area and an agricultural site. The wetland variation may be due to dynamic water levels that change the relative portion of water and vegetation at the surface. The agricultural site is likely influenced by crop growth and harvest. Because α values did not differ greatly across sites and no consistent values could be distinguished by land use, constant α values were considered to see if they provided reasonable estimates of R_N. Two constant α approaches were examined. One uses a single average α value determined from the mean measured values (0.141), and the other uses two α values determined from the mean measured values for land (0.149) and water (0.062).

The second approach considered to estimate α was to use literature values. Literature values typically were consistent with the measured data from Florida. However, literature values had a range (0.10 to 0.20) that exceeded the entire range observed across measured sites, excluding water. Given the wide range of literature values, the measured α were evaluated further.

The third approach uses satellite-derived α values. Data from the MODIS satellite-borne sensors can be interpreted to provide α data at 16-day, 500-m resolution (Salomon et al., 2006). However, because the MODIS data are not available prior to December 1999 and the period of interest for this study begins in 1995 MODIS α was not considered further. GOES α values were derived as an intermediate product of the GDM, and was produced for the entire period of study at the same temporal resolution as solar radiation. To assess the validity of the GOES estimate, we compared it to sites with measured α that had consistent land cover over a large surrounding area, so that land use variation within the grid would be minimal. After examining satellite images of all the radiation sites, three sites were identified that had relatively homogeneous land cover conditions over at least a 2 km footprint (rangeland, forest, and agricultural land covers). The GOES product was found to consistently underestimate α compared to measured α, however the seasonal patterns appeared to be consistent. Thus, we conducted a linear regression analysis using measured values from the three sites and estimated scaled GOES α for 2005. The original GOES α product was improved by an average regression equation [$y = 0.634x+0.0679$, where y is the scaled GOES α, x is the original GOES α, and where correlation coefficient (r^2) values were <0.2]. The corrected GOES α was more suitable for estimating R_N. However, as a whole, the GOES α tends to create a 6% positive bias, while the constant α approach only slightly underestimate the measured R_N values. Overall, the constant α values therefore provided better R_N estimates than did the GOES α estimates. The bimodal α approach gives $r^2 = 0.93$, a RMSE of 16.0 Wm^{-2}, and a 2% negative bias, suggesting a very minor reduction in performance using a constant α rather than site specific values.

Based on the comparison of several α approaches described above, constant α values were used for land (0.149) and water (0.062) in computation of PET. The SJRWMD, in collaboration with SFWMD and the USGS, classified all the 2 km grid pixels in Florida as either land or water (Fig. 5). Inland pixels were identified as water if 75% or more of the pixel contained water. The Atlantic Ocean, Gulf of Mexico, lagoons and bays were not classified as water in Fig. 5, and are coded as land with α values of 0.149. The calculated PET values use a land α for these regions.

6. Estimating reference ET

For the purpose of establishing uniform ET estimates and transferable crop coefficients, the ASCE Evapotranspiration in Irrigation and Hydrology Committee (ASCE-ET) recommends two standardized reference ET surfaces: (1) a short crop (similar to grass) and (2) a tall crop (similar to alfalfa). Also recommended is one standardized RET equation based on the Penman-Monteith equation (Penman, 1948; Monteith, 1965; Allen et al., 1998). As a part of the standardization, the "full" form of the Penman-Monteith equation and associated equations for calculating aerodynamic and bulk surface resistance were combined and reduced to a single equation having two constants (see Allen et al., 1998).

For a grass reference on a daily basis, the RET (RET_o in Eq. 15) method is given as

$$RET_o = \frac{0.408\Delta(R_N - G) + \gamma \dfrac{900}{T+273} u_2 (e_s - e_a)}{\Delta + \gamma(1 + 0.34u_2)}, \tag{15}$$

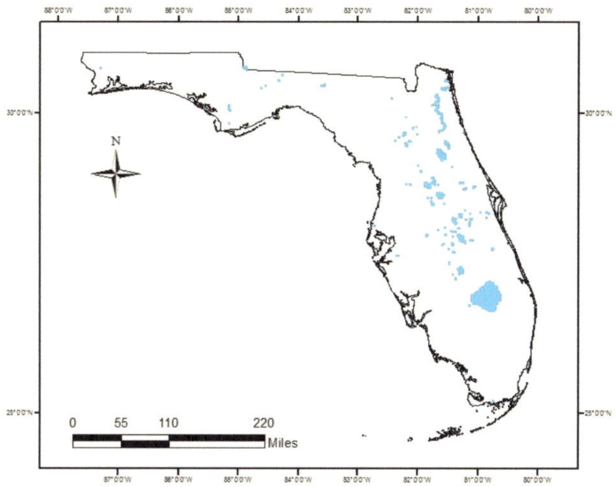

Fig. 5. Florida inland water bodies.

where the variables are as for Eqs. (4)-(8), T the average air temperature (°C), and while u_2 (ms^{-1}), e_a (kPa), and e_s–e_a (kPa) are defined below.

The standardized method used in this project is the short crop or grass reference calculated on a daily basis. This hypothetical reference surface has an assumed canopy height of 0.12 m, a constant bulk surface resistance of 70 sm^{-1}, and an assumed albedo of 0.23. The zero plane displacement height and roughness lengths are estimated as a function of the assumed crop height, so that aerodynamic resistance becomes a function of only the measured wind speed. The height for the temperature, humidity, and wind measurements is assumed to be 2 m. The latent heat of vaporization λ, is assigned a constant value of 2.45 $MJkg^{-1}$.

Wind speed values not collected at 2 m were converted to the 2-m height prior to calculating RET using the standard procedures outlined in the Food and Agriculture Organization of the United Nations Irrigation and Drainage Paper No. 56 (FAO56; Allen et al., 1998) according to the following equation:

$$u_2 = u_z \frac{4.87}{\ln(67.8z - 5.42)}, \tag{16}$$

where u_2 is the measured wind speed at z-meters above ground surface (ms^{-1}), in this case z=2, and z is the height of measurement above ground surface (m). The vapor pressure deficit requires the saturation vapor pressure, calculated in Eq. (7).

For this ASCE method, the daily mean saturation vapor pressure, e_s (kPa), is calculated as

$$e_s = \frac{e_s(T_{max}) + e_s(T_{min})}{2} \tag{17}$$

and the daily mean actual vapor pressure (kPa) is calculated from Eq. (11) (actual vapor pressure). R_N calculated using this ASCE method is the difference between net shortwave radiation (R_{nSW}) and net LW radiation (R_{nLW})

$$R_N = R_{nSW} - R_{nLW}. \tag{18}$$

Net shortwave radiation, R_{nSW}, is calculated the same as previously using insolation (R_s), with α=0.23:

$$R_{nSW} = (1-\alpha)R_s \,. \tag{19}$$

The R_{nLW} calculation is somewhat more involved. R_{nLW} is calculated as

$$R_{nLW} = \sigma\left[\frac{(T_{max,K})^4 + (T_{min,K})^4}{2}\right]\left(0.34 - 0.14\sqrt{e_a}\right)\left(1.35\frac{R_s}{R_{so}} - 0.35\right), \tag{20}$$

where $T_{max,K}$ and $T_{min,K}$ are the maximum and minimum absolute temperatures (K) during the 24-hour period, extraterrestrial radiation, R_a, is calculated as in Eq. (2) from Julian day and latitude, and outgoing solar radiation, R_{so}, is calculated from:

$$R_{so} = 0.75R_a \,. \tag{21}$$

Other terms are as defined previously.

7. Calibration influences on PET and RET, and ancillary meteorological data

Paech et al. (2009) provide a detailed analysis of the issues surrounding the calibration of the GOES insolation estimates. These issues include: (1) the change in GDM model bias with time (as a function of GOES visible sensor age since launch), and (2) less well known seasonal bias oscillations caused by sun angle, sun glint, and season aerosol concentrations (e.g., presence of the Saharan Air Layer over Florida in summer). However, one aspect that is particularly relevant for this analysis is how the GOES insolation errors, in terms of percentage, propagate into RET and PET estimates.

PET can be highly influenced by errors or variability in insolation because the computations are not offset by an aerodynamic component as with RET (Eq. 15). For example, in the Priestley-Taylor approach (Priestley & Taylor, 1972) used in this study (Eq. 4), PET is directly proportional to R_N, and because most of the variability in R_N is due to the insolation component, PET is particularly sensitive to that energy flux. From a preliminary analysis, it is estimated that a 10% error in insolation causes a 10-11% error in annual average PET and only a 6-7% error in annual average RET. Overall, monthly RET generally exceeds PET in the winter due to the often high winds during this season and the presence of wind speed in the aerodynamic term of the RET equation.

Lastly, the RET and PET calculations require ancillary daily meteorological data. Tables 1 and 2 summarize the required data and their sources. A quality assurance procedure was applied to measured data. A threshold analysis was applied to limit the maximum relative humidity to 100%. Temperature, relative humidity, and wind speed were assessed using graphical tools. The short periods having erroneous or missing values were replaced with an average of the previous and next day's values. Longer periods having erroneous or missing values were replaced with average recorded values using the remaining years' observations for that site and day. The gridded meteorological data are created from point station data. An inverse distance weighting method is used to interpolate daily point meteorological data to a 2 km grid scale on an annual basis for each meteorological variable. The five meteorological variables each require a separate interpolation.

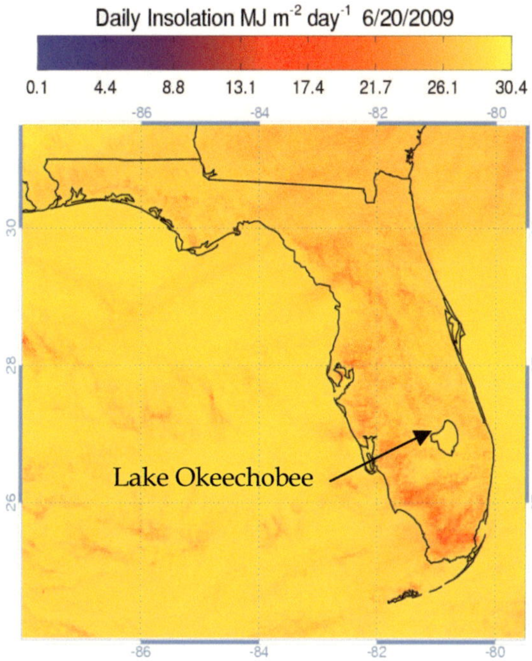

(a)

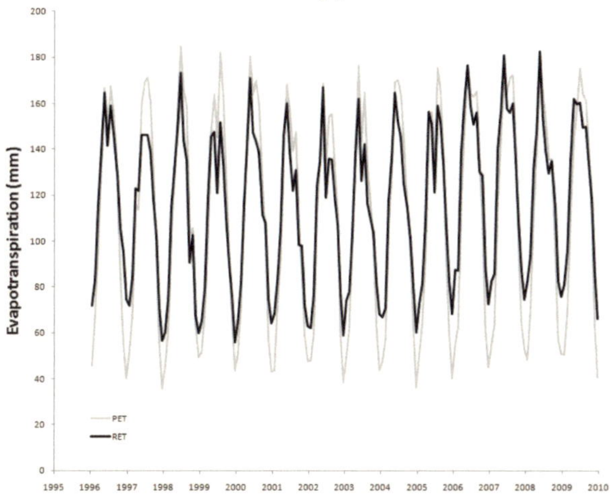

(b)

Fig. 6. (a): Daily insolation for 20 June 2009. (b): Estimated monthly PET and RET near Lake Starr in central Florida as estimated using the insolation data.

Figure 6a shows a map of the final product, 2 km daily insolation for one day (20 June 2009) developed from the calibrated GDM method using GOES satellite data. Note the spatial variation in the insolation in Fig. 6a resulting from differences in cloud cover across the State. The relatively clear-sky conditions over the ocean, near-coastal area, and the large inland Lake Okeechobee are features that produce persistent patterns in insolation. The variation of monthly PET and RET is seen in Figure 6b, exhibiting the substantial correlation between these two ET values. Also, the year-to-year variation in PET and RET is apparent, largely related to similar variations in insolation.

Model Input	PET_o	RET_o	Source
Solar Radiation (mean)	X	X	GOES
Air Temperature (min and max)	X	X	NOAA/NCDC, FAWN
Relative Humidity (min and max)	X	X	NOAA/NCDC, WMDs, FAWN
Wind Speed (mean)		X	NOAA/NCDC, WMDs, FAWN
Incoming Longwave Radiation	X		Calculated using Air Temperature, Relative Humidity, and Insolation
Outgoing Longwave Radiation	X		Calculated using Air Temperature
Albedo	X		SJRWMD R_N Network Values
Land or Water	X		GIS Landcover Analysis

Table 1. Required input data sources by method.

Source	Full Name	Source
NOAA/NCDC	National Oceanic and Atmospheric Administration (NOAA) National Climate Data Center	http://www.ncdc.noaa.gov/oa/ncdc.html
FAWN	Florida Automated Weather Network	http://fawn.ifas.ufl.edu/data/reports/
SJRWMD	St. Johns River Water Management District	http://sjr.state.fl.us/data.html
SWFWMD	Southwest Florida Water Management District	http://www.swfwmd.state.fl.us/data/
SFWMD	South Florida Water Management District: DBHydro	http://www.sfwmd.gov/dbhydroplsql/ show_dbkey_info.main_menu

Table 2. Web addresses for source data (effective November, 2007).

8. High-resolution ET products and applications across Florida

The spatially continuous PET and RET coverages of the entire State of Florida derived from GOES satellite data are presented and discussed in this section. Figure 7(a-d) illustrates an example for 2004: mean annual PET (Fig. 7a), mean July RET (Fig. 7b), mean annual RET (Fig. 7c), and insolation (Fig. 7d). Note that the continuity and high spatial resolution of these data are not attainable from ground-based observation networks. All data are plotted at 2 km resolution. The general north-to-south increase in PET and RET is seen in Figs. 7 (a and c), with the insolation map for July showing less variability as expected.

Figure 8(a-b) illustrates an example of daily PET and RET for 20 June 2009 over Florida. In this figure, the water bodies are seen clearly (see Fig. 5), especially in the PET map, because

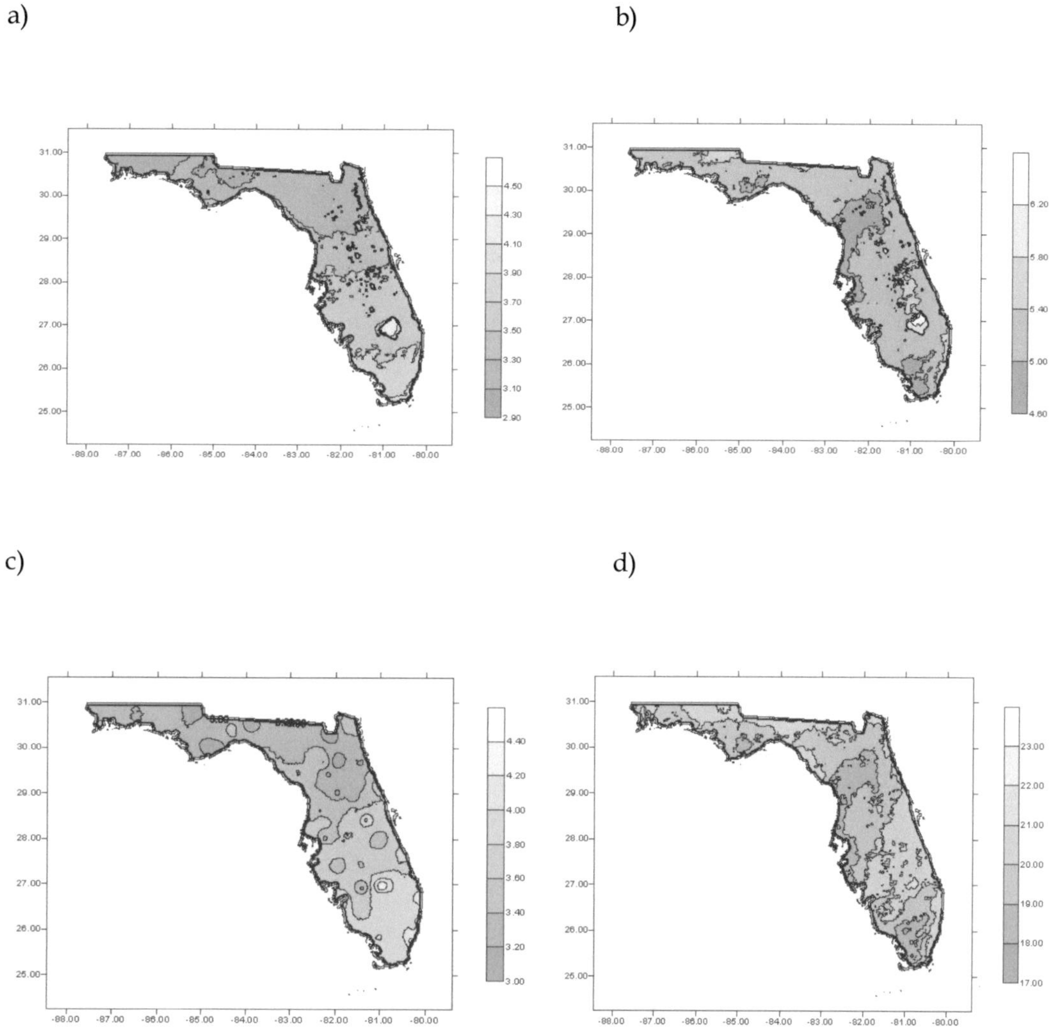

Fig. 7. (a) 2004 mean annual PET (mm day^{-1}), (b) mean July 2004 RET (mm day^{-1}), (c) 2004 mean annual RET (mm day^{-1}), (d) and insolation (MJ m^{-2} day^{-1}) for July 2004.

of the water/land variation in surface α used in the PET computation. In addition, the clear-sky conditions and associated high insolation (Fig. 6a) over the large inland Lake Okeechobee contribute to a distinct area of high PET and RET. The strong spatial correlation between insolation (Fig. 6a) and PET/RET (Fig. 8a-b) is indicative of the primary control of insolation on these ET values. The GOES insolation product is well suited to capture Statewide variations in this variable and the procedures outlined in this chapter present a viable approach to integrate this high-quality insolation product into computations of PET and RET.

This chapter has demonstrated how high resolution (2 km) insolation from GOES visible satellite data can be used to produce spatially uniform and continuous estimates of RET and PET over a statewide region. Florida, which experiences a wide range of ET conditions, has been a demonstration testbed for these products, which have proved to be valuable to the agricultural and hydrologic communities. These data also close the water budget by allowing for quality estimation of maximum evaporation and/or ET, which has been a difficult quantity to measure consistently over large regions. High-quality insolation data will be available from geostationary visible sensors for the long-term future, which allows any region within the view of a geostationary satellite (~55° N to ~55° S latitude) to develop a similar application.

Daily PET, RET insolation, and ancillary meteorological data for the State of Florida are processed and archived on an annual basis and are available at the USGS Florida Water Science Center Hydrologic Web Portal (hdwp.er.usgs.gov).

Use of trade, product, or firm names in this publication is for descriptive purposes only and does not imply endorsement by the U.S. Government.

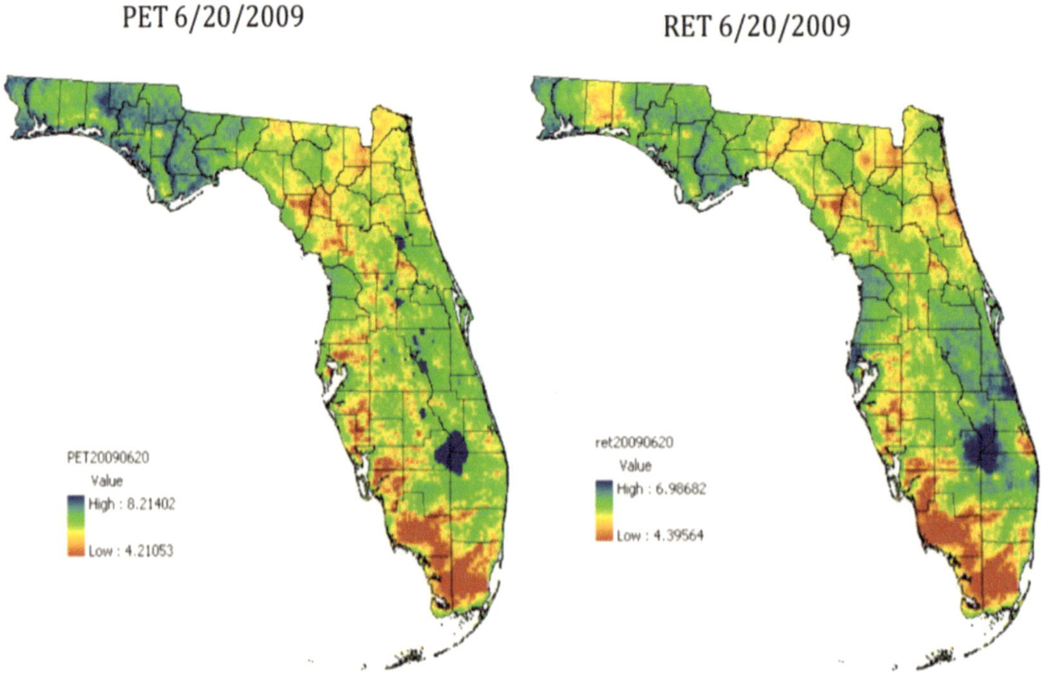

Fig. 8. Daily PET (mm day^{-1}; left) and RET (mm day^{-1}; right) for 20 June 2009. These images used the daily-integrated insolation as shown in Fig. 6a, and exemplify the high-resolution (2 km) detail that can be retrieved when GOES data are exploited in this manner.

9. References

Allen, R. G. (1996). Assessing integrity of weather data for use in reference evapotranspiration estimation. *J. Irrig. and Drain. Engrg.*, ASCE. 122, 97-106.

Allen, R. G.; Periera, L. S., Raes D. & M. Smith (1998). Crop evapotranspiration: Guidelines for computing crop requirements. Irrigation and Drainage Paper No. 56. Rome, Italy: FAO.

Allen, R. G.; I. A. Walter, R. L. Elliott, T. A. Howell, D. Itenfisu, M. E. Jensen, & R.L. Snyder (2005). The ASCE standardized reference evapotranspiration equation, American Society of Civil Engineers, Reston, Virginia.

Anderson, M. C.; W. L. Bland, J. M. Norman & G. R. Diak (2001). Canopy wetness and humidity prediction using satellite and synoptic-scale meteorological observations, *Plant Dis.*, 85, 1018–1026.

Anderson, M. C.; W. P. Kustas, & J. M. Norman (2003). Upscaling and downscaling – A regional view of the soil-plant-atmosphere continuum, *Agron. J.*, 95, 1408–1432.

Anderson, M. C.; J. M. Norman, J. R. Mecikalski, R. D. Torn, W. P. Kustas, & J. B. Basara (2004). A multiscale remote sensing model for disaggregating regional fluxes to micrometeorological scales, *J. Hydrometeor.*, 5, 343–363.

Brutsaert, W. (1982). *Evaporation into the Atmosphere – Theory, History and Application*. Kluwer Academic Publishers, Dordrecht.

Choi, M.; J. M. Jacobs, & W. P. Kustas (2008). Assessment of clear and cloudy sky parameterizations for daily downwelling longwave radiation over different land surfaces in Florida, USA, *Geophys. Res. Lett.*, 35, L20402, doi: 10.1029/2008GL035731.

Cosgrove, B. A.; D. Lohmann, K. E. Mitchell, P. R. Houser, E. F. Wood, J. Schaake, A. Robock, C. Marshall, J. Sheffield, L. Luo, Q. Duan, R. T. Pinker, J. D. Tarpley, R. W. Higgins, & J. Meng (2003). Realtime and retrospective forcing in the North American Land Data Assimilation Systems (NLDAS) project. *J. Geophys. Res.*, 108 (D22), 8842, doi:10.1029/ 2002JD003118.

Cosgrove, B. A.; D. Lohmann, K. E. Mitchell, P. R. Houser, E. F. Wood, J. C. Schaake, A. Robock, J. Sheffield, Q. Duan, L. Luo, R. W. Higgins, R. T. Pinker, & J. D. Tarpley (2003). Land surface model spinup behavior in the North American Land Data Assimilation System (NLDAS). *J. Geophys. Res.*, 108 (D22), 8845, doi:10.1029/2002JD003119.

Crawford, T. M. & C. E. Duchon (1999). An improved parameterization for estimating effective atmospheric emissivity for use in calculating daytime downwelling longwave radiation. *J. Appl. Meteorol.* 38, 474–480.

Darnell, W. L.; W. F. Staylor, S. K. Gupta, & F. M. Denn (1988). Estimation of surface insolation using sun-synchronous satellite data, *J. Climate*, 1, 820–835.

Dedieu, G.; P. Y. Deschamps, & Y. H. Kerr (1987). Satellite estimates of solar irradiance at the surface of the earth and of surface albedo using a physical model applied to meteosat data, *J. Climate Appl. Meteor.*, 26, 79–87.

Diak, G. R. & C. Gautier (1983). Improvements to a simple physical model for estimating insolation from GOES data, *J. Climate Appl. Meteor.*, 22, 505–508.

Diak, G. R.; W. L. Bland, & J. R. Mecikalski (1996). A note on first estimates of surface insolation from GOES-8 visible satellite data, *Agric. For. Meteor.*, 82, 219–226.

Diak, G. R.; M. C. Anderson, W. L. Bland, J. M. Norman, J. R. Mecikalski, & R. M. Aune (1998). Agricultural management decisions aids driven by real-time satellite data, *Bull. Amer. Meteor. Soc.*, 79, 1345–1355.

Douglas, E. M.; J. M. Jacobs, D. M. Sumner & R. L. Ray (2009). Potential evapotranspiration method selection for Florida land cover types. *J. Hydrology*. 373(3-4), 366-376.

Duffie, J. A. & W. A. Beckman (1980). *Solar Engineering of Thermal Processes*. John Wiley and Sons, New York, pp. 1-109.

Frouin, R.; C. Gautier, K. B. Katsaros, & J. Lind (1988). A comparison of satellite and empirical formula techniques for estimating insolation over the oceans, *J. Appl. Meteor.*, 27, 1016–1023.

Frouin, R. & B. Chertock (1992). A technique for global monitoring of net solar irradiance at the ocean surface. Part I: Model, *J. Appl. Meteor.*, 31, 1056–1066.

Gautier, C.; G. R. Diak, & S. Masse (1980). A simple physical model to estimate incident solar radiation at the surface from GOES satellite data, *J. Appl. Meteor.*, 19, 1007–1012.

Gautier, C.; G. R. Diak, & S. Masse (1984). An investigation of the effects of spatially averaging satellite brightness measurements on the calculation of insolation, *J. Climate Appl. Meteor.*, 23, 1380–1386.

Hoblit, B. C.; C. Castello, L. Liu, & D. Curtis (2003). Creating a seamless map of gage-adjusted radar rainfall estimates for the State of Florida, paper presented at EWRI World Water and Environmental Congress, Philadelphia, Pennsylvania, 23-26 June.

Jacobs, J.; J. Mecikalski, & S. Paech (2008). Satellite-based solar radiation, net radiation, and potential and reference evapotranspiration estimates over Florida. A Technical Report prepared for the State of Florida Water Management Districts. Available online at: hdwp.er.usgs.gov/ET/ GOES_FinalReport.pdf.

Jacobs, J. M.; D. A. Myers, M. C. Anderson, & G. R. Diak (2002). GOES surface insolation to estimate wetlands evapotranspiration, *J. Hydrol.*, 56, 53–65.

Jacobs, J. M.; M. C. Anderson, L. C. Friess, & G. R. Diak (2004). Solar radiation, longwave radiation and emergent wetland evapotranspiration estimates from satellite data in Florida, USA, *Hydrological Sci. J.*, 49, 461-476.

Mecikalski, J. M.; G. R. Diak, M. C. Anderson, & J. M. Norman (1999). Estimating fluxes on continental scales using remotely sensed data in an atmosphere-land exchange model, *J. Appl. Meteor.*, 38, 1352–1369.

Meng, C. J.; Pinker, R. T., Tarpley, D. J., & I. Laszlo (2003). A satellite approach for estimating regional land surface energy budget for GCIP/GAPP. *J. Geophys. Res.*, 108, D22, 8861, 10.1029/2002JD003088.

Monteith, J. L. (1965). Evaporation and environment. pp. 205-234. In G.E. Fogg (ed.) *Symposium of the Society for Experimental Biology, The State and Movement of Water in Living Organisms*, 19, Academic Press, Inc., NY.

Möser, W. & E. Raschke (1984). Incident solar radiation over Europe from METEOSAT data, *J. Climate Appl. Meteor.*, 23, 166–170.

Otkin, J.; M. C. Anderson, J. R. Mecikalski & G. R. Diak (2005). Validation of GOES-Based insolation estimates using data from the United States Climate Reference Network. *J. Hydrometeorol.*, 6, 475-640.

Oudin, L., C.; Michel, & F. Anctil (2005). Which potential evapotranspiration input for a lumped rainfall–runoff model? Part 1 – Can rainfall–runoff models effectively handle detailed potential evapotranspiration inputs? *J. Hydrology*, 303, 275–289.

Paech, S. J.; J. R. Mecikalski, D. M. Sumner, C. S. Pathak, Q. Wu, S. Islam, & T. Sangoyomi (2009). Satellite-based solar radiation in support of potential and reference evapotranspiration estimates over Florida: A 10-year climatology. *Water. Resources. Res.*, 45(6), 1328-1342.

Penman, H. L. (1948). Natural evaporation from open water, bare soil and grass. *Proc. Roy. Soc.* London A(194), S. 120-145.

Pinker, R. T. & J. A. Ewing (1985). Modeling surface solar radiation: Model formulation and validation, *J. Climate Appl. Meteor.*, 24, 389–401.

Pinker, R. T. & I. Laszlo (1992). Modeling surface solar irradiance for satellite applications on global scale, *J. Appl. Meteor.*, 31, 194–211.

Pinker, R. T.; R. Frouin, & Z. Li (1995). A review of satellite methods to derive surface shortwave irradiance, *Remote Sens. Environ.*, 51, 105–124.

Priestley, C. H. B. & Taylor, R. J. (1972). On the assessment of surface heat flux and evaporation using large-scale parameters. *Mon. Wea. Rev.*, 100, 81–92

Raphael, C. & J. E. Hay (1984). An assessment of models which use satellite data to estimate solar irradiance at the Earth's surface, *J. Climate Appl. Meteor.*, 23, 832–844.

Salomon, J. G.; C. B. Schaaf, A. H. Strahler, F. Gao, & Y. Jin. (2006). Validation of the MODIS bidirectional reflectance distribution function and albedo retrievals using combined observations from the Aqua and Terra platforms. *IEEE Transactions on Geoscience and Remote Sensing*, 44, 1555-1565, doi:10.1109/TGRS.2006.871564.

Schmetz, J. (1989). Towards a surface radiation climatology. Retrieval of downward irradiance from satellites, *Atmos. Res.*, 23, 287–321.

Sellers, W. D. (1965). *Physical Climatology*. University of Chicago Press, Chicago, Ill.

Stewart, J. B., C. J. Watts, J. C. Rodriguez, H. A. R. De Bruin, A. R. van den Berg, & J. Garatuza-Payan, 1999: Use of satellite data to estimate radiation and evaporation for northwest Mexico, *Agric. Water Manage.*, 38, 181–193.

Sumner, D. M. (2001). Evapotranspiration from a cypress and pine forest subjected to natural fires in Volusia County, Florida, 1998-99: U.S. Geological Survey WRIR 01-4245, 55 p.

Swancar, A.; T. M. Lee, & T. M. O'Hare (2000). Hydrogeologic setting, water budget, and preliminary analysis of ground-water exchange at Lake Starr, a seepage lake in Polk County, Florida: U.S. Geological Survey WRIR 00-4030, 65 p.

Tarpley, J. D. (1979). Estimating incident solar radiation at the surface from geostationary satellite data, *J. Appl. Meteor.*, 18, 1172–1181.

Vörösmarty, C. J.; C. A. Federer., & A. L. Schloss (1998). Potential evaporation functions compared on US watersheds: Possible implications for global-scale water balance and terrestrial ecosystem modeling. *J. Hydrology*, 207, 147–169.

Weymouth, G., & J. LeMarshall (1999). An operational system to estimate global solar exposure over the Australian region from satellite observations, *Aust. Meteor. Mag.*, 48, 181–195.

Part 3

Applications and Case Studies

11

A Calibration-Free Evapotranspiration Mapping (CREMAP) Technique

József Szilágyi[1,2], János Józsa[1] and Ákos Kovács[1]
[1]*Budapest University of Technology and Economics*
[2]*University of Nebraska-Lincoln*
[1]*Hungary*
[2]*USA*

1. Introduction

Obtaining spatially distributed accurate evapotranspiration (ET) estimates is crucial in most water balance calculations for the identification of mass and energy fluxes across the area of interest. While routine Bowen-ratio or eddy-covariance measurements of sensible and latent heat fluxes are typically representative of a horizontal scale of a few hundred meters (i.e., the so-called plot- or field-scale), many hydrologic models work at regional to continental to global scales. Rather than aggregating from the field scale, which may be difficult not only from a theoretical point of view but also from a practical one when long time-periods (years or decades) and large areas are considered, an ET-estimation technique that directly works at the regional scale, yet, at the same time, spatially distributed may be of high practical value.

An ET estimation method had been proposed by Bouchet (1963) almost half a century ago, now called the complementary relationship (CR) of evaporation which was subsequently formulated for practical regional-scale applications by Brutsaert & Stricker (1979), and Morton et al. (1985). The CR states that under constant available energy at the surface (Q_n) actual (E) and potential (E_p) evapotranspiration rates are complementary, their sum yielding twice the wet-environment (E_w) evapotranspiration rate, provided E_p is obtained by the Penman (1948) equation (Szilágyi & Józsa, 2009a) and E_w by the Priestley-Taylor equation (Priestley & Taylor, 1972). In other words

$$E = 2E_w - E_p \quad (1)$$

where

$$E_w = \alpha \frac{\Delta}{\Delta + \gamma} Q_n \quad (2)$$

with Δ being the slope of the saturation vapor pressure curve at the temperature of the air, γ ($\approx$ 0.67 hPa K^{-1}) the psychrometric constant, and α the Priestley-Taylor coefficient with typical values within the interval of 1.2 – 1.3. Penman defined E_p (1948) as

$$E_p = \frac{\Delta}{\Delta + \gamma} Q_n + \frac{\gamma}{\Delta + \gamma} f(u)(e^* - e) \quad (3)$$

where e and e* is the actual and saturation vapor pressure values (in hPa) at the temperature of the air, respectively, and the wind function, f(u) is defined as

$$f(u) = 0.26(1 + 0.54u_2) \quad (4)$$

with u_2 (m s^{-1}), the mean horizontal wind velocity, measured at 2-m height above the ground. With Eq. (4) E_p is given in mm d^{-1}, provided Q_n is also specified as water-depth equivalent in the same units.

While the Advection-Aridity (AA) model of Brutsaert & Stricker (1979) employs the above definitions for the different terms in Eq. (1), Morton et al. (1985) specify E_p and E_w differently in their WREVAP model, but in a way that Eq. (1) is still valid (Szilágyi & Józsa, 2008). They suggest Eq. (1) be employed for time-periods equal or longer than about a week (i.e., a minimum time-period of five days) for improved accuracy. This is so because the CR is based on an assumed dynamical equilibrium of the atmosphere and the underlying land surface, the latter influencing the humidity of the air through ET regulated by various feedbacks across the land-atmosphere interface. To attain such an equilibrium certainly takes some time in the wake of each weather front passing over the region.

In the past decade or so, Eq. (1) has been applied in several continental-scale studies ranging from specifying long-term mean ET across the conterminous US (e.g., Hobbins et al., 2001a,b) to identifying long-term trends in ET rates (Szilágyi, 2001), to investigating the global hydrologic cycle (Brutsaert & Parlange, 1998). While these studies yield regionally representative values of ET, none of them attempt to account for the spatial variability of ET within the region. Recently however Szilágyi & Józsa (2009b), employing the wet surface equation (Yeh & Brutsaert, 1971; Szilágyi & Józsa, 2009c), proposed such a disaggregation scheme utilizing Moderate Resolution Imaging Spectroradiometer (MODIS) data that have a nominal spatial resolution of about 1 km. Below a modified and simplified version of their ET mapping technique is explained with applications over two regions, Hungary in Central-Europe, and Nebraska in Central North America. To the best knowledge of the present authors no other calibration-free ET mapping technique exists that works at the watershed/regional scale and requires such a minimal and freely available input data as the current method to be discussed.

2. Description of the ET mapping approach

Spatial disaggregation of the regionally representative ET rates of Eq. (1), employing the AA (Brutsaert & Stricker, 1972) or WREVAP (Morton et al., 1985) models or the wet-surface equation (Yeh & Brutsaert, 1971; Szilágyi & Józsa, 2009c), is based on a linear transformation of the 8-day composited MODIS daytime surface temperature (T_s) values into actual ET rates (Szilágyi & Józsa, 2009b). Compositing is used for eliminating cloud effects by retaining the warmest pixel value in each 8-day period. See the MODIS website (modis.gsfc.nasa.gov) for more details of data collection and characteristics. The transformation requires the specification of two anchor points in the T_s – ET plane (Fig. 1). The first anchor point is defined by the spatially averaged daytime surface temperature, $\langle T_s \rangle$, and the corresponding regionally representative ET rate, E, from Eq. (1). The second anchor point results from a spatial averaging of the coldest pixel values, $\langle T_{sw} \rangle$, within the region and E_w of Eq. (2) out of consideration that the coldest pixels are the wettest evaporating at the wet environment evapotranspiration rate, E_w. The two points define the linear transformation of the T_s pixel values into ET rates for each 8-day period, or, as was done by Szilágyi & Józsa (2009b) and repeated here, for each month by employing monthly means. The resulting line is extended to the right, since in about half of the pixels ET is less than the regional mean, E. A monthly time-step is ideal since most of the watershed- or large-scale hydrologic models work at this time-resolution, plus a monthly averaging further reduces any lingering cloud effect in the 8-day composited T_s values.

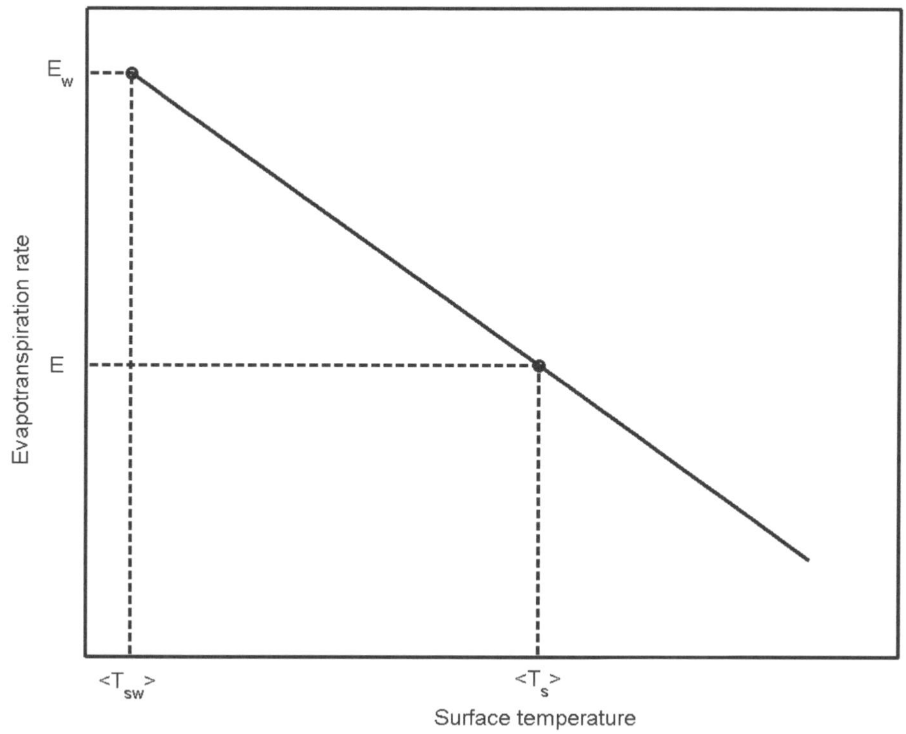

Fig. 1. Schematic representation of the linear transformation of the MODIS daytime surface temperature values into ET rates.

The linear transformation of Szilágyi & Józsa (2009b) can be derived through a generalization of the main idea that is behind the popular plot-scale ET-estimation models of SEBAL (Bastiaanssen et al., 1998) and METRIC (Allen et al., 2007). Namely, the vertical gradient of the air temperature near the surface can be assumed to be linearly dependent of the surface temperature (Bastiaanssen et al., 1998; Allen et al., 2007), thus the sensible heat (H) transfer across the land-atmosphere interface, provided changes in the aerodynamic resistance (r_a) among MODIS pixels are moderate, can also be taken a linear function of T_s. This can be so because under neutral atmospheric conditions (attained for time-steps a day or longer) r_a depends linearly on the logarithm of the momentum roughness height, z_{0m} (Allen et al., 2007), thus any change in z_{0m} between pixels becomes significantly dampened in the r_a value due to the logarithm. Consequently, the latent heat (LE) transfer itself becomes a linear function of T_s under a spatially constant Q_n term required by the CR, since the soil heat flux is generally negligible for intervals longer than a day, therefore $Q_n = H + LE$, from which $LE = mT_s + c$ follows, m and c being constants for the computational time step, i.e., a month here.

While a spatially constant Q_n term may, at first, seem an overly stringent requirement in practical applications due to spatial changes in vegetation cover as well as slope and aspect of the land surface, Q_n will change only negligibly in space provided the surface albedo also changes negligibly among the pixels over a flat or rolling terrain. The spatially constant Q_n (and to a lesser extent z_{0m}) requirement thus limits the spatial resolution the current method can be applied at, since by decreasing the pixel size, heterogeneity among the cells surely increases. Below we show that at least for the study regions, the MODIS pixel size of about 1 km ensures a largely constant Q_n value among the pixels due to only small observed spatial changes in the mean monthly surface albedo value among the MODIS cells.

3. Application of the CREMAP ET mapping method

8-day composited MODIS daytime surface temperature data were collected for Hungary and Nebraska over the 2000 - 2008 and 2000 - 2009 periods, respectively. The 8-day composited pixel values were averaged for each month to obtain one surface temperature per pixel per month.

3.1 Hungary

From the Hungarian Meteorological Service (HMS) 0.1-degree gridded mean monthly air temperature and specific humidity data were collected for the same period as well as sunshine duration values from 15 stations spread across the country which subsequently were interpolated to the same 0.1-degree grid. With these data, the linear transformation of mean monthly T_s values into monthly ET rates can be performed on a (MODIS) pixel-by-pixel basis. For validation purposes (a) 0.1-degree gridded precipitation (P) data from HMS; (b) stream discharge (R) values from the Water Resources Research Center (VITUKI), the University of West Hungary (NYME) and the Regional Water Authorities (RWA) for selected watersheds (Fig. 2); (c) eddy-covariance (EC) measurements of latent heat fluxes from HMS and the CarboEurope web-site (www.carboeurope.org) (Fig. 2); (d) lake evaporation measurements/estimates for Hungary's large shallow lakes (Balaton, Ferto, Velencei) from VITUKI and RWA were also collected.

The country due to its relatively small size was considered as one region which means that the linear transformation is valid for the whole country in the absence of elevation changes. However, with elevation, surface temperatures are expected to decrease independent of whether a pixel is wet or not. Therefore, the country was subdivided into three elevation zones: below 200 m, between 200 and 500 m, and above 500 m (Fig. 3). The anchor points of the linear transformation were calculated separately for the three zones each month by (a) calculating the spatial average of the MODIS-pixel surface temperatures within each zone and the corresponding regional evaporation rate, E, with the help of the WREVAP program for which the input variables were the spatially averaged values of the 0.1-degree gridded air temperature, specific humidity and sunshine duration values within the zone; (b) calculating the wet-environment evaporation rate, E_w, within the zone by Eq. (2) (with $\alpha = 1.26$) via the WREVAP-estimated Q_n value plus taking the mean of the coldest 30-50 (30 in the highest region) MODIS pixels from the middle part (i.e., 300-400 m and 550-650 m, for the latter it was considered that the area vanishes fast with elevation) of the elevation zone.

For the lowest zone no such middle part was defined due to the small change in elevation. As Szilágyi & Józsa (2009b) demonstrated, the method is not sensitive to the exact number of cold pixels employed. Note that only the zone-averaged values of the 0.1-degree gridded meteorological data are needed for defining the anchor points, thus no matching of the MODIS and the 0.1-degree grid is required.

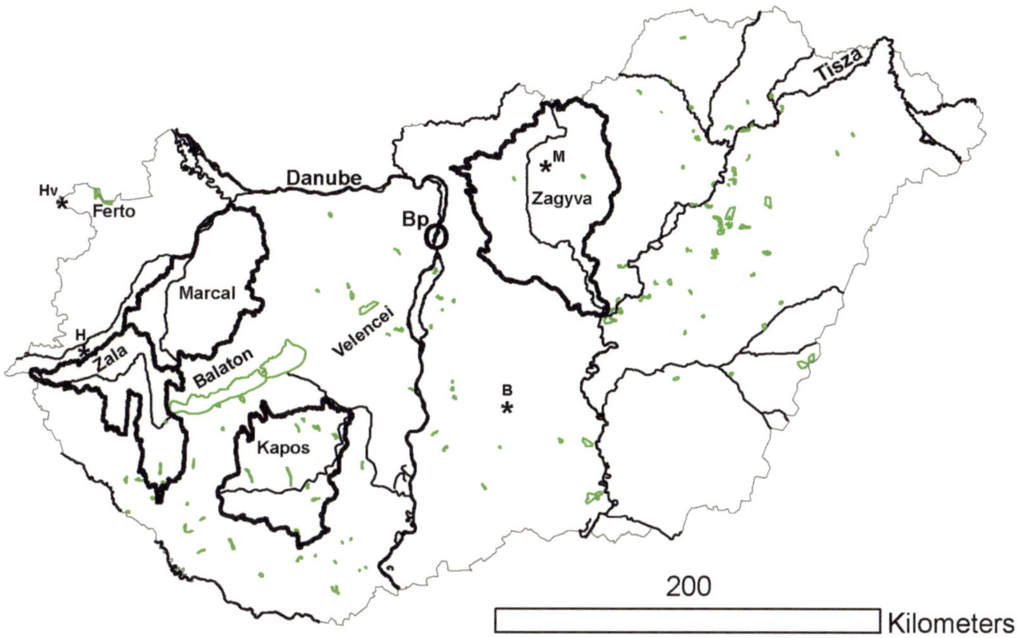

Fig. 2. Location of the watersheds [Kapos, Marcal, Zala, Zagyva and Hidegvizvolgy (Hv)] and eddy-covariance towers [Bugac (B), Matra (M), and Hegyhatsal (H)] used for validating the model estimates. The largest natural shallow lakes (denoted by green polygons, e.g., Balaton, Lakes Ferto and Velencei) of Hungary and the capital, Budapest (Bp) are also identified.

For avoiding possible sharp changes in the resulting pixel ET rates between the zones, the transformation equation was allowed to change linearly with pixel-elevation (z) between the limiting equations of the lower (l) and upper (u) zones. Mathematically,

$$ET(z) = \frac{(z_u - z)[m_l T_s(z) + c_l] + (z - z_l)[m_u T_s(z) + c_u]}{z_u - z_l} \tag{5}$$

where m and c are the parameters of the linear transformations by zone, obtained with the help of the anchor points. Here the reference elevations (z_u or z_l) are taken at 100, 350, and 600 meters.

Fig. 4 displays the linear transformations by months for the lowest zone (i.e., under 200 m). The line sections are bounded by the anchor points illustrated in Fig. 1 but during the calculations they are allowed to extend downward to accommodate for pixels warmer than the zonal-mean. Occasionally the transformation may yield negative pixel-ET values for very warm pixels which then were replaced by zeros. The lines, however, are bounded from above, i.e., in the few occasions (about half the number of the coldest points considered) when the pixel temperature is lower than $<T_{ws}>$, the corresponding ET value assigned is still

E_w, since at the 1-km scale this is the highest rate of ET achievable from energy considerations [i.e., the 1-km is the smallest scale for the validity of Eq. (2)].

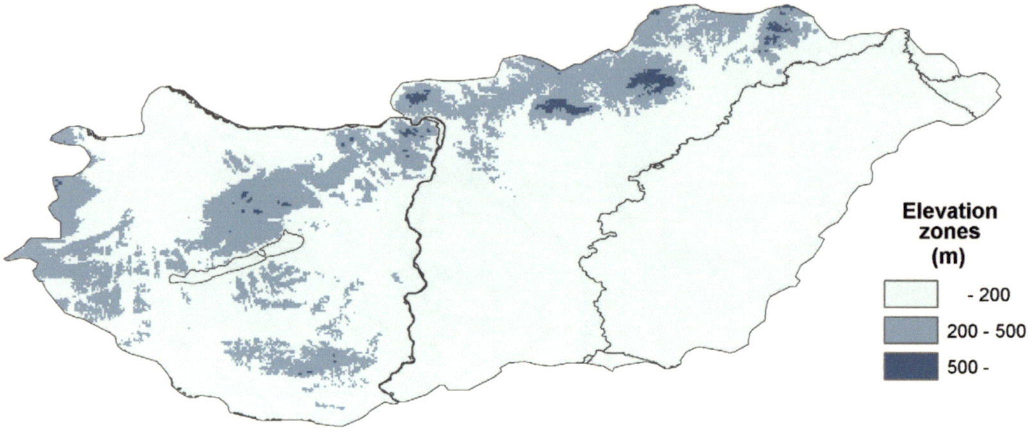

Fig. 3. The extent of the three elevation zones employed.

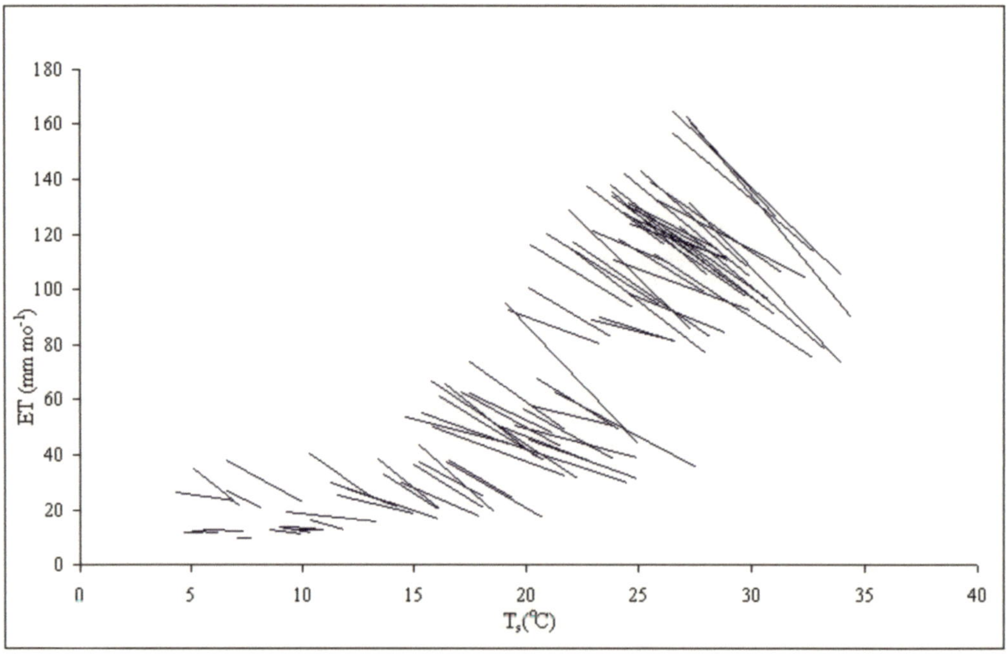

Fig. 4. Linear transformations for the lowest elevation zone in Hungary by month. Each line ends in the anchor points illustrated in Fig. 1. During the transformations the lines are extended downward to accommodate for pixels warmer than the zonal-mean.

The transformations were not performed for the winter months (December, January, and February) because then the ground may have patchy snow cover which violates the constant Q_n assumption since the albedo of snow is markedly different from that of the land. This is not a serious restriction on the general applicability of the model since in temperate climates ET is very limited in the winter months (with temperatures around 0 °C) anyway.

In the warm months the distribution of the albedo values among the MODIS cells is very narrow, with a standard deviation of about 1.6% and mean of 15%. This corroborates the working hypothesis that Q_n can be considered quasi-constant among the MODIS pixels. The only exception are the cells with open water. Open water surfaces have an albedo of around 5%, therefore the transformation-equation derived ET estimates are overwritten for such cells by the estimates of the lake evaporation module of WREVAP.
Table 1 displays the main characteristics of the ET validation sites and watersheds. Selection of the watersheds was limited mostly by data availability for the 2000-2008 period. Since the precipitation data (needed for back-calculating ET over the watershed as P – R) provided by HMS is restricted to Hungary, only those watersheds could be considered that are fully within the country's boundary.

Location	Land cover	Coordinates (°) [lat (N), long]	Elevation (m a.s.l.)	Area (km^2)
Hungary				
CarboEurope, Bugac (4-m tall EC tower)	Grassland	46.69, 19.6E	113	0.16
CarboEurope, Matra (3-m tall EC tower)	Vineyards/Orchards	47.85, 19.73E	300	0.09
Hegyhatsal (EC instr. at 82 m above ground)	Forest and cropland	46.95, 16.65E	248	67
Hidegvizvolgy catchm.	Forest	-	284 (mean)	6
Kapos catchment	Forest and cropland	-	184 (mean)	3210
Marcal catchment	Forest and cropland	-	185 (mean)	3042
Zagyva catchment	Forest and cropland	-	211 (mean)	4207
Zala catchment	Forest and cropland	-	195 (mean)	1528
Nebraska				
Sandhills, station #1 (2.68-m tall BR tower)	Sub-irrigated meadow	42.07, 101.47W	1087	0.07
station #2 (2.61-m tall BR tower)	Dry meadow	42.07, 101.44W	1081	0.07
station #3 (2.68-m tall BR tower)	Dunal upland grass	42.07, 101.37W	1086	0.07
Gothenburg (EC instr. at 7.8 m above canopy)	Riparian forest	40.92, 100.18W	783	0.61
Odessa (EC instrum. at 7.8 m above canopy)	Riparian forest	40.67, 99.22W	694	0.61
Mead, station #1 (4.5-m tall BR tower)	Irrigated maize	41.16, 96.47W	369	Varies < 0.2
station #2 (4.5-m tall BR tower)	Irrigated maize/soybean	41.16, 96.47W	369	Varies < 0.2
station #3 (4.5-m tall BR tower)	Non-irrigated maize/soybean	41.16, 96.47W	369	Varies < 0.2

Table 1. Validation sites and watersheds. EC: eddy covariance; BR: Bowen-ratio station. For the EC, BR sites, area means the estimated (100 times the height of the instrument above ground/canopy) footprint size.

3.2 Nebraska, USA

Mean annual precipitation, mean monthly maximum/minimum air and dew-point temperature values came from the PRISM database (www.prism.oregonstate.edu) at 2.5-min resolution. Mean monthly incident global radiation data at half-degree resolution were downloaded from the GCIP/SRB site (www.atmos.umd.edu/~srb/gcip/cgi-bin/historic.cgi). Due to its large size the state was subdivided into eight regions (Fig. 5) and the transformation equations calculated separately for each region, employing the coldest T_s value every month in each region and α = 1.2 in Eq. (2). For a smooth ET surface, the parameters of the transformation equations were linearly interpolated among all the cells.

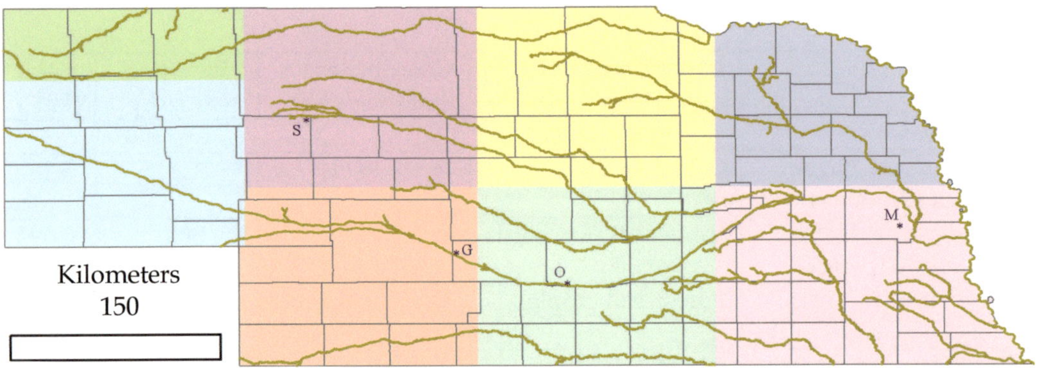

Fig. 5. Location of the validation sites (S: Sandhills; G: Gothenburg; O: Odessa; M: Mead) and the distribution of the eight regions.

As for Hungary, the transformations were performed only from March till November, each year. Due to the large size of the state, as well as the simultaneous presence of deep reservoirs and many small shallow lakes, CREMAP evaporation values were not corrected for open water surfaces. In a planned future online publication of the monthly maps, WREVAP-estimates of lake evaporation will be listed with the maps for each region: one estimate for a shallow, and a separate one for a deep open water body. See Table 1 for the characteristics of the validation sites.

4. Results and discussions

4.1 Hungary

Fig. 6 displays the spatial distribution of the ET rates as period-averaged (2000-2008) annual values. This is possible because we postulated that ET in the three winter months (with mean air temperatures around the freezing point) is negligible.

The driest continuous region is the sandy inter-fluvial plateau in south-central Hungary, the karstic-plateau just north of Lake Balaton, the urban conglomerate of Budapest and the southern side of the Matra mountain north-east of Budapest. The wettest regions are the mountains with good forest cover and ample precipitation in the north-eastern part of the country, the gallery-forests along the main rivers (the largest is Gemenc, a natural protection area along the Danube in the south-central part), the wetlands (Kis-Balaton, immediately south-west of Lake Balaton, as well as areas around Lake Ferto in north-western Hungary), and the forests in the south-western part of the country where annual precipitation is most abundant, in excess of 800 mm. The evaporation rates of the largest water bodies (about 890 mm) were calculated by the WREVAP program with the winter months included since lake evaporation -- provided the lakes do not freeze up -- is typically not negligible even in cold temperatures. WREVAP has been shown to yield accurate estimates of lake evaporation. These evaporation rates are typically larger than the T_s-obtained values by about 10%, caused mainly by the difference in the albedo values of the two surfaces. For the calculations the otherwise hard-to-locate WREVAP documentation with the original FORTRAN source code of Morton et al. (1985) can be downloaded from the following web-site: http://snr.unl.edu/szilagyi/szilagyi.htm.

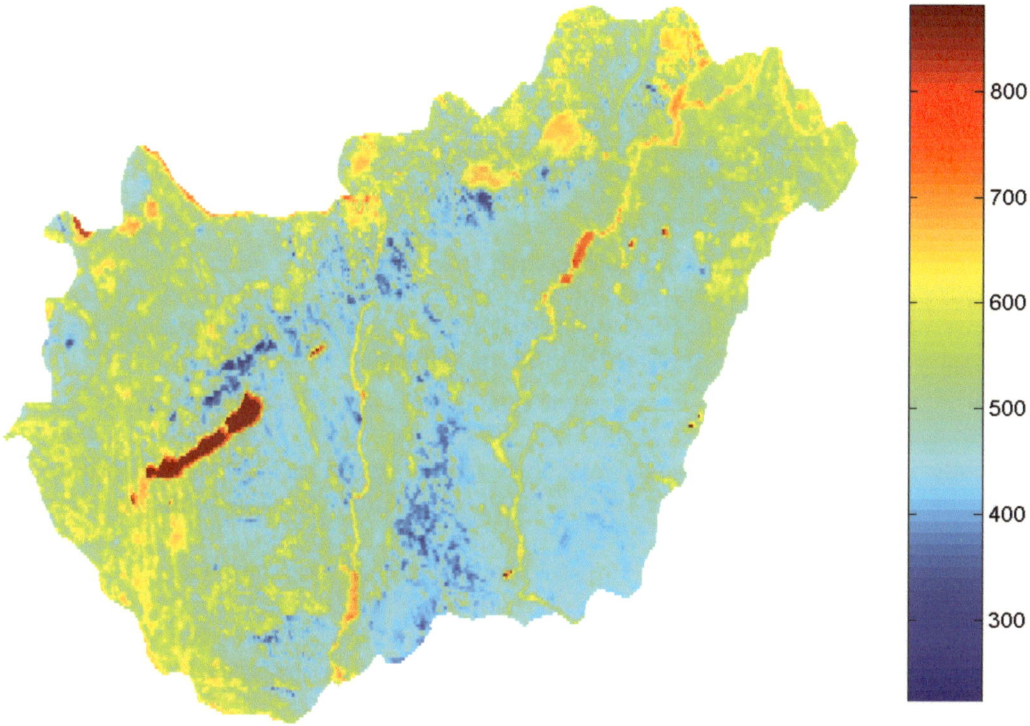

Fig. 6. Period-averaged (2000-2008) mean annual ET rates (mm) in Hungary. < ET > = 510 mm yr^{-1}.

Validation of the results were performed with the help of three eddy-covariance sites and five watersheds (Fig. 2 and Tables 1 & 2). While continuous data from the two CarboEurope sites (Nagy et al., 2007; Pinter et al., 2008) were available for each month of our study period, from the third site, Hegyhatsal (Barcza et al., 2009), continuous measurements were available only for selected months. Here the CREMAP ET values (averaged over 8 x 8 = 64 MODIS pixels around the tower) yield an unbiased estimates of the observed ET rates (Fig. 7), provided the Hegyhatsal measurements are multiplied by 1.78. This is necessary because the uncorrected eddy-covariance measurements represent a 44% systematic underestimation in comparison with energy balance closure. The Q_n values were obtained from two independent sources (a) locally measured data, and; (b) WREVAP-model estimates based on 0.1-degree gridded sunshine duration measurements. The two Q_n means are within 5% reinforcing the accuracy of the net radiation measurements at Hegyhatsal and the systematic underestimation of the EC instruments.

Location	Monthly (mm mo^{-1})				Annual (mm yr^{-1})				Period-averaged (mm yr^{-1})		
	mv	me	de	re (%)	mv	me	de	re (%)	mv	me	re (%)
Bugac	50	-0.7	16.1	-1.4	456	-29	28.3	-6.3	-		
Matra	43	1.4	14.1	3.3	401	4.4	29.2	1.1	-		
Hegyhatsal	55	0.1	7.9	0.2	-				-		
Hidegvizvolgy	-				-				620	18	2.9
Kapos	-				-				582	-56	-9.6
Marcal	-				-				559	-10	-1.8
Zagyva	-				-				531	-9	-1.7
Zala	-				-				564	11	1.9

Table 2. Error (estimated minus measured value) analysis of the ET estimates for Hungary. me: error mean; de: standard deviation of error; re: relative error (%) defined as the ratio of the error mean and the measurement mean (mv). Note that there is only one single period averaged value for each watershed, so there the mean results from one value.

A 44% underestimation by the EC technique is somewhat larger than the generally accepted 10-30% systematic undershoot (Baldocchi, 2003) well known for such systems. The unusually high location and placement (82 m above ground on a 117-m TV transmitter tower) of the instruments (Barcza et al., 2009) may have contributed to the increased undershoot.

At the watershed-scale, validation is possible only in a multi-year sense because even at an annual basis water balance closure of P – R yields a highly uncertain estimate of ET due to the typically unknown changes in water storage within the catchment in the form of soil moisture and groundwater (Szilágyi & Józsa, 2009b). Period-averaged ET estimates are within a few percent of the water-balance values, except at the Kapos catchment, where CREMAP undershoots water-balance ET by about 10%. As can be seen in Fig. 2, this catchment contains the highest number of reservoirs (in fact more than is discernible from the map) having typically elongated shapes nested in the rolling terrain which are narrow enough so that they would not show up clearly in the 1-km resolution MODIS pixels as low surface temperature areas. If the (unknown) total surface area of such reservoirs is just 2% of the drainage area, then it means about a 4% increase in watershed ET due to the almost doubling of the latent heat fluxes between open water surfaces and the land (see Fig. 6). Therefore, the real undershoot in land area ET by CREMAP is probably less severe.

Fig. 8 displays the distribution of the multi-year (2000-2006) precipitation recycling index, the ratio of the mean annual ET and precipitation rates. The corresponding period is two years shorter than the study period because of the availability of water balance data for the whole country, necessary to validate the index values. When the pixel values are aggregated over the country the index becomes 0.892 versus the officially derived value of 0.896 for the same period, a remarkable match.

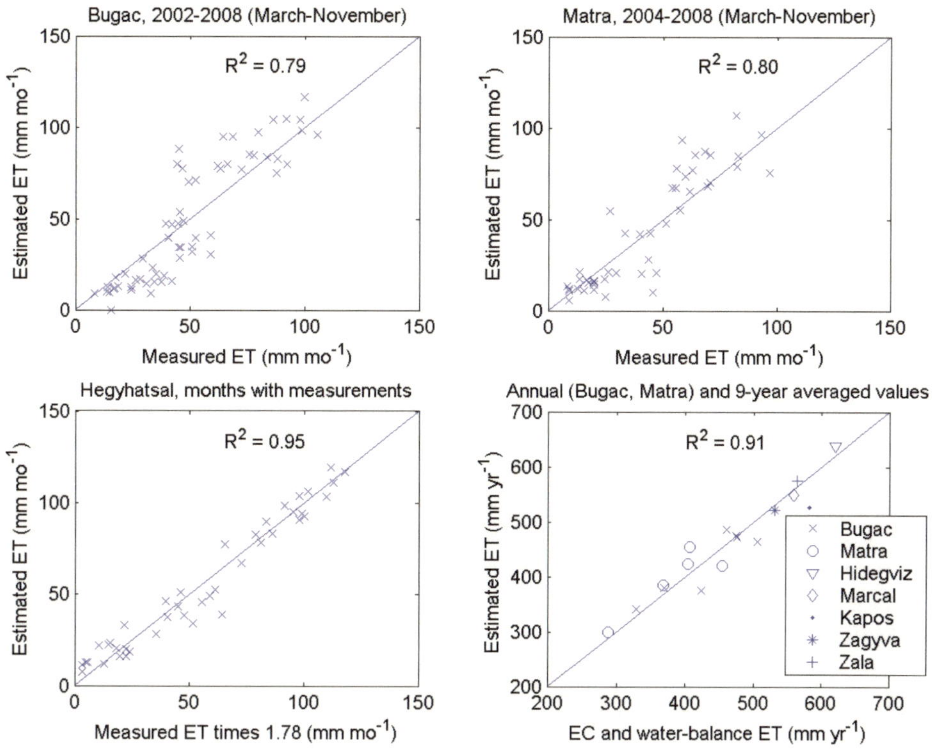

Fig. 7. Validation of the ET estimation method with eddy-covariance (Bugac, Matra and Hegyhatsal) and catchment water-balance data in Hungary.

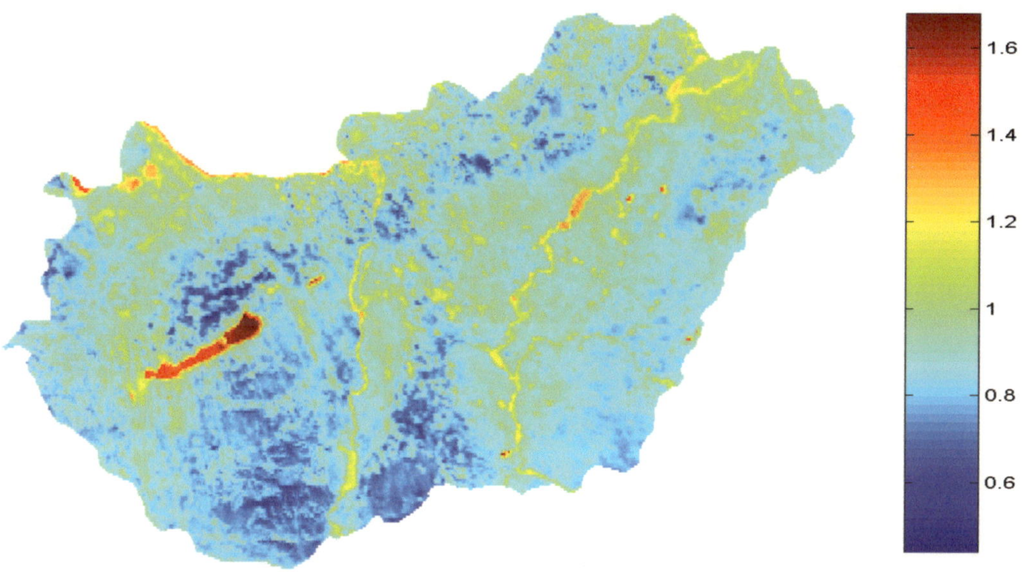

Fig. 8. Spatial distribution of the multi-year mean precipitation recycling index (ET/P) for 2000-2006 in Hungary. < ET/P > = 0.89.

To demonstrate the practical value of the 1-km resolution ET maps in regional water resources management, let us have a look at what ET rates the maps predict for forested areas in the central inter-fluvial sand plateau region of the country, an area in focus for water resources managers due to the long-term groundwater decline (e.g., Szilágyi & Vorosmarty, 1997) reported in the region. As seen in Fig. 8, some forested areas evaporate more (the index is larger than one) than the precipitation rate they receive. This is only possible if these forests create a depression in the groundwater table, to induce local groundwater flow directed toward them which supplies the difference in ET and precipitation. In fact, this has been reported by Major (1976, in Stelczer, 2000) from the region based on piezometer measurements, where they found that under the forest the groundwater table was on average 1 m deeper than in the surrounding non-forested areas. They calculated a mean annual ET rate of 712 mm for an actively growing black spruce community covering an area of 500 x 500 m. On average the forest consumed 130 mm more water annually than it received from precipitation. Our figures match these findings, yielding an average 620 mm annual ET for the forests in the area (this value may go up to 650 mm in certain pixels), which is about 80 mm more than the mean annual precipitation rate of the region. Similar conclusion have been drawn by Szilagyi & Vorosmarty (1997) from a complex regional, coupled surface water-groundwater balance model.

4.2 Nebraska, USA

The spatial distribution of the mean annual CREMAP ET values (2000-2009) in Nebraska is displayed in Fig. 9.

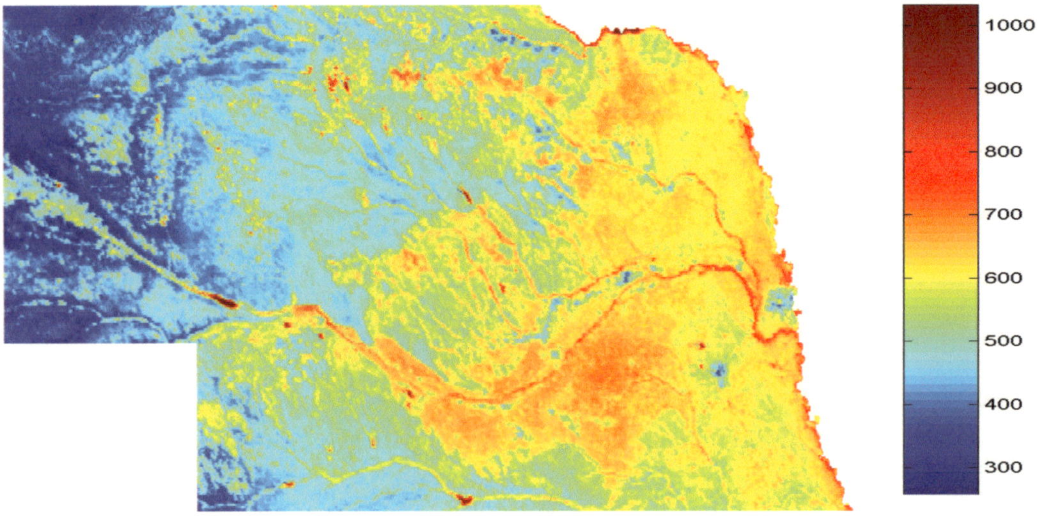

Fig. 9. Period averaged (2000-2009) mean annual ET rates (mm) in Nebraska, USA. < ET > = 531 mm yr^{-1}.

The high ET areas match closely the distribution of irrigation wells in the state (Fig. 10). The Sandhills region (Fig. 10) with its porous soil and grass-covered sand dunes is generally not suited for surface irrigation, thus the high ET areas within it correspond to shallow interdunal lakes prevalent mainly in the western and north-central part of the region. The river valleys with their shallow groundwater (most distinctly the Platte, flowing west to east with its tributaries in the middle, as well as the Missouri that forms the eastern boundary of the state and the Republican River in the southern part), also show up distinctly in the ET map, such as the two largest urban areas (i.e., Omaha and Lincoln) with their reduced evaporation rates in the south-east section of the state. The largest reservoirs on the streams with their high evaporation rates are clearly visible too. Precipitation has a strong gradient across the state, exceeding 800 mm per year in the eastern part, and decreasing to nearly 300 mm per year in the western portion of it.

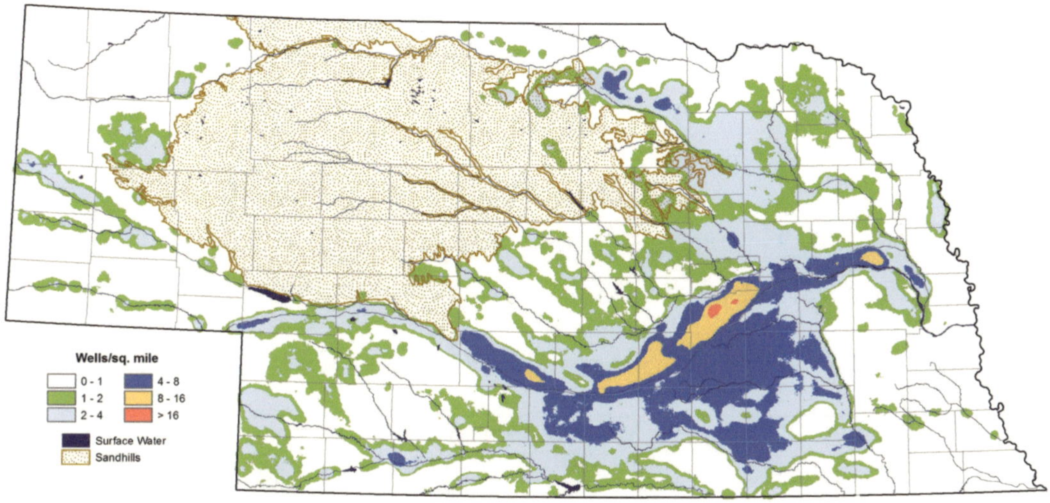

Fig. 10. Density of irrigation wells in Nebraska, 2009 (after Burbach & Korus, 2009).

The precipitation recycling index (ET/P) is displayed in Fig. 11. Again, the stream network is clearly visible. The Sandhills region has long been known as the main recharge area of groundwater in the state due to its porous soil. As Fig. 11 indicates, recharge is highly variable even within the Sandhills, its south-eastern portion (i.e., the largest continuous dark-blue area) generating the largest amount of recharge, probably because precipitation is most abundant there within its boundaries. It is also noteworthy to mention how sharply the south-eastern boundary of the Sandhills region is distinguishable in Fig. 11, with the southernmost part reaching the Platte River, in comparison with Fig. 10. As CREMAP does not use any soil or vegetation cover information, it is remarkable that from the ET distribution of the CREMAP estimates alone it is possible to map boundaries of different physical soil types (i.e., sand vs loess). The linear dark-blue feature east of the Sandhills between the Platte and its tributary, the Loup, is also made up of sand such as the area (called the Duncan dune field) just east of it, close to the confluence.

When the index is larger than unity, it means the area has a water supply other than precipitation over the same area. For rivers and lakes the additional water supply comes from the corresponding catchment area, for irrigated land, it is the ground or surface water used for irrigation. The largest index values are found in the western part of the state where precipitation is the scarcest and for irrigated areas the relative portion of irrigated water volumes is the largest when compared to precipitation, in order to satisfy the water demand of the crop.

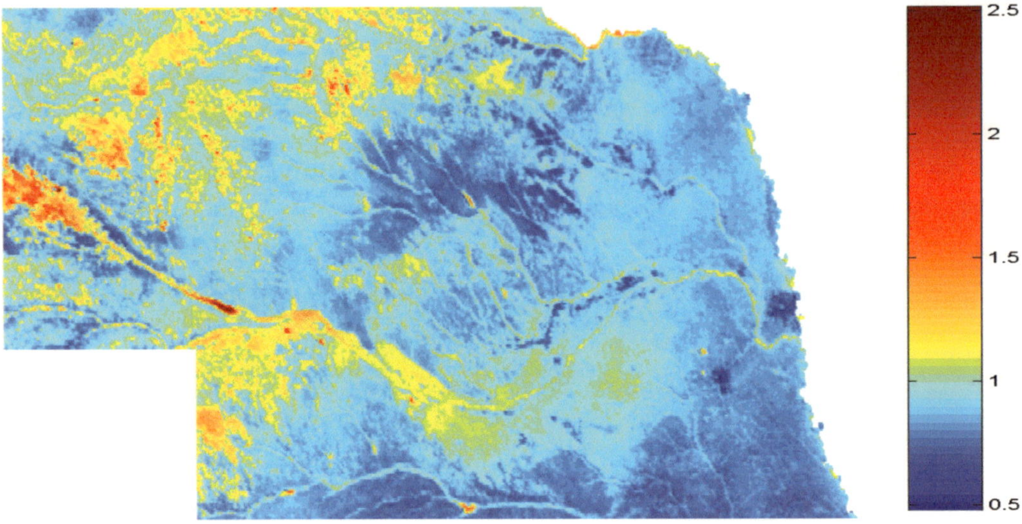

Fig. 11. The precipitation recycling index (ET/P) for 2000-2009 in Nebraska, USA. < ET/P > = 0.93.

Table 3 displays the results of the validation at the four locations listed in Table 1.

Location	Monthly (mm mo^{-1})				Annual (mm yr^{-1})			
	mv	me	de	re (%)	mv	me	de	re (%)
Sandhills, average of 3 sites	52	5.6	17.6	10.7	471	36	55.6	7.6
Gothenburg	68	10.1	19.5	14.9	663	38	38.4	5.7
Odessa	65	7.25	19.6	11.1	623	17	50.5	2.7
Mead, average of irrig. maize & rainfed sites	69	0.6	25.5	0.8	642	-34	68.6	-5.3

Table 3. Error (estimated minus measured value) analysis of the ET estimates for Nebraska. me: error mean; de: standard deviation of error; re: relative error (%) defined as the ratio of the error mean and the measurement mean (mv).

At the Sandhills and in the Platte riparian forest, the CREMAP values tend to overestimate the measured values, probably due to differences in footprints. At both locations the MODIS pixel is larger than the footprint area of the local measurements (Table 1), thus the MODIS pixels always include wetter inter-dunal valleys in the Sandhills, and wide, open channel portions, as well as irrigated crop areas around the riparian forest locations. Because of the high spatial variance in ET over the Sandhills (ET depends heavily on whether it is a typically wet inter-dunal valley or the much drier top of a sand dune) over relatively small scales, an average of the three sites was used for comparison with MODIS-cell derived ET estimates. The difference in the resulting ET estimates (i.e., measured vs estimated) is typically stronger in the summer months (Fig. 12) when the foot-print area of the local instruments decreases due to increased buoyancy of the air (Landon et al., 2009). At the riparian forest sites the uncorrected ET-flux measurements were multiplied by 1.13 in this study in order to close the energy balance, as discussed by Landon et al. (2009). From the two riparian sites the MODIS cell at Odessa had a better collocation with local instruments (i.e., the instrument is almost in the middle of the MODIS cell) and covered the riparian forest almost perfectly, explaining why MODIS estimates show less bias of the measured values (Table 3 and Fig. 12). At Mead, two of the plots (i.e., irrigated maize and rainfed maize/soyben) fell in one MODIS cell while the third covered only a fraction of the corresponding MODIS cell, thus the average of the two former plots were chosen for analysis.

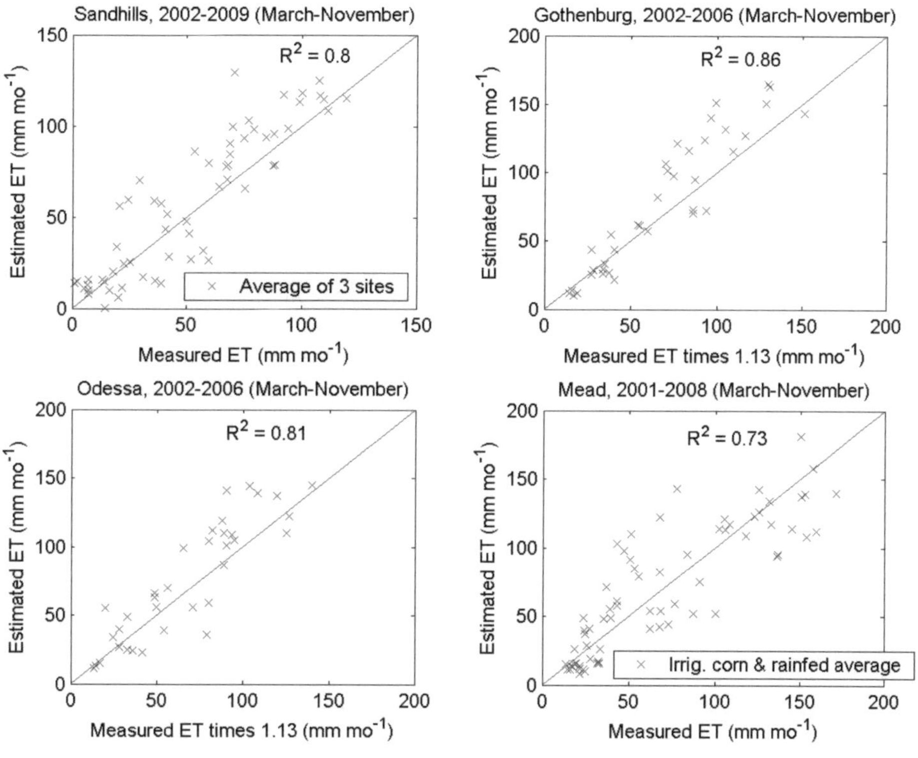

Fig. 12. Validation of the CREMAP estimates with Bowen-ratio (Sandhills and Mead) and eddy-covariance data. The EC values were multiplied by 1.13 to close the energy balance.

In Fig. 5 the smallest region in the north-western part of the country was chosen because it contains the Pine Ridge area, the most rugged part of the state. Over this area (and to a lesser degree, along the Niobrara River breaks in north-central Nebraska) the multi-year average of the CREMAP estimates is typically 15% larger than the same average of precipitation. The Pine Ridge (and the deep and wide Niobrara valley) creates increased large-scale turbulence of the air leading to a more effective heat transfer across the land/vegetation-atmosphere interface, especially that the predominant wind in the two regions is almost perpendicular to the axes of the ridge/valley. A similar problem has not been observed in Hungary because the country, due to its basin location (surrounded by the Alps and the Carpathian Mountains), is much less windy than the American prairie. Aided by an eco-region map of the state, the Pine Ridge and Niobrara River breaks could be identified in the MODIS cells, and the overlapping cell ET values corrected, so that their multi-year areal averages became constrained by precipitation.

5. Summary and conclusions

The present ET estimation model (CREMAP) is a modified and updated version of the ET estimation technique of Szilágyi and Józsa (2009b) and utilizes 1-km 8-day composited MODIS daytime surface temperature, T_s, and basic atmospheric data (mean air temperature and humidity as well as either sunshine duration or incident global radiation) to estimate the latent heat flux at a monthly time-step. The approach is based on a linear transformation of the mean monthly MODIS T_s pixel values into ET rates (Fig. 1). The anchor points of the transformation come from the (a) regional ET rate of the complementary relationship (Bouchet, 1963) of evaporation and the corresponding spatial mean of T_s; (b) Priestley-Taylor equation of wet environment ET and the corresponding spatial mean of the coldest points in the region. The resulting linear equation is valid always for the given computation interval. With each month a new transformation equation is obtained similarly (Fig. 4). For typical temperate climates, such as found in Hungary, the customary value of 1.26 for α is generally satisfactory in the Priestley-Taylor equation, the value of which to be reduced somewhat for drier climates (Szilágyi et al., 2009), such as the value of 1.2, applied for Nebraska. Changes in elevation can be accounted for by a linear interpolation of the parameters of the transformation equation (Eq. 5). A similar, 2-D linear interpolation can be applied for large study areas after a subdivision of them into smaller regions, as was performed for Nebraska. The advantage of the present method over that of Szilágyi and Józsa (2009b) is that the regional evaporation rate is calculated by the WREVAP model which is easier to employ than the wet-surface equation (Yeh & Brutsaert, 1971; Szilágyi & Józsa, 2009c) which needs daily maximum air temperature values, in addition of the data requirement of WREVAP, for the calculation of the daytime air temperature. The model was validated with eddy covariance (EC) and Bowen-ratio (BR) station measurements plus catchment-scale water-balance closure data. The validation, spanning almost three magnitudes in spatial scale (Table 1), indicated a favorable match between ET estimates and observations (Figs. 7 & 12; Tables 2 & 3). On a monthly basis the ET estimates resulted in an R^2 value of 0.8 – 0.9 with a minimum value of 0.73 and maximum of 0.95. During validation a problem arises from differences in foot-print areas between local measurements and the MODIS cell. As a consequence, more weight is to be given to validation results where the two foot-print sizes match, such as a) at Hegyhatsal (R^2 = 0.95) where the strong linear correspondence with measurements and high accuracy of the ET estimates result from high-tower mounted (82 m above ground) EC instruments, and; b) at the watersheds in Hungary, which thus yield improved statistics. At the rest of the EC and BR validation sites the instruments sample a smaller area than the size of the MODIS pixel. The overall strong correspondence with measured ET values and unbiased nature of the ET estimates are generally maintained at the annual and multi-annual level (Fig. 9; Tables 2 & 3), where seasonality effects are eliminated. On an annual and multi-year basis a typical value of R^2 is between 0.7 – 0.8 with a minimum of 0 (this happened at the riparian forest sites where inter-annual variance of ET is much reduced) and maximum of 0.91. The mean annual ET estimates remain well within 10% of the measured values.

The model is simple, requires only a minimum amount of easily accessible data, and calibration free. From the first author's personal website documentation of the WREVAP program with its FORTRAN source code is also accessible. The WREVAP program can be applied separately to estimate evaporation from open water surfaces. Typically mean annual open water evaporation is in excess of the CREMAP estimate by about 10%, mainly due to differences in albedo between the land and water.

Generally, CREMAP is expected to work correctly in areas where the CR is valid (Szilágyi et al., 2009). In its present form, it is not recommended to be applied in mountainous areas or in areas with strong surface-albedo and/or momentum roughness height changes at a scale in excess of 1-km. Also, due to its inherent assumptions (i.e., a spatially constant Q_n and aerodynamic resistance as well as the application of the Priestley-Taylor equation), it is not recommended to be applied at a spatial resolution finer than 1-km.

Research is currently undertaken to extend the method to conditions when the spatially constant Q_n requirement is violated. Employing high-resolution digital elevation models and MODIS surface albedo data, known changes in albedo, mean surface slope and aspect among the MODIS cells can be accounted for in the transformation equations. Further research is required to investigate the aerodynamic resistance altering effect of rolling-to-rugged terrain transitions and the inclusion of such effects in the modeling framework.

6. Acknowledgments

This work has been supported by the Hungarian Scientific Research Fund, #77364. We thank G. Antal, Z. Barcza, D. Billesbach, A. Clement, Z. Gribovszki, L. Haszpra, P. Kalicz, L. Sütheő, J. Szalai, USGS-Lincoln, G. Varga, & S. Verma for their data.

The views, conclusions, and opinions expressed in this study are solely those of the writers and not the University of Nebraska, state of Nebraska, or any political subdivision thereof.

7. References

Allen, R.; Tasumi, M. & Trezza, R. (2007). Satellite-based energy balance for mapping evapotranspiration with internalized calibration (METRIC)-model. *J. Irrig. Drainage Eng.*, 133(4), 380-394.

Baldocchi, D. (2003). Assessing the eddy covariance technique for evaluating carbon dioxide exchange rates of ecosystems: past, present and future. *Global Change Biol.*, 9, 1-14.

Barcza, Z.; Kern, A.; Haszpra, L. & Kljun, N. (2009). Spatial representativeness of tall tower eddy covariance measurements using remote sensing and footprint analysis. *Agric. Forest Meteor.*, 149, 795-807.

Bastiaanssen, W.; Menenti, M.; Feddes, R. & Holtslag, A. (1998). A remote sensing surface energy balance algorithm for land (SEBAL): 1. Formulation. *J. Hydrol.*, 212, 198–212.

Bouchet, R. (1963). Evapotranspiration reelle, evapotranspiration potentielle, et production agricole. *Annal. Agronom.*, 14, 543–824.

Brutsaert, W. & Parlange, M. (1998). Hydrologic cycle explains the evaporation paradox. *Nature*, 396(6706), 30.

Brutsaert, W. & Stricker, H. (1979). An Advection–Aridity approach to estimate actual regional evapotranspiration. *Water Resour. Res.*, 15, 443–449.

Burbach, M. & Korus, J. (2009). Density of active irrigation wells in Nebraska. Conservation & Survey Division, Institute of Agricultural and Natural Resources, University of Nebraska - Lincoln, http://snr.unl.edu/data/water/groundwatermaps.asp

Hobbins, M.; Ramirez, J. & Brown, T. (2001a). The complementary relationship in estimation of regional evapotranspiration: An enhanced advection-aridity model. *Water Resour. Res.*, 37(5), 1389–1403.

Hobbins, M.; Ramirez, J.; Brown, T. & Claessens L. (2001b). The complementary relationship in estimation of regional evapotranspiration: The complementary relationship areal evaporation and advection-aridity models. *Water Resour. Res.*, 37(5), 1367–1387.

Landon, M.; Rus, D.; Dietsch, B.; Johnson, M. & Eggemeyer, K. (2009). Evapotranspiration rates of riparian forests, Platte River, Nebraska, 2002-06. *USGS Report 2008-5228*.

Major, P. (1976). Groundwater balance investigations in flat lands. 2. Piezometer readings. (VITUKI report, in Hungarian), VITUKI, Budapest.

Morton, F.; Ricard, F. & Fogarasi, S. (1985). Operational estimates of areal evapotranspiration and lake evaporation - Program WREVAP. National Hydrological Research Institute Paper #24, Ottawa, Ontario, Canada.

Nagy, Z.; Pintér, K.; Czóbel, S.; Balogh, J.; Horváth, L.; Fóti, S.; Barcza, Z.; Weidinger, T.; Csintalan, Z.; Dinh, N.; Grósz, B. & Tuba, Z. (2007). The carbon budget of semi-arid grassland in a wet and a dry year in Hungary. *Agric. Ecosyst. Environ.*, 121, 21-29.

Penman, H. (1948). Natural evaporation from open water, bare soil, and grass. *Proc. Royal Soc. London*, A193, 120–146.

Pintér, K.; Barcza, Z.; Balogh, J.; Czóbel, S.; Csintalan, Z.; Tuba, Z. & Nagy, Z. (2008). Interannual variability of grasslands' carbon balance depends on soil type. *Community Ecol.*, 9(Suppl1), 43-48. doi: 10.1556/ComEc.9.2008.S.7.

Priestley, C. & Taylor, R. (1972). On the assessment of surface heat flux and evaporation using large-scale parameters. *Month. Weather Rev.*, 100, 81–92.

Stelczer, K. (2000). A vizkeszletgazdalkodas hidrologiai alapjai (Hydological bases of water resources management, in Hungarian). ELTE Eotvos Kiado, Budapest, Hungary.

Szilágyi, J. (2001). Modeled areal evaporation trends over the conterminous United States. *J. Irrig. Drainage Engin.*, 127(4), 196-200.

Szilágyi, J. & Vorosmarty, C. (1997). Water-balance modeling in a changing environment: reductions in unconfined aquifer levels in the area between the Danube and Tisza rivers in Hungary. *J. Hydrol. Hydromech.*, 45, 348-364.

Szilágyi, J. & Józsa, J. (2008). New findings about the complementary relationship based evaporation estimation methods. *J. Hydrol.*, 354, 171-186.

Szilágyi, J.; Hobbins, M. & Józsa, J. (2009). A modified Advection-Aridity model of evapotranspiration. *J. Hydrol. Engin.*, 14(6), 569-574.

Szilágyi, J. & Józsa, J. (2009a). Analytical solution of the coupled 2-D turbulent heat and vapor transport equations and the complementary relationship of evaporation. *J. Hydrol.*, 372, 61-67.

Szilágyi, J. & Józsa, J. (2009b). Estimating spatially distributed monthly evapotranspiration rates by linear transformations of MODIS daytime land surface temperature data. *Hydrol. Earth System Sci.*, 13(5), 629-637.

Szilágyi, J. & Józsa, J. (2009c). An evaporation estimation method based on the coupled 2-D turbulent heat and vapor transport equations. *J. Geophys. Res.*, 114, D06101. doi:10.1029/2008JD010772.

Yeh, G. & Brutsaert, W. (1971). A solution for simultaneous turbulent heat and vapor transfer between a water surface and the atmosphere. *Bound. Layer Meteorol.*, 2, 64–82.

12

Estimation of Reference Evapotranspiration using the FAO Penman-Monteith Method for Climatic Conditions of Poland

Leszek Łabędzki, Ewa Kanecka-Geszke, Bogdan Bak and Sandra Slowinska
Institute of Technology and Life Sciences
Poland

1. Introduction

Evapotranspiration is the combination of soil evaporation and crop transpiration. Weather parameters, crop characteristics, management and environmental factors affect evapotranspiration. Reference, potential and actual evapotranspiration are distinguished. These terms are commonly used, although some differences of their definitions can be found among researches.

The potential evapotranspiration of a given crop is defined as soil evaporation and plant transpiration under unlimited soil water supply and actual meteorological conditions. According to Brutseart (1982) the potential evapotranspiration is a maximum intensity of evapotranspiration from a large surface covered completely and homogeneously with actively growing plants under conditions of unlimited availability of soil water. Allen et al. (1998) call it the crop evapotranspiration under standard conditions and define as the evapotranspiration from disease-free, well-fertilized crops, grown in large fields, under optimum soil water conditions and achieving full production under the given climatic conditions.

The actual evapotranspiration is the amount of water transpired from plants and evaporated from soil surface under actual meteorological conditions and under non-optimal soil, biological, management and environmental conditions. It differs from the potential evapotranspiration due to soil water shortage or waterlogging, diseases, soil salinity, low soil fertility. According to Allen et al. (1998) the evapotranspiration from crops grown under management and environmental conditions that differ from the standard conditions defined for the potential evapotranspiration can be called the crop evapotranspiration under non-standard conditions.

The evapotranspiration from a reference surface is called the reference evapotranspiration and is denoted as ET_o. A large uniform grass (or alfalfa) field is considered worldwide as the reference surface. The reference grass crop completely covers the soil, is kept short, well watered and is actively growing under optimal agronomic conditions.

Reference evapotranspiration (ET_o) is an important agrometeorological parameter for climatological and hydrological studies, as well as for irrigation planning and management. There are several methods to estimate ET_o. The FAO Penman-Monteith (FAO PM) method has been considered as a universal standard to estimate ET_o (Allen et al., 1989, 1994, 1998;

Feddes & Lenselink, 1994; Smith, 1992). The method is recommended to use as the international standard for definition and estimation of reference crop evapotranspiration.
The reference evapotranspiration describes the evaporative demand of the atmosphere independently of crop type, crop development and management practices. As water is available sufficiently, soil factors do not affect ET_o. Relating evapotranspiration to a specific surface provides a reference to which evapotranspiration from other surfaces can be related. The evapotranspiration from cropped surfaces other than the reference surface can be calculated with the crop factor method using the crop coefficient (under standard conditions) and water stress coefficient (under soil water stress conditions). To estimate crop water requirements, one can relate potential evapotranspiration from the cropped soil with an optimum water supply to an estimated reference evapotranspiration by means of a crop coefficient.
ET_o values measured or calculated at different locations or in different seasons are comparable as they refer to the evapotranspiration from the same reference surface. The only factors affecting ET_o are climatic parameters. ET_o is an agro-climatic parameter and can be computed from weather data. ET_o expresses the evaporating power of the atmosphere at a specific location and time of the year and does not consider the crop characteristics and soil factors (Allen et al., 1998).
The FAO PM method considers many meteorological parameters related to the evapotranspiration process (net radiation, air temperature, vapor pressure deficit, wind speed) and has presented very good results when compared to data from lysimeters populated with short grass (Allen et al., 2006). Labedzki (1999) found a good agreement between the calculated FAO PM reference evapotranspiration and measured evapotranspiration in lysimeters covered with grass with standardized parameters under climatic conditions of central Poland. The linear correlation coefficient was r=0.997 and the standard error of estimation SEE=0.45 mm day^{-1}. Allen et al. (1989, 1994) obtained similar values (r=0.82-0.99 and SEE=0.09-0.77 mm day^{-1}) for lysimeter measurements carried out in different locations in USA.
There are many researches who carried out the analysis of the FAO PM method and the reference evapotranspiration calculated with this method from the point of view of its spatial distribution (e.g. Buttafuoco et al., 2010; McVicar et al., 2007; Song et al., 2010), its accuracy and sensivity (e.g. Allen et al., 2005; Buttafuoco et al., 2010; Fitzmaurice & Beswick, 2005), in relation to other methods of calculating reference evapotranspiration (e.g. Maulé et al., 2006; Paltineanu et al., 1999; Sumner &. Jacobs, 2005).
The objective of this study was to evaluate the performance of the FAO PM method to estimate ET_o under climatic conditions of Poland. The analysis of temporal and spatial variability of the PM reference evapotranspiration is contained as well as the application of the reference evapotranspiration to the determination of agro-climatic regions and climatic water balance in Poland.

2. Climate of Poland

Poland is situated in the Great European Plain between the Baltic Sea and the Carpathian mountains. It is located in a transitory temperate climate zone, influenced by a mild oceanic climate from the west and a dry continental climate from the east. The geographical location of Poland facilitates the occurrence of a great diversity of weather and climatic conditions. The major atmospheric systems shaping the weather conditions in Poland are the yearlong

Icelandic Low and Azorian High in the west and, to a lesser extent, the seasonal Asian High (Wos, 1999). Climatic conditions in Poland are characterized by a considerable variability in weather during long periods of time (years) as well as short periods (days, weeks, months). The annual precipitation amounts to 600 mm. It reaches 350 mm on the average during the vegetation period (April - September). In this period precipitation ranges from 300 mm in the central part to 400 mm in the northern Poland and to 600 mm in the southern Poland (Fig. 1). Taking into account precipitation tendencies in the period 1891-2000, it can be distinguished, in general, three main regions: northern - with an increasing precipitation tendency, central - with a decreasing precipitation tendency and the southern one - without any significant tendency. The clearest tendencies are visible between individual stations. The statistically lowest difference of 55 mm should be accepted as the value characteristic for the total area of Poland (Zawora & Ziernicka, 2003).
Due to the shortage of precipitation and the increase trend of temperature, drought frequency has increased, particularly during the last decade. The dry regions of Poland are almost the entire central region, as well as north-western and mid-eastern parts. These are the regions most threatened by droughts with the annual rainfall amount often less than 300 mm. The most frequent and most severe droughts occur in this area, which sometimes experiences extremely long periods without rain. It is commonly assumed that droughts appear in Poland once every 4-5 years (Labedzki, 2007). Drought usually begins in western Poland, moves through the central part and eventually reaches eastern side. Droughts have become a severe problem in Polish agriculture. They are the main reason for decline in crop yields.

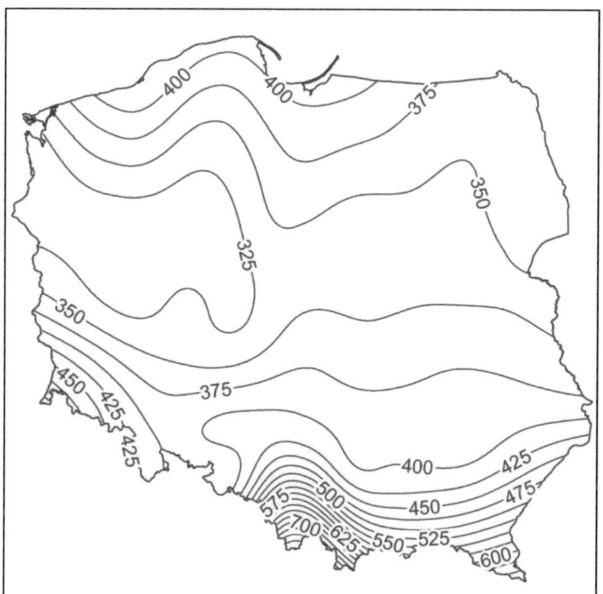

Fig. 1. Precipitation [mm] in the growing season (Apr-Sep), the mean values in 1970-2004

The annual mean temperature is +7.5°C. Averages of temperature in Poland in the 20th century varied from 6°C to over 9°C, with tendency for 7-year periodicity and revealed an increasing trend, which means the warming up reaching 0.9°C per 100 years (Kozuchowski & Zmudzka, 2003). The mean temperature amounts from -2.5°C in January to 18°C in July.

The maximum temperature often rises to above 30°C in summer. According to Kejna et al. (2009) the distribution of Tmin is meridional, with high values in the west and on the Baltic coast (4–5°C), and lowest in the east (2.5°C). In the period 1951–2005 a statistically significant increase of Tmin occurred in north-west Poland, reaching 0.2–0.3°C per 10 years. The highest mean Tmax are in the west of Poland (up to 12°C) and they decrease eastwards and on the coast (10.5°C). In most of the country they increase (up to 0.26°C/10 years), though the changes in south-east Poland are not significant. The number of days with Tmin<0°C varied from less than 100 days on the Baltic coast to over 140 days in the east of Poland. Their frequency drops almost throughout the country, and by as much as 4.3 days per 10 years in the north. The frequency of days with Tmin<-10°C in Poland oscillated between 10 on the coast to 40 in the east, showing a significant growth only in the south (by 2–3 days/10 years). Days with Tmin<-20°C are sporadic (1–8) and their frequency grows in south-east Poland (by 1.2 days/10 years). Days with Tmin<-30°C are very rare (less than 1). Hot days (Tmax>25°C) occur from 15–20 times a year on the coast and 25 in the south to over 30 times in the centre of the Poland. Throughout country the number of hot days increases by up to 4.6 days/10 years. Similarly, the number of very hot days (Tmax>30°C) has increased to 1.4 days in the Pomeranian Lake District (north-western Poland). Their frequency ranges from 2 to 6 days per year. Extremely hot days (Tmax>35°C) are very sporadic in Poland (less than 1 day per year) and their frequency has also slightly increased.

3. Calculation of reference evapotranspiration ET_o

3.1 Method

The general form Penman-Monteith equation (Allen et al., 1998) is:

$$\lambda\, ET = \frac{\Delta\,(R_n - G) + \rho c_p\,(e_s - e_a)/r_a}{\Delta + \gamma\,(1 + r_c/r_a)} \quad (1)$$

where ET is evapotranspiration, R_n is the net radiation, G is the soil heat flux, (e_s-e_a) is the vapour pressure deficit of the air, γ is the psychrometric constant, Δ is the slope of the vapour pressure curve, λ is the latent heat of vaporization, c_p is the specific heat of moist air at constant pressure, ρ is the air density, r_a is the aerodynamic resistance to turbulent heat and vapor transfer, r_c is the surface resistance.

This equation can be applied to different vegetation and soil moisture conditions. By defining the reference crop as a actively growing and well watered green grass surface with a height of 0.12 m having a surface resistance of 70 $s \cdot m^{-1}$ and an albedo of 0.23, the FAO Penman-Monteith method to estimate ET_o can be derived as (Allen et al., 1998):

$$ET_o = \frac{0.408\,\Delta(R_n - G) + 900\,\gamma u_2(e_s - e_a)/(T + 273)}{\Delta + \gamma\,(1 + 0.34\,u_2)} \quad (2)$$

where:

ET_o - reference evapotranspiration [mm day^{-1}],
R_n - net radiation [MJ m^{-2} day^{-1}],
G - soil heat flux density [MJ m^{-2} day^{-1}],
T - mean daily air temperature at 2 m height [°C],

u_2 - wind speed at 2 m height [m s^{-1}],
e_s - saturation vapour pressure [kPa],
e_a - actual vapour pressure [kPa],
e_s-e_a - saturation vapour pressure deficit [kPa],
Δ - slope of the vapour pressure curve [kPa °C^{-1}],
γ - psychrometric constant [kPa °C^{-1}].

The time step with which ET_o is calculated depends on the purpose of the calculation, the required accuracy and the time step of the meteorological data available. Several data in Equation (2) are not measured directly and they should be computed. The method of computation of such data also depends on the meteorological data available.

For the purpose of this study ET_o was calculated with the ten-day time step. Allen et al. (1998) write that, notwithstanding the non-linearity in the Penman-Monteith equation and some weather parameter methods, mean ten-day weather data can be used to compute the mean ten-day values for the reference evapotranspiration. The value of the reference evapotranspiration calculated with mean ten-day weather data is similar to the average of the daily ET_o values calculated with daily average weather data.

Using Equation (2) the mean ten-day values of ET_o are calculated in the growing period (April – September) of 1970-2004 for the mean ten-day values of R_n, T, u_2, e_s and e_a. Then the ten-day sums of ET_o are calculated by multiplying the mean ten-day values by 10 or 11. The monthly and the whole period sums are computed as the sums of the ten-day sums.

The net radiation R_n is the difference between the incoming net shortwave radiation (R_{ns}) and the outgoing net longwave radiation (R_{nl}):

$$R_n = R_{ns} - R_{nl} \tag{3}$$

The net shortwave radiation R_{ns} resulting from the balance between incoming and reflected solar radiation is calculated as:

$$R_{ns} = (1 - a) R_s \tag{4}$$

where:
R_{ns} - net solar shortwave radiation [MJ m^{-2} day^{-1}],
a - albedo = 0.23 for the hypothetical grass reference crop [dimensionless],
R_s - incoming solar radiation [MJ m^{-2} day^{-1}].

Because the solar radiation R_s was not measured, it was calculated with the Angstrom formula which relates solar radiation to extraterrestrial radiation and relative sunshine duration, in the form with the constants according to Allen et al. (1998):

$$R_s = (0.25 + 0.50\, S/S_o)\, R_a \tag{5}$$

where:
S - actual duration of sunshine [hour],
S_o - maximum possible duration of sunshine or daylight hours [hour],
R_a - extraterrestrial radiation [MJ m^{-2} day^{-1}].

The extraterrestrial radiation R_a and the maximum possible daylight hours S_o for a specific day of the month and a given latitude are taken from Sun Tables (1974).

The outgoing net longwave radiation R_{nl} is calculated from the equation (Allen et al., 1998):

$$R_{nl} = \sigma T^4 (0.34 - 0.14\sqrt{e_a})(1.35\, R_s/R_{so} - 0.35) \tag{6}$$

where:
R_{nl} - net longwave radiation [MJ m^{-2} day^{-1}],
σ - Stefan-Boltzmann constant [4.903 10^{-9} MJ K^{-4} m^{-2} day^{-1}],
T – mean ten-day temperature [K],
R_{so} – clear-sky radiation [MJ m^{-2} day^{-1}], calculated as:

$$R_{so} = 0.75\ R_a \tag{7}$$

Saturation vapour pressure e_s was calculated from the relationship:

$$e_s = 0.611\exp[17.27T/(T+237.3)] \tag{8}$$

where:
e_s - saturation vapour pressure [kPa],
T - mean ten-day air temperature [°C].
Actual vapour pressure e_a was calculated as:

$$e_a = e_s\ RH/100 \tag{9}$$

where:
e_a - actual vapour pressure [kPa],
RH - mean ten-day relative humidity [%].
The slope of the relationship between saturation vapour pressure and temperature is given by:

$$\Delta = 4098\ e_s\ /(T + 237.3)^2 \tag{10}$$

where:
Δ - slope of the vapour pressure curve [kPa °C^{-1}],
T – mean ten-day temperature air temperature [°C].
Because soil heat flux G is small compared to R_n, when the surface is fully covered by vegetation and the calculation time step is ten days, the estimation of G can be ignored and assumed to be zero in Equation (2). Allen et al. (1998) state that the soil heat flux beneath the grass reference surface is relatively small for that time period.
The sums of reference evapotranspiration ET_o in each month and in the whole growing period from April to September in the years 1970-2004 were established at 40 meteorological stations located in different regions of Poland. The statistical analysis was performed to get the values of mean, minimum, maximum, standard deviation, variability coefficient and median.
Statistical parameters characterizing the temporal variability of ET_o are determined using ET_o averaged over 40 stations in each year.
Statistical parameters characterizing the spatial variability of ET_o are determined using ET_o averaged over 35 years at each station. The gamma probability distributions were fitted to the multi-year series of the sums of ET_o in the growing season at each station, checking goodness-of-fit by using the χ^2-Pearson test. On that basis the sums of ET_o at the probability of exceeding of 0.2, 0.5 and 0.8 were estimated.
The sums of reference evapotranspiration in the growing season were used to distinguish the agro-climatic regions of Poland with the *k*-means clustering algorithm.
The same sums of ET_o were used to calculate the climatic water balance in the growing period. Climatic water balance is the difference between precipitation sum P and reference evapotranspiration ET_o calculated according to the Penman-Monteith formula:

$$CWB = P - ET_o \quad (11)$$

where:

CWB - climatic water balance [mm],

P - precipitation [mm],

ET_o - reference evapotranspiration according to the Penman-Monteith method [mm].

The maps of the spatial distributions of ET_o (the sums in the growing season at the probability of 0.2, 0.5 and 0.8, the mean monthly sums) and of CWB (the sums in the growing season) were created using the Regularized Spline Radial Basic Function method with the Spatial Analyst Module of ArcView GIS 9.1.

3.2 Data

The ten-day records of meteorological parameters (air temperature, air humidity, sunshine hours, wind velocity and precipitation) came from 40 meteorological stations of the Institute of Meteorology and Water Resources and of the Institute of Technology and Life Sciences (Tab. 1, Fig. 2).

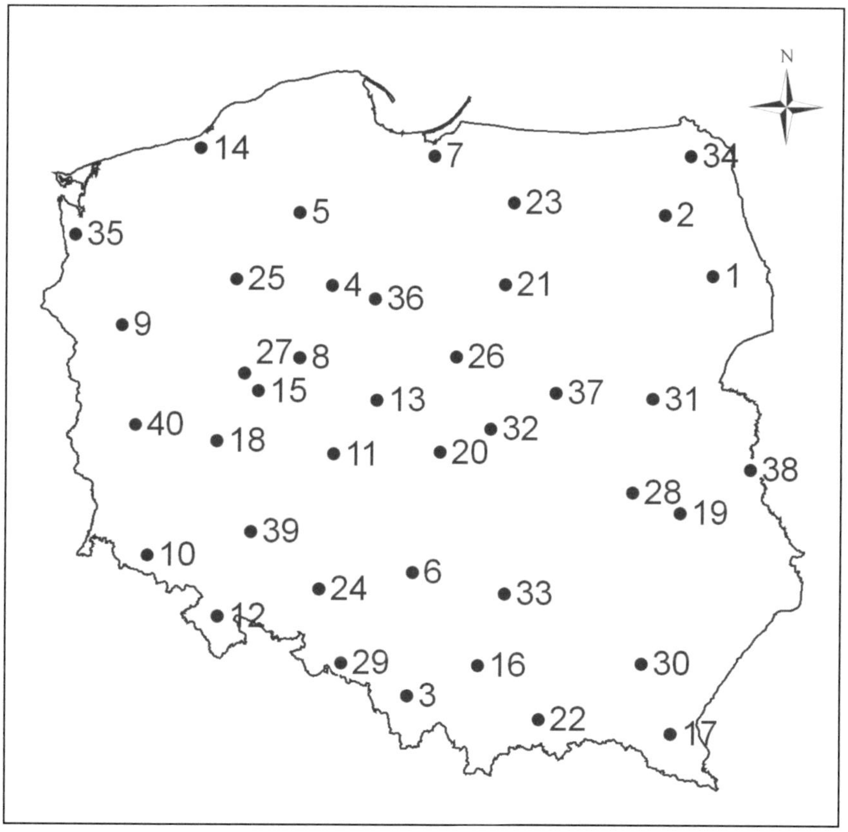

Fig. 2. Geographical location of the studied meteorological stations in Poland (1 - the number of the station from Tab. 1)

No	Station	Altitude (m a.s.l.) *	Latitude	Longitude
1	Białystok	139	53°13'	23°10'
2	Biebrza	117	53°39'	22°36'
3	Bielsko-Biała	398	49°48'	19°00'
4	Bydgoszcz	46	53°08'	18°01'
5	Chojnice	173	53°42'	17°33'
6	Częstochowa	261	50°49'	19°06'
7	Elbląg	38	54°10'	19°26'
8	Gniezno	110	52°33'	17°34'
9	Gorzów	65	52°44'	15°15'
10	Jelenia Góra	342	50°54'	15°48'
11	Kalisz	140	51°44'	18°05'
12	Kłodzko	316	50°26'	16°39'
13	Koło	95	52°12'	18°40'
14	Koszalin	33	54°12'	16°09'
15	Kórnik	77	52°15'	17°06'
16	Kraków	209	50°04'	19°57'
17	Lesko	386	49°28'	22°20'
18	Leszno	93	51°51'	16°35'
19	Lublin	171	51°14'	22°34'
20	Łódź	184	51°44'	19°24'
21	Mława	141	53°06'	20°21'
22	Nowy Sącz	292	49°37'	20°41'
23	Olsztyn	133	53°46'	20°25'
24	Opole	176	50°40'	17°58'
25	Piła	72	53°08'	16°45'
26	Płock	62	52°33'	19°40'
27	Poznań	86	52°25'	16°50'
28	Puławy	142	51°25'	21°58'
29	Racibórz	190	50°05'	18°13'
30	Rzeszów	200	50°06'	22°03'
31	Siedlce	146	52°11'	22°16'
32	Skierniewice	129	51°57'	20°09'
33	Skroniów	256	50°38'	20°16'
34	Suwałki	165	54°08'	22°57'
35	Szczecin	1	53°24'	14°37'
36	Toruń	69	53°03'	18°35'
37	Warszawa	106	52°09'	20°59'
38	Włodawa	175	51°38'	23°33'
39	Wrocław	116	51°06'	17°05'
40	Zielona Góra	182	51°56'	15°30'

* m a.s.l. metres above mean sea level

Table 1. Geographical location of the studied meteorological stations

4. Reference evapotranspiration ET_o

4.1 Temporal variability of ET_o

The monthly and the growing season sums of ET_o, averaged over 40 stations, characterise the multi-year mean reference evapotranspiration in Poland. In particular months it ranges from 54 mm in September to 110 mm in July (Tab. 2). It means the mean daily reference evapotranspiration from almost of 2 mm day^{-1} to 4 mm day^{-1}, respectively. The highest monthly sum (157 mm) occured in July 1994 and the lowest (42 mm) in September 2001.
In the whole vegetation period, from April to September, ET_o amounts to 520 mm. Taking into account the average precipitation in this period equal to 350 mm, water deficit can be expected in Poland, on average. During 1970-2004, the lowest whole-country mean ET_o (422 mm) occured in 1980 and the highest ET_o (615 mm) in 1992 (Tab. 2, Fig. 3). The 1980 was a very wet year with the highest precipitation in most regions of Poland, whereas in 1992 the extreme drought occured all other the country, caused by very low precipitation, high air temperature and high insolation. The close relationship between reference evapotranspiration and precipitation is a characteristic feature of a transitory temperate climate. Reference evapotranspiration had rather high temporal variability in the multi-year period, with the variability coefficient of 12-15% for the monthly sums and 9% for the sums in the growing season. It is the other special feature of climate in Poland.

Parameter	Period						
	Apr	May	June	July	Aug	Sep	Apr-Sep
Mean [mm]	61	97	104	110	96	54	520
Minimum [mm]	50	78	80	77	75	42	422
Maximum [mm]	83	116	134	157	123	71	615
Median [mm]	59	96	102	107	96	52	516
Standard deviation [mm]	8	11	13	17	12	8	46
Variability coefficient [%]	13	12	13	15	13	14	9

Table 2. Statistics of temporal variability of ET_o averaged over 40 stations

ET_o changed in a wide range in 1970-2004. The amplitude (maximum-minimum) of the ET_o sums in the growing season was the lowest in southern Poland (135 mm) and the highest in central Poland (316 mm). It shows that meteorological conditions were more stable in southern Poland than in central.
The statistically significant linear increasing trend of the growing season spatially averaged reference evapotranspiration was determined in the analyzed period 1970-2004 (Fig. 3). The trend shows the increase of 27 mm per 10 years and is an effect of the increasing trends of temperature, sunshine hours and solar radiation observed in the last 50 years in Poland.

4.2 Spatial variability of ET_o

The monthly and the growing season sums of ET_o, averaged over 35 years, characterise the spatial variability of the multi-year mean reference evapotranspiration. The minimum multi-year average ET_o in the growing season equals to 469 mm occured in the station no. 10, located in the south-western part of Poland. The maximum sum of ET_o, average in time, equals to 566 mm, was observed in the central part (the station no. 8).

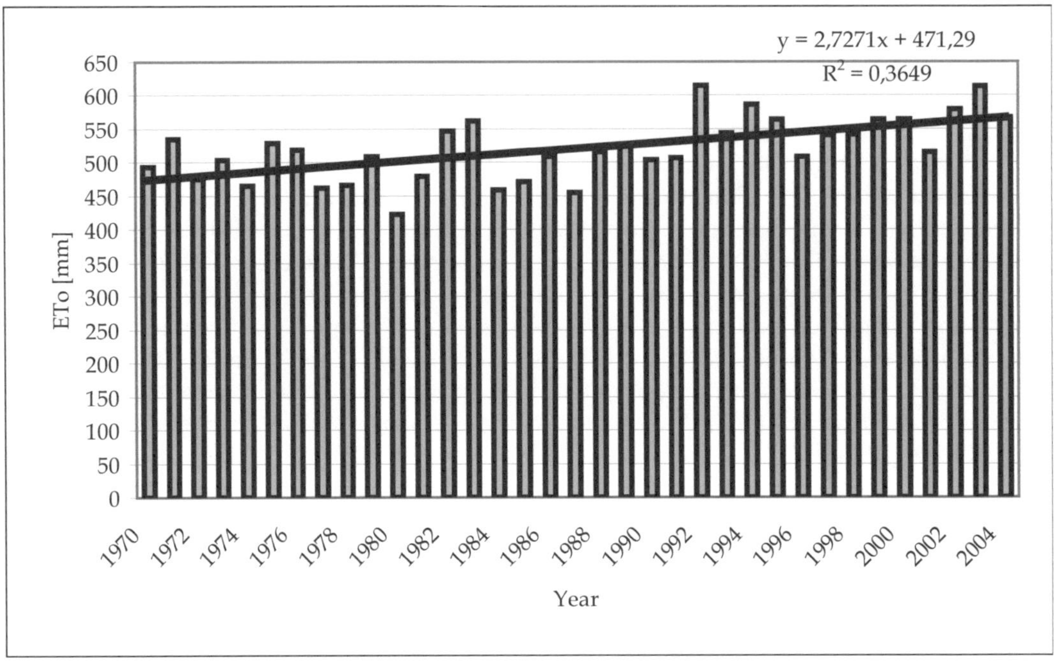

Fig. 3. Sums of ET_o in the growing season (Apr-Sep) in 1970-2004, averaged over 40 stations

Similar to the spatially averaged ET_o (Tab. 2), the monthly sums of ET_o, averaged in time, were the highest in July (120 mm) in the central Poland and the lowest in September (45 mm) in the north-eastern part (Tab. 3).

The spatial variability of reference evapotranspiration (the variability coefficient 5-7%) was lower than the temporal variability. It means the bigger differentation of the reference evapotranspiration in the years than in the space.

Parameter	Period						
	Apr	May	June	July	Aug	Sep	Apr-Sep
Mean [mm]	61	97	104	110	96	54	520
Minimum [mm]	53	85	92	98	86	45	469
Maximum [mm]	66	105	113	120	107	62	566
Median [mm]	62	97	103	110	97	55	524
Standard deviation [mm]	3	5	6	6	6	4	28
Variability coefficient [%]	6	5	5	5	6	7	5

Table 3. Statistics of spatial variability of ET_o averaged over 35 years

The absolute minimum reference evapotranspiration sum in the growing season at all stations and in all years amounted to 366 mm and was determined in 1980 in the north-eastern part of Poland (station no. 2) (Tab. 4). The absolute maximum was 732 mm in 1982 in the central Poland (station no. 27).

Period	Abs min	Abs max
April	31	105
May	46	138
June	67	161
July	62	180
August	51	161
September	30	168
April-September	366	732

Table 4. Absolute minimum and maximum sums of reference evapotranspiration ET_o [mm] at 40 stations in 1970-2004

The values of reference evapotranspiration determined for the probability of exceeding of 0.2, 0.5 and 0.8 can be referred to the conditions of warm and dry, average and cold and wet weather, respectively. The probability of 0.2 means that evapotranspiration can be higher than or equal to a given amount. Practically it means that evapotranspiration at the probability of 0.2 can be expected every five years, at the probability of 0.5 - every two years and at the probability of 0.8 - eight years in a decade.
A spatial differentiation of the reference evapotranspiration is observed in Poland (Fig. 4, 5, 6). The growing season sum of the reference evapotranspiration, calculated with the Penman-Monteith method, ranges in the area of the country from 480 to 560 mm in an average year (Fig. 4), 500 to 620 mm in a hot and dry year (Fig. 5) and 460 to 520 mm in cold and wet year (Fig. 6).

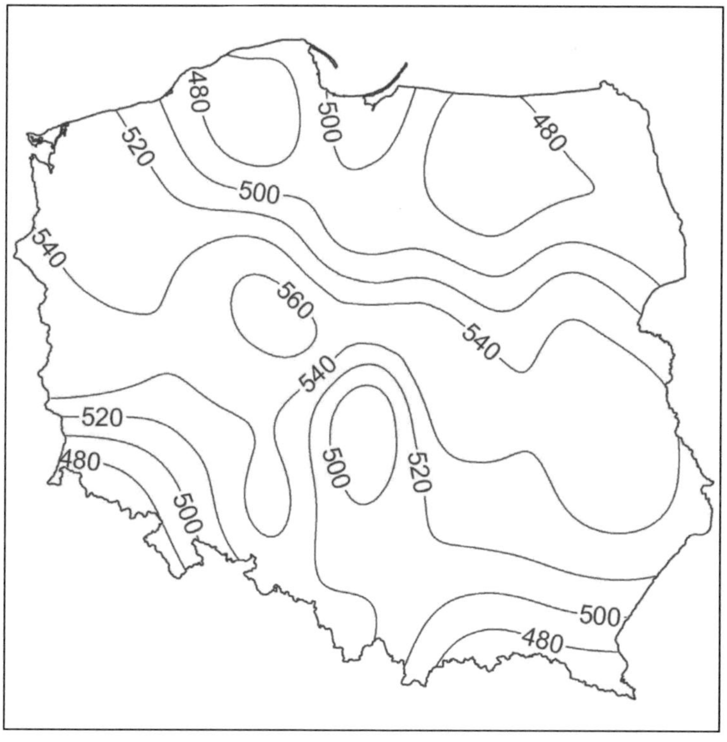

Fig. 4. Reference evapotranspiration ET_o [mm] in the growing season (Apr-Sep) at the probability 0.5

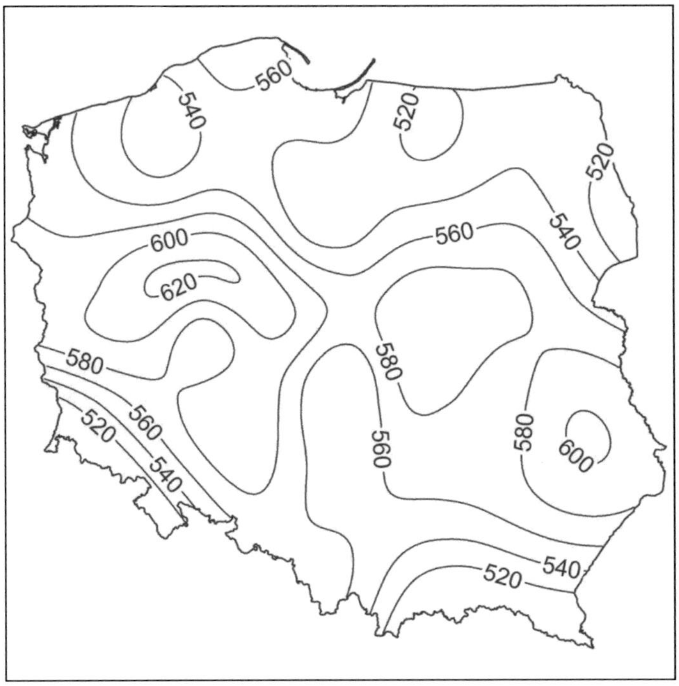

Fig. 5. Reference evapotranspiration ET_o [mm] in the growing season (Apr-Sep) at the probability 0.2

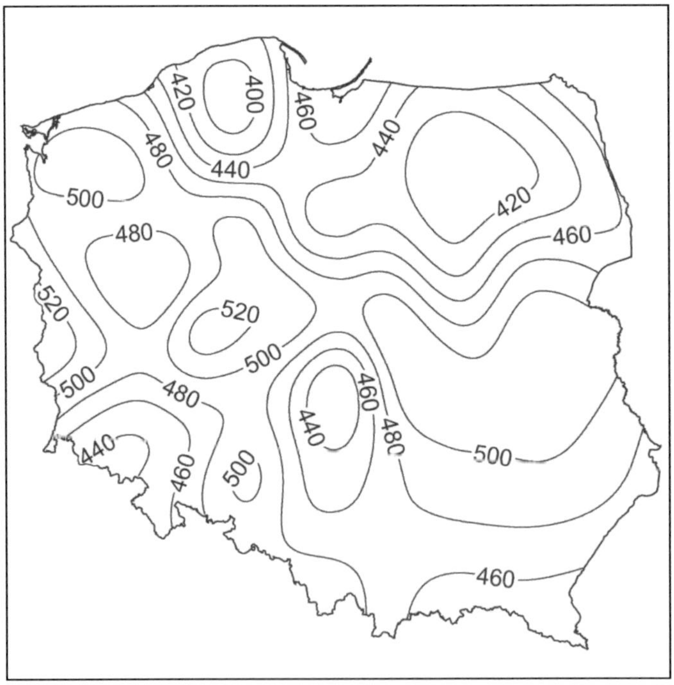

Fig. 6. Reference evapotranspiration ET_o [mm] in the growing season (Apr-Sep) at the probability 0.8

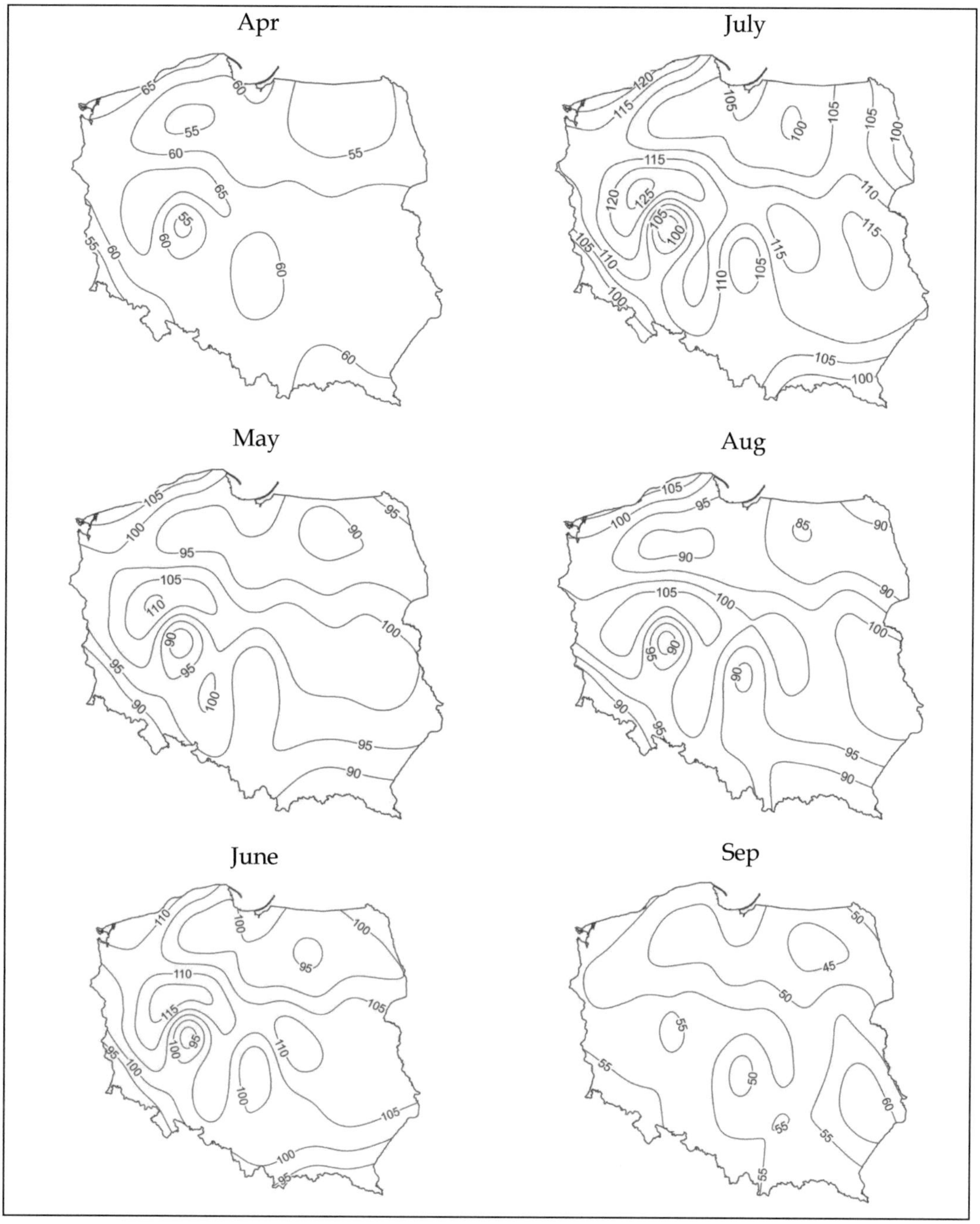

Fig. 7. Mean monthly sums of reference evapotranspiration ET_o [mm]

Generally the higher reference evapotranspiration were recorded in central Poland from west to east. The highest values occurred in the small area with the stations nos. 4, 8, 11, 13, 15, 25 and 27. This is the driest region of Poland with the lowest rainfall which often is less than 300 mm in a year. The region is most threatened by meteorological droughts. In connection with high reference evapotranspiration the region is threatened by agricultural droughts. Although no significant decrease in the annual precipitation is predicted in Central Europe due to global climate change, considering the forecasted increase in temperature, possible increase in reference evapotranspiration and water shortage is very likely in the future in that region.

Similar to the sum of the reference evapotranspiration in the growing season, the highest values of the monthly sums occurred in the mid-western part of Poland, reaching 125 mm in July in that region (Fig. 7). The lowest ET_o is observed in April and September, being less twice than in July.

5. Agro-climatic regions

According to the sums of reference evapotranspiration in the growing season the agro-climatic regions of Poland have been distinguished using cluster analysis. The *k*-means clustering algorithm has been applied for grouping the meteorological stations of similar reference evapotranspiration.

Cluster analysis is an exploratory data analysis tool which aims at sorting different objects into groups in a way that the degree of association between two objects is maximal if they belong to the same group and minimal otherwise. The *k*-means method is one of different clustering methods and it produces exactly *k* different clusters of the greatest possible distinction. The goal of the method is to minimize variability within clusters and maximize variability between clusters. The similarity rules apply maximally to the members of one cluster and minimally to members belonging to the rest of the clusters. The significance test evaluates the between group variability against the within-group variability when computing the significance test for the hypothesis that the means in the groups are different from each other. In *k*-means clustering, the algorithm moved stations in and out of groups (clusters) to get the most significant ANOVA results (Romesburg, 2004).

Taking into account the range of variability of ET_o in the area of Poland as well as the regionalization of climatic conditions, three clusters ($k = 3$) were determined *a priori*, that are to be as distinct as possible. The performed analysis showed very different means. The magnitude of the *F* value from the analysis of variance indicates a very significant discrimination between clusters ($p<0.0000001$).

The analysis allowed to classify reference evapotranspiration occurring in the area of Poland into the meaningful piles. On that basis the three agro-climatic regions can be distinguished in Poland:

A - with high reference evapotranspiration (high evaporative demand of the atmosphere),

B - with medium reference evapotranspiration (medium evaporative demand of the atmosphere),

C - with low reference evapotranspiration (low evaporative demand of the atmosphere).

The distinguished regions are shown in Fig. 8 and the values of ET_o in these regions in Tab. 5. Into the A region 14 stations were classified, into the B region - 14 and into the C region - 12. The mean ET_o in each region is calculated as the mean of all stations classified into the region. The maximum and minimum ET_o are the maximum and minimum multi-year (1970-2004) average values recorded at any station located in the region.

The A region is characterized by the spatially and temporally averaged reference evapotranspiration according to the Penman-Monteith method equal to 550 mm from April to September, the B region – 520 mm and the C region – 490 mm. It should be realized that

ET_0 in a give year and at a given station can be of the magnitude that matches another class (region). This should not be the base for changing the classification and the affiliation of a station to that class.

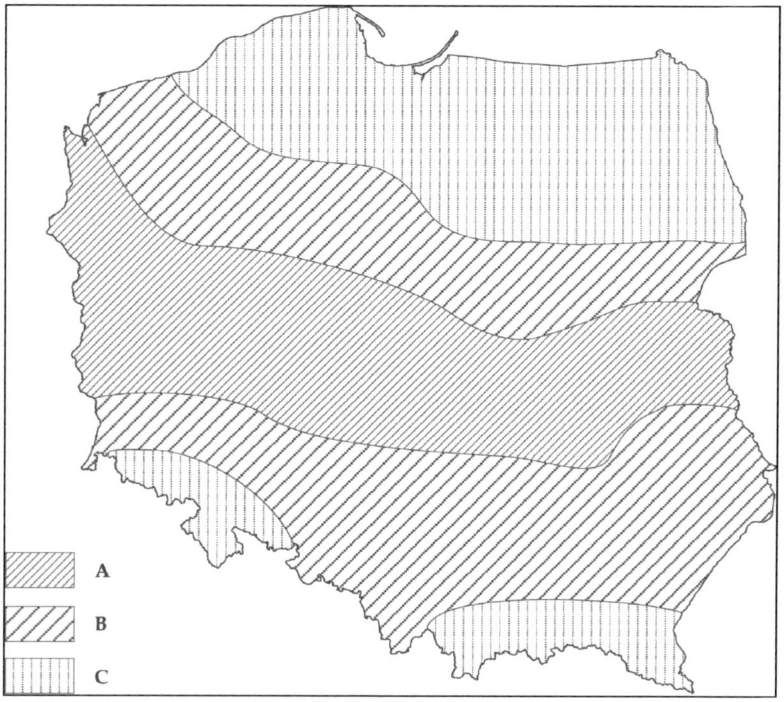

Fig. 8. Agro-climatic regions: A - high ET_0, B - medium ET_0, C - low ET_0

Parameter	Agro-climatic region		
	A	B	C
Class of ET_0	high	medium	low
Mean ET_0 [mm]	550	520	490
Max ET_0 [mm]	566	546	514
Min ET_0 [mm]	536	489	469

Table 5. Reference evapotranspiration ET_0 in the agro-climatic regions of Poland

6. Climatic water balance

Reference evapotranspiration can been applied to a wide variety of research problems in the field of agrometeorology and agricultural water management. Besides of the very popular method of calculating crop evapotranspiration and crop water use with crop factors, reference evapotraspiration can be used to compute the climatic water balance. The climatic water balance one can interpret as the indicator of climate dryness.

The climatic water balance CWB was calculated using the measured precipitation and the calculated ET_0 at 40 stations and in 35 years (1970-2004). The spatial distribution of CWB (Fig. 9) shows its differentiation in Poland, depending on the agrometeorological conditions.

The mean multiannual values of this balance in the growing season (April to September) were negative at most examined stations. In the most part of the country reference evapotranspiration exceeds precipitation. The climatic water balance, averaged over all years and all stations, was equal to -144 mm in the growing season.
The higher positive values of CWB (0-100 mm) were found in the southern Poland, the region situated near the Karpaty mountains. Precipitation exceeds reference eapotranspiration in that region. In the north part of Poland – near the Baltyk sea and in the north-east part of the country the values of climatic water balance were negative but higher than –100 mm. The values of CWB less than –150 mm were in the central Poland, from west to east. The lowest values of the balance (less than –200 mm) were on the area from the north-west to the central part. The highest risk of agro-climatic water deficit can be expected in these regions.

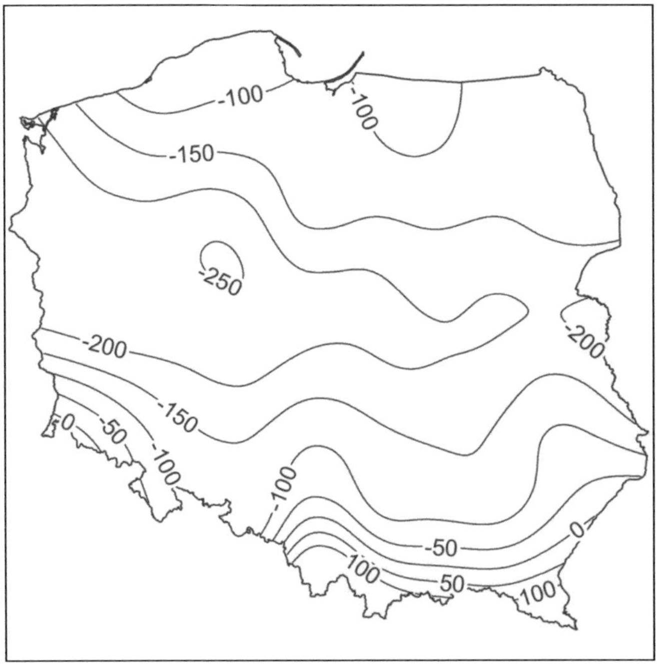

Fig. 9. Climatic water balance [mm] in the growing season (Apr–Sep), mean in 1970-2004

For the operational, real-time estimation of climatic water balance in different regions and to get the comparable evaluation of precipitation deficit or surplus, the standardized climatic water balance S-CWB can be used (Łabedzki & Bąk, 2004). After the performance of standardization one can analyse the actual situation at different stations and the evaluation of actual agro-climatic conditions is comparable. The standardized climatic water balance is an index indicating climatic water balance lower or higher than the mean value or the median if the probability distribution of climatic water balance is not normal. This method allows the CWB distribution at the station to be represented by a mathematical cumulative probability function. One can then tell what is the probability of CWB being less than or equal to a certain amount. The probability of CWB being less than or equal to the median is 0.5. The probability shows how often one can expect a particular value of CWB.

To categorize and evaluate the deficit or surplus of precipitation in relation to reference evapotranspiration, CWB should be compared with the boundaries of different agro-climatic conditions classes. Following the example of the classification of the standardized precipitation index SPI (McKee et al., 1993; Paulo & Pereira, 2006; Vermes, 1998), seven classes can be distinguished (Tab. 6). Because S-CWB are standardized values, the threshold for particular classes corresponds to the definite probabilities of non-exceeding a given value. For example a threshold S-CWB for extremely dry period (S-CWB $\leq$ -2.0) corresponds to the climatic water balance (with the lower) with a probability of occurrence 2%. S-CWB = -1.0 corresponds to the climatic water balance (with the lower) with a probability of occurrence 16%.

Anyway S-CWB needs a careful interpretation. It is a relative index being the standardized deviation from the median and indicating climatic water balance lower or higher than the median for the period and location under consideration. In the S-CWB methodology the average climatic water balance is treated as normal, which often means water deficit for agriculture. This index characterises only climatic water deficit or surplus.

S-CWB	Agro-climatic conditions	Cumulative probability
≤ -2.00	extremely dry	0.0228
-1.99 to -1.50	very dry	0.0668
-1.49 to -1.00	moderately dry	0.1587
-0.99 to 1.00	normal	0.8413
0.99 to 1.50	moderately wet	0.9332
1.49 to 2.00	very wet	0.9772
> 2.0	extremely wet	1

Table 6. Classification of agro-climatic conditions according to standardized climatic water balance S-CWB

7. Conclusion

1. In this study the performance of the FAO Penman-Monteith method to estimate reference evapotranspiration under climatic conditions of Poland is presented. Using ten-day weather data from 40 weather stations from 1970-2004, the PM reference evapotranspiration was calculated. The analysis of temporal and spatial variability of the PM reference evapotranspiration was made. In addition, the reference evapotranspiration was used to determine the agro-climatic regions and climatic water balance in Poland.
2. The sum of reference evapotranspiration in the growing season from April to September exceeds 500 mm in most area of Poland. Averaged over 35 years and 40 stations it amounts to 520 mm. The monthly sums of reference evapotranspiration, averaged over 40 stations, ranges from 54 mm in September to 110 mm in July.
3. The spatial variability of reference evapotranspiration was lower than the temporal variability. It means the bigger differentation of the reference evapotranspiration among years than among stations.
4. As a general rule, reference evapotranspiration isolines as yielded by the Penman-Monteith method are traced almost west-east. They showed a characteristic pattern, by decreasing from the central part to the northern and southern part of Poland. The

highest reference evapotranspiration were recorded in the region with the lowest rainfall.

5. The spatial distribution of reference evapotranspitration in the area of Poland is similar to the spatial distribution of precipitation.
6. Using cluster analysis three agro-climatic regions were distinguished in Poland differing significantly in reference evapotranspiration.
7. The statistically significant increase in reference evapotranspiration was determined in 1970-2004 due to the increase in temperature.
8. In most of the country reference evapotranspiration exceeds precipitation and this is the reason of water shortage for crops. This tendency is expected to grow with the foreseen climate warming.
9. Future research will be addressed to further develop an operational system for assessing reference evapotranspiration with the FAO Penman-Monteith method and potential and actual evapotranspiration in a regional and the entire country scale. Future studies will consider the ability of forecasting reference evapotranspiration to forecast meteorological and agricultural droughts and to improve the management of irrigation systems. Of particular interest is which method of forecasting is better - based on synoptic and mezoscale models or statistical models.

8. References

Allen, R.G.; Clemmens, A.J.; Burt, C.M.; Solomon, K. & O'Halloran, T. (2005). Prediction accuracy for projectwide evapotranspiration using crop coefficients and reference evapotranspiration. *Journal of Irrigation and Drainage Engineering,* Vol. 131, Issue 1, 24-36

Allen, R.G.; Jensen, M.E.; Wright, J.L. & Burman, R.D. (1989). Operational estimate of reference evapotranspiration. *Agronomy Journal*, No. 81, 650-662

Allen, R.G.; Pereira, L.S.; Raes, D. & Smith, M. (1998). Crop evapotranspiration - Guidelines for computing crop water requirements. *FAO Irrigation and Drainage Paper*, No. 56, FAO, Rome

Allen, R.G.; Pruitt,W.O.; Wright, J.L.; Howell, T.A.; Ventura, F.; Snyder, R.; Itenfisu, D.; Steduto, P.; Berengena, J.; Yrisarry, J.B.; Smith, M.; Pereira, L.S; Raes, D.; Perrier, A.; Alves, I.; Walter, I. & Elliott, R. (2006). A recommendation on standardized surface resistance for hourly calculation of reference ET_o by the FAO56 Penman-Monteith method. *Agricultural Water Management*, No. 81, 1-22

Allen, R.G.; Smith, M.; Perrier, A. & Pereira, L.S. (1994). An update for the definition of reference evapotranspiration. *ICID Bull.* Vol. 43, No. 2, 1-34

Buttafuoco, G.; Caloiero, T. & Coscarelli, R. (2010). Spatial uncertainty assessment in modelling reference evapotranspiration at regional scale. *Hydrol. Earth Syst. Sci. Discuss.*, No. 7, 4567-4589, DOI:10.5194/hessd-7-4567-2010

Brutseart, W. (1982). *Evaporation into the atmosphere: theory, history, and applications,* Kluwer Academic Publishers, ISBN 90-277-1247-6, Dordrecht, The Netherlands

Feddes, R.A. & Lenselink, K.J. (1994). Evapotranspiration. In: *Drainage Principles and Applications. ILRI Publication*, No. 16, 145-173, Wageningen

Fitzmaurice, L. & Beswick, A. (2005). Sensitivity of the FAO56 crop reference evapotranspiration to different input data. Technical Report QNRM05484, ISBN 1 74172 060 5, Department of Natural Resources and Mines, Queensland, Australia

Kejna, M.; Arazny, A.; Maszewski, R.; Przybylak, R.; Uscka-Kowalkowska, J. & Vizi, Z. (2009). Daily minimum and maximum air temperature in Poland in the years 1951-2005. *Bulletin of Geography*, No. 2, 35-56, Nicolaus Copernicus University, Torun, Poland

Kozuchowski, K. & Zmudzka, E. (2003). 100-year series of areally averaged temperatures and precipitation totals in Poland. In: *Man and Climate in the 20th Century. Studia Geograficzne*, No. 75, 117-122, Wroclaw University, Poland

Labedzki, L. (1999). Usability of the Penman-Monteith equation for calculating reference and grassland evapotranspiration. Wiadomosci IMUZ, Vol. XX, No. 2, 89-101 (*in Polish, summary in English*)

Labedzki, L. (2007). Estimation of local drought frequency in central Poland using the standardized precipitation index SPI. *Irrigation and Drainage*, Vol. 56, Issue 1, 67-77. DOI: 10.1002/ird.285

Labedzki, L. & Bak, B. (2004). Standardized climatic water balance as drought index. *Acta Agrophysica*, Vol. 3(1), 117-124 (*in Polish, summary in English*)

Maulé, C.; Helgalson, W.; McGinn, S. & Cutforth, H. (2006). Estimation of standardized reference evapotranspiration on the Canadian Prairies using simple models with limited weather data. *Canadian Biosystems Engineering*, Vol. 48, 1.1-1.11

McKee, T.B.; Doesken, N.J. & Kleist, J. (1993). The relationship of drought frequency and duration to time scales. Proc. 8th Conf. Applied Climatology, 17-22 January 1993, Anaheim, California, 179-184

McVicar, T.R.; van Niel, T.G.; Li, L.; Hutchinson, M.F.; Mu, X.. & Liu, Z. (2007). Spatially distributing monthly reference evapotranspiration and pan evaporation considering topographic influences. *Journal of Hydrology*, Vol. 338, Issue 3-4, 196-220. DOI: 10.1016/j.jhydrol.2007.02.018

Paltineanu, C.; Panoras, A.G.; Mavroudis, I.G & Louisakis, A. (1999). Estimating reference evapotranspiration and irrigation water requirements in the Gallikos river basin, Greece. *International Agroghisics*, Vol. 13, 49-62

Paulo, A.A. & Pereira, L.S. (2006). Drought concepts and characterization. Comparing drought indices applied at local and regional scales. *Water International*, Vol. 31, No. 1, 37-49

Romesburg, H. C. (2004). *Cluster Analysis for Researchers*. Lulu Press, North Carolina.

Smith, M. (1992). Report on the expert consultation on revision of FAO methodologies for crop water requirements. *Land and Water Development Division*, FAO, Rome

Song, Z.W.; Zhang, H.L.; Snyder, R.L.; Anderson, F.E. & Chen, F. (2010). Distribution and Trends in Reference Evapotranspiration in the North China Plain. *Journal of Irrigation and Drainage Engineering*, Vol. 136, Issue 4, 240-247

Sumner D.M. & Jacobs, J.M. (2005). Utility of Penman–Monteith, Priestley–Taylor, reference evapotranspiration, and pan evaporation methods to estimate pasture evapotranspiration. *Journal of Hydrology*, Vol. 308, 81–104. DOI:10.1016/j.jhydrol.2004.10.023

Sun Tables. (1974). Institute of Meteorology and Water Management, Warsaw

Vermes L., 1998. How to work out a drought mitigation strategy. An ICID guide. Guidelines for water management. Bonn: DVWK, No. 309

Wos, A. (1999). *Climate of Poland*. Wydawnictwo Naukowe PWN, Warszawa (*in Polish)*

Zawora, T. & Ziernicka, A. (2003). Precipitation variability in time in Poland in the light of multi-annual mean values (1891-2000). In: *Man and Climate in the 20th Century. Studia Geograficzne*, No. 75, 123-128, Wroclaw University, Poland

13

Application of a New Web-Based Tool(*CropWaterUse*) for Determining Evapotranspiration and Irrigation Requirements of Major Crops at Three Locations in Queensland

José O. Payero[1, 2], Dhananjay Singh[1]
Graham Harris[1], Simon Vriesema[1], Jenelle Hare[1]
Lance Pendergast[1] and Yash Chauhan[1]
[1]*Agri-Science Queensland, Department of Employment Economic Development & Innovation (DEEDI) 203 Tor St, PO Box 102, Toowoomba, Qld 4350*
[2]*The University of Queensland, Queensland Alliance for Agricultural and Food Innovation (QAAFI), St. Lucia, QLD 4072 Australia*

1. Introduction

Decreased water availability in many areas has created the need to make more efficient use of limited water resources. To maximize production and profits, growers need to make decisions regarding planting date, crop type, planted area, and irrigation management, which are affected by the amount of water available. Therefore, they need to know how much water is needed to grow a particular crop, which they currently estimate in most cases based on previous experience. However, crop water requirements can vary significantly, among other things, with crop type, season, location, planting date, and available water, which could make it inappropriate to extrapolate previous experiences to future planning. Also, occasionally growers may want to grow unfamiliar crops, which can further complicate planning. Growers and consultants need reliable estimates of crop water requirements to make irrigation decisions aimed at improving water use efficiency and profits. A current challenge is that growers' perception of crop water requirements for particular crops and corresponding irrigation scheduling practices are based on historical weather patterns. These weather patterns are likely to change in the future if climate change and climate variability predictions prove to be true (Howden et al., 2007; IPCC, 2007), leading to changes in planting times and crop growth patterns. Growers could benefit from new tools to help them anticipate and adapt to the effects of these potential changes.

In the last two decades there have been considerable advances in automatic weather station networks across the world and in developing improved procedures to calculate crop water use from weather data (Allen et al., 1998). However, the fact remains that most growers still do not use weather data to make irrigation decisions. Part of the reason is that the information is not easily accessible or it is not available in the required format. In Australia, for example, daily weather data and the calculated grass-reference evapotranspiration (ETo) are available online from the Australian Bureau of Meteorology (http://www.bom.gov.au/silo/) in two types of datasets, the *Data Drill* and the *Patched Point* datasets. The *Data Drill* dataset provides interpolated data for any location in Australia and the *Patched Point* dataset combines interpolated data with Bureau of Meteorology measurements. Even though ETo data are available, for this information to be useful to growers it needs to be transformed into crop evapotranspiration (ETc) for specific crops and locations, or even further transformed into estimates of soil water status for particular fields. Growers, in general, do not have the knowledge and/or the tools to make these transformations. Therefore, there is a need to make ETc values derived from weather data accessible to growers. The availability of internet services and on-line weather databases provides new opportunities to develop and maintain web-based decision support tools for growers.

In general, growers need to make both strategic (planning) and tactical (day-to-day) irrigation and cropping system decisions. The strategic decisions concern questions such as: what crops to grow, how much area of each crop to grow, when to plant each crop, how much area of crop to irrigate and how much to grow as dryland, etc. If water is the main limiting factor, as it is in many parts of Australia, it is important to know how much water they would need to grow specific crops in a given area and planting day, which will determine how much area can be planted according to the amount of water available. The tactical decisions involve day-to-day irrigation scheduling, which determine timing and amount of irrigation to be applied to each crop during each irrigation event. Tools can be developed that use weather data to assist growers make both strategy and tactical decisions based on daily weather data. There is a long history of development of computer programs that use weather data to aid in irrigation scheduling, some recent and some dating back to the time when personal computers first became available (Abourached et al., 2007; Brown et al., 2010; Car et al., 2008; Chapman et al., 2008; Chauhan et al., 2011; Chopart et al., 2007; Cull, 1979; Davidson et al., 1998; Evett and Lascano, 1993; Fox et al., 1993; Howell et al., 1995; Inman-Bamber and Attard, 2005; Jensen, 1969; Kincaid and Heermann, 1974; McKay et al., 2009; Raes et al., 2009; Richards et al., 2008; Steduto et al., 2009; Thysen and Detlefsen, 2006).

Unfortunately, however, historically growers have been slow to adopt these tools for practical decision-making. There have been many reasons for that. In the early days, one of the main reasons was that weather data were not readily available for the different locations. Also, weather data were collected with manual weather stations, which meant that someone would have to physically travel to the station site every day to record the data. Therefore, there could be a long time lag between data collection and the time the data got to the end user, especially for stations located in remote areas. Another issue limiting the uptake of decision-support tools in the early days was the need for manual data input, which was time-consuming and tedious. With the development of electronic weather stations and the internet, some of these early issues can be overcome. However, some barriers to adoption of this technology still remain. First, there is the issue of computer illiteracy of many of the older growers who are predominantly the ones making decisions on the farm. Also, the lack of internet connectivity in many rural areas can be a problem. Additionally, there are issues relating to the way many decision-support tools have been conceived and designed. Many are just too complex for the normal grower since they have been designed as research tools, targeting scientists rather than growers, as is the case of many mechanistic crop growth

models. Others have been developed as user-friendly tools, but training and on-going support for growers have been lacking. In other cases, tools have been developed for just one crop, which become of limited use for growers that have to deal with a variety of crops. Other tools have been developed for a given region, therefore cannot be applied to other regions without considerable modifications. The lack of local validation can also be an issue affecting accuracy of many tools, which contributes to failing to gain the trust of local growers.
Despite all these issues, with recent advances in technology, there are still opportunities to develop tools that would respond to the need of many growers, if an adequate process for development and adoption is followed. This chapter reports on a new web-based tool (*CropWaterUse*) recently developed in Australia by Agri-Science Queensland to help growers and crop consultants determine, among other things, daily and seasonal crop evapotranspiration and irrigation requirements, assuming full irrigation (no crop stress). The chapter presents a description of *CropWaterUse* and illustrates its application by using it to simulate evapotranspiration and irrigation requirements of several major crops grown in three different environments (cool, mild, and hot) in Queensland, Australia.

2. Description of *CropWaterUse*

CropWaterUse is a web-based tool designed to help growers and crop consultants make strategic decisions. Although it was designed for growers and crop consultants, it could have application for policy makers and water resource planning agencies that need to create policies for regulating water use and allocating water resources. The information generated by *CropWaterUse* can also be useful for irrigation engineers when designing irrigation systems, especially for unfamiliar crops and locations. It has been designed to be user-friendly and to require minimum inputs. It can assist growers in making strategic water management decisions by allowing them to easily answer the question of how much water would be needed to grow a crop at a given planting day and location. Users can compare scenarios by changing inputs. The tool can then be used to make side-by-side comparisons of different scenarios, which allow the user to answer "what if" questions. It uses historical weather data and, therefore, is not intended to be used as an irrigation scheduling tool, which would require daily real-time data. *CropWaterUse* is freely available online at http://cropwateruse.dpi.qld.gov.au.

2.1 System architecture

CropWaterUse is designed with a client-server multi-tiered architecture consisting of three tiers: client-side (presentation), server-side (logic), and server-side (data) as shown in Figure 1. It was developed using the technologies and tools shown in Table 1. All presentation logic is handled on the client-side tier. The server-side application logic interacts with a server-side SQL database that has been developed using MSSQL 9.0. Figure 2 shows the logic flowcharts for accessing the system and performing analysis.

2.2 Website statistics

CropWaterUse is capable of monitoring website statistics using the Google Analytics suite of web tools. It has the ability to provide website statistics visits using various breakdowns, including: Geo-locations, visitor loyalty, browser capabilities and traffic sources. For example, Figure 3 shows some of the website statistics produced from October 2009 to

September 2010. It shows 223 visits to the website from 17 countries, with most of the visitors coming from Australia. Since the website consists of several pages, the 223 visits have resulted in 1000 page views.

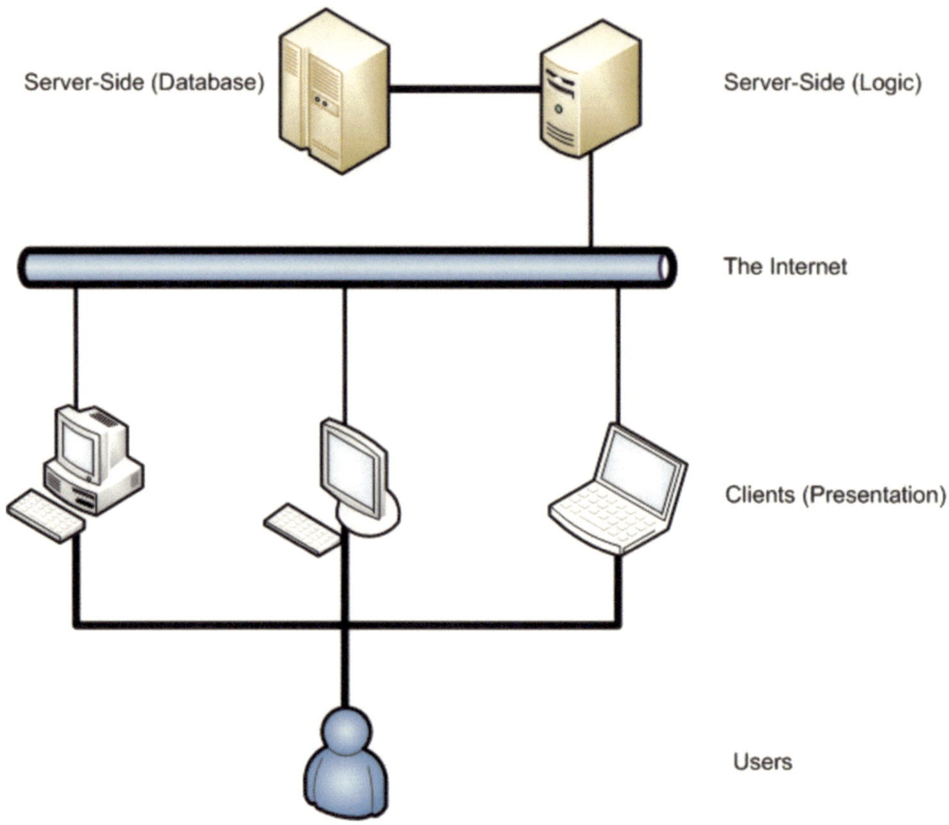

Fig. 1. Three-tier system architecture used to design *CropWaterUse*

Tool/language	**Version/comments**
Visual Studio.NET	2008
ASP.Net and C#	3.5 SP1
SQL Server	2005
TeeChart 3.0.NET	3.0 March Build
Developer Express Components ASP.NET	v.9.3+
Javascript	

Table 1. Technologies and tools used to develop *CropWaterUse*

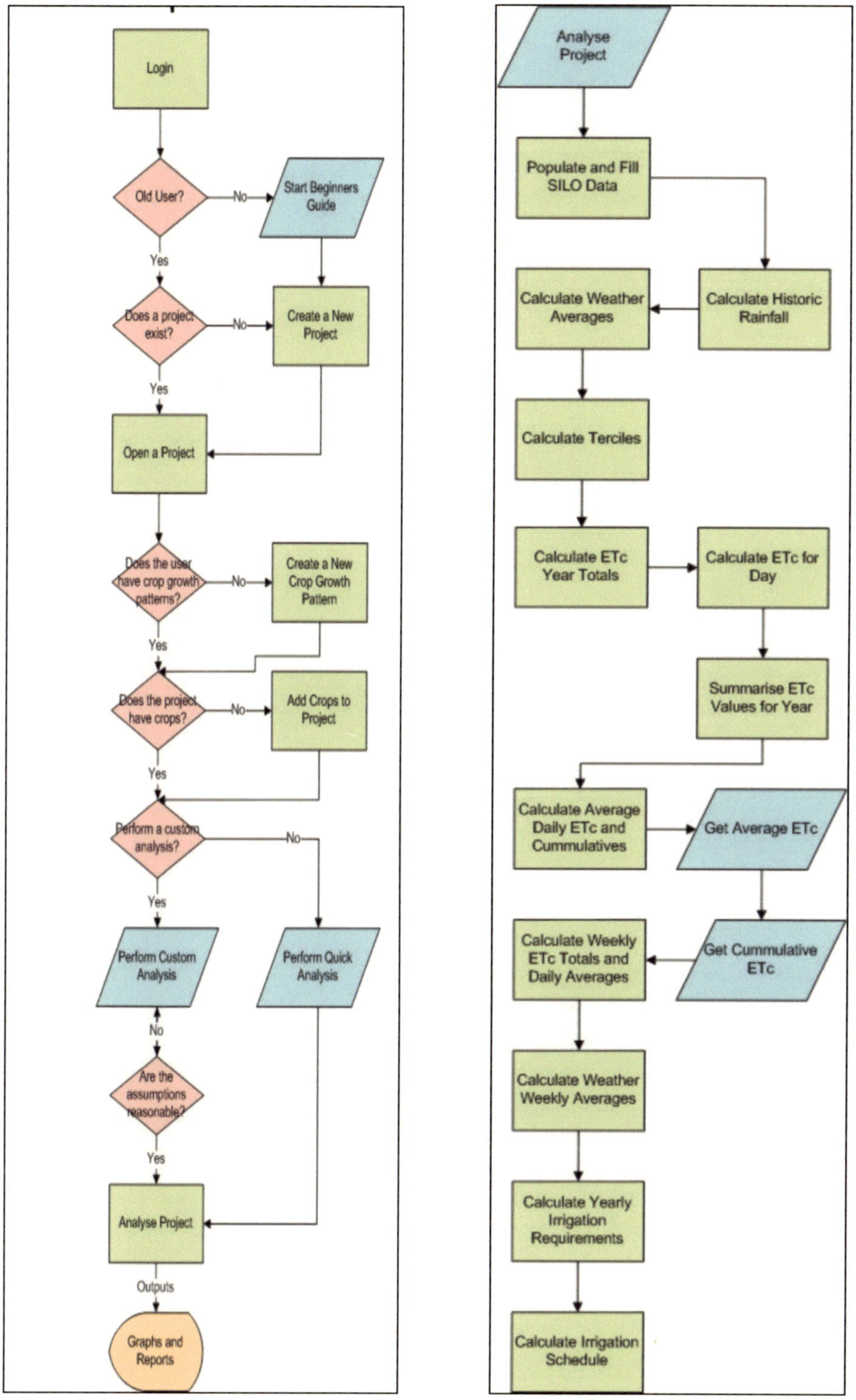

Fig. 2. Logic flowcharts for accessing the system and performing analysis in *CropWaterUse*

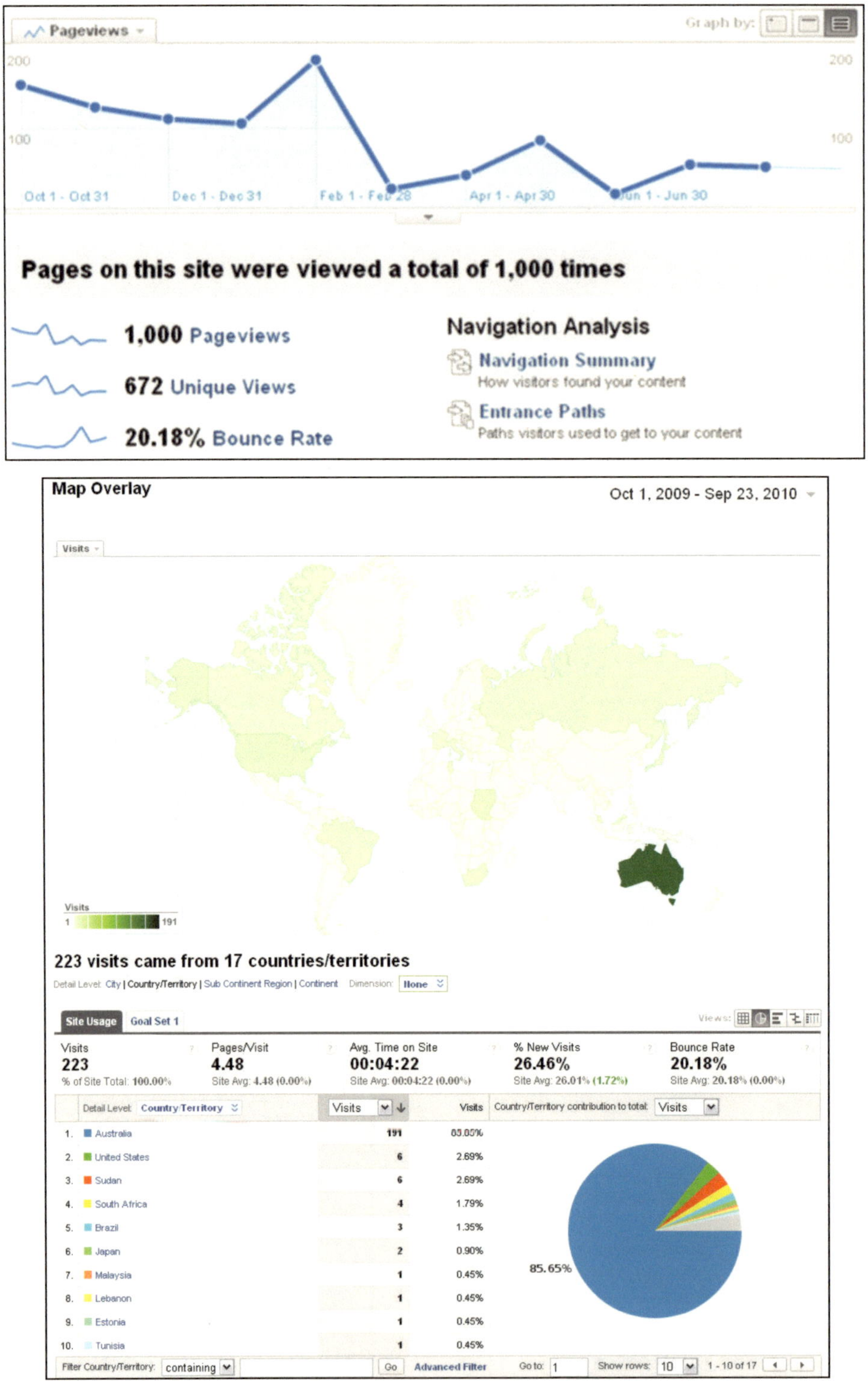

Fig. 3. Website statistics of *CropWaterUse* during October 2009 to September 2010

2.3 Inputs and analysis

Before using the system for the first time, users need to create an account, which is done online by answering a few questions. Then the system automatically sends the login information to the specified email address. After login, the next step is for the user to create a project to perform an analysis. A project consists of crops that are added to the project to create an analysis scenario. Each scenario specifies a crop type (such as cotton, corn…), which has a sowing date, location, and crop growth pattern. The crop growth pattern is the crop coefficient (Kc) curve, that is, the Kc values and the lengths of growth stages (LGS). For instance, cotton planted on 15 October at Dalby would be a scenario that can be added to the project and cotton planted on 1 November at the same location would be a different scenario. In each project, users can specify as many scenarios as needed. Projects can be saved and can be re-used or edited later and the user can also save many different projects. When an analysis is performed, outputs for each scenario in the project are produced, which allow side-by-side comparison of the different scenarios. Currently, users can choose crop growth patterns from a variety of broad-acre and horticultural crops and from many locations across Queensland and New South Wales, Australia. The number of locations is currently limited by the available weather data, but could easily be expanded by including weather data from other locations. The historical weather record for the locations available in *CropWaterUse* currently spans from 1957 to 2008.

Once scenarios are created, the user can perform either a quick analysis or a custom analysis to generate outputs. The quick analysis uses all the years in the historical weather record, an irrigation efficiency of 75%, and a soil water deficit of 75 mm to trigger irrigation events. The custom analysis allows the user to change these assumptions. However, a minimum of 10 consecutive years of weather data record needs to be included in the custom analysis to be able to generate statistics. Both types of analyses assume that the soil profile is full at sowing and that the crop is never under water stress. To facilitate its use, *CropWaterUse* includes extensive help facilities, including written step-by-step instructions, a wizard, and even an online video tutorial.

2.4 Calculations

CropWaterUse uses historical weather data to calculate daily values of crop evapotranspiration (ETc). ETc for a crop with no water stress is calculated using the FAO-56 single crop coefficient procedure as (Allen et al., 1998):

$$ETc = Kc \times ETo \quad (1)$$

where, ETc = crop evapotranspiration (mm day^{-1}), Kc = crop coefficient (unitless) and ETo= grass-reference evapotranspiration (mm day^{-1}).

The user can specify a crop growth pattern by inputting Kc values for the initial, mid and end periods (Kc ini, Kc mid, and Kc end), and the LGS for the initial, crop development, mid-season, and late season development stages. The definitions of these stages provided by FAO-56 (Allen et al., 1998) are given in Table 1.

Growth Stage	Definition
Initial	Planting to 10% ground cover
Crop Development	10% ground cover to effective full cover
Mid-Season	Effective full cover to start of maturity
Late Season	start of maturity to harvest or full senescence

Table 2. Definitions of crop development stages given by FAO-56 (Allen et al., 1998)

For each crop, *CropWaterUse* provides default Kc values, which have been taken from FAO-56, and LGS for combinations of three generic growing environments (Cool, Mild and Hot) and crop maturity groups (Early, Medium, and Late). The LGS values were determined by local experience and by conducting simulations with the APSIM crop growth model (Keating et al., 2003). Also, lysimeter and eddy covariance measurements of ETc are currently underway to determine Kc and LGS values for local crops, which will be incorporated into *CropWaterUse* as new information becomes available. Although default values are provided for Kc and LGS, advanced users have the flexibility to change these values. In fact, the tool allows users to create new crops or new crop varieties by inputting adequate Kc and LGS values, if known. The tool calculates daily Kc values by linear interpolation between the specified values for the initial, mid and late season stages.

The historical weather data is taken from the Enhanced Meteorological Dataset available from SILO, which is part of the Department of Environmental and Resources Management (DERM) (Jeffrey et al., 2001). The enhanced dataset is interpolated from data collected by the Australian Bureau of Meteorology. *CropWaterUse* takes ETo values directly from SILO, which are calculated with the FAO-56 Penman-Montheith method (Allen et al., 1998) as:

$$\mathrm{ETo} = \frac{0.408\Delta(R_n - G) + \gamma \dfrac{900}{T+273} u_2\,(e_s - e_a)}{\left[\Delta + \gamma\,(1 + 0.34u_2)\right]} \tag{2}$$

where, ETo = grass-reference evapotranspiration (mm day^{-1}), Δ = slope of the saturation vapor pressure versus air temperature curve (kPa °C^{-1}), Rn = net radiation at the crop surface (MJ m^{-2} day^{-1}), G = soil heat flux density (MJ m^{-2} day^{-1}), T = mean daily air temperature at 2 m height (°C), u_2 = mean daily wind speed at 2 m height (m s^{-1}), e_s = saturation vapour pressure (kPa), e_a = actual vapour pressure (kPa), e_s – e_a = saturation vapour pressure deficit (kPa) and γ = psychrometric constant (kPa °C^{-1}). Since actual measurements of wind speed and solar radiation are not available for many sites in Australia, SILO calculates ETo using a fixed value of u_2 = 2.0 m s^{-1} and solar radiation data derived from observed cloud oktas. An evaluation of errors associated with these assumptions has been reported by Fitzmaurice and Beswick (2005).

CropWaterUse calculates daily ETc values for every day in the specified historical weather record. This information is used to calculate statistics (such as averages and measures of variability) and seasonal totals. Daily rainfall and ETc values are used to conduct a daily soil water balance to obtain an estimate of irrigation requirements, number of irrigations, and timing of each irrigation event. An estimate of water losses (runoff + deep percolation) is calculated from the soil water balance, considering the efficiency of rain and irrigation.

Growing seasons are grouped according to seasonal rainfall and are classified as "DRY", "AVERAGE", and "WET", for which separate statistics for seasonal ETc, rainfall and irrigation requirements are produced. Seasons are classified into "DRY", "AVERAGE", and "WET" using long-term average of rainfall data, rather than using the specified analysis period. This is done to ensure that short term analyses do not portray years incorrectly against the long-term averages. The historical lowest to highest and determining the 33% and 66% indexes of the sorted values. In the output the years are colour-coded to indicate "DRY", "AVERAGE", and "WET" seasons.

2.4 Outputs

The system produces a series of graphical and tabular outputs that show side-by-side comparison of the different scenarios. Users are not limited to the number of scenarios they can run at one time and the graphical displays automatically expand and contract to accommodate the number of scenarios. The outputs can be printed or saved in a variety of formats (ie., xls, pdf, doc, png), which facilitates further manipulation of the data by the user. Although graphical and tabular outputs are created and displayed separately to facilitate online viewing, a compiled summary report is also produced that includes all the graphs and tables.

To illustrate some of the calculations and outputs of *CropWaterUse*, we performed a quick analysis comparing cotton (summer crop) planted on 12 November at Oakey (Queensland, Australia) and wheat (winter crop) planted on 6 June at the same location. Figure 4 shows the calculated average of daily evapotranspiration (ETc) as a function of days after sowing (DAS) for each of the two scenarios. Figure 4 only shows the average ETc for each DAS including all the years in the historical weather record. It, therefore, does not show the variability between years, which can be quite significant. Figure 5 shows the same information as Figure 4, but plotted as a function of calendar day, illustrating the fact that the growing seasons for the two crops do not overlap, since wheat is a winter crop while cotton is a summer crop. It also shows that the average peak daily ETc is about 5 mm day^{-1} and 6 mm day^{-1} for wheat and cotton, respectively. This information is vital for determining irrigation system capacity. Figure 6 shows the daily cumulative average ETc for each crop, which indicates that at that location cotton uses an average of about 670 mm of seasonal ETc while wheat only uses about 400 mm.

Figure 7 shows boxplots comparing the seasonal ETc, effective rain, and irrigation requirement for both scenarios. Boxplots are useful because they show the data divided into quartiles, which provides a measure of central tendency (such as the median), a measure of variability from season to season, and the magnitude of extreme events. Figure 7 shows that even though cotton at this location uses an average of around 670 mm, for a given season the actual value can vary between about 460 to 750 mm, the lower value probably corresponding to a very wet season and the highest value to a very dry season. It also shows a similar variability in seasonal ETc for wheat. The rainfall boxplots indicate that on average that location receives more rain during the summer than during the winter crop, with similar variability among seasons for the two crops. Because it depends on both ETc and rain, the irrigation requirement inherits even more variability from season to season than ETc and rain. Figure 7 shows that at that location, cotton requires anywhere from 300 mm to 750 mm of irrigation while wheat requires from about 100 to about 520 mm. This is a huge variability, which highlights the need to employ irrigation scheduling techniques that allow the grower to adapt irrigation management to changing weather conditions. In this regards, there is a need for further development of weather forecasting technology, for the short and medium time scale, and the integration of weather forecast information into irrigation decision-support tools.

In addition to the graphical outputs, *CropWaterUse* produces tabular outputs. Figure 8 shows a sample summary output produced by *CropWaterUse* for the cotton and wheat scenarios at Oakey. The summary shows the seasonal statistics, including the minimum, maximum, average, and standard deviation for the calculated variables for each scenario. These outputs also show the assumptions that were used to generate those numbers, such as the analysis period, crop, location, irrigation efficiency, growth pattern (Kc and LGS), soil water deficit to trigger irrigation and sowing date.

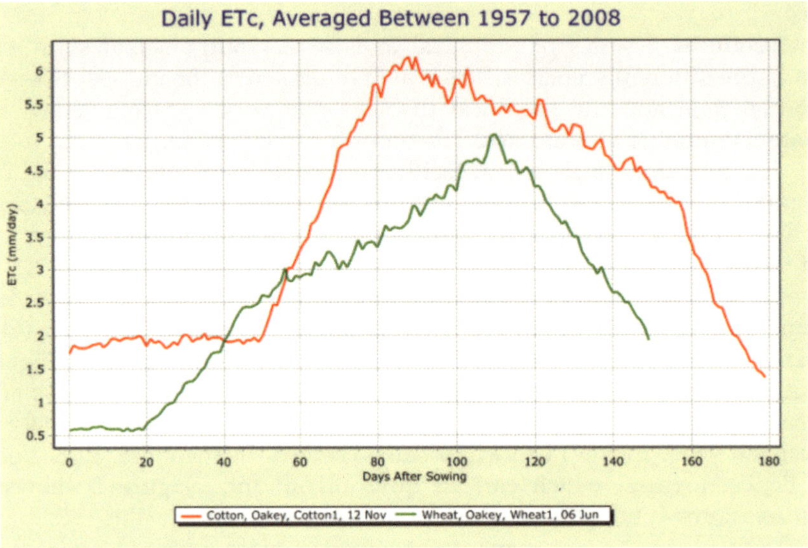

Fig. 4. Calculated average daily crop evapotranspiration as a function of days after sowing for cotton and wheat at Oakey

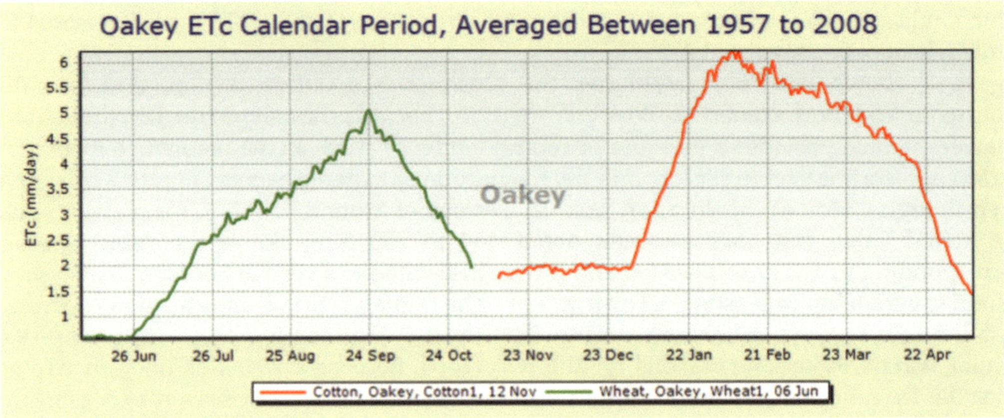

Fig. 5. Calculated average daily crop evapotranspiration as a function of calendar day for cotton and wheat at Oakey

Figure 9 shows a sample weekly weather and crop water use output produced by *CropWaterUse*. It shows the averages of the weather data that was used to calculate ETo and ETc for each week of the growing season for each scenario. Figure 10 shows a sample of the irrigation requirement output for each year included in the analysis. It includes seasonal ETc, rainfall, rainfall losses, final soil water deficit (at the end of the season), irrigation demand, and irrigation required (taking into account the irrigation efficiency). Figure 11 shows a sample output for irrigation timing for each year in the analysis period. The last column in the table shows the number of irrigations required during each year. Years for the irrigation requirement and irrigation timing outputs are colour-coded to represent "DRY", "AVERAGE", and "WET" seasons. Separate summary statistics are shown for the "DRY", "AVERAGE", and "WET" seasons (Figure 12).

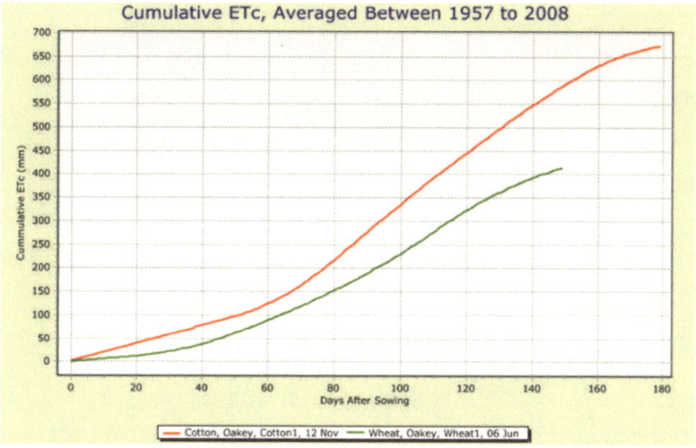

Fig. 6. Calculated average daily cumulative crop evapotranspiration as a function of days after sowing for cotton and wheat at Oakey

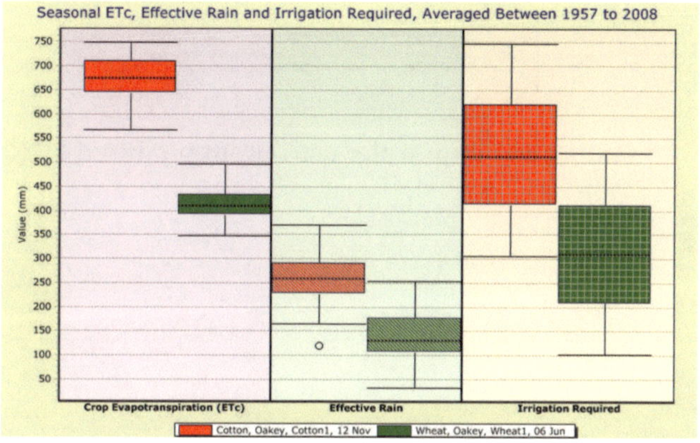

Fig. 7. Boxplot of seasonal crop evapotranspiration, effective rain, and irrigation requirement for cotton and wheat at Oakey

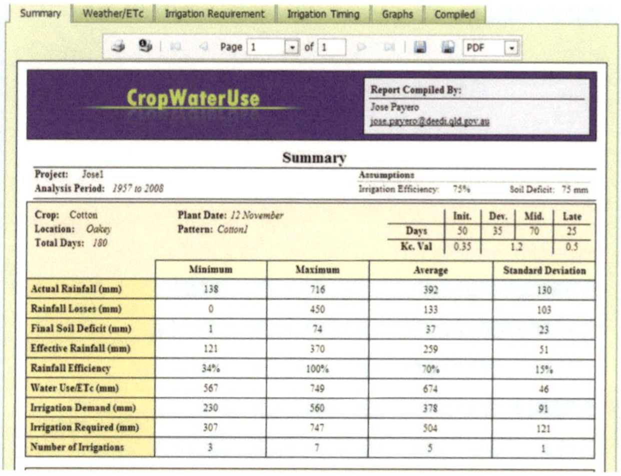

Summary

Project: Jose1
Analysis Period: *1957 to 2008*

Assumptions
Irrigation Efficiency: 75% Soil Deficit: 75 mm

Crop: Cotton **Plant Date:** *12 November*
Location: *Oakey* **Pattern:** *Cotton1*
Total Days: *180*

	Init.	Dev.	Mid.	Late
Days	50	35	70	25
Kc. Val	0.35	1.2		0.5

	Minimum	Maximum	Average	Standard Deviation
Actual Rainfall (mm)	138	716	392	130
Rainfall Losses (mm)	0	450	133	103
Final Soil Deficit (mm)	1	74	37	23
Effective Rainfall (mm)	121	370	259	51
Rainfall Efficiency	34%	100%	70%	15%
Water Use/ETc (mm)	567	749	674	46
Irrigation Demand (mm)	230	560	378	91
Irrigation Required (mm)	307	747	504	121
Number of Irrigations	3	7	5	1

Fig. 8. Sample summary output produced by *CropWaterUse*

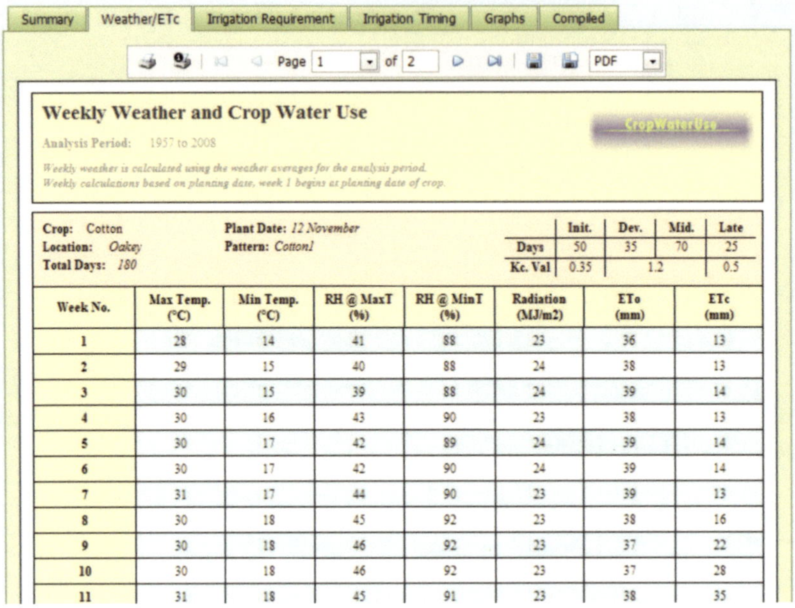

Weekly Weather and Crop Water Use

Analysis Period: 1957 to 2008

Weekly weather is calculated using the weather averages for the analysis period.
Weekly calculations based on planting date, week 1 begins at planting date of crop.

Crop: Cotton
Location: *Oakey*
Total Days: *180*
Plant Date: *12 November*
Pattern: *Cotton1*

	Init.	Dev.	Mid.	Late
Days	50	35	70	25
Kc. Val	0.35	1.2		0.5

Week No.	Max Temp. (°C)	Min Temp. (°C)	RH @ MaxT (%)	RH @ MinT (%)	Radiation (MJ/m2)	ETo (mm)	ETc (mm)
1	28	14	41	88	23	36	13
2	29	15	40	88	24	38	13
3	30	15	39	88	24	39	14
4	30	16	43	90	23	38	13
5	30	17	42	89	24	39	14
6	30	17	42	90	24	39	14
7	31	17	44	90	23	39	13
8	30	18	45	92	23	38	16
9	30	18	46	92	23	37	22
10	30	18	46	92	23	37	28
11	31	18	45	91	23	38	35

Fig. 9. Sample Weekly weather and crop water use output produced by *CropWaterUse*

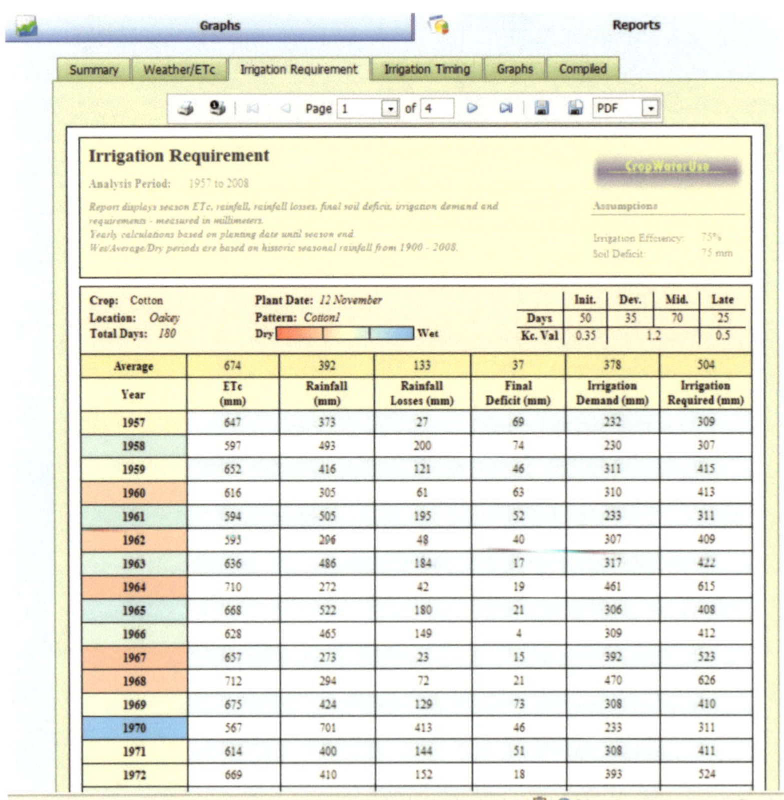

Irrigation Requirement

Analysis Period: 1957 to 2008

Report displays season ETc, rainfall, rainfall losses, final soil deficit, irrigation demand and requirements - measured in millimeters.
Yearly calculations based on planting date until season end.
Wet/Average/Dry periods are based on historic seasonal rainfall from 1900 - 2008.

Assumptions
Irrigation Efficiency: 75%
Soil Deficit: 75 mm

Crop: Cotton
Location: *Oakey*
Total Days: *180*
Plant Date: *12 November*
Pattern: *Cotton1*
Dry — Wet

	Init.	Dev.	Mid.	Late
Days	50	35	70	25
Kc. Val	0.35	1.2		0.5

Average	674	392	133	37	378	504
Year	**ETc (mm)**	**Rainfall (mm)**	**Rainfall Losses (mm)**	**Final Deficit (mm)**	**Irrigation Demand (mm)**	**Irrigation Required (mm)**
1957	647	373	27	69	232	309
1958	597	493	200	74	230	307
1959	652	416	121	46	311	415
1960	616	305	61	63	310	413
1961	594	505	195	52	233	311
1962	593	206	48	40	307	409
1963	636	486	184	17	317	422
1964	710	272	42	19	461	615
1965	668	522	180	21	306	408
1966	628	465	149	4	309	412
1967	657	273	23	15	392	523
1968	712	294	72	21	470	626
1969	675	424	129	73	308	410
1970	567	701	413	46	233	311
1971	614	400	144	51	308	411
1972	669	410	152	18	393	524

Fig. 10. Sample irrigation requirement output produced by *CropWaterUse*

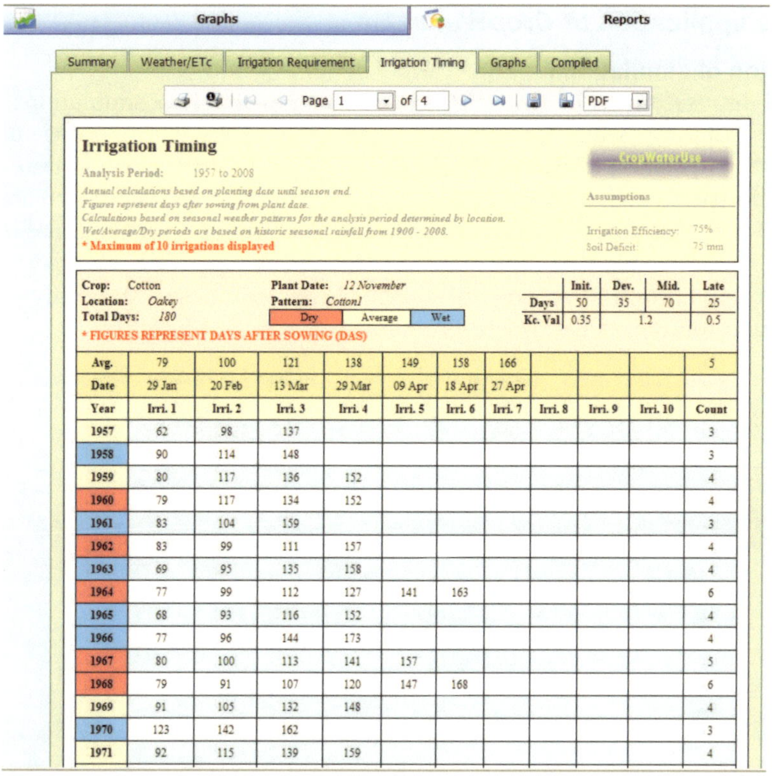

Irrigation Timing

Analysis Period: 1957 to 2008

Annual calculations based on planting date until season end.
Figures represent days after sowing from plant date.
Calculations based on seasonal weather patterns for the analysis period determined by location.
Wet/Average/Dry periods are based on historic seasonal rainfall from 1900 - 2008.
*** Maximum of 10 irrigations displayed**

Assumptions
Irrigation Efficiency: 75%
Soil Deficit: 75 mm

Crop: Cotton
Location: *Oakey*
Total Days: *180*
Plant Date: *12 November*
Pattern: *Cotton1*
Dry | Average | Wet

	Init.	Dev.	Mid.	Late
Days	50	35	70	25
Kc. Val	0.35	1.2		0.5

* FIGURES REPRESENT DAYS AFTER SOWING (DAS)

Avg.	79	100	121	138	149	158	166				5
Date	29 Jan	20 Feb	13 Mar	29 Mar	09 Apr	18 Apr	27 Apr				
Year	Irri. 1	Irri. 2	Irri. 3	Irri. 4	Irri. 5	Irri. 6	Irri. 7	Irri. 8	Irri. 9	Irri. 10	Count
1957	62	98	137								3
1958	90	114	148								3
1959	80	117	136	152							4
1960	79	117	134	152							4
1961	83	104	159								3
1962	83	99	111	157							4
1963	69	95	135	158							4
1964	77	99	112	127	141	163					6
1965	68	93	116	152							4
1966	77	96	144	173							4
1967	80	100	113	141	157						5
1968	79	91	107	120	147	168					6
1969	91	105	132	148							4
1970	123	142	162								3
1971	92	115	139	159							4

Fig. 11. Sample irrigation timing output produced by *CropWaterUse*

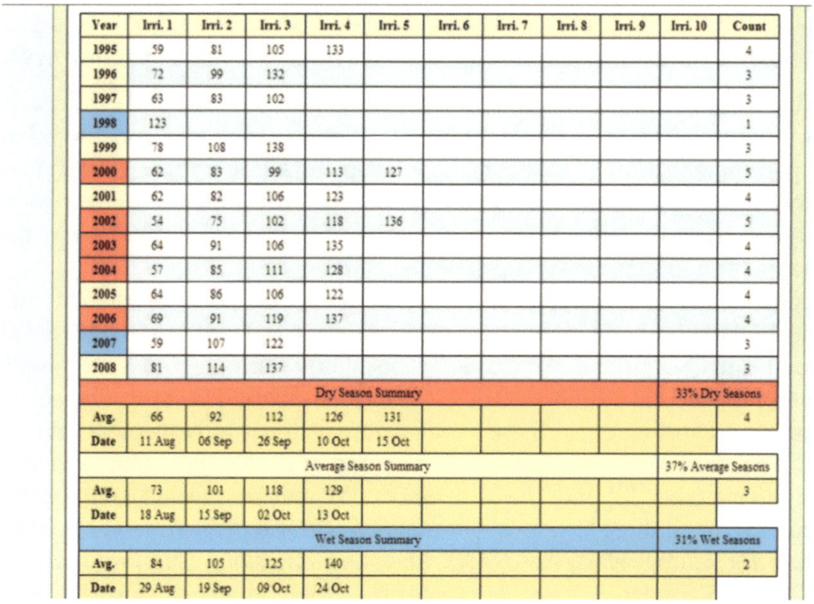

Year	Irri. 1	Irri. 2	Irri. 3	Irri. 4	Irri. 5	Irri. 6	Irri. 7	Irri. 8	Irri. 9	Irri. 10	Count
1995	59	81	105	133							4
1996	72	99	132								3
1997	63	83	102								3
1998	123										1
1999	78	108	138								3
2000	62	83	99	113	127						5
2001	62	82	106	123							4
2002	54	75	102	118	136						5
2003	64	91	106	135							4
2004	57	85	111	128							4
2005	64	86	106	122							4
2006	69	91	119	137							4
2007	59	107	122								3
2008	81	114	137								3
Dry Season Summary										33% Dry Seasons	
Avg.	66	92	112	126	131						4
Date	11 Aug	06 Sep	26 Sep	10 Oct	15 Oct						
Average Season Summary										37% Average Seasons	
Avg.	73	101	118	129							3
Date	18 Aug	15 Sep	02 Oct	13 Oct							
Wet Season Summary										31% Wet Seasons	
Avg.	84	105	125	140							2
Date	29 Aug	19 Sep	09 Oct	24 Oct							

Fig. 12. Sample summary of average irrigation timing for Dry, Average and Wet seasons produced by *CropWaterUse*

3. Example application of *CropWaterUse*

3.1 Description of simulations

To illustrate the usefulness of *CropWaterUse* we conducted two simulations (studies) for hypothetical cropping scenarios. In the first study, simulation runs using default assumptions for irrigation efficiency of 75% and irrigation trigger of 75 mm of soil water deficit was compared with improved irrigation efficiency of 85% and increased soil water deficit of 100 mm. Also, comparison of variety and planting date combinations including early maturing/early planted and late maturing/late planted summer grain crops, including sorghum, corn, soybeans, mungbeans and sunflowers in relatively cooler and hotter climatic conditions of Dalby and Emerald, respectively (Fig. 13).

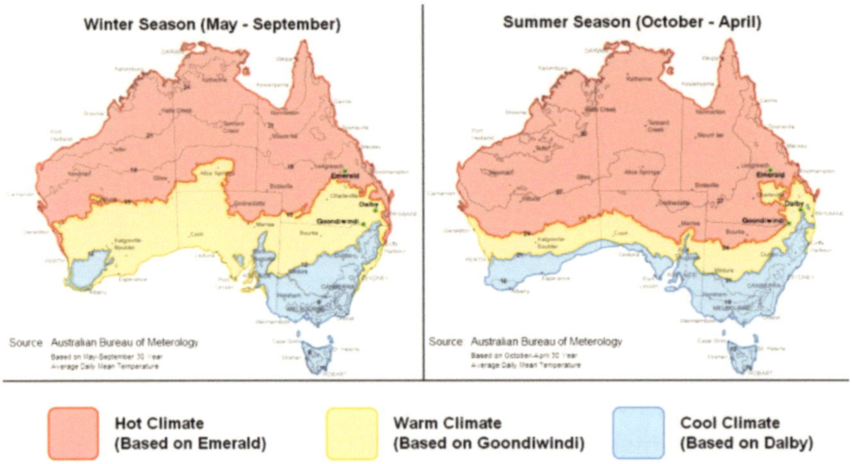

Fig. 13. Relatively hotter, warmer and cooler climatic regions in Australia based on seasonal minimum and maximum temperatures in Emerald, Goondiwindi and Dalby in Queensland.

Early planting of sorghum, corn and sunflower was parameterized at 30 Sept, whereas for soybeans and mungbeans at 30 Oct. Late planting of all crops was considered at 30 Jan. In this study, parameterisation of maturity type, crop duration, critical growth stages (initial, development, mid and late) and Kc value required for only 3 growth stages (Kc ini, Kc mid, and Kc end) are given in Table 3. These values can be changed in the tool (user defined) if they are known for a location and crop type.

In the second study we conducted simulations of evapotranspiration and irrigation requirements of early and late-planted cotton (summer crop) and wheat (winter crop) grown in three locations in Queensland, Australia. The locations were selected to represent cool, mild, and hot environments and included Dalby, Goondiwindi and Emerald. The current planting windows for cotton and wheat at these locations are October to November and May to June, respectively. We compared crop water use patterns and irrigation requirements for cotton planted early (September 15) and late (November 15), and for wheat planted early (April 15) and late (June 15) for all three locations. Simulations with *CropWaterUse* were performed using default values for rainfall efficiency (75%), irrigation efficiency (75%) and trigger irrigation deficit (75 mm). Parameterization of maturity type, length of growth stages (initial, development, mid and late) and Kc values (Kc ini, Kc mid, and Kc end) for cotton and wheat at three regions (Dalby, Goondiwindi, Emerald) is given in Table 4.

		Maturity	Crop	Growth stages (days)				Kc		
Crops	Regions	type	Duration (days)	Initial	Development	Mid	Late	Kc ini.	Kc mid	Kc end
Sorghum	Dalby	Early	120	20	30	40	30	0.40	1.20	0.50
		Late	130	20	35	45	30	0.40	1.20	0.50
	Emerald	Early	110	20	30	35	25	0.40	1.25	0.50
		Late	120	20	30	45	25	0.40	1.25	0.50
Maize	Dalby	Early	125	20	30	30	25	0.40	1.20	0.50
		Late	135	20	35	40	30	0.40	1.20	0.50
	Emerald	Early	115	20	20	25	20	0.60	1.25	0.50
		Late	120	20	25	30	25	0.60	1.25	0.50
Soybean	Dalby	Early	130	20	25	60	25	0.40	1.20	0.50
		Late	140	20	30	65	25	0.40	1.20	0.50
	Emerald	Early	95	15	15	50	15	0.40	1.20	0.50
		Late	105	15	20	55	15	0.40	1.20	0.50
Mungbean	Dalby	Early	85	15	20	25	25	0.40	1.15	0.30
		Late	95	15	25	30	25	0.40	1.15	0.30
	Emerald	Early	65	10	15	20	20	0.40	1.15	0.30
		Late	70	10	15	25	20	0.40	1.15	0.30
Sunflower	Dalby	Early	130	30	20	60	20	0.40	1.20	0.50
		Late	140	30	25	60	25	0.40	1.20	0.50
	Emerald	Early	95	15	15	50	15	0.40	1.20	0.50
		Late	105	15	20	55	15	0.40	1.20	0.50

Table 3. Parameterization of maturity type, crop duration, critical growth stages (initial, development, mid and late) and Kc value required for only three growth stages (Kc ini, Kc mid, and Kc end) used for this study

		Maturity	Crop	Growth stages (days)				Kc		
Crops	Regions	type	Duration (days)	Initial	Development	Mid	Late	Kc ini.	Kc mid	Kc end
Cotton	Dalby	Early	180	50	35	70	25	0.35	1.20	0.50
	Goondiwindi	Early	170	45	30	70	25	0.35	1.20	0.50
	Emerald	Early	130	35	25	65	20	0.35	1.20	0.50
Wheat	Dalby	Early	140	15	30	55	40	0.30	1.15	0.40
	Goondiwindi	Early	140	15	30	55	40	0.30	1.15	0.40
	Emerald	Early	115	10	25	45	35	0.30	1.15	0.40

Table 4. Parameterization of maturity type, crop duration, critical growth stages (initial, development, mid and late) and Kc value (Kc ini, Kc mid, and Kc end) for cotton and wheat at three regions

3.2 Results of simulations

3.2.1 First Study: early maturity/early planted vs. late maturity/late planted summer grain crops

Crop ETc decreased significantly (about 100 mm on average across crops and locations) when agronomic conditions were changed from early maturity/early planted to late maturity/late planted systems for both (cooler and hotter) climatic conditions (Fig. 14). Sorghum, soybeans, maize and sunflower had significantly greater seasonal ETc than mungbeans, which had the lowest (340 mm). Sunflower, mungbeans and soybeans had significantly greater seasonal ETc at Dalby than at Emerald, whereas Sorghum had higher ETc at Emerald than at Dalby (Fig. 14).

At Dalby, early maturing/early planted grain crops appeared to receive considerably more rainfall than the late maturity/late planted grain crops (Table 5). On average, early maturity/early planted crops received about 90 mm more rain than the late maturity/late planted crops. Among early maturity/early planted crops, mungbeans received the least rainfall, somewhere between 40 mm and more than 100 mm less than the other crops, whereas this difference with late maturity/late planted crop was less than 50 mm for mungbeans (Table 5).

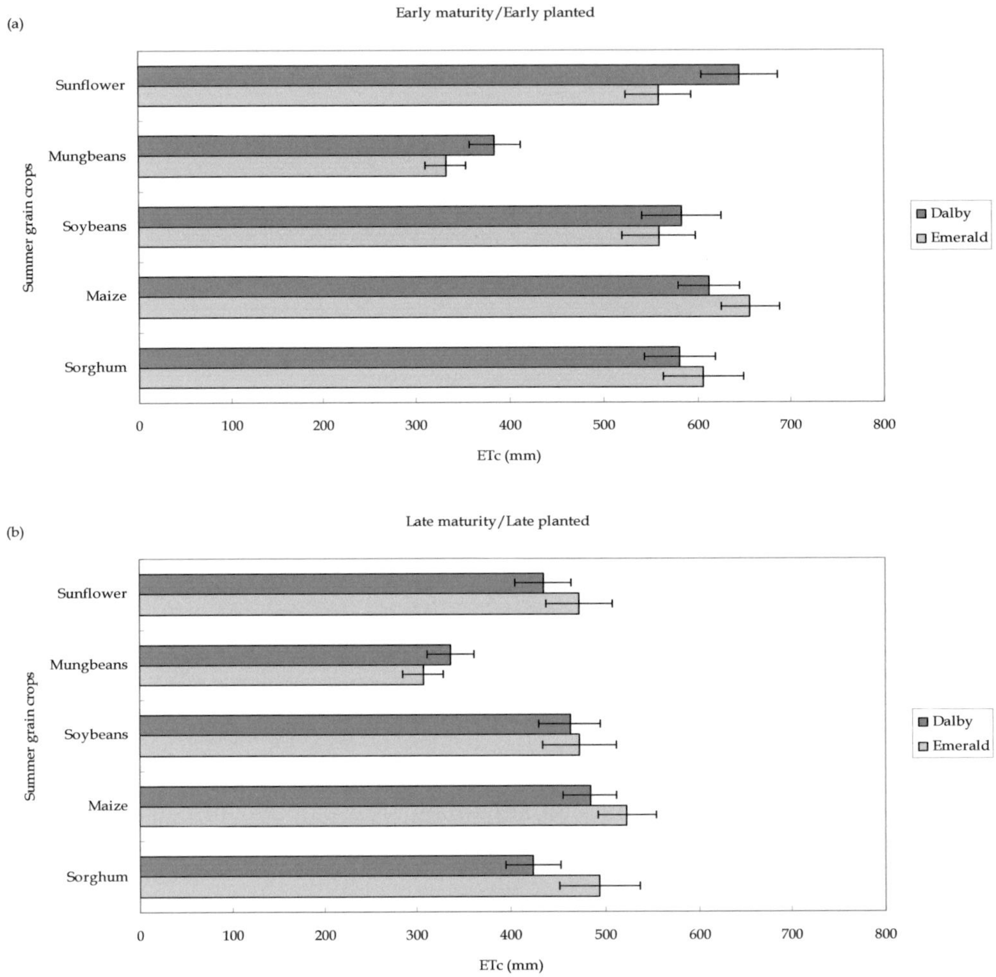

Fig. 14. Range of crop evapotranspiration (ETc) as estimated by the *CropWaterUse* tool using historical weather data (1957-2008) for "fully irrigated" sunflower, mungbeans, soybeans, maize and sorghum for various scenarios (a) early planting, early maturing varieties, and (b) late planting, late maturing varieties in cooler (at Dalby) and hotter (at Emerald) climatic conditions in Queensland. Bar indicates +/- standard deviation

Runoff+deep drainage losses of rainfall water varied between 60 mm and 100 mm for early maturity/early planted crops, whereas this loss was around 70 mm to 85 mm for late maturity/late planted crops. Rainfall efficiency was between 70% and 80% for both scenarios at Dalby.

Increasing irrigation efficiency by 10% (from 75% to 85%) and soil water deficit about 25 mm (from 75 mm to 100 mm) resulted in one less irrigation event required for all crops whether

	Summer grain crops in cooler climate (Dalby)				
Planting/Irrigation Parameters	Sorghum	Maize	Soybeans	Mungbeans	Sunflower
Early maturing, early planted					
Rainfall (mm)	299 (97)	265 (82)	343 (107)	227 (76)	333 (110)
Runoff+deep drainage (mm)	73 (56)	61 (50)	112 (78)	65 (49)	97 (70)
Rainfall Efficiency (%)	78 (13)	79 (14)	70 (16)	74 (15)	73 (14)
Irrigation Requirement (mm)					
At 75% eff., 75 mm deficit	437 (116)	343 (111)	532 (126)	260 (86)	505 (122)
At 85% eff., 100 mm deficit	367 (112)	289 (97)	431 (115)	203 (87)	407 (111)
Number of irrigation					
At 75% eff., 75 mm deficit	4 (1)	3 (1)	5 (1)	3 (1)	5 (1)
At 85% eff., 100 mm deficit	3 (1)	2 (1)	4 (1)	2 (1)	3 (1)
Late maturing, late planted					
Rainfall (mm)	209 (97)	203 (95)	219 (101)	172 (90)	219 (101)
Runoff+deep drainage (mm)	75 (77)	74 (78)	77 (76)	68 (71)	84 (81)
Rainfall Efficiency (%)	71 (21)	70 (22)	71 (19)	69 (23)	68 (21)
Irrigation Requirement (mm)					
At 75% eff., 75 mm deficit	346 (87)	336 (90)	386 (94)	268 (80)	360 (86)
At 85% eff., 100 mm deficit	294 (81)	280 (91)	324 (84)	219 (85)	300 (84)
Number of irrigation					
At 75% eff., 75 mm deficit	3 (1)	3 (1)	4 (1)	3 (1)	4 (1)
At 85% eff., 100 mm deficit	2 (1)	2 (1)	3 (1)	2 (1)	3 (1)

Table 5. Pattern of rainfall, losses through runoff+deep drainage, irrigation requirement and number of irrigation as estimated from the *CropWaterUse* tool (using historical weather data from 1957-2008) for a range of scenarios, early maturing/early planted and late maturing/late planted, for sorghum, maize, soybeans, mungbeans and sunflower in cooler climatic conditions at Dalby, Queensland. Values in parenthesis indicate standard deviation

planted early or late. However, late maturity/late planted sorghum, soybeans and sunflower required one irrigation less than the early maturity/early planting, there was no change in the number of irrigations for maize and mungbeans. Across the growing conditions, soybeans and sunflower needed 3-5 irrigations, whereas sorghum, maize and mungbeans required 2-4 irrigations for the relatively cooler conditions at Dalby.

At Emerald, there was only about 10 mm difference in rainfall between early maturity/early planted and late maturity/late planted scenarios (Table 6). On average, Emerald received about 56 mm less rainfall water than Dalby in the growing season stretching from October to April. Runoff+deep drainage losses for early planted crops ranged between 30 and 70 mm, whereas losses were between 70 and 90 mm for late planted crops. Improving irrigation efficiency by 10% (from 75% to 85%) and increasing water deficit by 25 mm (from 75 mm to 100 mm) decreased the number of irrigations for all crops for both scenarios, except for mungbeans, which required a minimum of 2 irrigations. Sorghum and maize at Emerald required one extra irrigation for various scenarios (early vs. late planting, unimproved irrigation efficiency/smaller deficit vs. improved irrigation efficiency/larger water deficit) than Dalby (Table 6).

In the first study, seasonal ETc for various summer grain crops as influenced by planting decision and irrigation management practices were compared between relatively cooler and hotter climatic regions in Queensland. While using the *CropWaterUse* tool, we tested two strategies to understand the irrigation water requirements. One strategy was to plant early in the season with a quick-maturing variety, this may reduce the number of irrigations, and another strategy was to slightly improve the irrigation efficiency and delay the irrigation with a slightly greater soil water deficit.

	Summer grain crops in hotter climate (Emerald)				
Planting/Irrigation Parameters	Sorghum	Maize	Soybeans	Mungbeans	Sunflower
Early maturing, early planted					
Rainfall (mm)	259 (102)	170 (85)	206 (90)	168 (74)	206 (90)
Runoff+deep drainage (mm)	68 (66)	34 (43)	48 (57)	50 (49)	48 (57)
Rainfall Efficiency (%)	63 (22)	85 (15)	81 (16)	75 (18)	81 (16)
Irrigation Requirement (mm)					
At 75% eff., 75 mm deficit	515 (119)	428 (100)	498 (114)	259 (76)	498 (114)
At 85% eff., 100 mm deficit	427 (105)	357 (93)	431 (108)	200 (79)	431 (108)
Number of irrigation					
At 75% eff., 75 mm deficit	5 (1)	4 (1)	5 (1)	2 (1)	5 (1)
At 85% eff., 100 mm deficit	4 (1)	3 (1)	4 (1)	2 (1)	4 (1)
Late maturing, late planted					
Rainfall (mm)	216 (115)	186 (96)	193 (100)	157 (85)	193 (100)
Runoff+deep drainage (mm)	94 (26)	76 (76)	83 (76)	72 (74)	83 (76)
Rainfall Efficiency (%)	63 (22)	67 (22)	64 (21)	63 (25)	64 (21)
Irrigation Requirement (mm)					
At 75% eff., 75 mm deficit	455 (103)	421 (94)	442 (98)	253 (67)	442 (98)
At 85% eff., 100 mm deficit	388 (103)	340 (93)	364 (97)	205 (740	364 (97)
Number of irrigation					
At 75% eff., 75 mm deficit	5 (1)	4 (1)	4 (1)	2 (1)	4 (1)
At 85% eff., 100 mm deficit	3 (1)	3 (1)	3 (1)	2 (1)	3 (1)

Table 6. Pattern of rainfall, losses through runoff+deep drainage, irrigation requirement and number of irrigations as estimated using *CropWaterUse* for a range of scenarios, early maturing/early planted and late maturing/late planted, for sorghum, maize, soybeans, mungbeans and sunflower in hotter climatic conditions at Emerald, Queensland. Values in parenthesis indicate standard deviation

The ETc was about 100 mm more for the early maturity/early planted than the late maturity/late planted crops at both cooler and hotter climatic conditions (Fig. 14). This may be due to increasing evaporative demand with increasing temperature and greater in-crop rainfall between the Oct to Jan growing period for early planted crops (Table 5 and 6). Whereas the effective growing period for the late planted crops between Feb and May would have cooler conditions towards maturity of the crops, it was though expected that the hotter conditions at Emerald would lead to greater ETc than the cooler conditions at Dalby. But opposite to this, ETc for sunflower and soybeans was greater at Dalby than at Emerald, particularly for early maturity/early planted crops (Fig. 14). However, for the late maturity/late planted crop, ETc at Emerald appeared to be slightly greater than at Dalby, but differences were not significant.

Hotter conditions at Emerald compared with Dalby may increase total soil evaporation but not necessarily total crop transpiration, as hotter conditions may also advance crop maturity

(Muchow, Sinclair and Bennett 1990; Boote and Sinclair 2006), thus reducing the duration of crop growth significantly. Secondly, significantly less in-crop rainfall at Emerald than Dalby may have also contributed to reduce ETc at Emerald, particularly for early maturity/early planted crops. Rainfall differences between Dalby and Emerald were minimal for the late maturity/late planted crops, this may have resulted in slightly greater ETc for late maturity/late planted crops with hotter conditions at Emerald than at Dalby (Fig. 14).

Differences in seasonal conditions with greater rainfall, higher temperature or reduced maturity, had variable impact on the number of irrigations required for various crops. Sorghum and corn and to some extent mungbeans required one less irrigation at Dalby than at Emerald for early or late planted and with improved or unimproved irrigation efficiency and increased irrigation deficit. There was no change in number of irrigations required for soybeans and sunflower due to seasonal differences between Dalby and Emerald (Table 4 and 5).

In addition to seasonal conditions, planting time and maturity of crop can also influence the amount of in-crop rainfall, directly influencing the amount and frequency of irrigations, particularly in the cooler climatic conditions at Dalby. The early maturity/early planted crops planted in September, with a growing season of 3-4 months received about 90 mm more rainfall than the late planted crops planted in January (Table 5). This resulted in one less irrigation for most of the summer grain crops, saving water and increasing profitability. Anecdotal evidence also suggests that early-planted sorghum crops, in particular, produce higher yields than late-planted sorghum in southeast and southwest Queensland. On the other hand, there was minimal difference in the amount of in-crop rainfall between early and late planted crops at Emerald (Table 6). It should also be noted that factors such as crop vigour, soil fertility and soil depth would also influence the number of irrigations. For example, newer varieties of sorghum do not have as much crop vigour as older varieties, requiring less irrigation.

Seasonal conditions, planting time and maturity of a crop/variety may interact and complicate irrigation requirement and scheduling decisions, whereas slightly improved irrigation efficiency and increased soil water deficit to trigger irrigations are likely to have a significant impact on reducing the number of irrigations. For example, improved irrigation efficiency and increasing soil water deficit slightly resulted in one less irrigation for most of the crops, subjected to variable planting, maturity or seasonal conditions (Table 5 and 6).

3.2.2 Second study: early and late planted cotton and wheat at three locations in Queensland

The average rate of irrigation water use in Queensland has been estimated (ABS., 2007/08) at 3.1 ML/ha and 4.9 ML/ha for wheat and cotton, respectively (Figure 15). Changing the planting strategy in order to adapt to future climatic change (predicted for earlier planting of these crops) is likely to result in change in irrigation water use due to expected changes in in-crop rainfall and ETc.

For the summer crop (cotton), the simulation results showed that ETc would increase if the planting of cotton is brought forward to an earlier planting in September from November at all 3 locations (Table 7). The increase in ETc with cotton planted in mid September would be due to greater overall increase in temperature over the duration of the cotton season. Since there was not much difference in the in-crop seasonal rainfall, an additional irrigation would be needed for the early-planted cotton at all sites (Table 7).

For the winter crop (wheat), in contrast, early planting would considerably decrease ETc (Table 8), decreasing the number of irrigations from 4 to 3 at Dalby and Goondiwindi, and from 5 to 3 at Emerald. This analysis, however, does not consider the impact of early planting on crop yield, the danger of frost, or the practicality of early planting.

Among the three locations, Goondiwindi has the highest and Emerald has the lowest irrigation demand for cotton. On the other hand, Emerald has the highest irrigation demand for wheat. On average, cotton requires about 350 mm more water than wheat. The loss of water through runoff and deep drainage is also 80 mm more for cotton compared to wheat.

In summary, earlier planting of wheat at these three locations would result in less demand on irrigation water, whereas earlier planting of cotton would increase irrigation demand. This type of simulation can be performed using *CropWaterUse* in just a few minutes, and it is the type of information that can be used by growers for long-term planning.

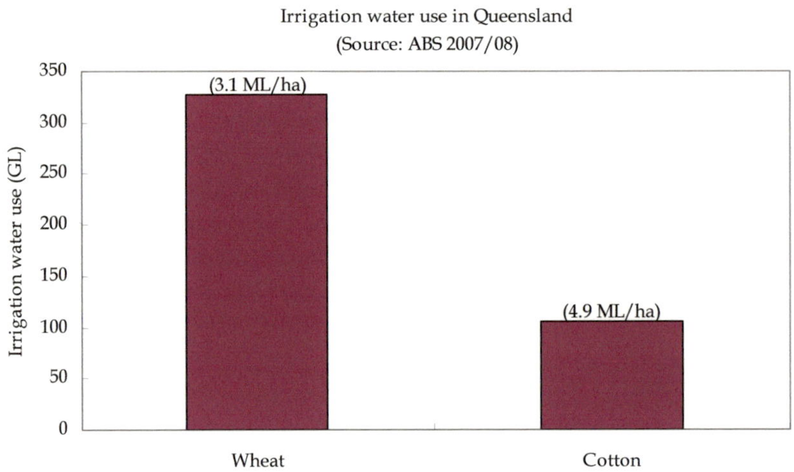

Fig. 15. Total and per hectare (ha) irrigation water use for wheat and cotton in Queensland, Australia

Parameters	Dalby		Goondiwindi		Emerald	
	Sep 15	Nov 15	Sep 15	Nov 15	Sep 15	Nov 15
Evapotranspiration (mm)	834 (50)	732 (45)	862 (48)	774 (48)	705 (42)	661 (48)
Rainfall (mm)	424 (126)	386 (129)	357 (135)	347 (144)	291 (116)	362 (139)
Runoff+Drainage (mm)	133	136	96	104	85	131
Irrigation Demand (mm)	543 (107)	482 (84)	601 (116)	531 (108)	499 (85)	430 (110)
No. of Irrigation	7 (1)	6 (1)	8 (1)	7 (1)	6 (1)	5 (1)
Dry seasons (%)	33	39	29	33	27	33
Normal seasons (%)	35	27	35	31	39	39
Wet seasons (%)	31	33	37	35	33	27

Table 7. Comparing water requirement for early (September) and late (November) planted cotton at Dalby, Goondiwindi, and Emerald, based on historical weather data (1957-2008). Values in brackets are one standard deviation

Parameters	Dalby		Goondiwindi		Emerald	
	Apr 15	Jun 15	Apr 15	Jun 15	Apr 15	Jun 15
Evapotranspiration (mm)	331 (27)	482 (50)	332 (78)	495 (53)	362 (25)	466 (33)
Rainfall (mm)	161 (81)	205 (76)	178 (94)	194 (84)	117 (82)	93 (75)
Runoff+Drainage (mm)	31	57	40	45	22	10
Irrigation Demand (mm)	201 (66)	334 (81)	194 (61)	346 (74)	267 (56)	383 (73)
No. of Irrigation	3 (1)	4 (1)	3 (1)	4 (1)	3 (1)	5 (1)
Dry season (%)	35	27	35	33	29	38
Normal season (%)	33	40	35	33	33	31
Wet season (%)	33	33	31	35	38	31

Table 8. Comparing water requirement for early (April) and late (June) planted wheat at Dalby, Goondiwindi, and Emerald, based on historical weather data (1957-2008). Values in brackets are one standard deviation

4. Conclusions

Water scarcity is becoming one of the major challenges agricultural production in Australia and in many other parts of the world is facing. Competition for limited water resources is increasing between agricultural, domestic, industrial and environmental uses. Environmental use has especially become a formidable competitor for water resources in the last two decades, as more and more regulations are put in place to protect water courses and habitats for endangered species. Since agriculture is usually the major user of fresh water resources diverted from rivers and pumped from groundwater, agricultural producers will likely face increasing social pressures to increase crop water productivity. These social pressures will be eventually translated into regulations that further restrict the use of water for agricultural production. At the same time that irrigation water is becoming scarce and more costly for growers, human population continues to grow. Therefore, agriculture will have to sustain more people, with about the same land area and with less water. Because of this, to maximize profits and sustain an increasing population with limited water, growers will have to become more creative on how they allocate water among competing enterprises and how they manage water within each enterprise on a day-to-day basis. This is a formidable task, which will demand using our ingenuity to develop new technologies and tools to assist growers make better decisions in regards to allocation of resources within the farm, including land, water, nutrients, chemicals, labour, machinery, capital, etc. New and improved tools and skills will be needed to assist growers in managing individual resources, such as water, and also for integrating, harmonising, and optimising the use of all the resources that contribute to agricultural production as a whole system, rather than as collection of disparate components.

In this chapter we have presented the description and application of *CropWaterUse,* a new web-based tool that has been developed in Australia to help growers plan the use of water, one of the many components of the agricultural production system. The system was developed with feedback from many growers and crop consultants, mainly located in the cotton producing areas of eastern Australia. Growers and consultants were involved in the development and testing of *CropWaterUse,* with the aim of developing a system that answered questions that were relevant to them and that, at the same time, was user-friendly. Since its release in October 2009, the system has had a positive uptake by producers, consultants, and even policy makers. To promote uptake, the system has been supported by extension and research personnel located in different agricultural areas of Queensland, including Dalby, Emerald, Kingaroy, Gatton, and Toowoomba. To continuously improve the tool, there is also ongoing research aimed at improving the way we calculate ETc for different crops and as new research knowledge is gained, it will be incorporated into *CropWaterUse*. In this regards, *CropWaterUse* will also serve as an effective extension tool since research information is immediately made available to end users, avoiding the usual time lag that occurs within the research-extension-adoption process.

Since it has been designed as a simple tool, it is recognised that *CropWaterUse* has some limitations. One of the limitations is that it does not deal with crop stress. This is an important limitation for dryland and deficit-irrigation production. Another important limitation of *CropWaterUse* is that it uses LGS, needed to construct the crop growth patterns (i.e. Kc curves), in "days" rather than "cumulative growing degree days (CGDD)." That can have an important impact on season length when comparing the same crop grown in areas with significantly different weather conditions. We tried and overcome this limitation by providing default LGS values for three growing environments and three maturity groups for each crop. However, better information on Kc curves based on CGDD's rather than "days" is needed for the different crops and growing environments. Another limitation is that it does not deal with different plant populations or with different planting configurations, such as the different types of skip-row configurations that are commonly used by Australian growers.

Despite its limitations, *CropWaterUse* can be a useful planning tool for growers and crop consultants, if used for the intended purpose and with appropriate Kc and weather data. A more comprehensive new tool is now under development that will complement the capabilities of *CropWaterUse*. The new tool will be aimed at assisting growers in day-to-day irrigation scheduling, rather than long-term planning. Therefore, it will use daily near-real-time weather data, rather than just historical data. It will use the dual Kc approach detailed in FAO-56, rather than the single Kc used in *CropWaterUse*. The dual Kc considers the effect of crop stress and the impact of rain and irrigation on evaporation from the soil surface. It will also integrate economic analysis and weather forecasting into the day-to-day irrigation decision-making process.

5. Acknowledgements

The authors would like to acknowledge Agri-Science Queensland, Department of Employment, Economic Development & Innovation (DEEDI), for funding the development of *CropWaterUse*. We also would like to acknowledge the contributions of a the many growers and crop consultants who provided valuable feedback during the development and testing of *CropWaterUse*.

6. References

Abourached, C.; Hillyer, C.; Sayde, C.; English, M. & Bush, J. (2007). A web-based advisory service for optimum irrigation management, *Proceedings of the 2007 ASABE Annual International Meeting* , 17-20 June. ASABE, Paper No. 072253, Minneapolis, Minnesota, pp. 1-12.

ABS. (2007/08). Australian Bureau of Statistics. www.abs.gov.au/.

Allen, R.G.; Pereira, L.S.; Raes, D. & Smith, M. (1998). Crop *evapotranspiration-Guidelines for computing crop water requirements*. FAO Irrigation and drainage paper 56. FAO, Rome.

Boote K.J. & Sinclair T.R. (2006). Crop physiology: Significant discoveries and our changing perspective on research. *Crop Science*, 46: 2270-2277.

Brown, P.D.; Cochrane, T.A. & Krom, T.D. (2010). Optimal on-farm irrigation scheduling with seasonal water limit using simulated annealing. Agricultural Water Management, 97: 892-900.

Car, N.J.; Christen, E.W.; Hornbuckle, J.W. & Moore, G.A. (2008). A web services-supported, calendar-based, irrigation decision support system, Irrigation Australia Ltd., *Proceeding of the 2008 Australian Irrigation Conference*. Irrigation Australia, Ltd, Melbourne, pp. 10.

Chapman, M.; Chapmam, L. & Dore, D. (2008). *National audit of on-farm irrigation information tools*. Final report prepared for the Australian Goverment Department of the Environment, Water, Heritage and the Arts by RuralPlan Pty. Ltd.

Chauhan, Y.S.; Wright, G.C.; Holzworth, D.; Rachaputi, R.C.N. & Payero, J.O. (2011). AQUAMAN-a web-based decision support system for irrigation scheduling of peanuts. *Irrigation Science*: in press.

Chopart, J.L.; Mezino, M.; Aure, F.; Le Mezo, L.; Mete, M.; & Vauclin, M. (2007). OSIRI: A simple decision-making tool for monitoring irrigation of small farms in heterogeneous environments. Agricultural Water Management, 87:128-138.

Cull, P.O. (1979). *Irrigation scheduling of cotton by computer*. Dissertation Thesis, University of New England, 295 pp.

Davidson, J.I.; Bennett, C.T.; Tyson, T.W.; Baldwin, J.A.; Beasley, J.P.; Bader, M.J.; & Tyson, A.W. (1998). Peanut irrigation management using EXNUT and MOISNUT computer programs. Peanut Science, 25:103-110.

Evett, S.R. & Lascano, R.J. (1993). ENWATBAL.BAS: A mechanistic evapotranspiration model written in compiled BASIC. *Agronomy Journal*, 83(3): 763-772.

Fitzmaurice, L. & Beswick, A.R. (2005). *Sensitivity of the FAO56 crop reference evapotranspiration to different input data*, Technical Report, The Australia Bureau of Meteorology and the Queensland Department of Natural Resources and Mines, pp. 16 (available at http://www.longpaddock.qld.gov.au/silo/pet/index.html, 29 Sep 2010).

Fox, F.A.; Slack, D.C.; Clark, L.J. & Scherer, T.F. (1993). AZSCHED-Software for real-time irrigation scheduling, *Proceedings of the Am. Soc. Civil Eng. Nat. Conf. Proc. on Irrig. and Drain. Eng., Management of Irrigation and Drainage Systems: Integrated Perspectives*, pp. 891-898.

Howden, S.M.; Soussana J.F.; Tubiello, F.N.; Chhetri, N.; Dunlop, M. & Meinke, H. (2007). Adapting agriculture to climate change. PNAS 104: 19691-19696. Doi. 10.1073/pnas.0701890104.

Howell, T.A.; Johnson, K. & Dusek, D.A. (1995). Use of the SCS-Scheduler program--Southern High Plains. In: L. Ahuja, J. Leppert, K. Rojas and E. Seely (Editors), Workshop on Computer Applications in Water Management. Colorado State University, Fort Collins, Colorado, pp. 109-113, & 265.

IPCC. (2007). Climate Change 2007: Synthesis Report, Contribution of Working Groups I, II and III to the fourth assessment Report of the Intergovernmental Panel on Climate Change, Geneva, Switzerland.

Inman-Bamber, G. & Attard, S. (2005). Inventory of Australian software tools for on-farm water management. Technical Report No. 02/05, Cooperative Research Centre for Irrigation Futures.

Jeffrey, S.J.; Carter, J.O.; Moodie, K.B. & Beswick, A.R. (2001). Using spatial interpolation to construct a comprehensive archive of Australian climate data. *Environmental Modelling & Software*, 16: 309-330.

Jensen, M.E. (1969). Scheduling irrigations using computers. *Journal of Soil and Water Conservation*, 24(8): 193-195.

Keating, B.A.; Carberry, P.S.; Hammer, G.L.; Probert, M.E.; Robertson, M.J.; Holzworth, D.; Huth, N.I.; Hargreaves, J.N.G.; Meinke, H.; Hochman, Z.; McLean, G. ; Verburg, K.; Snow, V.; Dimes, J.P.; Silburn, M.; Wang, E.; Brown, S.; Bristow, K.L; Asseng, S.; Chapman, S.; McCown, R.L.; Freebairn, D.M.; & Smith, C.J. (2003). An overview of APSIM, a model designed for farming systems simulation. European Journal of Agronomy, 18:267-288.

Kincaid, D.C. & Heermann, D.F. (1974). Scheduling irrigations using a programmable calculator. USDA-ARS Report No. ARS-NC-12. 10 pp.

McKay, A.; Denby, C.; Prince, R.; Calder, T. & Ellement, D. (2009). Scheduling vegetable irrigation using the web. *Irrigation Australia Journal*, 24(1): 17.

Muchow, R.C.; Sinclair, T.R. & Bennett J.M. (1990). Temperature and solar radiation effects on potential maize yield across locations. *Agronomy Journal*, 82: 338-343.

Raes, D.; Steduto, P.; Hsiao, T.C. & Fereres, E. (2009). *AquaCrop-The FAO crop model to simulate yield response to water*. Food and Agriculture Organization of the United Nations, Rome, Italy.

Richards, Q.D.; Bange, M.P. & Johnston, S.B. (2008). HydroLOGIC: An irrigation management system for Australian cotton. *Agricultural Systems,* In press, doi:10.1016/j.agsy.2008.03.009: 10.

Steduto, P., Raes, D.; Hsiao, T.C.; Fereres, E.; Heng, L.; Izzi, G.; & Hoogeveen, J. (2009). AquaCrop: A new model for crop prediction under water deficit conditions. Options Mediterraneennes, Series A: 285-292.

Thysen, I. & Detlefsen, N.K. (2006). Online decision support for irrigation for farmers. *Agricultural Water Management*, 86: 269-276.

14

Estimates of Evapotranspiration and Their Implication in the Mekong and Yellow River Basins

Maichun Zhou
College of Water Conservancy and Civil Engineering, South China Agricultural University, Guangzhou 510642, China

1. Introduction

Potential evapotranspiration (PET) is generally considered to be the maximum rate of evaporation from vegetation-covered land surfaces when water is freely available and is primarily determined by meteorological controls (McVicar et al., 2007; Lhomme, 1997; Granger, 1989). It is a key input to hydrological models. Evapotranspiration can be directly measured by lysimeters or eddy correlation method but expensively and practically only in research over a small plot for a short time. The pan evaporation has long records with dense measurement sites. To apply it in hydrological models, however, first, a pan coefficient, K_p, then a crop coefficient, K_c, must be multiplied. Due to the difference on sitting and weather conditions, K_p is often expressed as a function of local environmental variables such as wind speed, humidity, upwind fetch, etc. A global equation of K_p is still lack. The values of K_c from the literature are empirical, most for agricultural crops, and subjectively selected.

When the environment is water-limited, PET is difficult to measure and therefore is usually estimated as an area-average by the model. If the environment is energy-limited, then the measurement of actual evapotranspiration provides a measure of PET (Donohue et al., 2007). Among the many evaporation models available, the Penman equation (Penman, 1948) and derivation of this equation are preferred, with two extensions being widely employed: (a) the Penman-Monteith equation (P-M) (Monteith, 1965); and (b) the Shuttleworth-Wallace model (S-W) (Shuttleworth and Wallace, 1985). The P-M equation was even standardized as FAO-24 (Doorenbos and Pruitt, 1992) and FAO-56 (Allen et al., 1998) for the reference evapotranspiration of a hypothetical crop. In contrast to many empirical PET formulations of uni- or bi-meteorological variables (e.g. air temperature only of Thornthwaite, 1948; air temperature and solar radiation of Priestley and Taylor, 1972), the P-M and S-W extensions of Penman equation use air temperature, solar radiation, humidity, wind speed, vegetation dynamics, and implicitly consider the influence of feedbacks among forcing meteorological variables, vegetation and evaporation, therefore is physically-based. This chapter compares use of the P-M equation (i.e. the FAO-56 standardized form) and the S-W model over the Mekong and Yellow River basins, representing humid and semi-arid regions, respectively, at a monthly time-step for the period from 1981 to 2000.

2. Methods

2.1 Penman-Monteith (P-M) estimate of reference crop evaporation

The P-M equation treats the vegetation canopy as a single uniform cover or "big-leaf". In its FAO-56 standardized form, the evaporation of a hypothetical crop is estimated (Allen et al., 1998), and hereafter referred to as a reference evapotranspiration (RET) as in Equation (1). The hypothetical crop is assumed: closely resembling an extensive surface of green grass of uniform height (0.12 m), actively growing (canopy resistance of 70 s m-1), completely shading the ground (albedo of 0.23) and with adequate water.

$$RET = \frac{0.408\Delta(R_n - G) + 900\gamma u_2 (e_s - e_a)/(273+T)}{\Delta + \gamma(1+0.34u_2)} \tag{1}$$

where RET is in mm day^{-1}, R_n is the net radiation above vegetation while G is the soil heat flux both (MJ m^{-2} day^{-1}), e_s and e_a are the saturation and actual vapour pressures respectively both (kPa), T is the mean air temperature (°C), u_2 is the wind speed at 2 m height (m s^{-1}), Δ is the curve slope of relationship between saturation vapour pressure and air temperature (kPa °C^{-1}), and γ is the psychrometric constant (kPa °C^{-1}).

Over a large basin, the big leaf assumption is rarely valid. There are often many vegetation types co-existent, and always some parts or periods where or when the vegetation is not "closed". Both the soil surface and the vegetation leaves evaporate or transpire moisture to the atmosphere and their relative importance changes dynamically as the vegetation develops. The ideal approach is that applicable at all times and places and able to reflect the changes of surface conditions. The S-W model meets this criterion.

2.2 Shuttleworth-Wallace (S-W) model

As an extension of the P-M equation, the S-W model considers dual sources, namely transpiration from vegetation and evaporation from underlying soil (Shuttleworth and Wallace, 1985):

$$\lambda ET = C_c ET_c + C_s ET_s \tag{2}$$

where ET is the total evapotranspiration (mm day^{-1}), λ is the latent heat of water vapourization (MJ kg^{-1}), ET_c and ET_s are equivalent to transpiration and evaporation by applying the P-M model to "closed" canopy and bare substrate respectively (MJ m^{-2} day^{-1}), C_c and C_s are weighting coefficients as functions of resistances. The formulation of all terms in Equation (2) is given as:

$$ET_c = \frac{\Delta(R_n - G) + \left[(24\times3600)\rho c_p (e_s - e_a) - \Delta r_a^c (R_n^s - G)\right] / (r_a^a + r_a^c)}{\Delta + \gamma\left[1 + r_s^c/(r_a^a + r_a^c)\right]} \tag{3}$$

$$ET_s = \frac{\Delta(R_n - G) + \left[(24\times3600)\rho c_p (e_s - e_a) - \Delta r_a^s (R_n - R_n^s)\right] / (r_a^a + r_a^s)}{\Delta + \gamma\left[1 + r_s^s/(r_a^a + r_a^c)\right]} \tag{4}$$

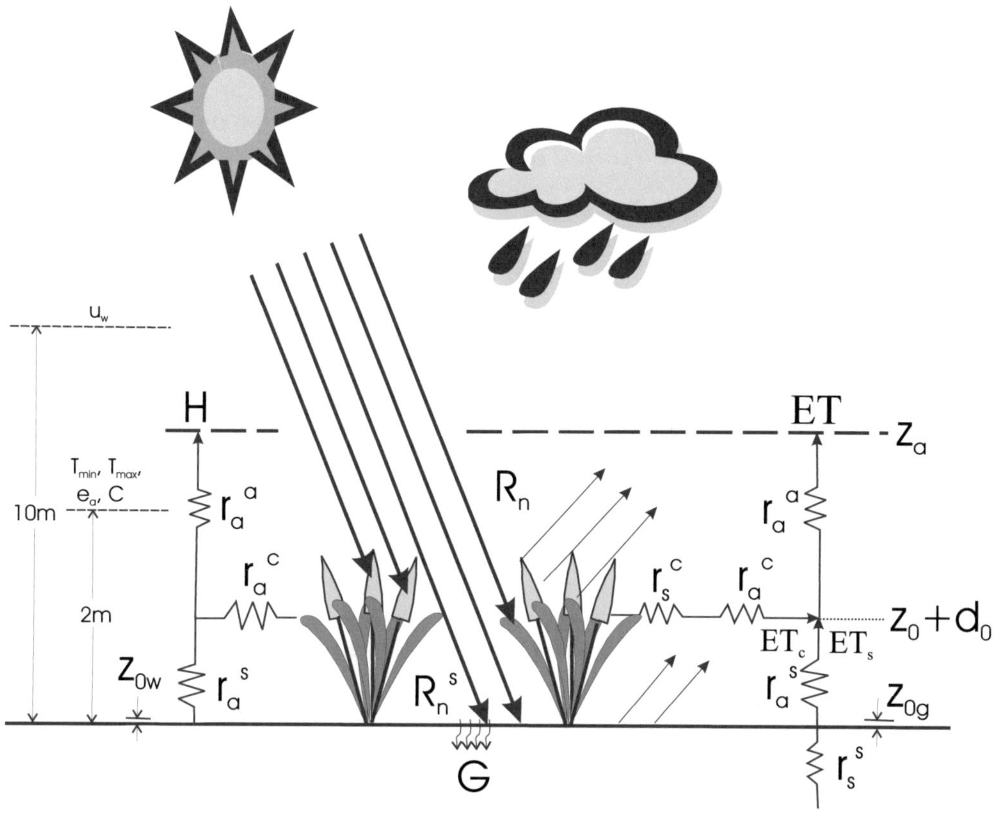

Fig. 1. Schematic diagram of the S-W model. From right to left, r_s^c and r_a^c bulk resistances of canopy stomatal and boundary layer respectively, r_a^s and r_a^a aerodynamic resistances from soil to canopy and from canopy to reference height respectively, r_s^s soil surface resistance, z_{0g} ground roughness length, $z_0 + d_0$ effective height of canopy source, z_a reference height (= h_c + 2 m), h_c vegetation height, R_n and R_n^s net radiations above canopy and to soil surface respectively, G soil heat flux, ET_c transpiration from canopy, ET_s evaporation from soil, ET total evapotranspiration, H sensible heat, T_{min} and T_{max} daily min and max air temperatures respectively, e_a actual vapour pressure, C cloud cover, u_w wind speed at z_w height, h_w observation height for other meteorological variables (usually, z_w = 10 m and h_w = 2 m), z_{0w} ground roughness length at weather station.

$$C_c = \frac{1}{1+(R_c R_a)/\left[R_s(R_c+R_a)\right]} \tag{5}$$

$$C_s = \frac{1}{1+(R_s R_a)/\left[R_c(R_s+R_a)\right]} \tag{6}$$

$$R_a = (\Delta+\gamma)r_a^a \tag{7}$$

$$R_c = (\Delta+\gamma)r_a^c + \gamma r_s^c \tag{8}$$

$$R_s = (\Delta + \gamma) r_a^s + \gamma r_s^s \tag{9}$$

where R_n^s is the net radiation over soil surface (MJ m-2 day-1), ρ is the mean air density (kg m-3), c_p is the specific heat of moist air (MJ kg-1 °C-1), r_s^c and r_a^c are the bulk stomatal and boundary layer resistances of canopy respectively, r_a^s and r_a^a are the aerodynamic resistances between soil and canopy and between canopy and reference height respectively, r_s^s is the surface resistance of soil, all five resistances are in s m-1, other notations are the same as in Equation (1). Figure 1 shows the sensible and latent heat transfer structure of S-W model.

The evaporation from water surface is estimated by substituting the aerodynamic resistance of Penman wind speed function and $r_s = 0$ into the P-M (Shuttleworth, 1993):

$$ET = \frac{\Delta}{\Delta + \gamma} 0.408 (R_n - G) + \frac{\gamma}{\Delta + \gamma} 2.624 (1 + 0.536 u_2)(e_s - e_a) \tag{10}$$

Stannard (1993) and Federer et al. (1996) compared a number of models, including the P-M and the S-W, and found that they give very different prediction. The research of Stannard (1993) and Vorosmarty et al. (1998) shows that hydrological modeling is sensitive to the PET methods, higher in humid regions, and the S-W model performs best. Furthermore, the interception plays an important role in water cycle. Only the S-W model is applicable to the evaporation from interception (Federer et al., 1996). The S-W model, however, is highly complex with many parameters and demands a great deal of data on the meteorology and the land surface characteristics. Most previous work has been focusing on the model validation and comparison with some specific cover types over small experimental catchments in a short time (e.g. Iritz et al., 1999 among others) or in the water balance model at a continent (Vorosmarty et al., 1998). Its application to a large basin for a long term is still lack. In this chapter, first, the S-W model is developed only using parameter values from the literature. Neither experimental measurement nor calibration is introduced. Second, all input data are publicly available, so that it can be applied to the data-poor or ungauged basins, particularly to the large basins. Third, using this method, the spatial distribution of potential evapotranspiration is estimated for a long term over the Mekong and Yellow River basins

2.3 Model parameterization

2.3.1 Climate-related parameters

In Equations from (1) to (10), parameters λ, e_s, Δ, ρ, c_p and γ are directly related to the climatic variables (Shuttleworth, 1993; Allen et al., 1998):

$$\lambda = 2.501 - 0.002361 T_s \tag{11}$$

$$e^o(T) = 0.6108 \exp\left(\frac{17.27T}{T + 237.3}\right) \tag{12}$$

$$e_s = \frac{e^o(T_{\max}) + e^o(T_{\min})}{2} \tag{13}$$

$$\Delta = \frac{4098\left[0.6108\exp\left(\frac{17.27T}{T+237.3}\right)\right]}{\left(T+237.3\right)^2} \tag{14}$$

$$\rho = \frac{P}{T_{kv}R} \tag{15}$$

$$P = 101.3\left(\frac{293-0.0065z}{293}\right)^{5.26} \tag{16}$$

$$\gamma = \frac{c_p P}{\varepsilon\lambda} \tag{17}$$

where T_s is the temperature of water surface (°C), substituted with the daily mean air temperature (T) for simplicity, equal to the arithmetic average of $T_{\max}$ and $T_{\min}$, the daily maximum and minimum air temperatures (°C), $e^o(T)$ is the saturation vapour pressure (kPa) at T, Equation (15) is based on the ideal gas law, P is the atmospheric pressure (kPa), assuming 20°C for a standard atmosphere, T_{kv} is the air virtual temperature (K), T_{kv} = 1.01 (273 + T), R is the specific gas constant, 0.287 kJ kg^{-1} K^{-1}, z is the elevation above sea level (m), ε is the ratio of the molecular weight of water vapour to that of dry air.
By substituting c_p = 1.013×10^{-3} MJ kg^{-1} °C^{-1} (a value under average atmospheric conditions), λ = 2.45 MJ kg^{-1} (when the temperature is about 20°C) and ε = 0.622 into Equation (17), the psychrometric constant is approximated as:

$$\gamma = 0.665\times10^{-3}P \tag{18}$$

2.3.2 Aerodynamic resistances

The aerodynamic resistances, r_a^s and r_a^a, are derived by integrating the eddy diffusion coefficients within and above the canopy (K-theory). Shuttleworth and Wallace (1985) applied the K-theory to the complete canopy cover and to the bare substrate soil separately and obtained the final by linearly interpolating between the two limits in terms of LAI, while Shuttleworth and Gurney (1990) applied the K-theory directly to the sparse vegetation canopy through an assumed "preferred" height therefore the interpolation with LAI is unnecessary. Furthermore, Shuttleworth and Wallace (1985) used the "preferred" equations to estimate the roughness length and zero plane displacement of the canopy while Shuttleworth and Gurney (1990) derived them from a second-order closure theory and considered arguably that the latter are superior to the former. In this chapter, the formulation of Shuttleworth and Gurney (1990) is used and the stability effects are ignored to avoid the iterative running of the model, as expressed by Equations (19) and (20). Hess (1998) showed an acceptable estimate for the long-term monthly PET by supposing the sunshine and wind stationary using the P-M method.

$$r_a^s = \frac{h_c\exp(n)}{nK_h}\left[\exp\left(-nz_{0g}/h_c\right)-\exp\left\{-n\left(Z_0+d_p\right)/h_c\right\}\right] \tag{19}$$

$$r_a^a = \frac{1}{\kappa u_*}\ln\left(\frac{z_a - d_0}{h_c - d_0}\right) + \frac{h_c}{nK_h}\left[\exp\left\{n\left[1-\left(Z_0 + d_p\right)/h_c\right]\right\} - 1\right] \quad (20)$$

where h_c is the vegetation height (m), n is the eddy diffusivity decay constant of the vegetation, K_h is the eddy diffusion coefficient at the top of canopy (m² s⁻¹), z_{0g} is the roughness length of ground (m), varying with the vegetation type, Z_0 is the "preferred" roughness length (= 0.13 h_c) (m), d_p is the "preferred" zero plane displacement (= 0.63 h_c) (m), κ is von Karman's constant (κ = 0.41), u_* is the friction velocity (m s⁻¹), z_a is the reference height (m), 2 m above vegetation, d_0 is the zero plane displacement of canopy (m). All items are parameterized as follows (Monteith, 1973; Choudhury and Monteith, 1988; Shuttleworth and Gurney, 1990; Federer et al., 1996):

$$K_h = \kappa u_*\left(h_c - d_0\right) \quad (21)$$

$$u_* = \kappa u_a / \ln\left\{\left(z_a - d_0\right)/z_0\right\} \quad (22)$$

$$d_0 = \begin{cases} h_c - z_{0c}/0.3 & LAI \ge 4 \\ 1.1h_c \ln\left[1 + \left(c_d LAI\right)^{1/4}\right] & LAI < 4 \end{cases} \quad (23)$$

$$z_0 = \min\left\{0.3\left(h_c - d_0\right),\ z_{0g} + 0.3h_c\left(c_d LAI\right)^{0.5}\right\} \quad (24)$$

$$z_{0c} = \begin{cases} 0.13h_c & h_c \le 1 \\ 0.139h_c - 0.009h_c^{\,2} & 1 < h_c < 10 \\ 0.05h_c & h_c \ge 10 \end{cases} \quad (25)$$

$$c_d = \begin{cases} 1.4\times 10^{-3} & h_c = 0 \\ \left[-1 + \exp\left(0.909 - 3.03 z_{0c}/h_c\right)\right]^4 / 4 & h_c > 0 \end{cases} \quad (26)$$

$$n = \begin{cases} 2.5 & h_c \le 1 \\ 2.306 + 0.194h_c & 1 < h_c < 10 \\ 4.25 & h_c \ge 10 \end{cases} \quad (27)$$

where u_a is the wind speed at the reference height (m s⁻¹), z_0 is the roughness length of canopy while z_{0c} is that for a "closed" canopy (m), c_d is the mean drag coefficient for individual leaves. Equations (23) and (24) combine two cases: closed canopy (LAI ≥ 4) and sparse growing vegetation, using the "preferred" values or related to LAI in the second-order closure theory. Equation (25) uses the "preferred" values but differentiates the canopy from short and tall vegetation types and linear interpolated between. Equation (26) uses the value of open water surface for bare surface (Brutsaert, 1982, pp. 118). Wilson and Shaw (1977) assumed c_d as 0.2 for maize leaves, Jones (1992) gave a typical value in the range between 0.03 and 0.6, whereas Shuttleworth and Gurney (1990) set it to 0.07. Equation (26) gives 0.05 around for short vegetation (h_c < 1 m), but 0.412 for tall vegetation (h_c > 10 m). Equation (27) uses the typical values to represent for short vegetation and tall vegetation, and linear interpolation between: 2.5 of agricultural crop (Monteith, 1973; Uchijima, 1976) and 4.25 of pine forest (Brutsaert, 1982, pp. 106).

The wind speed observed at the weather stations is converted to the reference height using a logarithmic profile in that the internal boundary layer heights over the weather ground and canopy surface are matched and a step change in surface roughness from z_0 to z_{0w} is assumed (Brutsaert, 1982, pp. 59 and pp. 167; Federer et al., 1996):

$$u_a = u_w \frac{\ln(z_b/z_{0w})}{\ln(z_b/z_0)} \frac{\ln[(z_a - d_0)/z_0]}{\ln(z_w/z_{0w})} \tag{28}$$

where u_w is the wind speed observed at the weather station (m s^{-1}). The height of observation, according to the CRU data (New et al., 1999), z_w is set to 10 m. Over weather station ground, zero plane displacement is assumed zero, the roughness length, z_{0w}, is assumed as 0.005m (Federer et al., 1996), and the height of internal boundary layer, z_b (m), is estimated (Brutsaret, 1982, pp. 165) as:

$$z_b = 0.334 F_w^{0.875} z_{0w}^{0.125} \tag{29}$$

where F_w is the fetch at weather station and assumed F_w = 5000 m.

2.3.3 Bulk stomatal and boundary layer resistances

The bulk stomatal resistance of canopy is affected not only by LAI but also by the environmental variables. It is often expressed as the form (Jarvis, 1976):

$$r_s^c = \frac{r_{ST\min}}{LAI_e \prod_i F_i(X_i)} \tag{30}$$

where LAI_e is the effective LAI, X_i is any environmental variable upon which stomatal response depends, $F_i(X_i)$ is the stress function of X_i, $0 \le F_i(X_i) \le 1$, $r_{ST\min}$ represents the minimal stomatal resistance of individual leaves under optimal conditions (s m^{-1}), for instance, the soil moisture was near field capacity, the temperature was between 17°C and 25°C, the vapor pressure deficit at 2 m level was below 5 hPa and the global radiation was over 400 W m^{-2} (Tourula and Heikinheimo, 1998). In case one or more of stress functions reach zero, r_s^c is given maximum 50000 s m^{-1}, corresponding to the molecular diffusivity of water vapour through leaf cuticula (Tourula and Heikinheimo, 1998).

Shuttleworth and Wallace (1985) expressed the bulk stomatal resistance as $r_s^c = r_{ST}/2LAI$, where r_{ST} is the mean stomatal resistance (of amphistomatous leaves) and LAI is the total leaf area index, and took r_{ST} 400 s m^{-1} for a fully grown agricultural crop (LAI = 4), resulted r_s^c 50 s m^{-1}. Gardiol et al. (2003) used $r_s^c = r_s/2LAI_e$, where $r_s/2$ was considered as the mean stomatal resistance of amphistomatous leaves and LAI_e is equal to $LAI/2$ when $LAI \ge 4$. Gardiol et al. (2003) took r_s 294 s m^{-1} for the corn crop, assuming LAI = 4, resulted r_s^c 73.5 s

m-1. Considering the general crop in Shuttleworth and Wallace (1985) but a specific crop in Gardiol et al. (2003), their equations are actually equivalent. Brisson et al. (1998) used $r_s^c = r_{ST}\sigma_c/LAI$, where σ_c is the shielding factor. If the same value of r_{ST} is used in both Shuttleworth and Wallace (1985) and Brisson et al. (1998), σ_c is equivalent to 1/2. However, Brisson et al. (1998) used r_{ST} 250 s m-1 for the maize crop and expressed the shielding factor $\sigma_c = 0.5LAI + 1$. Considering the actually adopted large value for r_{ST} and another expression for σ_c, the equation of Brisson et al. (1998) is also equivalent to that of Shuttleworth and Wallace (1985). Unifying these equations, the bulk stomatal resistance can be expressed $r_s^c = r_{ST}/LAI_e$, where the use of LAI_e means only the upper leaves of canopy are active in heat and vapour transfer due to illumination-induced stomatal closure deep in the canopy (Allen et al., 1993). It should be noted that r_{ST} takes implicitly accounts for the environmental effects. If explicitly, Equation (30) is used. LAI_e is calculated equal to actual LAI for $LAI \le 2$, $LAI/2$ for $LAI \ge 4$ and 2 for intermediate LAI (Gardiol et al., 2003).
For the environmental stress functions, CO_2 concentration is ignored and others are outlined as follows:

$$F_1(S) = (dS)/(c+S) \tag{31}$$

$$F_2(D) = \begin{cases} 1-0.409D & \textit{for short vegetation} \\ 1-0.238D & \textit{for tall vegetation} \end{cases} \tag{32}$$

$$F_3(T) = \begin{cases} 1 & T \ge 298 \\ 1-1.6\times10^{-3}(298-T)^2 & 273 < T < 298 \\ 0 & T \le 273 \end{cases} \tag{33}$$

$$F_4(\theta) = \begin{cases} 1 & \theta \ge \theta_f \\ \dfrac{\theta-\theta_r}{\theta_f-\theta_r} & \theta_f < \theta < \theta_r \\ 0 & \theta \le \theta_r \end{cases} \tag{34}$$

As shown in Equation (31), the hyperbolic solar radiation function is used (Jarvis, 1976; Stewart, 1988; Stewart and Gay, 1989), where S is the incoming photo-synthetically active radiation (PAR) (in the range of 0.4-0.72 microns) flux (W m-2), d = 1 + c /1000, c = 100 for forests and 400 for crops. For simplicity, the net radiation rather than the solar irradiation is used (Monteith, 1995) and averaged for the day time in daily simulation. The coefficient of vapour pressure deficit function, Equation (32), is represented for short vegetation obtained from the Konza Prairie in Kansas (FIFE data) (Stewart and Gay, 1989), and for tall vegetation by the coniferous forest from the Hydrologic Atmospheric Pilot Experiment/Modelisation du Bilan Hydrique (HAPEX-MOBILHY) data (Noilhan and Planton, 1989; also see Lhomme et al., 1998), where $D = e_s - e_a$. When the air temperature, T, is higher than 25oC, the stomatal openness is not limited; less than 0oC, stomatal closes completely; and varies linearly between, as shown in Equation (33) where T in K. The stress function of soil moisture content is shown in Equation (34), where θ is the soil moisture content in root

zone, θ_f is the field capacity below which the transpiration is stressed, θ_r is the residual soil moisture content. For PET estimation, the soil moisture is assumed at field capacity.
The bulk boundary layer resistance of canopy is calculated equivalent to all leaf boundary layers in parallel (Shuttleworth and Wallace, 1985; Brisson et al., 1998):

$$r_a^c = r_b \sigma_b / LAI \tag{35}$$

where σ_b is the shielding factor, taking 0.5 (Shuttleworth and Wallace, 1985; Brisson et al., 1998), r_b is the boundary layer resistance of individual leaf and is obtained as (Shuttleworth and Gurney, 1990):

$$r_b = \frac{100}{n}\left(\frac{w}{u_h}\right)^{1/2}\left[1-\exp(-n/2)\right]^{-1} \tag{36}$$

where r_b is in s m^{-1}, w is the canopy characteristic leaf width (m), u_h is the wind speed at the top of canopy (m s^{-1}), estimated using Equation (28) by substituting h_c for z_a, and n is the eddy diffusivity decay constant (see Equation (27)).

2.3.4 Surface resistance of substrate soil

Monteith (1981) proposed a two-layer model to simulate the soil evaporation, namely a drying soil layer overlaying a wet soil layer. The soil surface resistance is interpreted the resistance for the water vapour to diffuse through the dry top layer from the "evaporative front", the level of wet soil layer. It has been observed that a dry top layer developed quickly after a rainfall event in one day or even in a few hours, but very thin not extending a few centimeters (Hiller, 1980, p. 122; van de Griend and Owe, 1994; Lund and Soegaard, 2003). This is also supported by the finding of Stannard (1993) that the surface resistance was unrelated to the shallow soil moisture (at a depth of 0.15 m). Not only the soil physical parameters (such as moisture, vapour pressure and temperature gradients, matric forces, pore diameter, etc.) but also the air turbulence at the soil atmosphere interface affect this rapid process. It is most difficult, if not impossible, to develop the very detailed models which require a very high vertical resolution on the order of 1 mm close to the soil-atmosphere interface in order to describe the continuity of water fluxes properly. Consequently, to simplify the problems, Choudhury and Monteith (1988) calculated the soil surface resistance using the depth of upper dry soil, Equation (37), while van de Griend and Owe (1994) using the topsoil moisture, Equation (38):

$$r_s^s = \tau l / (p D_v) \tag{37}$$

$$r_s^s = 10 \exp\left[\alpha_\theta \left(\theta_{\min} - \theta\right)\right] \quad \theta \le \theta_{\min} \tag{38}$$

where l is the depth of upper dry layer (m), p is the porosity, τ is the tortuosity factor, D_v is the molecular diffusion coefficient for water vapour (m^2 s^{-1}), θ is the soil moisture content (percent by volume) in the top 1 cm, and $\theta_{\min}$ is an empirical minimum value above which the soil is able to deliver vapour at a potential rate, α_θ is a coefficient. The value of 10 in Equation (38) represents the resistance to molecular diffusion across the water surface itself.

Due to the lack of research on the parameters (τ, p, D_v, $\theta_{\min}$ and α_θ) for different soil types and the difficulty to determine the variables (l and θ), it is not practical at the moment to apply Equations (37) and (38) to large basins. Therefore, r_s^s is set to 500 s m-1 for PET estimation, as Shuttleworth and Wallace (1985) and Federer et al. (1996) did. This value is reasoned by that (a) Fuchs and Tanner (1968) suggested between 1000 and 2500 s m-1 for a sandy soil with a dry surface layer of 1.5 cm; (b) Camillo and Gurney (1986) found that a resistance between 100 and 600 s m-1 significantly improved the performance of a bare soil evaporation model; (c) Kondo et al. (1990) suggested that the surface resistance can exceed 4000 s m-1 for a very dry soil layer of 2 cm depth; (d) Mahfouf and Noilhan (1991) reported r_s^s could vary between 0 and several thousands second per meter; (e) when the wilting moisture θ_r = 4.1% (Rawls and Brakensiek, 1985) is used for θ, Equation (38) gives r_s^s 486 s m-1 where van de Griend and Owe (1994) used $\theta_{\min} = \theta_{field}/2 = 15\%$ and α_θ was calibrated as 0.3563 for a fine sandy loam; (f) Lund and Soegaard (2003) found that on the first day after rain the soil resistance increases from 0 in the morning to about 500 s m-1 by the end of the day in a millet field with a sandy loam; (g) when θ_r = 4.1 % and θ_{sat} = 45.3 % (Rawls and Brakensiek, 1985) are used, the empirical equation of Sun (1982) gives r_s^s 417 s m-1 for the sandy loam.

2.3.5 Net radiations over vegetation canopy, substrate soil surface and soil heat flux

Net radiation over vegetation canopy As a combination method, the reliability of S-W model is largely dependent on the accuracy of net radiation. However, the measurement of net radiation over large areas is impracticable but can be estimated from the solar radiation:

$$R_n = R_{ns} - R_{nl} \tag{39}$$

$$R_{ns} = (1-\alpha) R_{solar} \tag{40}$$

$$R_{nl} = \sigma\left(\frac{T_{\max,k}^4 + T_{\min,k}^4}{2}\right)\left(0.34 - 0.14\sqrt{e_a}\right)\left(1.35\frac{R_{solar}}{R_{solar}^0} - 0.35\right) \tag{41}$$

where R_{solar}, R_{solar}^0 and R_{ns} are the solar, clear-sky solar and net solar radiations respectively, R_{nl} is the net loss of energy in longwave radiation to the atmosphere, α is the albedo of land surface, σ is the Stefan-Boltzmann constant (= 4.903×10^{-9} MJ K-4 m-2 day-1), $T_{\max,k}$ and $T_{\min,k}$ are the max and min atmospheric temperatures in Kalvin, $T_{\max,k} = T_{\max}$ + 273.16 and $T_{\min,k} = T_{\min}$ + 273.16. All radiation terms are in MJ m-2 day-1. R_{solar} depends on the Julian date, the latitude location and the cloud cover condition (see Appendix A). In Equation (41), the first term is based on the Stefan-Boltzmann law, second term is the net emissivity between the atmosphere and the land surface, and third term is the correctness factor for the cloud cover (Shuttleworth, 1993; Allen et al., 1998). The land surface albedo is related to vegetation LAI (Uchijima, 1976) as:

$$\alpha = \alpha_m - (\alpha_m - \alpha_s)\exp(-0.56LAI) \tag{42}$$

where α_m and α_s are the albedo corresponding to the "closed" canopy and the bare soil, respectively. For PET estimation, α_s = 0.1 is used for the wet bare soil (Shuttleworth, 1993) and α_m is from the literature and changes with the vegetation types.

Radiation flux over substrate soil surface The radiation reaching soil surface, R_n^s, can be calculated using a Beer's law relationship of the form

$$R_n^s = R_n \exp(-C_r LAI) \tag{43}$$

where C_r is the extinction coefficient of the vegetation for net radiation. Monteith (1973) arbitrarily chose 0.7. Denmead (1976) gave 0.5-0.7 for the wheat and was cited by Kelliher et al. (1995). Lafleur and Rouse (1990) reported the value in a range from 0.3 to 0.8 and Ross (1975) from 0.3 to 0.6. Brisson et al. (1998) obtained C_r = 0.25 ± (0.001) for a well watered soybean crop by calibration, which is less than the value generally admitted (0.25 instead of 0.4, Shuttleworth and Wallace, 1985; Braud et al., 1995), due to the soil surface water status. If the soil dries out, its temperature and albedo increase, leads to a decrease in soil net radiation. Both Stannard (1993) and Mo et al. (2004) took 0.5. It is convenient to ignore variation in C_r which occurs in response to structural differences of vegetation, thus, C_r = 0.5 is also used here.

Soil heat flux The heat conduction into the substrate is commonly taken about 30% of R_n^s (Stannard, 1993). However, it is found too much in our monthly simulation. Accordingly, the equation of Allen et al. (1998) is used:

$$G = 0.07(T_{i+1} - T_{i-1}) \tag{44}$$

where T_{i-1} and T_{i+1} are the mean air temperatures in previous and next months respectively (°C).

2.3.6 Vegetation parameters

Except the soil surface resistance, all other resistances and each component of radiation are related to the vegetation parameters such as LAI, height and leaf width. Vegetation morphology changes dynamically with the environmental conditions (e.g. the prolonged water stress) and seasons. Probably, only the satellite can efficiently provide a frequent measurement to the time-varying vegetation in a long term and over a lager area.

Leaf area index (LAI) The LAI is used not only intensively in S-W parameterization but also in interception estimation. The LAI for each vegetation class can be derived from NOAA-AVHRR NDVI through FPAR (Myneni and Williams, 1994; Sellers et al., 1994; Andersen et al., 2002). Here the SiB2 method is used (Sellers et al., 1996):

$$SR = \frac{1 + NDVI}{1 - NDVI} \tag{45}$$

$$FPAR = FPAR_{\min} + (FPAR_{\max} - FPAR_{\min}) \frac{(SR - SR_{\min})}{(SR_{\max} - SR_{\min})} \tag{46}$$

$$LAI = (1 - F_{cl}) LAI_{\max} \frac{\ln(1 - FPAR)}{\ln(1 - FPAR_{\max})} + F_{cl} LAI_{\max} \frac{FPAR}{FPAR_{\max}} \tag{47}$$

where SR is the simple ratio of hemispheric reflectance for the NIR (near-infrared) light to that for the visible light, $FPAR$ is the fraction of photo-synthetically active radiation, F_{cl} is the fraction of clumped vegetation, $SR_{\min}$ and $SR_{\max}$ are SR with 5% and 98% of NDVI population. The values of F_{cl}, NDVI at 5% and 98% population are adopted from SiB2 for all vegetation types (NDVI at 5% setting to 0.039 globally, F_{cl} and NDVI at 98%, see Table 1). $FPAR_{\min}$ = 0.001 and $FPAR_{\max}$ = 0.950 consider the satellite-sensed NDVI saturation (Sellers et al., 1996). $LAI_{\max}$ is the maximum LAI when the vegetation develops fully, prescribed for each vegetation class by referring to the literature.

Vegetation height The vegetation height is calculated by differentiating the annual and perennial vegetation. For perennial vegetation, the height is assumed constant; but for the annual, the height grows with the LAI in a linear relationship, although an exponential relationship has also been used by some investigators (e.g. Brisson et al., 1998):

$$h_c = \begin{cases} 0 & LAI_{\max} = 0 \ \ (\textit{water surface, bare soil, etc.}) \\ h_{c\min} + \left(h_{c\max} - h_{c\min}\right)\dfrac{LAI}{LAI_{\max}} & LAI_{\max} \neq 0 \end{cases} \tag{48}$$

where $h_{c\min}$ and $h_{c\max}$ are the minimum and maximum vegetation heights prescribed by referring to the literature. For the perennial vegetation such as the forests, shrubs, etc., the same value is prescribed to $h_{c\min}$ and $h_{c\max}$.

Leaf width The leaf width is also treated by differentiating the annual and perennial vegetation. For annual vegetation, it is calculated using an exponential relation with LAI (Farahani and Bausch, 1995):

$$w = \begin{cases} w_{\max} & \textit{for perennial vegetation} \\ w_{\max}\left[1 - \exp(-0.6LAI)\right] & \textit{for annual vegetation} \end{cases} \tag{49}$$

2.3.7 Prescribed parameters

In S-W model parameterization, we need to prescribe a number of threshold parameters for vegetation canopy and roughness length of substrate soil surface. They are listed in Table 1 for IGBP (Global International Geosphere-Biosphere Programme) land cover classification by referring to the literature and our general knowledge, mainly from

1. α_m: Brutsaert (1982), Shuttleworth (1993), Fennessey and Kirshen (1994)
2. $LAI_{\max}$: Vorosmarty et al. (1998) (3.6 for shortgrass, 4.1 for tall grass, 5.2 for woodland, 5.0 for deciduous forest, 4.4 for evergreen forest and 1.6 for alpine), Andersen et al. (2002) (7 for evergreen broad leaf forest and 5 for other vegetation), Vourlitis et al. (2002) (6 for tropical rain forest and 1 for savanna), and so on
3. h_c, $w_{\max}$, F_{cl} and $NDVI_{98\%}$: Sellers et al. (1996) and Mo et al. (2004)
4. $r_{ST\min}$ (s m^{-1}): McNaughton and Black (1973) (75 for Douglas fir forest), Korner et al. (1979) (90-100 for barley crop), Calder et al. (1986) (120 for rain forest), Verma et al. (1986) (75-160 for deciduous forests), Rowntree (1991) (40-60 for growing crop, 80-130 for forests and 60-200 for grassland), Allen (1994) (100 for grass), Huntingford (1995) (82.6-125 for savanna), Kelliher et al. (1995) (80-90 for crop, 125 for grassland, 160-220 for forests), Vorosmarty et al. (1998) (90 for cultivated lands and 190 for forest and savanna), and Shen et al. (2002) (128 for wheat crop) among many other investigators

5. z_{0g} : Federer et al. (1996)
6. A number of parameters consisting of above items: Noihan and Planton (1989), Liang et al. (1994), Dunn and Mackay (1995), Raupach (1995), Mauser and Schadlich (1998)

Code	Type	α_m	LAI_{max}	h_c (m)	w_{max} (m)	F_{cl}	$r_{ST\cdot min}$ (s m^{-1})	$NDVI_{98\%}$	z_{0g} (m)
1	Evergreen needleleaf forests	0.16	5.5	17.0-17.0	0.001	1.0	150	0.689	0.020
2	Evergreen broadleaf forests	0.20	7.0	30.0-30.0	0.05	0.0	150	0.611	0.020
3	Deciduous needleleaf forests	0.15	3.3	17.0-17.0	0.001	1.0	150	0.689	0.020
4	Deciduous broadleaf forests	0.18	7.0	25.0-25.0	0.08	0.0	150	0.721	0.020
5	Mixed forests	0.17	5.7	20.0-20.0	0.04	0.5	150	0.721	0.020
6	Closed shrublands	0.20	4.6	1.5- 1.5	0.01	0.0	150	0.674	0.020
7	Open shrublands	0.15	3.0	1.0- 1.0	0.01	1.0	100	0.674	0.020
8	Woods savannas	0.20	1.5	0.8- 0.8	0.01	0.5	180	0.611	0.020
9	Savannas	0.25	0.9	0.1- 0.8	0.01	0.8	120	0.674	0.020
10	Grasslands	0.23	1.8	0.05-0.8	0.01	0.0	115	0.674	0.010
11	Permanent wetlands	0.10	6.0	0.05-1.0	0.01	0.0	65	0.674	0.010
12	Croplands	0.20	7.0	0.0- 0.8	0.01	0.0	90	0.674	0.005
13	Urban and built-up	0.18	----	------	-----	0.0	0	0.674	0.020
14	Cropland/natural vegetation mosaic	0.20	6.5	0.1- 0.8	0.01	0.5	120	0.674	0.010
15	Permanent snow and ice	0.70	----	------	-----	0.0	0	0.674	0.001
16	Barren or sparsely vegetated	0.15	0.3	0.05-0.8	0.01	1.0	120	0.674	0.001
17	Water bodies	0.08	----	------	-----	0.0	0	0.674	0.001

Code is IGBP number for vegetation type, α_m is maximum albedo (full cover), LAI_{max} is maximum leaf area index, h_c is vegetation height (a function of LAI), w_{max} is maximum vegetation leaf width, F_{cl} is fraction of clumped vegetation, $r_{ST\ min}$ is minimum stomatal resistance of individual leaf, z_{0g} is roughness length of substrate ground. Perennial vegetations: codes from 1 to 8, and annual vegetations: others; tall vegetations: codes from 1 to 5, and short vegetations: others.

Table 1. Land cover threshold parameters, NDVI at 98% population, and ground roughness length of substrate surface based on the IGBP classification from the literature

3. Study region description

The Mekong River, see Figure 2, is the longest river in Southeast Asia and the 12th longest river in the world, with a length of 4 800 km, a drainage area of 805 604 km^2 (WRI et al., 2003), and with an annual runoff of 475 × 10^9 m^3. It originates on the Tibetan Plateau and flows southwards through China, Myanmar, Laos, Thailand, Cambodia, and Vietnam before it discharges into the South China Sea. The upper Mekong River (1 600 km, from the Tibetan

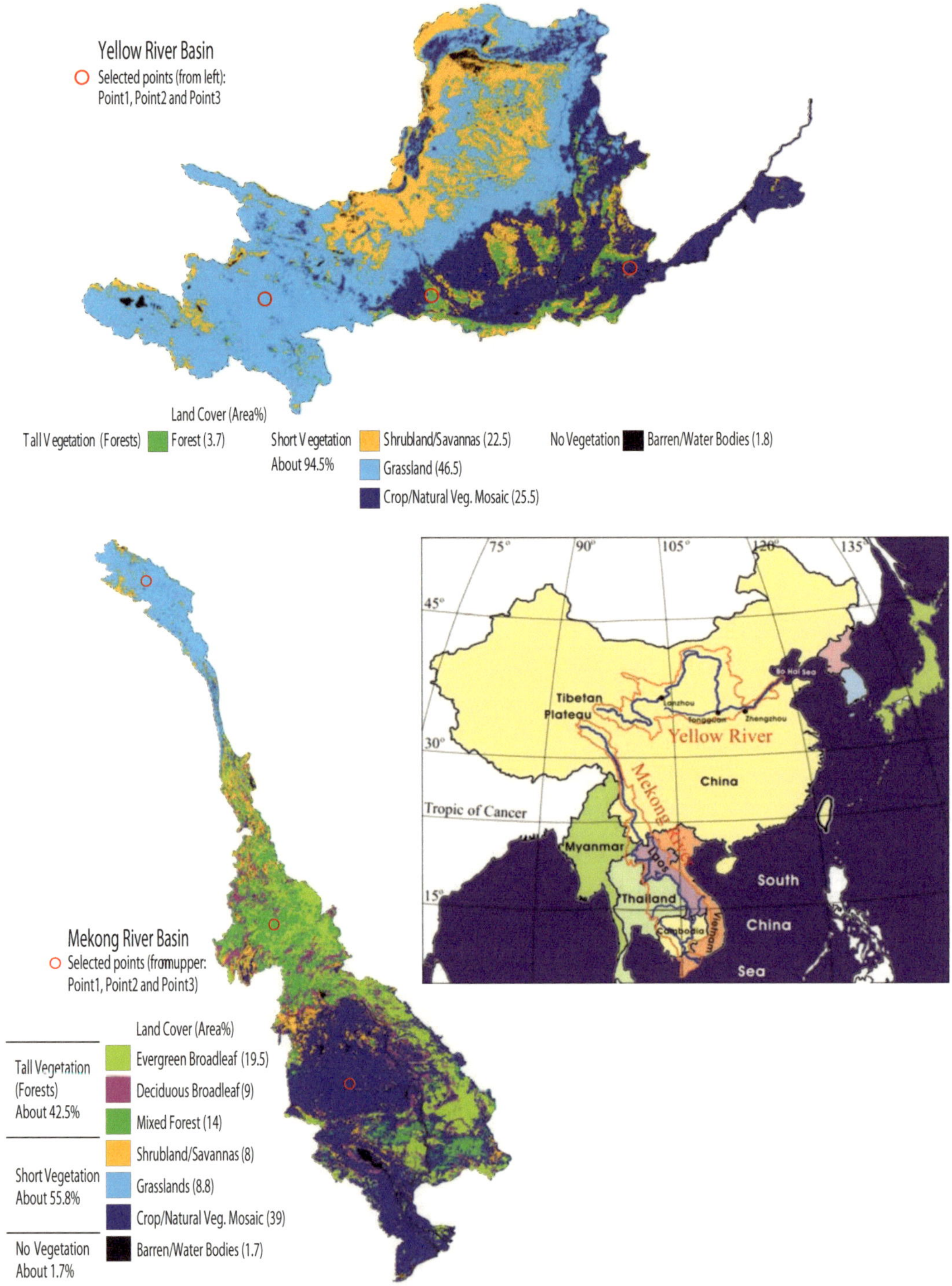

Fig. 2. Mekong and Yellow basins in Asian monsoon region: main streams, IGBP land cover, and selected points.

Plateau to the Thailand-Myanmar border), called the Lancang River in China, drops rapidly by about 4 500 m in a series of large mountain ranges. After exiting China and entering the Golden Triangle (an area of around 350 000 km^2 overlapping mountains of four countries: Myanmar, Laos, Vietnam, and Thailand), it is called the Lower Mekong River. The Lower Mekong River is gentle and most of its reaches are navigable. The climate of upper Mekong River basin has a high-mountainous cold weather with mean annual precipitation from about 300 mm at its source on the Tibetan Plateau to 1600 mm before entering the Golden Triangle. The lower Mekong River is situated in the tropics and is dominated by two distinct monsoons: the southwest from Indian Ocean from mid-May to mid-October with frequent rainfall, and the northeast from China from mid-October to April with a dry spell. The mean annual rainfall in the lower Mekong River ranges from 1 000 mm in northeast Thailand to more than 3 200 mm in the mountainous regions in Laos (Kite, 2001), and around 85-90% of it falls during the rainy season.

The Yellow River, see Figure 1, is the second longest river in China, with a length of 5 464 km, draining an area of 794 712 km^2, and with an annual runoff of 58×10^9 m^3. The Yellow River is divided into upper (2 119 km, from the source to Lanzhou), middle (2 571 km, from Lanzhou to Zhengzhou) and lower (774 km, from Zhengzhou to the Bo Hai Sea). The upper river winds around a series of large mountain ranges on the eastern Tibetan Plateau falling more than 3 300 m and basin average elevation about 4 000 m. After Lanzhou, the river makes a large northern loop through the alluvial plains and the Loess Plateau to Tongguan, picking up more than 90% of its silt load (3 kg m^{-3} at Lanzhou increasing to 35 kg m^{-3} at Tongguan). The lower river is narrow and flows within levees and dikes which have been constructed over the past 2000 years. Sediment deposition from the highly erosive Loess Plateau has continuously increased the height of the river bed. In places, the river bottom is 20 m above the surrounding land surface. The Yellow River basin has an arid and semi-arid continental monsoon climate. In the upper basin, the temperature is a complicated function of the elevation and low throughout the year. In the middle basin, the temperature decreases from south to north and from east to west, and is affected by local mountains and deserts. In the lower basin, the climate is dry and windy in spring, hot and wet in summer, dry in fall, and moderately cold and dry in winter. The annual precipitation is between 200-650 mm over the basin, being large in the lower basin and in the southern portion of the upper and middle basins.

4. Data sources

In order to apply the P-M equation and S-W model, topographic data, characteristics of land cover and meteorological data are required.

4.1 Topographic data

The P-M equation and S-W model use a DEM (Digital Elevation Model) to calculate the parameters of atmospheric pressure, mean air density and psychrometric constant. The Hydro1K DEM was downloaded (http://lpdaac.usgs.gov/gtopo30/hydro/index.html) and clipped to the basins defined by manual digitization from the DCW (Digital Chart of the World) (http://www.maproom.psu.edu/dcw; Danko, 1992). It is a hydrologically correct DEM developed at the Data Center of USGS EROS (Earth Resources Observation System, U.S. Geological Survey) (Verdin and Greenlee, 1996). Original topography at 1-km resolution was averaged on to an 8-km grid, i.e. the same resolution as NDVI (Normalized Difference Vegetation Index, see following).

4.2 Land cover

The IGBP (International Geosphere-Biosphere Programme) land cover dataset is used to characterize the basin land surface (http://edcwww.cr.usgs.gov/landdaac/glcc/glcc.html) (see Figure 2). It was derived from 1-km AVHRR data (Advanced Very High Resolution Radiometer) spanning April 1992 through March 1993 by the Data Center of USGS EROS, the University of Nebraska-Lincoln and the Joint Research Centre of the European Commission (Loveland et al., 2000). The IGBP categories the global land covers into 17 classes. To show the prevailing types, some minor land covers are classified into the similar types in Figure 2. In the Mekong River basin, the most common land cover is forests (~42.5%), the second most common is croplands or a mosaic of cropland and natural vegetation (~39%). Most of remainder is shrubland or grassland (each ~8%). In the Yellow River basin, the most common land cover is grassland (~47%), the second most common is croplands or a mosaic of cropland and natural vegetation (~26%), while the forested area is very small (< 4%). Most of remainder is shrubland or savanna (~23%). The original 1-km spatial resolution was converted to an 8-km grid by assigning the most common land cover to each 8-km cell. While the aggregation from 1-km to 8-km changes the proportion of land cover types to some extent, the general spatial distribution remains.

4.3 NDVI data

Monthly NOAA-AVHRR maximum NDVI composite data at 8-km grid resolution (ftp://daac.gsfc.nasa.gov/data/avhrr/global_8km/; Tucker et al., 2005) was obtained from July 13, 1981 to September 21, 2001, except for a period with missing data from September to December 1994. The NDVI data for the basins were transferred into Lambert Azimuthal Equal Area projection. The monthly NDVI data are assumed to represent the value for the middle day of the month. The average monthly NDVI data from 1981 to 2000 are used during the period when data were missing.

4.4 Meteorological data

The required meteorological data include air temperature, relative humidity, radiation and wind speed. The CRU (Climate Research Unit, University of East Anglia in UK) TS 2.0 dataset provides monthly time series of mean air temperature, diurnal air temperature range, precipitation, cloud cover, and actual vapour pressure from 1901 to 2000, and mean wind speed from 1961 to 1990 globally on a 0.5-degree grid (New et al., 1999 and 2000). The wind speed was mainly measured at 10 m height (New et al., 1999). CRU TS 2.0 dataset was interpolated as a function of latitude, longitude, and elevation (TBASE 5-min latitude–longitude global DEM at http://www.ngdc.noaa.gov/seg/topo/topo.shtml) from station data using thin-plate splines (Hutchinson, 1995) for the 1961-1990 climatic normals and angular distance–weighted for the monthly anomalies relative to the 1961–1990 mean in which the influence of elevation was ignored. The climatic variables were separated into two categories: primary and secondary (New et al., 1999). The primary data variables include mean air temperature, the diurnal air temperature range and precipitation, which were constructed directly from station observations. The secondary variables used include cloud cover and vapour pressure, which were constructed by merging station observations (where available) with synthetic data derived from the gridded primary variables. In the synthetic data, cloud cover was related to the diurnal air temperature range, and the actual vapour pressure to the daily minimum air temperature (New et al., 2000). The CRU data sets were extracted for the basins and transferred into Lambert Azimuthal Equal Area projection at 8-km resolution without interpolation.

All the climatic variables change significantly throughout the year and are very non-uniformly distributed across the basins. Figure 3 shows the general characteristics of six monthly climatic variables: maximum and minimum air temperatures (T_{max} and T_{min} in °C), precipitation (P in mm/month), actual vapour pressure (e_a in kPa), cloud cover (C in tenth), and wind speed (u in m s^{-1}). They were spatially and temporally averaged over the whole basins and from 1981 to 2000 (T_{max}, T_{min}, P, e_a and C) or from 1961 to 1990 (u). The Mekong River basin spans from 33.5°N to 9.0°N, crossing from the cold mountainous climate in the Tibetan Plateau to the hot tropical climate in the Indochina Peninsula. Although the Yellow River basin also originates from the Tibetan Plateau, it generally flows at constant latitude. In the Mekong River basin, a large proportion is located in the tropics, whereas almost the whole Yellow River basin sits in North China. Their climate patterns are very different. In the two basins, the coldest points are both located in the Tibetan Plateau, as shown by the similar monthly change of the lower dash lines in Figures from 3(a) to 3(d). The maximum point temperature occurs at the center of middle Yellow River and at the center of Lower Mekong, respectively. The basin-spatially averaged temperature (solid lines in Figures from 3(a) to 3(d)) and maximum cell temperature (upper dash lines in Figures from 3(a) to 3(d)) change similarly. In another words, basin-spatial average T_{max} and cell T_{max} peak in March and April in the Mekong River basin (Figure 3(a)), and in June and July in the Yellow River basin (Figure 3(b)). In the Mekong River basin, the cell T_{min} (upper dash line in Figure 3 (c)) peaks almost constantly from April to October (basin rainy season) but the basin-spatial average T_{min} peaks from June to August due to the weight of cold region in Tibetan Plateau. In the Yellow River basin, both cell T_{min} and basin-spatial average T_{min} peaks in July, the same as in the case of its T_{max}. Most precipitation occurs from May to October in both basins (Figures 3(e) and 3(f)). Accordingly the vapour pressure and cloud cover are high during this time (Figures from 3(g) to 3(j)). The general pattern of relative humidity is governed by the southeast rainy monsoon from the Indian Ocean from mid-May to mid-October in the Mekong River basin, and by the rainy monsoon from the Indian and Pacific oceans during summer months in the Yellow River basin. Basin-wide mean relative humidity is more constant throughout the year in the Mekong (Figures 3(g)), with much larger variation being seen in the Yellow River basin (Figures 3(h)). The almost zero minimum relative humidity in both basins is located in the Tibetan Plateau. The monthly distribution of wind speed, u, is uni-modal for the minimum, weekly bi-modal for the basin-spatial average and the maximum both in the Mekong and Yellow River basins due to the two distinct monsoons, with a strong peak in April and a weaker peak in November, see Figures 3(k) and 3(l).

To compute the solar radiation, the cloudiness, C (tenth), is converted into the relative sunshine duration using the relationship of Doorenbos and Pruitt (1992), as shown in Table 2 and fitted in Equation (50).

$$\frac{n}{N} = 0.9234 - 0.048C - 0.0042C^2 \tag{50}$$

where n and N are actual and maximum possible sunshine duration (hour day^{-1}).

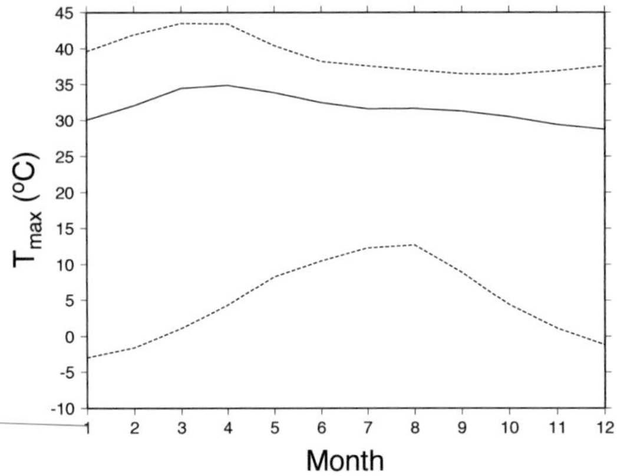

(a) T_{max} over Mekong

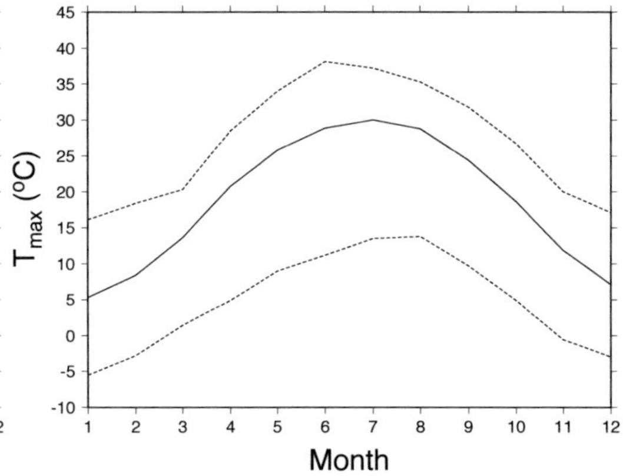

(b) T_{max} over Yellow

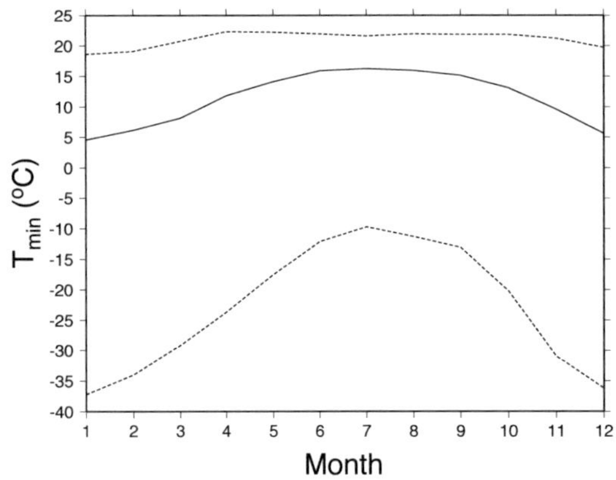

(c) T_{min} over Mekong

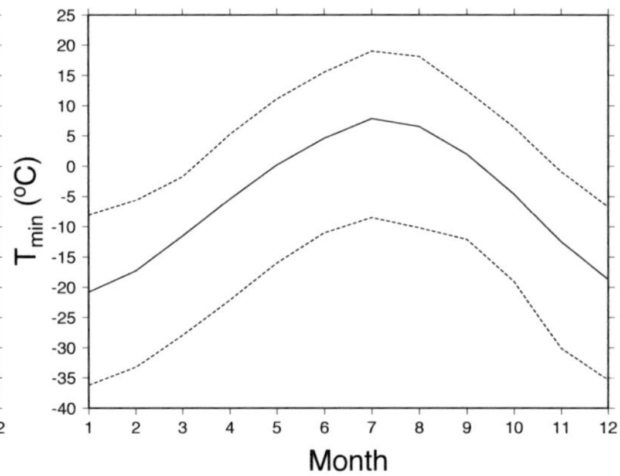

(d) T_{min} over Yellow

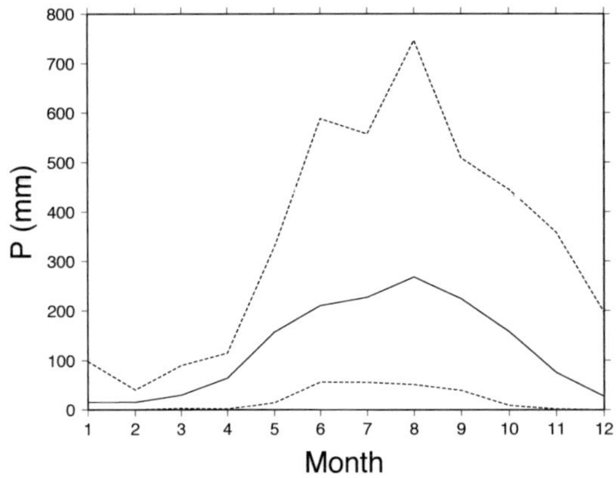

(e) P over Mekong

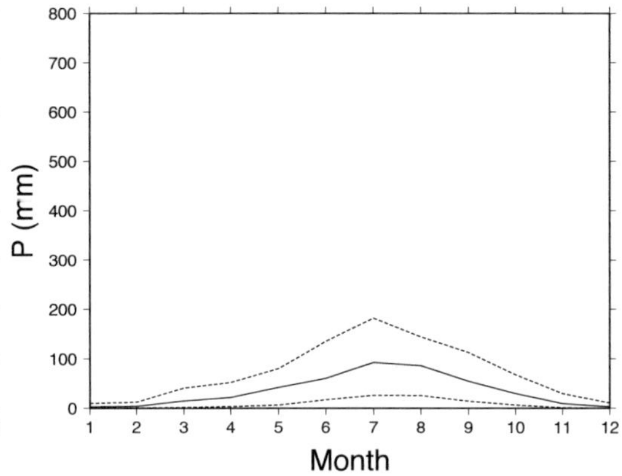

(f) P over Yellow

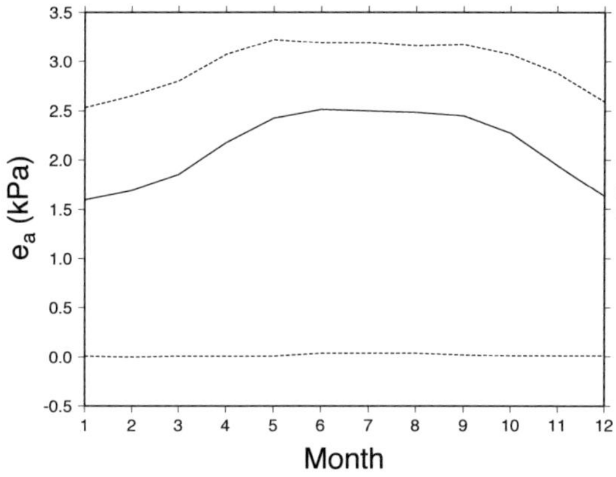

(g) e$_a$ over Mekong

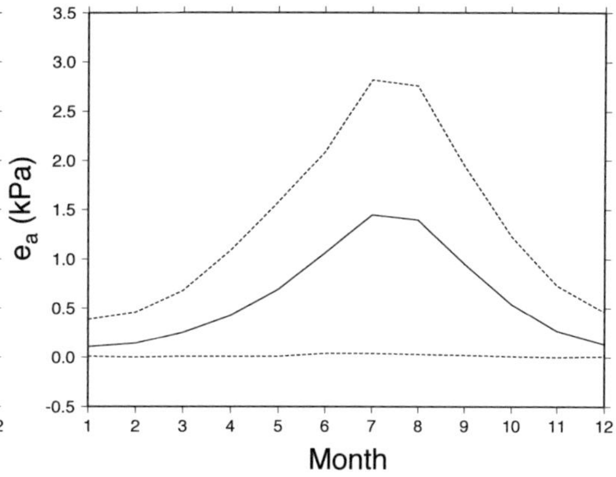

(h) e$_a$ over Yellow

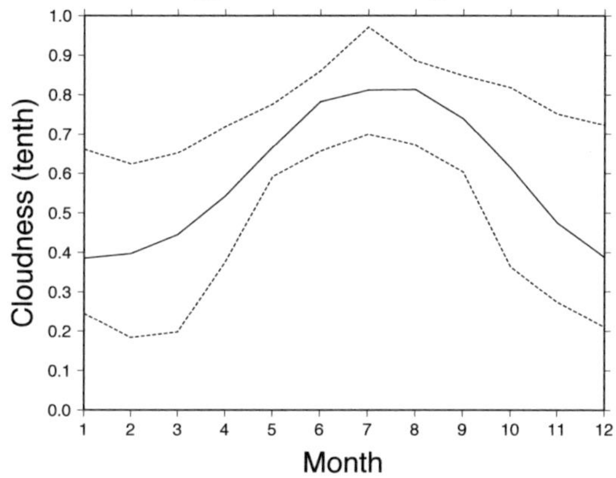

(i) C over Mekong;

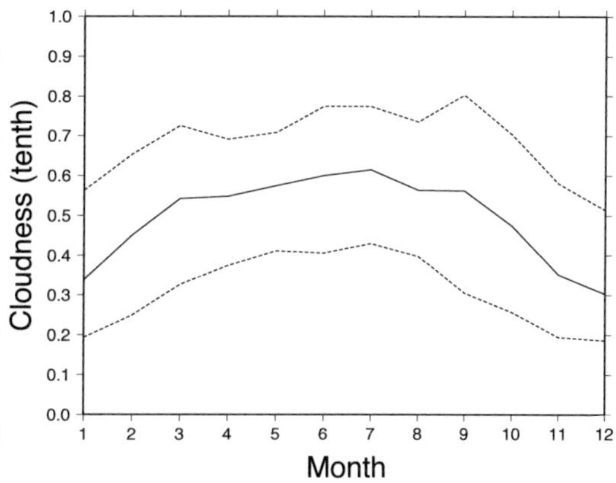

(j) C over Yellow

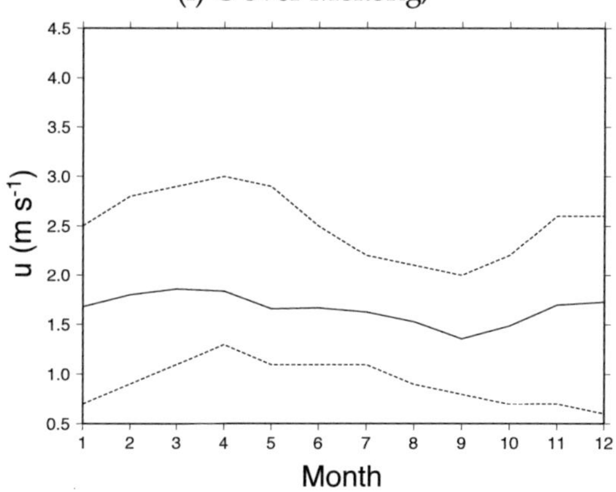

(k) u over Mekong

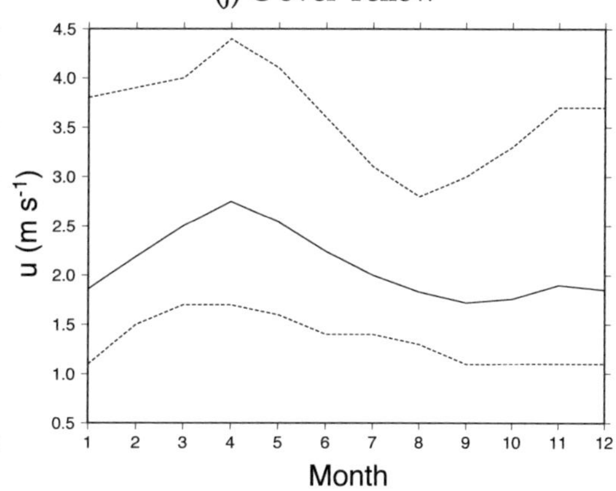

(l) u over Yellow

Fig. 3 Monthly climatological variables: T_{max}, T_{min}, P, e_a and C are averaged during 1981-2000, and u is averaged during 1961-1990; The spatial mean (solid line), the spatial minimum and maximum (dash lines) are shown.

Cloudiness (tenth)	0	1	2	3	4	5	6	7	8	9	10
n/N	0.95	0.85	0.8	0.75	0.65	0.55	0.5	0.4	0.3	0.15	0

n/N is the ratio of actual sunshine duration to maximum possible hours of sunshine.

Table 2. Relationship between cloud cover and sunshine duration (source: Doorenbos and Pruitt, 1992)

The 1961-1990 monthly mean wind speed data are used for whole 1981-2000 simulation period, we acknowledge that changes in RET and PET due to recently reported decreases in terrestrial tropical and mid-latitude near-surface wind speeds (McVicar et al. 2008) need to be accounted for when considering longer temporal extents. The wind speed at reference height (2 m height above the ground in FAO-56 equation and 2 m height above the vegetation canopy in S-W model) is converted using Equation (28).

5. Results and discussion

5.1 Comparison of spatial distributions of PET and RET and their interrelationships

The basin-average RET and PET, shown in Figure 4 for the period from 1981 to 2000, both reflect a similar variability and trend in climate. The good linearity between RET and PET, see Figure 5, means that about 70~80% of the inter-annual variability in basin-average PET can be explained by variations in climate, the remaining 20~30% being determined by vegetation diversity and dynamics which are incorporated in the S-W model but not in the FAO-56 P-M method. However, the spatial distributions of RET and PET are strikingly different, see Figure 6. The RET reflects the three climatic patterns over the individual basins: upper, middle and lower. On the other hand, these climatic patterns are not so clearly displayed in the PET distribution, but the effect of vegetation is more obvious. To clearly show the difference between RET and PET, the spatial distributions of average annual (RET-PET) and (RET-PET)/PET are displayed in Figure 7. Over the Tibetan Plateau of Mekong River basin where the actual vegetation of grassland has a lower LAI than the hypothetical reference crop, RET is estimated higher than PET because a lower LAI means more soil surface uncovered and the resistance of soil surface used in PET is much higher than the resistance of reference crop (500 m s^{-1} vs. 70 m s^{-1}). Over the Lower Mekong River basin where the large forest area has a higher LAI than the hypothetical reference crop, RET is lower than PET because a high LAI means soil surface covered well and the combined resistance of forest and soil surface is small (high LAI reducing the bulk stomatal resistance of forest canopy significantly, see Equation (30)). The vegetation morphology (forest being tall and grass being short) and LAI also affect the land surface albedo, then R_n. The estimate of RET is lower 120 mm/year (or only 1.7%) than PET basin-spatially (Figures 7(a) and 7(c)). Over the Yellow River basin, because there are few forests (see Figure 2) and the short vegetation generally has a lower LAI (due to water shortage), RET is estimated to be much higher than PET, basin-spatially 300 mm/year (or about 50%) (Figures 7(b) and 7(d)).

Three specific locations representing typical vegetation types were investigated in greater detail, namely, grass (Point 1), forest (Point 2) and crop (Point 3), see Figure 2. The aggregation of resolution from 1-km to 8-km does not change the vegetation types of these points. The year-to-year changes in both RET and PET are similar at the grass and forest points in the Mekong, and at all three points in the Yellow, that is when RET increases, in most cases PET also increases, and *vice versa*, see Figure 4. But at the crop point over the Mekong, the variation of amplitude in annual PET is not so large as in annual RET during 1985 to 1996, probably because the grass and forests are natural but the crop is also affected by irrigation, cultivation and human maintenance, with a heavy weight in the Mekong but a light weight in the Yellow comparing with the climate controls. However, the inter-annual changes are larger in PET than in RET at the three points in the Mekong and at the grass and forest points in the Yellow. Reasons are anticipated that: (a) vegetation types amplified or diminished the climatic effects on PET (e.g. the forest vs. the hypothetical crop, or the grass in the Tibetan Plateau vs. the hypothetical crop); (b) vegetation dynamics as a feedback of forcing meteorological controls changed the climatic effects on PET (see the following section).

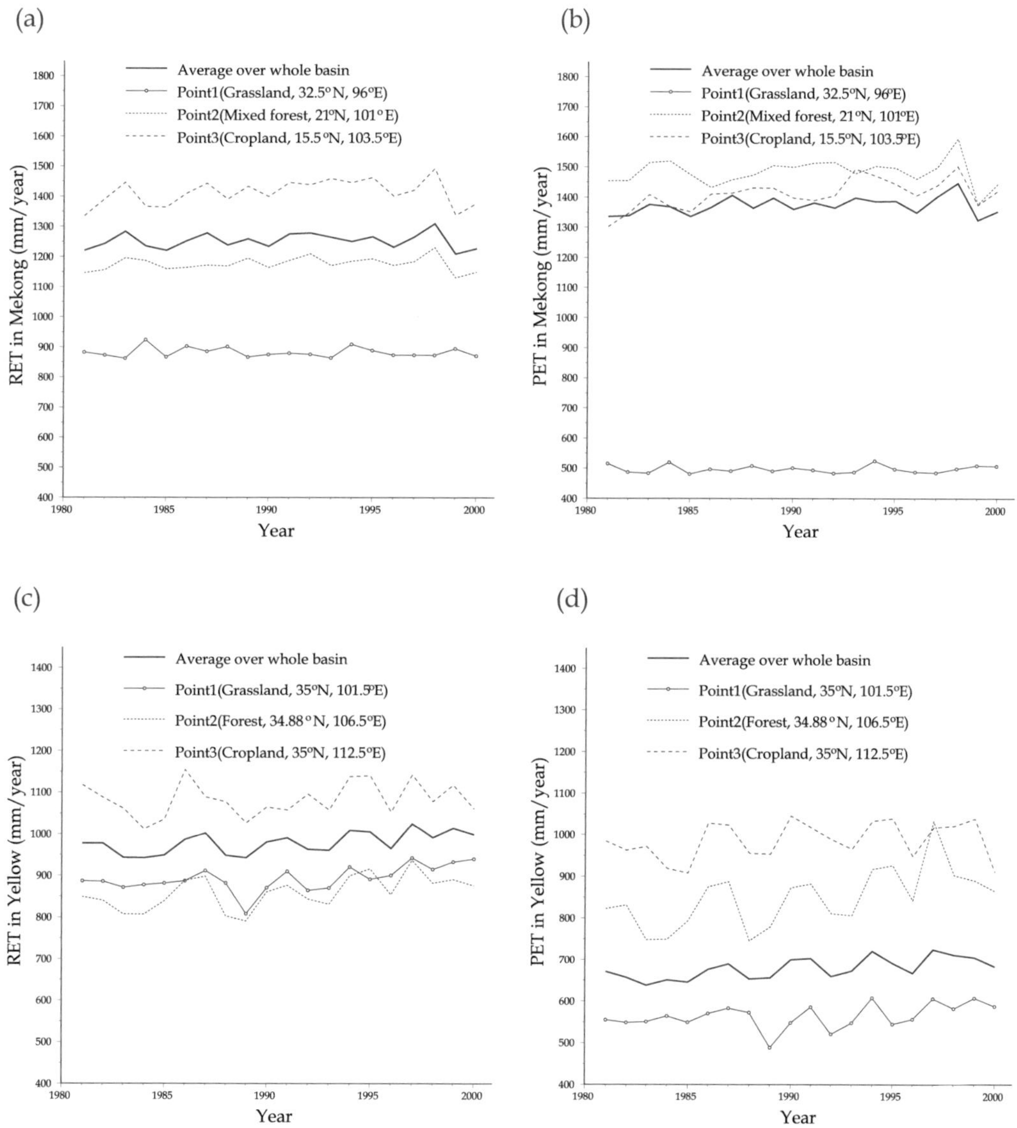

Fig. 4. Year-to-year change of annual RET and PET over whole basins and at three selected points in 1981-2000: (a) RET in Mekong; (b) PET in Mekong; (c) RET in Yellow; (d) PET in Yellow.

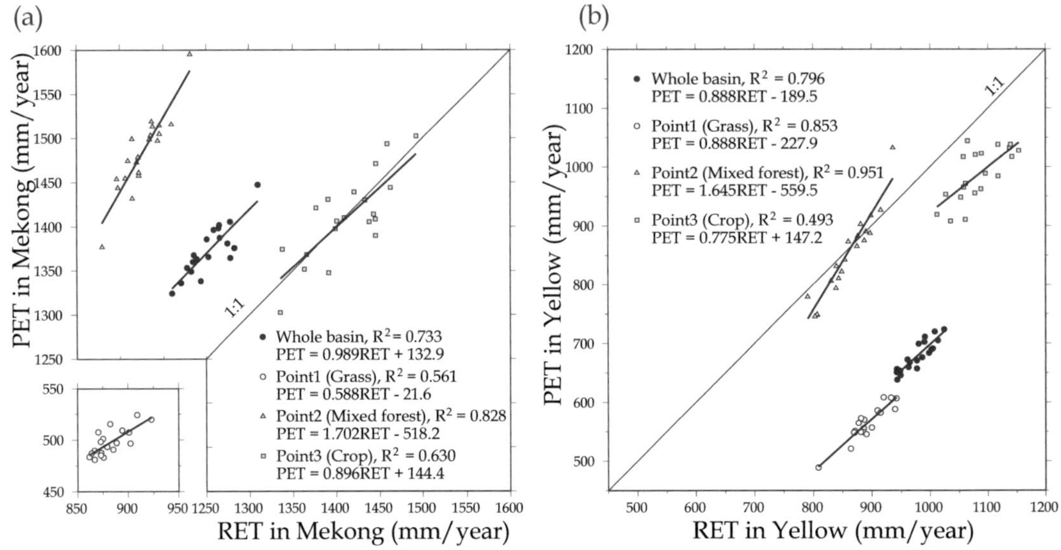

Fig. 5. Relationship between annual RET and PET over whole basins and at three selected points in 1981-2000: (a) RET vs. PET in Mekong; (b) RET vs. PET in Yellow.

5.2 Potential evapotranspiration for vegetation interception

By the S-W model, the potential evapotranspiration from interception, PET_0, is estimated with $r_s^c = r_s^s = 0$. In Mekong River basin, the ratio of PET_0/PET is 1.63 in average, with 2.90 for grassland, 1.74 for forests and 1.32 for cropland. PET_0 is much higher than the interception storage capacity, I_{max}. For example, PET_0 is about 7.3 mm day^{-1} in average in Mekong River basin but I_{max} is only 1.4 mm from $I_{max} = C_{int}$ LAI when 7.0 is used for LAI_{max} with the forests (see Table 1) where C_{int} is the interception coefficient and C_{int} = 0.2 mm is used (Dickinson, 1984; De Ridder and Schayes, 1997; Vazquez and Feyen, 2003; Mo et al., 2004), assuming one rainfall occurs one day. The interception storage controls its actual evaporation. In Yellow River basin, the ratio of PET_0/PET is 2.65 in average (much larger than in Mekong River basin because of the poor vegetation in favor of PET_0 but not PET relatively), with 2.69 for grassland, 3.40 for forests and 1.98 for cropland.

5.3 Implication of cyclical S-W estimates over Yellow River basin

PET at the forest and crop points in the Yellow River basin changes cyclically (Figure 4(d)), driven by the seasonal precipitation effect on the vegetation development. To show this, the relationship between seasonal vegetation LAI and seasonal precipitation is analyzed over a large area (i.e. the whole Loess Plateau in the middle reaches of Yellow River, gray shaded area in Figure 8) in order to avoid that the point vegetation is easily affected by local and surrounding factors (e.g. runoff from neighbouring grids). The forest point is located inside the Loess Plateau and the crop point close to the Loess Plateau. The vegetation LAI over the Loess Plateau in the middle Yellow River is estimated from NDVI data in Equations from (45) to (47). The regional average precipitation is calculated from the weather station observation (black points in the gray shaded area of Figure 8) using the Thiessen polygon method. As shown in Figure 9(c), the point vegetation LAI in warm and wet seasons (LAI_{wet}) changes similarly to the Loess Plateau vegetation LAI_{wet} but not exactly and with

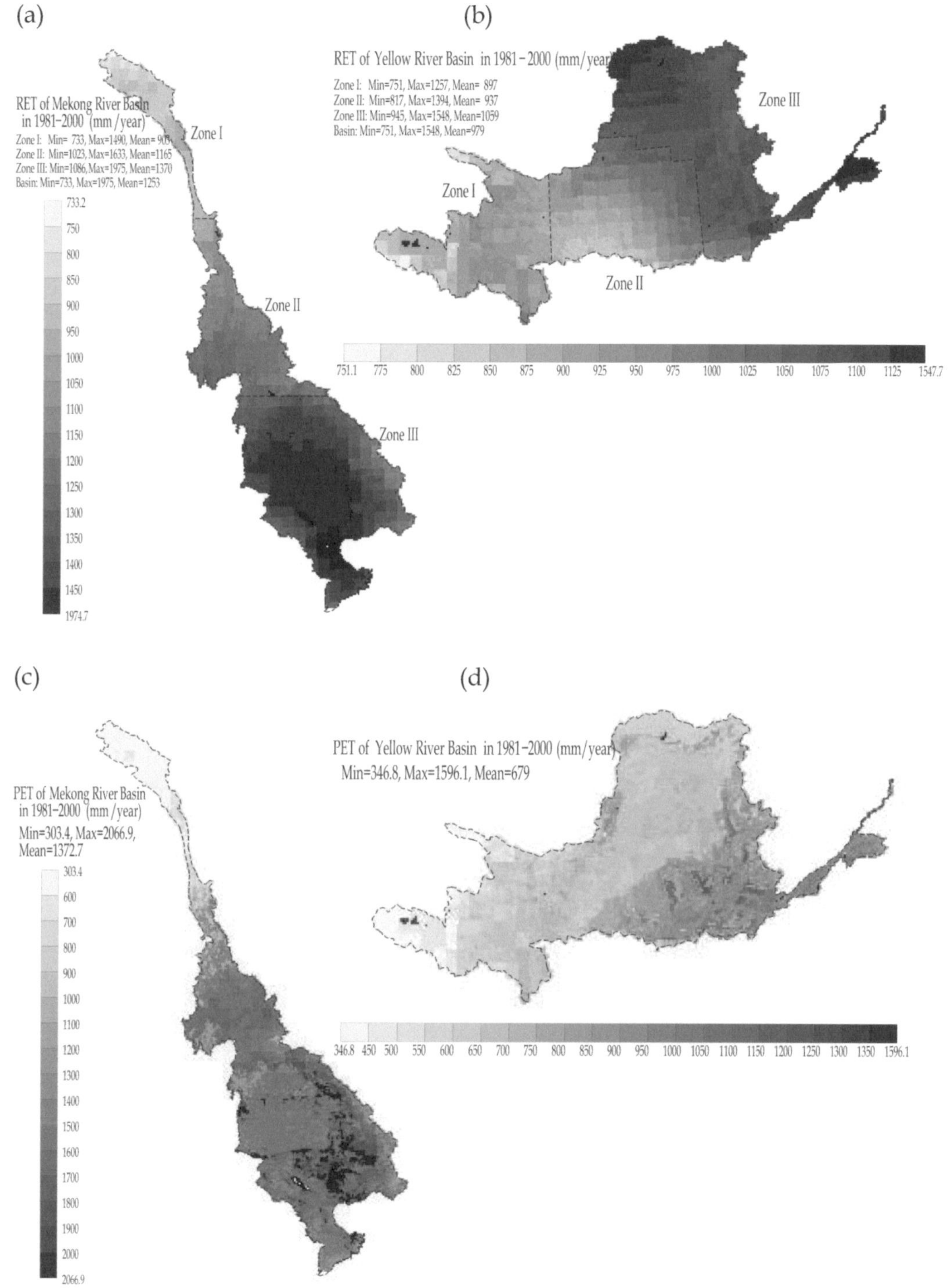

Fig. 6. Spatial distribution of annual RET and PET averaged in 1981-2000: (a) RET in Mekong; (b) RET in Yellow; (c) PET in Mekong; (d) PET in Yellow.

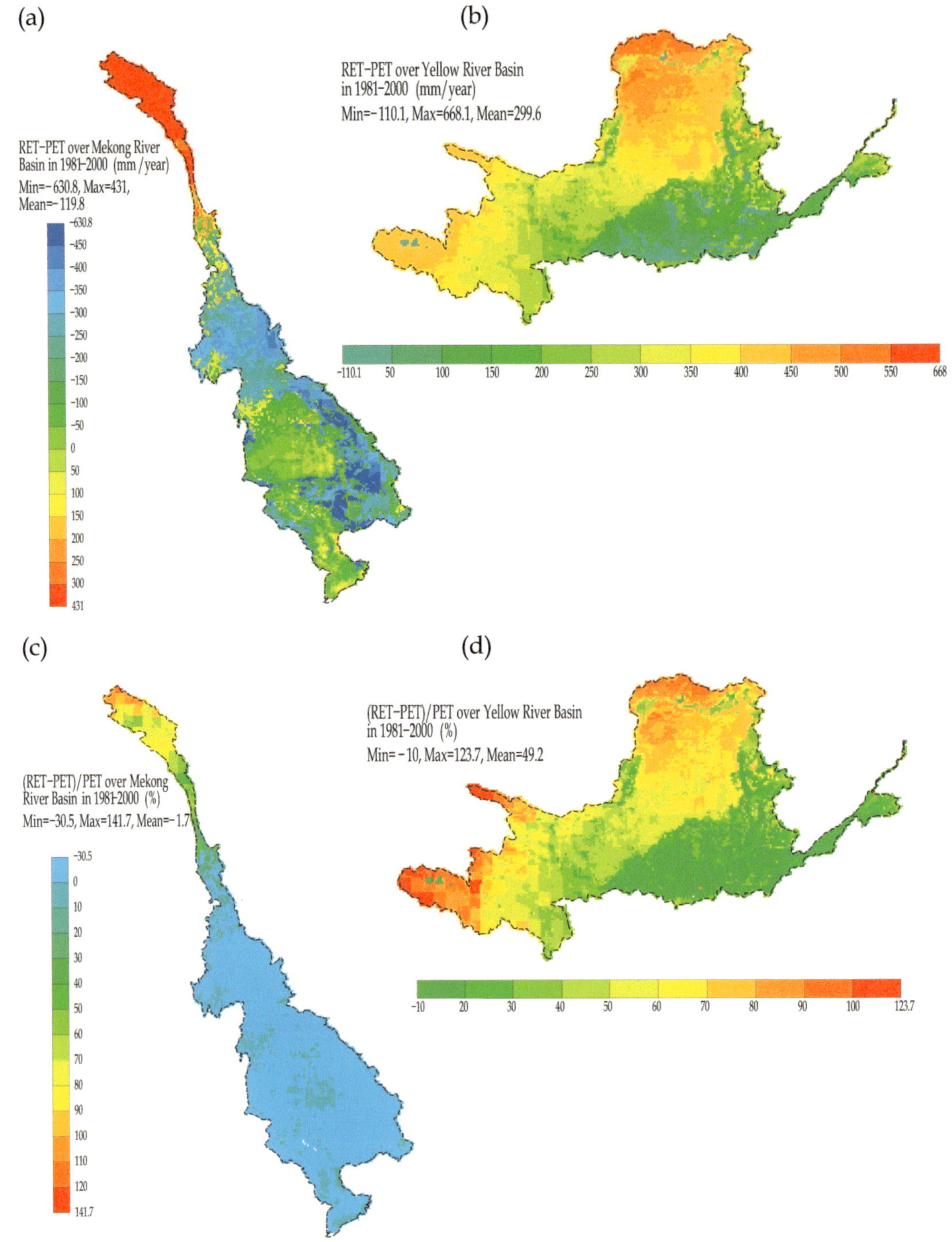

Fig. 7. Spatial distribution of the difference between annual RET and PET averaged in 1981-2000: (a) RET-PET in Mekong; (b) RET-PET in Yellow; (c) (RET-PET)/PET in Mekong; (d) (RET-PET)/PET in Yellow.

large amplitude because the points are located in the south of the Plateau (more precipitation). Comparing LAI_{wet} of the Loess Plateau in Figure 9(c) with the precipitation in Figures from 9(d) to 9(f), the vegetation LAI is not much related to the precipitation in the same period (P_{wet}) but well related to the antecedent precipitation in cold and dry season (P_{dry}). The cold and dry season refers to October in previous year to April the following year while the warm and wet season is from May to September. The regression R^2 is 0.5157 for LAI_{wet} vs. P_{dry} (Figure 9(a)) but 0.0228 for LAI_{wet} vs. P_{wet} (Figure 9(b)). Comparing Figure 9(c) with Figure 9(d), LAI_{wet} and P_{dry} have similar change trends, but LAI_{wet} and P_{wet} (or P_{wet}+P_{dry}) do not have this close relationship (Figure 9(c) vs. Figure 9(e), or Figure 9(c) vs. Figure 9(f)). This kind relationship can only be explained that: due to small potential evapotranspiration in the dry and cold season, in the form of light rainfall or snowfall, most of precipitation infiltrates to increase soil water in favor of root growing, seed germination and new bud development in spring; whereas in the warm and rainy season, because the rainfall intensity is usually heavy and the loess soil is often crusted, most of the precipitation loads flushes into the rivers and the little infiltration is consumed quickly by the strongly potential evapotranspiration. Table 3 shows that: when the years were separated into two periods, 1980's and 1990's, in average their LAI_{wet} has no difference because their P_{dry} is almost the same although their P_{wet} is different. Now we see these specific years with dry winter or wet winter. Figures 9(c) and 9(d) and Table 3 show that: in the years of 1982, 1985, 1999 and 2000, a dry antecedent winter experienced with P_{dry} less than 80 mm, their LAI_{wet} is at a very low level; whereas in the years of 1984, 1990, 1994 and 1998, a wet antecedent winter experienced with P_{dry} larger than 110 mm, their LAI_{wet} is at a high level. Precipitation in winter (P_{dry}) primarily controls vegetation condition in summer. Precipitation in summer (P_{wet}) is a secondary controlling factor, e.g. in 1985 LAI_{wet} being a bit higher due to high P_{wet} even though low P_{dry} in this dry-winter year, and in 1994 LAI_{wet} being a bit lower due to low P_{wet} even though high P_{dry} in this wet-winter year. Under a given forcing meteorological setting, high LAI means high PET. Over the middle Yellow River basin, only in summer vegetation LAI is at a bit high level. In winter, it is at a very low level.

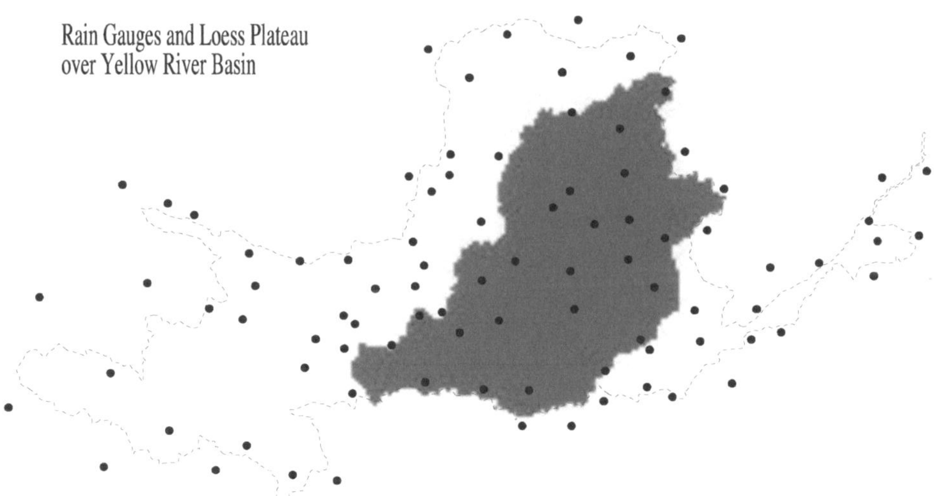

Fig. 8. Loess Plateau in the middle reaches of Yellow River (gray shaded) and rain gauges (black points)

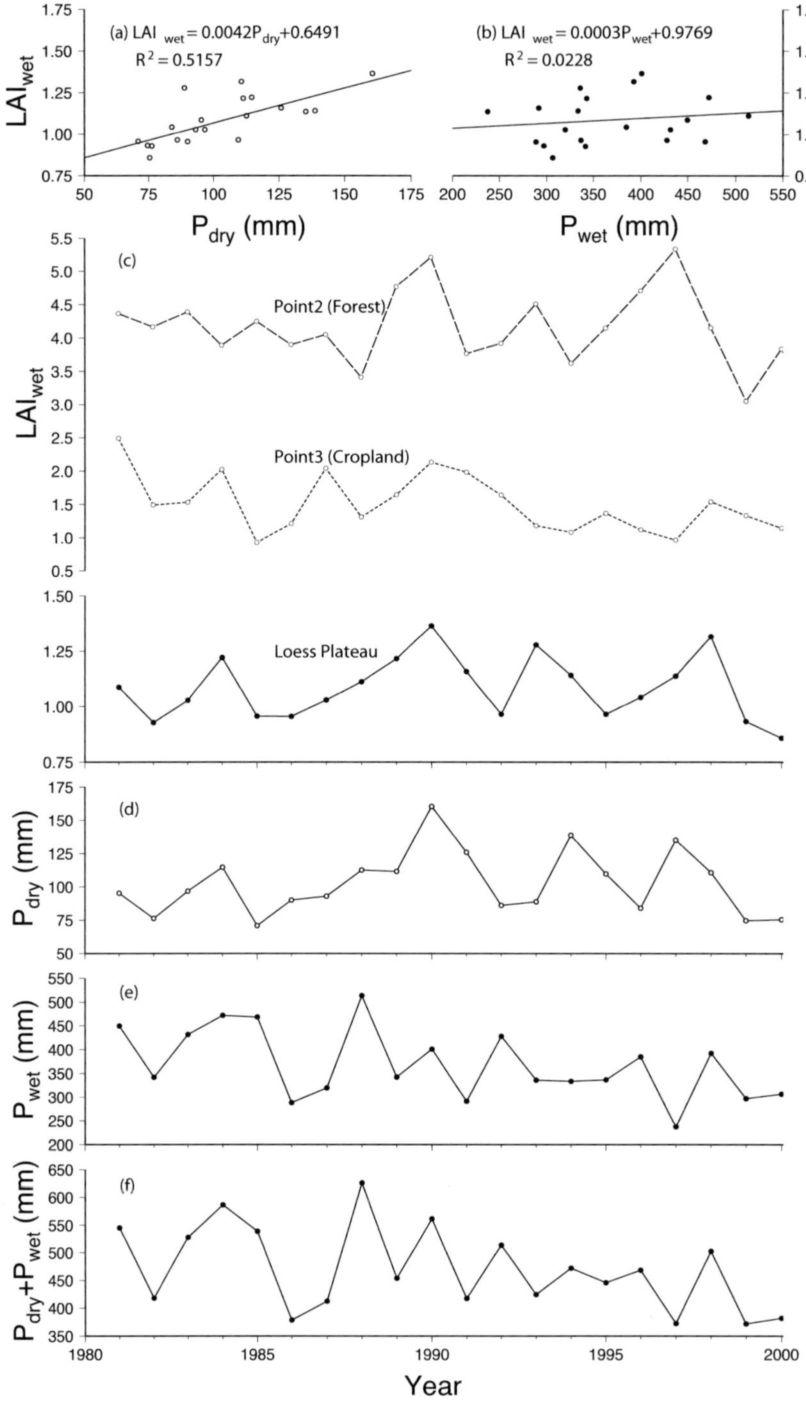

Fig. 9. Vegetation LAI_{wet} and precipitation during 1981-2000 over Loess Plateau: (a) regression relationship between LAI_{wet} and P_{dry}; (b) regression relationship between LAI_{wet} and P_{wet}; (c) LAI_{wet}; (d) P_{dry}; (e) P_{wet}; (f) $P_{dry}+P_{wet}$.

Year	LAI_{wet}	P_{dry} (mm)	P_{wet} (mm)
1981-1990	1.09	102.1	402.8
1991-2000	1.08	102.9	334.3
1982 (dry winter)	0.928	76.2	341.6
1985 (dry winter)	0.956	70.9	468.3
1999 (dry winter)	0.932	74.6	297.2
2000 (dry winter)	0.858	75.4	306.6
1984 (wet winter)	1.221	114.6	472.0
1990 (wet winter)	1.365	160.4	401.0
1994 (wet winter)	1.141	138.7	333.3
1998 (wet winter)	1.316	110.7	392.5

Table 3. Seasonal vegetation LAI and precipitation over Loess Plateau

5.4 Comparison between PET, RET and pan evaporation over Mekong River basin

The pan evaporation, E_{pan}, is available at 80 observation sites (8 in Laos, 58 in Thailand, 9 in Cambodia and 5 in Vietnam) from the Lower Mekong Hydrologic Yearbook (Interim Committee for Investigations of the Lower Mekong Basin, from 1964 to 1988). The pan type is Class A.

Figure 10 shows the comparison between annual PET, RET and E_{pan} where average monthly NDVI in period from 1981 to 2001 are used for estimates of PET and RET before 1981. As expectation, for both PET and RET, the points deviate from the 1:1 line when compared with E_{pan} (Figures 10(a) and 10(b)). Particularly for PET, the points are more scattered in total but also more aggregated with the vegetation types. This is because that the S-W method, different from the FAO-56, not only uses the climatic data but also considers the vegetation types. The less but some aggregation of the points with the vegetation type in Figure 10(b) is the result of their close location. For example, most of the forest and savannas points are located between 17.5°-20.5°N and therefore more aggregated in Figure 10(b) than the crop points which are located in a much large area. The 1:1 line seems go through the center of water surface points in Figures 10(a) and 10(b) but not representative with only one observation site (Sakon Nakhon in Thailand). The scatter is primary, especially along E_{pan} axis, and aggregation is secondary in Figures 10(a) and 10(b). It is difficult to fit a relationship between PET and E_{pan} or between RET and E_{pan} because PET or RET is an area estimate while E_{pan} is a point observation which is seriously affected by local micro-climate (e.g. often over a patch of lawn), the pan installation, its operation and maintenance. Even though these, at the crop points, PET and RET are estimated about 0.806 and 0.771 of E_{pan} respectively in average, larger than 0.7 reported by Beven (2001).

In contrast with E_{pan}, PET and RET have a good relationship, as show in Figure 10(c). The PET is larger at both forest and crop points but smaller at the savannas than RET, mainly due to the difference of their albedo (Table 1), apart due to their canopy structure (e.g. the vegetation heights and LAI). Over water surface, the points are parallel to the 1:1 line. The very small difference is resulted from the use of different equations to convert the wind speed to 2 m height.

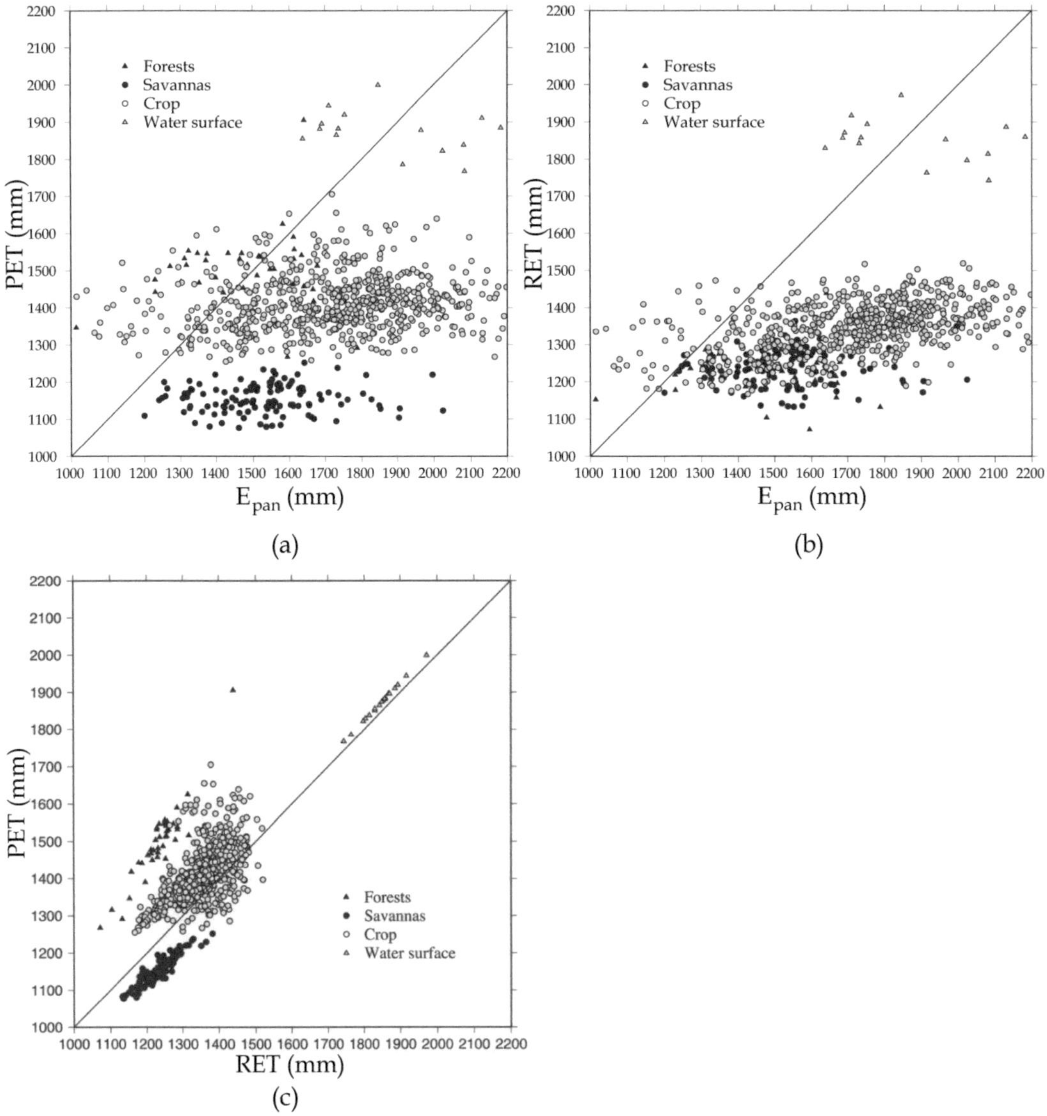

Fig. 10. Comparison among PET, RET and E_{pan} over Mekong River basin: (a) PET vs. E_{pan}, (b) RET vs. E_{pan}, and (c) PET vs. RET

5.5 Empirical relationship between FAO-56 P-M and S-W

The facts that use of the FAO-56 method is simple, and that there is a linearity between annual RET and PET over the basins and at the selected points encourage exploration for a similar relationship between monthly RET and PET using a more robust statistical analysis. In addition to the vegetation type, the LAI is the only factor used in the S-W model but not used in the FAO-56 method. Therefore, a multiple regression analysis of SPSS software was made by categorizing the vegetation types. Because the power function was found to have good R^2 for all land covers and it satisfies PET = 0 when RET = 0, the dependent variable was used as ln(PET) and independent variables were used as IGBP land cover, ln(RET), LAI and CRU climatic variables. In order to reduce data amount and not lose their statistical attributes, not all cell data of these variables were used but randomly sampled for regression analysis over the individual basins. Of all climatic variables, only use of month precipitation and daily mean temperature in the Mekong River basin, and only use of daily mean

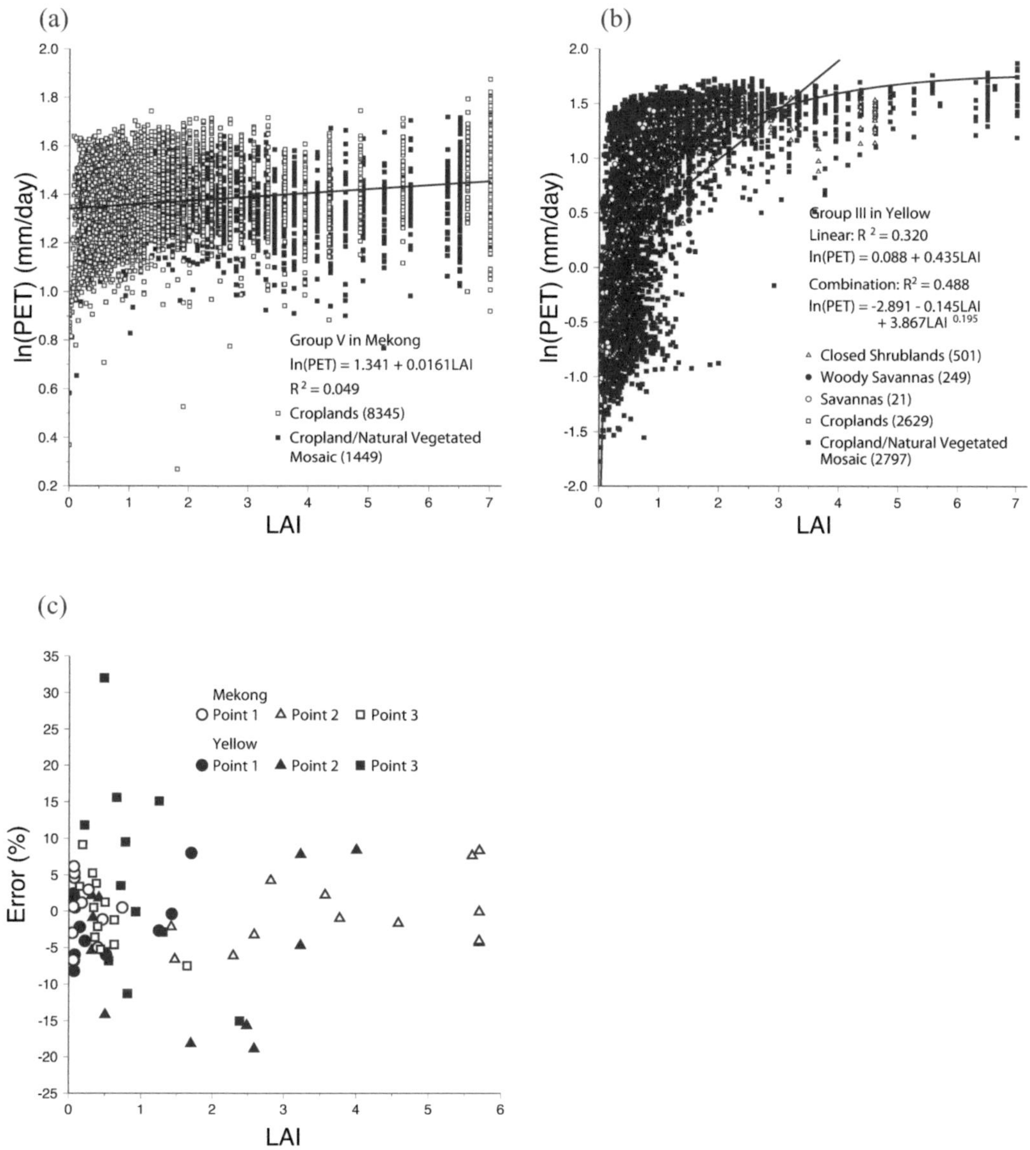

Fig. 11. Scatter plot between logarithmic PET and LAI (a) for vegetation Group V in Mekong; (b) for vegetation Group III in Yellow; and (c) prediction errors of multiple regression on three selected points corresponding to LAI.

temperature in the Yellow River basin improved the value of R^2 for some vegetation groups in the regression noticeably. Other variables gave only marginal improvement. A linear relationship seems able to describe the trend between LAI and ln(PET) in the Mekong River basin, e.g. in the case of vegetation Group V as shown in Figure 11(a). In the Yellow River basin, this relationship seems to be a combination of linear and power functions of LAI, e.g. in the case of vegetation Group III as shown in Figure 11(b). The numbers in the legend brackets in Figure 11 are the data amount sampled in the individual basins during 1981-2000 (also see "Samples" column for the same vegetation groups in Table 4). Therefore, the regression was made by further incorporating LAI in the Mekong and a combination of LAI and LAI^c in the Yellow, where c was determined by applying the SPSS "Nonlinear Regression" procedure with a function of $\ln(PET) = a + b \times LAI^c$ for each group. The multiple regression results were listed in Table 4. For the three points considered in the individual basins, the prediction errors, equal to {(PET predicted from the regression relationship between PET vs. RET and/or other variables) – (PET estimated using the S-W model)}/(PET estimated using the S-W model), were shown in Figure 11(c) for January to December in randomly selected years between 1981 and 2000 when NDVI data are available, to avoid the figure become a mess with too many data. Generally, by incorporating LAI, the quality of the prediction is improved to some extent, but quite large errors persist which are unrelated to LAI. We are therefore forced to the conclusion that no accurate statistical relationship exists between PET and RET because we were unable to fit a good function to account for the strongly nonlinear relationship between PET and LAI.

Category	Land cover (IGBP code)	Samples	ln(PET) = b0 + b1 × ln(RET) + b2 × Var + b3 × LAI						
			Var	b0	b1	b2	b3	R^2	σ
Group I	Evergreen Needleleaf Forest (2)	4811	P	0.03504	1.115	-3.185 E-04	0.02318	0.855	0.089
	Deciduous Broafleaf Forest (4)	1359							
	Mixed Forest (5)	3077							
Group II	Closed Shrublands (6)	536	T	-0.363	0.759	0.02186	0.06151	0.852	0.127
Group III	Woody Savannas (8)	613	No	-0.0416	0.879		0.125	0.921	0.042
Group IV	Open Shrublands (7)	123	T	-0.328	0.658	0.03964	0.123	0.989	0.061
	Grasslands (10)	2456							
	Barren or Sparsely Vegetated (16)	2							
Group V	Croplands (12)	8345	No	0.233	0.811		0.0351	0.782	0.063
	Cropland/Natural Vegetation Mosaic (14)	1449							
Group VI	Water Bodies (17)	229	No	0.002	1.007			0.999	0.002

P: monthly precipitation (mm $month^{-1}$); T: daily mean temperature (°C); No: no climate variable significantly involved, i.e. not to improve R^2 obviously; σ: standard error (mm day^{-1}).

(a) Mekong River basin

Category	Land cover (IGBP code)	Samples	$\ln(PET) = b0 + b1 \times \ln(RET) + b2 \times T + b3 \times LAI + b4 \times LAI^c$							
			b0	b1	b2	b3	b4	c	R^2	σ
Group I	Deciduous Needleleaf Forest (2)	2	-0.955	0.802	0.0453	0.0230	-0.457	0.220	0.977	0.136
	Deciduous Broafleaf Forest (4)	462								
Group II	Mixed Forest (5)	182	-0.634	0.907	0.0397	0.0328	-0.187	0.379	0.976	0.130
Group III	Closed Shrublands (6)	501	-1.142	0.753	0.0368	0.0087	-0.668	0.195	0.979	0.120
	Woody Savannas (8)	249								
	Savannas (9)	21								
	Croplands (12)	2629								
	Cropland/Natural Vegetation Mosaic (14)	2797								
Group IV	Open Shrublands (7)	4229	-0.908	0.820	0.0232	0.192	-0.473	0.152	0.980	0.107
	Grasslands (10)	11735								
	Barren or Sparsely Vegetated (16)	110								
	Urban and Built-up (13)	7								
Group V	Water Bodies (17)	76	0.048	0.986					1.000	0.006

(b) Yellow River basin

Table 4. Multiple regression between PET (mm day^{-1}), RET (mm day^{-1}), climatic variables and LAI

5.6 Limited validation of S-W estimates with available field data

In the O Thom I watershed (area 137 km^2, elevation from 46 to 273 m, located at 105°28′E, 12°44′N) of Mekong River basin in Kompong Thom province of central Cambodia, Nobuhiro et al. (2008) measured the evapotranspiration during two distinct sampling periods in 2003 and 2004. The vegetation type is a relatively undisturbed evergreen broadleaf forest (mean tree height in the upper crown layer 27.2 m, maximum tree height 45.1 m). For the field measurement, they established a 60 m high meteorological observation tower. By using the heat balance method (incorporating the Bowen ratio) (Httori, 1985), the daily evapotranspiration levels were estimated to be 4.3 (minimum 3.0 and maximum 5.5) mm day^{-1} during 20-28 October 2003 (late rainy season), and 4.6 (minimum 2.0 and maximum 5.7) mm day^{-1} during 1-9 March 2004 (middle dry season). Arguably, the average of the two values is used as a level indicator of annual evapotranspiration, equal to 4.45 mm day^{-1}. During the observation, the water table measured shallow at a depth less than 1 m. The soil moisture was very wet. These values of measured evapotranspiration can be thought to represent the potential one. At the same site, IGBP land cover is also evergreen broadleaf forest, both in original 1-km resolution version and in the aggregated 8-km resolution version. Because field observation items by Nobuhiro et al. (2008) were not enough to apply the FAO-56 method and the S-W model, here RET and PET estimated using the data described in the section of "DATA SOURCES" are compared, averaged in the period 1981-2000. During this period, average PET was estimated to be 7.2 (from 6.3 to 7.9) mm day^{-1} in March, 4.3 (from 3.7 to 5.1) mm day^{-1} in October, and 5.3 (from 5.1 to 5.6) mm day^{-1} for the whole year. Average RET was estimated to be 5.0 (from 4.5 to 5.4) in March, 3.4 (from 3.1 to 3.8) mm day^{-1} in October, and 3.94 (from 3.78 to 4.12) mm day^{-1} for the whole year. Obviously, comparing to the estimates of RET, which are a lot lower than the field measurement of the evapotranspiration due to the grass hypothetical crop in FAO-56 method, the estimates of PET are closer to the field measurement. The Yellow River basin is located in a semi-arid region. It is difficult to find a watershed where the LAI of vegetation cover is at a high level and soil moisture approaches field capacity for a period long enough to carry out an experiment for potential evapotranspiration measurement. Even though there is a heavy rainfall (high intensity in a short time), which happens in summer (see Figures 9(d) and 9(e), Table 3), most of it flushes into the rivers as a surface flow and the very little infiltration (due to the crusted soil surface) would be exhausted quickly by the high potential evapotranspiration. However, at the crop point (Point 3), the largest annual PET estimated from the S-W model corresponds to the most severe drought year in the last century (OSFCDRH and NWRHI, 1997).

Another way to validate the estimate of PET is applying the water balance equation to a wet basin, i.e. $\overline{ET} = \overline{P} - \overline{Q}$, where $\overline{P}$ is the precipitation, $\overline{Q}$ is the stream flow, $\overline{ET}$ is the actual evapotransipiration, all basin-averaged annual values (Donohue et al., 2007). When the basin is wetted constantly and spatially, the environment is energy-limited and $\overline{ET}$ can be a represent of PET. However, our database is not enough to support this analysis.

6. Conclusions

This chapter provides realistic estimates of potential evapotranspiration as inputs to drive hydrological modeling of large river basins, in particular for the poorly monitored or ungauged regions. For this, two extensions of P-M equation, FAO-56 method and S-W model are comparatively investigated over the Mekong and Yellow River basins, representing the humid and semi-arid Asian monsoon regions.

a. Although both are extensions of P-M equation, they use different assumptions, parameterization, and data sources to simulate different evaporation mechanisms. The estimates of RET and PET over the Mekong and Yellow River basins are spatially very different. The estimate of FAO-56 P-M is a good integrated climatic index which is able to reflect the temporal changes and spatial distribution of climate across the basins, and S-W estimate reflects the climate variability and the vegetation distribution and development.
b. Preliminary investigation into the relationship between FAO-56 P-M and S-W estimates suggest that investigating an empirical relationship between them may be worthwhile. Several attempts were made to do this, but ultimately doing so proved illusive. Large predictive errors exist because of the strong nonlinearity and scatter between PET and the LAI of the vegetation.
c. Available relevant field data are very scarce, and the only available data suggests the S-W estimate may be more realistic.

Consequently, use of the S-W model, albeit more complex, is recommended because of its more robust physical basis and because it successfully accounts for the effect of changing land surface conditions on PET. The estimated PET derived in this manner can be used to

provide a direct input to hydrological models without the need to use empirical pan or crop coefficients, which would be required if pan evaporation or RET were used.
Another finding is that: vegetation condition in summer is primarily controlled by the regional antecedent precipitation in the preceeding cold and dry seasons over the Loess Plateau in the middle reaches of the Yellow River.

Appendix A. Solar radiation

The extraterrestrial radiation is estimated from the solar constant and declination, which are the functions of the location latitude and the date in the year, expressed as follows (Shuttleworth, 1993; Allen et al., 1998).

Inverse relative distance Earth-Sun:

$$d_r = 1 + 0.033\cos\left(\frac{2\pi}{365}J\right) \tag{A1}$$

where J is Julian day in the year, corresponding to the middle day of the interval.

Solar declination:

$$\delta = 0.409\sin\left(\frac{2\pi}{365}J - 1.39\right) \text{ in rad} \tag{A2}$$

Sunset hour angle:

$$\omega_s = \arccos\left[-\tan(\varphi)\tan(\delta)\right] \text{ in rad} \tag{A3}$$

where φ is the location latitude in radians.

Extraterrestrial radiation in a day:

$$R_a = \frac{24(60)}{\pi} G_{sc} d_r \left[\omega_s \sin(\varphi)\sin(\delta) + \cos(\varphi)\cos(\delta)\sin(\omega_s)\right] \tag{A4}$$

where R_a is in MJ m^{-2} day^{-1} and G_{sc} is the solar constant, G_{sc} = 0.082 MJ m^{-2} min^{-1}.

Solar radiation over vegetation canopy:

$$R_{solar} = \left(a_s + b_s \frac{n}{N}\right) R_a \tag{A5}$$

where n/N is calculated using Equation (50), the parameters a_s and b_s are set to 0.25 and 0.50 respectively when their measurements are lack.
The clear-sky solar radiation is calculated as:

$$R_{solar}^{0} = \left(0.75 + 2 \times 10^{-5} z\right) R_a \tag{A6}$$

where z is the elevation above sea level (m)

7. Acknowledgements

We are pleased to acknowledge the financial support of the President Scientific Funds of South China Agricultural University through the research project 'Development of a distributed watershed hydrological model and integrated management of water resources' (7600-K07050), and of the Department of Education of Guangdong Province through the research project 'Non-point source pollution of Hanjiang River basin and the distributed hydrological and environmental simulation' (2008-86), and of the Scientific Innovation Funds of Department of Water Resources of Guangdong Province through the research project 'Application research of distributed hydrological model and data assimilation in Hanjiang River water resources and water quality' (2009).

8. References

Allen, R. G., Pereira, L. S., Raes, D. and Smith, M. (1998) Crop evapotranspiration – guidelines for computing crop water requirements. FAO Irrigation and Drainage Paper, No. 56, FAO, Rome.

Allen, R. G., Smith, M., Perrier, A. and Pereira, L. S. (1993) Updated reference evapotranspiration definition and calculation procedures, Revision of FAO Methodologies for Crop Water Requirements. 36 pp.

Allen, S. J. (1994) An update for calculation of the reference evapotranspiration. ICID Bull., 43(2), 35-91.

Andersen, J., Dybkjaer, G., Jensen, K. H., Refsgaard, J. C. and Rasmussen, K. (2002) Use of remotely sensed precipitation and leaf area index in a distributed hydrological model. J. Hydrol., 264, 34-50.

Beven, K. J. (2001) Rainfall-Runoff Modelling, the Primer. John Wiley, Chichester (England), 360 pp.

Braud, I., Dantas-Antonino, A. C., Vauclin, M., Thony, J. L. and Ruelle, P. (1995) A simple soil-plant-atmosphere transfer model (SiSPAT) development and field verification. J. Hydrol., 166, 213-250.

Brisson, N., Itier, B., L'Hotel, J. C. and Lorendeau, J. Y. (1998) Parameterisation of the Shuttleworth-Wallace model to estimate daily maximum transpiration for use in crop models. Ecol. Model., 107, 159-169.

Brutsaert, W. (1982) Evaporation into the Atmosphere. D. Reidel, Dordrecht, Holland, 299 pp.

Calder, I. R., Narayanswamy, M. N., Sirinivasalu, N. V., Darling, W. G. and Lardner, A. J. (1986) Investigation into the use of deuterium as a tracer for measuring transpiration from eucalyptus. J. Hydol., 84, 345-351.

Camillo, P. J. and Gurney, R. J. (1986) A resistance parameter for bare soil evaporation models. Soil Sci., 141, 95-106.

Choudhury, B. J. and Monteith, J. L. (1988) A four-layer model for the heat budget of homogeneous land surfaces. Quart. J. Royal Meteorol. Soc., 114, 373-398.

Danko, D. M. (1992) The digital chart of the world. GeoInfo Systems 2, 29-36.

De Ridder, K. and Schayes, G. (1997) Radiative transfer in the IAGL land surface model. J. Appl. Meteorol., 36, 12-21.

Denmead, O. T. (1976) Temperate cereals. In: J. L. Monteith (Editor), Vegetation and the Atmosphere, Vol. 2. Academic Press, 31 pp.

Dickinson, R. E. (1984) Modeling evapotranspiration for three-dimensional global climate models. In: J. E. Hanson, T. Takahashi (Editor), Climate Processes and Climate Sensitivity. Amer. Geophys. Union, 58-72.

Donohue, R. J., Roderick, M-. L. and McVicar, T. R. (2007) On the importance of including vegetation dynamics in Budyko's hydrological model. Hydrol. Earth System Sci. 11, 983-995.

Doorenbos, J. and Pruitt, W. O. (1992) Crop water requirements. FAO Irrigation and Drainage. Paper, No. 24, FAO, Rome, 144 pp.

Dunn, S. M. and Mackay, R. (1995) Spatial variation in evapotranspiration and the influence of land use on catchment hydrology. J. Hydrol., 171, 49-73.

Farahani, H. J. and Bausch, W. C. (1995) Performance of evapotranspiration models for maize - bare soil to closed canopy. Trans. ASAE, 38, 1049-1059.

Federer, C. A., Vorosmarty, C. J. and Fekete, B. (1996) Intercomparison of methods for potential evapotranspiration in regional or global water balance models. Water Resour. Res., 32, 2315-2321.

Fennessey, N. M. and Kirshen, P. H. (1994) Evaporation and evapotranspiration under climate change in New England. J. Water Resour. Plann. Manage., 120(1), 48-69.

Fuchs, M. and Tanner, C. B. (1968) Evaporation from a drying soil. J. Appl. Meteorol., 6, 852-857.

Gardiol, J. M., Serio, L. A. and Maggiora, A. I. D. (2003) Modeling evapotranspiration of corn (Zea mays) under different plant densities. J. Hydrol., 217, 188-196.

Granger, R. J. (1989) An examination of the concept of potential evaporation. J. Hydrol. 111, 9-19.

Hattori, S. (1985) Explanation on derivation process of equations to estimate evapotranspiration and problems on the application to forest stand. Bull. Forestry and Forest Products Research Institute 332, 139-165 (in Japanese).

Hess, T. M. (1998) Trends in reference evapotranspiration in the North East Arid Zone of Nigeria, 1961-91. J. Arid Envir., 38, 99-115.

Hiller, D. (1980) Applications of Soil Physics. Academic, San Diego, Clifo., 385 pp.

Huntingford, C. (1995) Non-dimensionalisation of the Penman-Monteith model. J. Hydrol., 170, 215-232.

Hutchinson, M. F. (1995) Interpolating mean rainfall using thin plate smoothing splines. Int. J. Geogr. Inform. Systems, 9, 385–403.

Interim Committee for Investigations of the Lower Mekong Basin, from 1964 to 1988.

Iritz, Z., Lindroth, A., Heikinheimo, M., Grelle, A. and Kellner, E. (1999) Test of a modified Shuttleworth-Wallace estimate of boreal forest evaporation. Agri. Forest Meteorol., 98-99, 605-619.

Jarvis, P. G. (1976) The interpretation of the variation in leaf water potential and stomatal conductance found in canopies in the field. Phil. Trans. Royal Soc. London, B 273, 593-610.

Jones, H. G. (1992) Plants and microclimate: a quantitative approach to environmental plant physiology, 2nd ed. Cambridge University Press, New York, 428 pp.

Kelliher, F. M., Leuning, R., Raupach, M. R. and Schulze, E. –D. (1995) Maximum conductances for evaporation from global vegetation types. Agri. Forest Meteorol., 73, 1-16.

Kite, G. (2001) Modeling the Mekong: hydrological simulation for environmental impact studies. J. Hydrol., 253, 1-13.

Kondo, J., Saigusa, N. and Sato, T. (1990) A parameterization of evaporation from bare soil surface. J. Appl. Meteorol., 29, 383-387.

Korner, G., Schell, J. A. and Bauer, H. (1979) Maximum leaf diffusive conductance in vascular plants. Photosynthesis, 13(1), 45-82.

Lafleur, P. M. and Rouse, W. R. (1990) Application of an energy combination model for evaporation from sparse canopies. Agri. Forest Meteorol., 49, 135-153.

Lhomme, J. -P. (1997) Towards a rational definition of potential evaporation. Hydrol. Earth System Sci., 1, 257-264.

Lhomme, J. –P., Elguero, Chehbouni, A. and Boulet, G. (1998) Stomatal control of transpiration: Examination of Monteith's formulation of canopy resistance. Water Resour. Res., 34(9), 2301-2308.

Liang, X., Lettenmaier, D. P., Wood, E. F. and Burges, S. J. (1994) A simple hydrologically based model of land surface water and energy fluxes for general circulation models. J. Geophy. Res., 99(D7), 14415-14428.

Loveland, T. R., Reed, B. C., Brown, J. F., Ohlen, D. O., Zhu, J., Yang, L., and Merchant, J. W. (2000) Development of a Global Land Cover Characteristics Database and IGBP DISCover from 1-km AVHRR Data. Int. J. Remote Sensing, 21, 1303-1330.

Lund, M. R. and Soegaard, H. (2003) Modelling of evaporation in a sparse millet crop using a two-source model including sensible heat advection within the canopy. J. Hydrol., 280, 124-144.

Mahfouf, J. F. and Noilhan, J. (1991) Comparative study of various formulations of evaporation from bare soil using in situ data. J. Clim. Appl. Meteorol., 30(9), 1354-1365.

Mauser, W. and Schadlich, S. (1998) Modelling the spatial distribution of evapotranspiration on different scales using remote sensing data. J. Hydrol., 212-213, 250-267.

McNaughton, K. G. and Black, T. A. (1973) A study of evapotranspiration from a Douglas Fir forest using the energy balance approach. Water Resour. Res., 9, 1579-1590.

McVicar, T. R., Van Niel, T. G., Li, L. T., Hutchinson, M. F., Mu, X. M. and Liu, Z. H. (2007) Spatially Distributing Monthly Reference Evapotranspiration and Pan Evaporation Considering Topographic Influences. J. Hydrol., 338, 196-220.

McVicar, T. R., Van Niel, T. G., Li, L. T., Roderick, M. L., Rayner, D. P., Ricciardulli, L. and Donohue, R. J. (2008) Wind speed climatology and trends for Australia, 1975-2006: Capturing the stilling phenomenon and comparison with near-surface reanalysis output. Geophy. Res. Letters, 35, L20403.

Mo, X., Liu, S., Lin, Z. and Zhao, W. (2004) Simulating temporal and spatial variation of evapotranspiration over the Lushi basin. J. Hydrol., 285, 125-142.

Monteith, J. L. (1965) Evaporation and environment. Symp. Soc. Exp. Bio. XIX, 205-234, Cambridge University Press.

Monteith, J. L. (1973) Principles of environmental physics. Edward Arnold, London, 214 pp.

Monteith, J. L. (1981) Evaporation and surface temperature. Quart. J. Royal Meteorol. Soc., 107, 1-27.

Monteith, J. L. (1995) Accommodation between transpiring vegetation and the convective boundary layer. J. Hydrol., 166(3-4), 251-263.

Myneni, R. B. and Williams, D. L. (1994) On the relationship between FPAR and NDVI. Remote Sensing Environ., 49, 200-211.

New, M., Hulme, M. and Jones, P. (1999) Representing twentieth-century space-time climate variability. Part I: Development of a 1961-90 mean monthly terrestrial climatology. J. Climate, 12, 829-856.

New, M., Hulme, M. and Jones, P. (2000) Representing twentieth-century space-time climate variability. Part II: Development of 1901-96 monthly grids of terrestrial surface climate. J. Climate, 13, 2217-2238.

Nobuhiro, T., Shimizu, A., Kabeya, N., Tamai, K., Ito, E., Araki, M., Kubota, T., Tsuboyama, Y. and Chann, S. (2008) Evapotranspiration during the late rainy season and middle of the dry season in the watershed of an evergreen forest area, central Cambodia. Hydrol. Process., 22(9), 1281-1289.

Noilhan, J. and Planton, S. (1989) A simple parameterization of land surface process for meteorological models. Mon. Weather Rev., 17, 536-549.

OSFCDRH (Office of State Flood Control and Drought Relief Headquarters) and NWRHI (Nanjing Water Resources & Hydrology Institute) (1997) China's Flood and Drought Disaster. Water Resources and Hydropower Press, Bejing (in Chinese).

Penman, H. L. (1948) Natural evaporation from open water, bare soil and grass. Proc. Royal Soc. London, A 193, 120-146.

Priestley, C. H. B. and Taylor, R. J. (1972) On the assessment of surface heat flux and evaporation using large-scale parameters. Monthly Weather Review, 100, 81-92.

Raupach, M. R. (1995) Vegetaion-atmosphere interaction and surface conductance at leaf, canopy and regional scales. Agri. Forest Meteorol., 73, 151-179.

Rawls, W. J. and Brakensiek, D. L. (1985) Prediction of soil water properties for hydrologic modeling. Watershed Management in the Eighties, ASCE, pp. 293-299.

Ross, J. (1975) Radiative transfer in plant communities. In: J. L. Monteith (Editor), Vegetation and the Atmosphere. Academic Press, London, pp. 13-55.

Rowntree, P. R. (1991) Atmospheric parameterization schemes for evaporation over land: basic concepts and climate modeling aspects. In: T. J. Schmugge, J. -C. Andre (Editors), Land Surface Evaporation: Measurement and Parameterization, p. 5-29.

Sellers, P. J., Los, S. O., Tucker, C. J., Iustice, C. O., Dazlich, D. A., Collatz, G. J. and Randall, D. A. (1996) A revised land surface parameterization (SiB2) for atmospheric GCMs. Part II: the generation of global fields of terrestrial biophysical parameters from satellite data. J. Climate, 9, 706-737.

Sellers, P. J., Tucker, P. J., Collatz, G. J., Los, S. O., Justice, C. O., Dazlich, D. A. and Randall, D. A. (1994) A global 1 degree by 1 degree NDVI data set for climate studies. Part 2: the generation of global fields of terrestrial biophysical parameters from NDVI. Int. J. Remote Sensing, 15(17), 3519-3545.

Shen, Y., Kondoh, A., Tang, C., Zhang, Y., Chen, J., Li, W., Sakura, Y., Liu, C., Tanaka, T. and Shimada, J. (2002) Measurement and analysis of evapotranspiration and surface conductance of a wheat canopy. Hydrol. Process., 16, 2173–2187.

Shuttleworth, W. J. (1993) Evaporation. In: D.R. Maidment (Editor), Handbook of Hydrology. McGraw-Hill, New York, pp. 4.1-4.53.

Shuttleworth, W. J. and Gurney, R. J. (1990) The theoretical relationship between foliage temperature and canopy resistance in sparse crops. Quart. J. Royal Meteorol. Soc., 116, 497-519.

Shuttleworth, W. J. and Wallace, J. S. (1985) Evaporation from sparse crops - an energy combination theory. Quart. J. Royal Meteorol. Soc., 111, 839-855.

Stannard, D. I. (1993) Comparison of Penman-Monteith, Shuttleworth-Wallace, and Modified Priestley-Taylor evapotranspiration models for wildland vegetation in semiarid rangeland. Water Resour Res., 29(5), 1379-1392.

Stewart, J. B. (1988) Modelling surface conductance of pine forest. Agri. Forest Meteorol., 43, 19-35.

Stewart, J. B. and Gay, L. W. (1989) Preliminary modeling of transpiration from the FIFE site in Kansas. Agri. Forest Meteorol., 48, 305-315.

Sun, S. F. (1982) Moisture and heat transport in a soil layer forced by atmospheric conditions. M. Sc. Thesis, Dept. of Civil Engineering, University of Connecticut, 72 pp.

Thornthwaite, C. W. (1948) An approach toward a rational classification of climate. Geograophical Rev., 38, 55-94.

Tourula, T. and Heikinheimo, M. (1998) Modelling evapotranspiration from a barley field over the growing season. Agri. Forest Meteorol., 91, 237-250.

Tucker, C. J., Pinzon, J. E., Brown, M. E., Slayback, D., Pak, E. W., Mahoney, R., Vermote, E. and El Saleous, N. (2005) An extended AVHRR 8-km NDVI data set compatible with MODIS and SPOT vegetation NDVI data. Int. J. Remote Sensing, 26(20), 4485-5598.

Uchijima, Z. (1976) Maize and Rice. In: J. L. Monteith (Editor), Vegetation and the Atmosphere, Vol. 2. Academic Press, New York, pp. 33-64.

van de Griend, A. A. and Owe, M. (1994) Bare soil surface resistance to evaporation by vapor diffusion under semiarid conditions. Water Resour. Res., 30(2), 181-188.

Vazquez, R. F. and Feyen, J. (2003) Effect of potential evapotranspiration estimates on effective parameters and performance of the MIKE SHE-code applied to a medium-size catchment. J. Hydrol., 270, 309-327

Verdin, K. L. and Greenlee, S. K. (1996) Development of continental scale digital elevation models and extraction of hydrographic features. In: Proceedings, Third International Conference/Workshop on Integrating GIS and Environmental Modeling, Santa Fe, New Mexico, January 21-26, 1996. National Center for Geographic Information and Analysis, Santa Barbara, California.

Verma, S. B., Baldocchi, D. D., Anderson, D. E., Matt, D. R. and Clement, R. J. (1986) Eddy fluxes of CO_2, water vapour, and sensible heat over a deciduous forest. Boundary-Layer Meteorol., 36, 71-91.

Vorosmarty, C. J., Federer, C. A. and Schloss, A. L. (1998) Potential evaporation functions compared on US watersheds: possible implications for global-scale water balance and terrestrial ecosystem modeling. J. Hydrol., 207, 147-169.

Vourlitis, G. L., Filho, N. P., Hayashi, M. M. S., Nogueira, J. S., Caseiro, F. T. and Campelo, J. H. (2002) Seasonal variations in the evapotranspiration of a transitional tropical forest of Mato Grosso, Brazil. Water Resour. Res., 38(6), 1094,

Wilson, R. and Shaw, R. (1977) A higher order closure model for canopy flow. J. Appl. Meteorol., 16, 1197-1205.

WRI, IUCN and IWMI, the Ramsar Convention Bureau, 2003. The Watersheds of the World_CD. World Resources Institute, Washington, DC. http://www.waterandnature.org/eatlas

SUGGESTED READINGS

Zhou, Maichun, Ishidaira, Hiroshi and Takeuchi, Kuniyoshi (2006) Estimating the potential evapotranspiration over the Yellow River basin by considering the land cover characteristics. IAHS RedBook No. 303 (Predictions in Ungauged Basins: Promises and Progress, edited by M. Sivapalan), p. 214-225, Proceedings of symposium S7 held during the Seventh IAHS Scientific Assembly at Foz do Iguassu, Brazil, April 2005.

Zhou, M. C., Ishidaira, H., Hapuarachchi, H. P., Magome, J., Kiem, A. S. and Takeuchi, K. (2006) Estimating potential evapotranspiration using Shuttleworth-Wallace model and NOAA-AVHRR NDVI data to feed a distributed hydrological model over the Mekong River Basin. Journal of Hydrology, 327, 151-173.

Zhou, M. C., Ishidaira, H. and Takeuchi, K. (2007) Estimation of potential evapotranspiration over Yellow River basin: reference crop evaporation or Shuttleworth-Wallace? Hydrological Processes, 21(14), 1860-1874.

Zhou, M. C., Ishidaira, H. and Takeuchi, K. (2008) Comparative study of potential evapotranspiration and interception evaporation by land cover over Mekong basin. Hydrological Processes, 22(9), 1290-1309.

Zhou, Maichun, Ishidaira, Hiroshi, Takeuchi, Kuniyoshi and Gao, Yongtong (2009) Evapotranspiration in the Mekong and Yellow River basins. Hydrological Sciences Journal, 54(3), 623-638.

15

Responses of Energy Budget and Evapotranspiration to Climate Change in Eastern Siberia

Hotaek Park, Takeshi Yamazaki, and Takeshi Ohta
Research Institute for Global Change, JAMSTEC
Tohoku University
Nagoya University
Japan

1. Introduction

The structure and functioning of the Arctic terrestrial ecosystems are greatly sensitive to climate change. Evidence continues to mount that warming experienced in the Arctic region during the past few decades has been affecting the structure and functioning of the Arctic terrestrial ecosystems (Oechel et al., 2000; Serreze et al., 2000). Observations indicate an increase in the number of shrubs in tundra regions (Chappin et al., 1995) and an increase in early greening of the Arctic vegetations (Buermann et al., 2003). Changes in the structure and phenology of vegetation act to modify energy and water budgets, because all components in the Arctic are interrelated through a network of linkages, feedbacks, and multi-dependent interactions. In fact, expansion of shrub cover has its own positive feedback on climate because of the lower albedo of shrubs compared to tundra, and consequently earlier snowmelt than snow-covered tundra (Chapin et al., 2005). In this manner, a change in one variable in a part of the system can initiate a cascade of regional effects and have global ramifications.

Soil moisture is the most important factor that links climate and vegetation. Climate influences the soil moisture via evapotranspiration (ET). Winter is the period of the lowest temperature in the Arctic, and hence, the low saturation of water vapour leads to less evaporation. Soil-freezing also controls plant water uptake. Snowmelt during the early spring releases the stored ice water causing higher soil moisture, subsequently increasing ET. Conversely, higher levels of ET cause a decrease in soil moisture content, resulting in soil water deficits. This soil water deficit in turn controls ET, stomatal conductance, and photosynthesis. ET rates at high-latitude are lower than at low-latitude. According to a recent review (Park et al., 2008), ET rates in eastern Siberia during the growing season were less than 3 mm day^{-1} whereas the corresponding rate at mid and low latitudes was 1–6 mm day^{-1}. The low ET rates in high latitudes can cause an extremely large seasonal variability in soil moisture. In Lena and Kolyma watersheds, for example, ET during the summer has often exceeded precipitation, resulting in a negative water balance (Park et al., 2008). The higher rates of ET during the summer subject the vegetation to frequent soil water stress. Unless the stress is settled by an increase in precipitation, summer water balances will become increasingly negative (Rouse, 2000). The active layer depth (ALD) increases during the summer. Plant roots penetrate beyond the immediate dry surface, providing a link to subsurface moisture. Therefore, the Arctic vegetation may not experience an immediate change in ET, and may ultimately experience less inter-annual variability of ET. In dry conditions, the increased ALD works as a buffer for transpiration. However, if the stage were reached when the water balance is considerably negative, this important interactive mechanism would no longer be active. In these manners, the Arctic presents a number of unique features that strongly influence their energy and water balances and their feedback to climatic and ecological processes.

Recently, changes such as early snowmelt (Nijssen et al., 2001), permafrost reduction (Jorgenson et al., 2006), and increase in runoff (Peterson et al., 2002) have been observed in the Arctic. These hydro-climatic changes influence the phenological and physiological functions of the Arctic ecosystems, thereby altering the exchange of radiation, water, and energy with the atmosphere. These changes may ultimately affect local, regional, and global climates (Bonan et al., 1992), which may in turn induce changes in the composition and distribution of vegetation. Global climate models (GCMs) also project increases in temperature and precipitation in the Arctic regions as global warming proceeds (IPCC, 2007). The increased temperature would lengthen snow-free periods, warm permafrost, and increase leaf area index (LAI), thus enhancing photosynthesis. Moreover, the high precipitation could also increase liquid water storage as the permafrost is warmed. Their interactions re-enforce both positive and negative feedback loops on ET, thus promoting ET. A summary of simulations by five GCMs indicates that ET rates from the major terrestrial Arctic watersheds will increase by a maximum of 10% in the years 2071-2090 relative to the 1981–2000 reference period (ACIA, 2005). The future projections by GCMs as well as historical field surveys (e.g., NOWES, BOREAS, NOPEX, and GAME-Siberia) are shedding new light on the interactions between the Arctic ecosystems and the climate system and suggest that the Arctic is highly sensitive to climate change. However, the issue of how the changes in air-surface energy and water exchange rates in eastern Siberia are influenced by the uncertainties of susceptibility and vulnerability of the boreal and Arctic ecosystems to future climatic changes remains unresolved. We attempt to resolve this issue by applying a land surface model (LSM) to two periods, 1986–2004 (Current) and 2081–2100 (Future). We also address the key factors affecting changes in their rates.

2. Model description

The land surface model used was developed by Yamazaki (2001), with the addition of a snow interception process (Yamazaki et al., 2007), and was then expanded to continental scale over eastern Siberia (Park et al., 2008). The LSM is a physically based process model dealing with thermal and hydrologic processes in an atmosphere-vegetation-snow-soil system. The major characteristics of LSM are briefly described. The vegetation canopy is divided into a 'crown space' and 'trunk space' (i.e., without leaves); the crown space is in turn subdivided into two layers. The energy budget is solved for the radiative and energy fluxes between the atmosphere and the two crown layers, considering both the heat storage within the canopy and the water or snow storage on the leaves. A multi-layer snow submodel is used to calculate the processes of snow accumulation and melting. The snow submodel calculates the profiles of snow temperature, density, and water content, incorporating depth hoar formed by the temperature gradient in snow layers. The model also calculates heat and water fluxes in soil layers. In permafrost regions, soil seasonally freezes and thaws. The fusion of heat in frozen soil is estimated using a method that assumes the heat capacity to be large within a small temperature range near the freezing point. The effects of soil ice content on water and energy fluxes have not yet been incorporated. The snow-free condition energy balance equation that incorporates a surface resistance of ground and understory vegetation is applied to the soil surface and forest floor. Such a model structure makes it possible to evaluate the contribution of understory ET to the total ET rate. The LSM successfully simulated the understory ET rates at the different vegetation types (Yamazaki et al., 2004) and in eastern Siberia (Park et al., 2008).

The input data for the model simulation are daily air temperatures, humidity, precipitation, solar radiation, downward longwave radiation, and wind speed. The LSM calculates hourly water and energy fluxes using several empirical models and assumptions with the daily forcing data. The method that interpolates hourly meteorological variables based on the daily ones was well described in Park et al. (2008). Leaf stomatal conductance is estimated by an empirical model (Jarvis, 1976) composed of a function of solar radiation, air temperature, vapour pressure deficit, and soil moisture. Parameter values with each function are derived by aggregating the data of all the single leaves without classification of the vegetation type.

3. Dataset preparation

An area of 40–72°N, 90–180°E was selected as the study area covering eastern Siberia, which included the Lena and Kolyma watersheds. Inhomogeneous topography and land cover may produce a very high spatial variation in meteorological variables. The climate data of 66 stations within the study area are available. The station data during 1986–2004 were interpolated to the grid of 0.5° x 0.5° using a minimum curvature technique (Park et al., 2008). To generate the data sets for predictions of climate change for Future (i.e., 2081–2100), outputs from 11 GCMs based on a scenario (i.e., SRES A1B) of IPCC (2007) were selected. The climate outputs from each GCM were linearly interpolated to 0.5° resolution for the use of LSM, and the interpolated values of 11 GCMs were averaged to all grid cells. The averaged data sets were used as the forcing data of the LSM for the future simulations.

4. Model simulation

Leaf area index is closely correlated to land surface processes. LAI is known to be a large source of uncertainty in the specification of the global time-series fields. The current LSM does not consider vegetation dynamics, implying that it cannot express the spatiotemporal variations of vegetation structure under the future climatic changes, as well as their effects on land surface processes. This can seriously affect the calculation of land surface processes. Therefore, it is assumed that the prescribed vegetation type at each grid does not change under the future climate. Instead, LSM considers the seasonal and inter-annual variations of LAI. The variations of LAI are estimated by a modified phenological model (Park et al., 2008) as

$$\mathrm{LAI} = LAI_{\max,i} \times \left(T_{\min,i} \times VPD_i \times Photo_i\right) \tag{1}$$

where $LAI_{\max,i}$ is a maximum LAI over grid i determined from June to August based on MODIS image data; T_{min}, the minimum temperature indicator; *VPD*, the vapour pressure deficit indicator; and *Photo*, the photoperiod indicator. The three indicators in eq. (1) are respectively bounded between 0 (inactive) and 1 (unconstrained) and their equations can be found in Jolly et al. (2005). The scenario SRES A1B estimates the air CO_2 concentration to reach 700 ppm in 2100. The increased CO_2 concentration will cause higher photosynthesis, so that LAI will also be increased. To reflect the effect of CO_2 concentration on LAI, $LAI_{max,i}$ in Future is assumed to 1.5 times higher than that in current. In the period of each calculation, the LAI is calculated as a 10-day moving average for all grids.

The LSM also takes into consideration eight soil types, assuming them to be constant with time. The treatment of the initial and boundary layer values is an important problem in the model simulation. In the calculation of current, the initial water content of soil layers was set to 0.3 for all grids. The soil surface and lower boundary layer temperature was set to the air temperature of the first day and annual mean air temperature, respectively. The temperature of soil layers was interpolated linearly between the surface and the lower boundary layer temperature. A spin-up computation was performed over five calculation years for the stabilization of soil temperature and water. The two modelling periods of this study are not continuous. Thus, values of individual variables at the final time step of the calculation period were used as the initial values of the next modelling period.

5. Results and discussion

5.1 Spatial fields

5.1.1 Air temperature and precipitation

The mean fields of air temperature and precipitation for the two periods over eastern Siberia are compared in Fig. 1. Air temperature exhibits considerable spatial variation between the two periods. The comparison also illustrates the significant increase in the temperature in Future across all areas of eastern Siberia, as compared to Current. In particular, it is found that the temperature increment migrates in a northward direction. In the years 1986–2004, the regions with temperature <-10°C are distributed over a wide area. However, this area is significantly decreased in Future. Above all, the increase in temperature is remarkable in regions with >140°E and >65°N, where it is classified as tundra under the current climate. In those regions, temperature in Future is increased by a minimum of 5°C as compared to Current.

The increase in temperature in the Arctic had already been addressed by previous observations and simulations (IPCC, 2007). Data coverage during the early part of the twentieth century is sparse, especially at high latitudes. During the twentieth century, however, averaged annual temperatures have exhibited an overall increase of approximately 1°C, but with large variability (Serreze and Barry, 2005). The temperature in eastern Siberia has risen by 1–2°C on an annual basis during the past few decades (Dolman et al., 2008). Moreover, the five ACIA-designated GCMs projected a warming in the annual mean air temperature of approximately 4–5 °C by 2080 (ACIA, 2005).

Indeed, precipitation exhibits similar spatial distributions in the two periods (Fig. 1). However, the absolute amount of precipitation obviously increased in Future as compared to Current. The increase was especially significant in the regions with >140°E, similar to what was observed for air temperatures. The Future precipitations in northern regions also indicated significant increases. A comparison of precipitation between Current and Future shows spatially large differences in ranges of approximately 50–200 mm.

Based on analyses from 1900 through 2002, annual precipitation, averaged over terrestrial regions over the 55–85°N band, exhibited a general increasing trend (Serreze and Barry, 2005). In general, GCMs had projected modest increases in precipitation by the end of the

21st century. The five ACIA-designated GCMs projected that precipitation in the major terrestrial Arctic watersheds will increase by approximately 10% in the years 2071-2090 relative to the 1981-2000 baseline (ACIA, 2005).

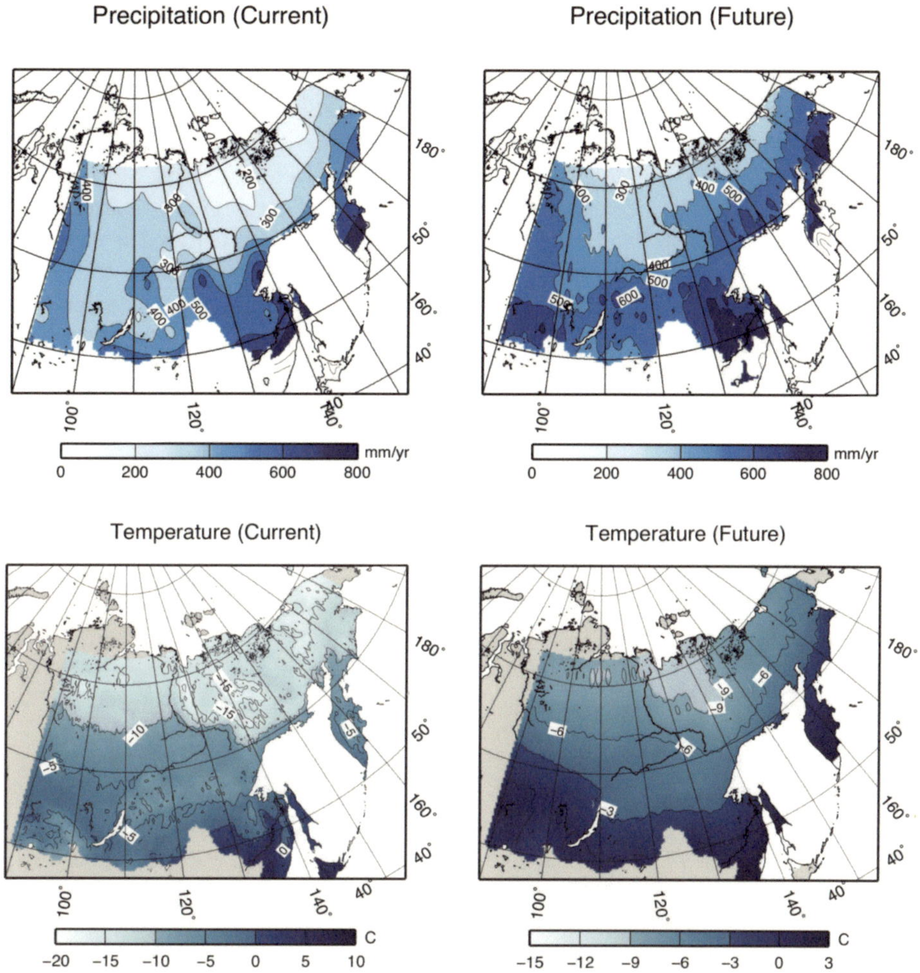

Fig. 1. Comparison of spatial distribution of mean annual precipitation (top) and mean annual air temperature (bottom) between Current (1986-2004, left-hand side panel) and Future (2081-2100, right-hand side panel).

5.1.2 Energy budget and evapotranspiration

Partitioning of net radiation (R_n) into sensible (H), latent heat (LE), and ground heat (G) fluxes reflects the different properties of vegetation and climate over the surface. Figure 2 illustrates the spatial distribution of averaged annual energy budget components in the two periods. Fig. 2 also shows the largest increase in R_n in forested area in Future as compared to Current. Consequently, in Future, LE and H also increased in the same area. LE is a major component of the energy budget during summer in Current as well as Future, especially in the boreal forest. A clear contrast in LE between the two periods exists along the boreal forest-tundra transition. In Future, the forest-tundra transition line migrates further northward than Current. This means an increase in evaporative rates in tundra regions. The increase in *LE* in tundra regions may be closely related to the increased air temperature, precipitation (Fig. 1), and LAI. Warming would cause increases in productivity and leaf areas, thus increasing *LE* (Eugster et al., 2000). Field surveys (Sturm et al., 2001) and satellite observations (Jia et al., 2003) had shown the increase in the biomass within ecosystems in tundra regions. The increased biomass, especially LAI, is a likely contributor to higher evaporative rates. LAI in Lena watershed was 2.0 in Current and 2.8 in Future.

A notable fact is that *LE* dominates the energy balance in northern regions (i.e., tundra). The absolute value of *LE* in the tundra region is lower than that in forest areas, but *LE* in Future is larger than that in Current. Relative to the differences among biomes, Arctic tundra is remarkably homogeneous in land surface parameters, with low values for canopy height, surface roughness, LAI, etc. However, measurements showed that significant differences exist among tundra ecosystem types in surface energy partitioning and moisture exchange (Eugster et al., 2000). Therefore, climatically induced changes in ecosystem structures can greatly affect regional energy exchanges. This is because during summer, large areas of the Arctic tundra and boreal zone are actually heat and moisture sources rather than sink deposits.

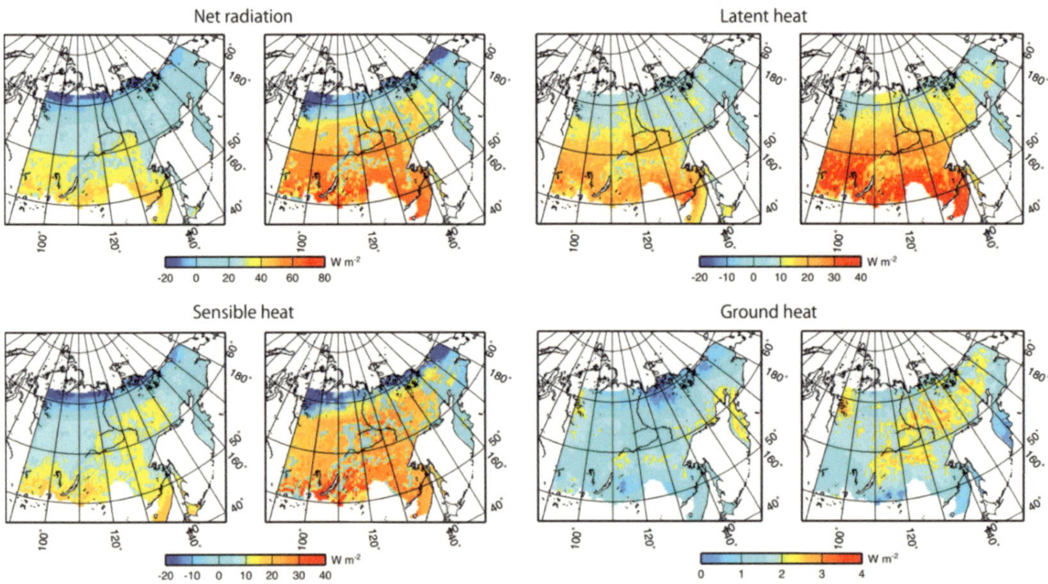

Fig. 2. Comparison of spatial distribution of averaged annual energy budget components between Current (left-hand side panel) and Future (right-hand side panel)

The spatial difference in *H* is as large as that of *LE* (Fig. 2); *H* in Future is larger than that in Current. The increase in *H* is majorly correlated to the increased R_n. Warming increases LAI, enhancing the absorption of radiation with low albedo. As a result, the temperature gradient between canopy surface and the overlying air increases, thereby resulting in higher *H*. Potential shifts of vegetation in the future would also have various climate feedbacks. According to a simulation dealing with vegetation change, transition from moist to shrub tundra increased near-surface temperature by 3.5°C during summer because of increased *H* (Chappin et al., 2000). Higher LAI also increases surface roughness, which decreases aerodynamic resistance, generating turbulence and ultimately higher *H*. The large *H* over boreal forests is likely to contribute to the diurnal growth of a very deep planetary boundary layer (PBL) (McNaughton and Spriggs, 1989). It is known that PBL in Siberia does not grow to the depths observed in Canada (Hollinger et al., 1995) because of the existence of the Siberian High Pressure zone. However, the formed boundary layers entrain a substantial amount of dry air, which makes it difficult to humidify the air. This response can have a negative feedback on stomatal conductance and a positive feedback on *H* and PBL growth (McNaughton and Spriggs, 1989).

As compared to forests, low leaf area in tundra regions results in more radiation reaching the ground surface, and thus, higher values of *G*. *G* in Future is slightly higher than in Current, in the tundra regions as well as in forests (Fig. 2). The increase in G in Future is not as large as the accompanying temperature increase. This can be attributed to soil surface shading caused by the increased LAI. Eastern Siberia is covered with permafrost. A considerable amount of *G* is consumed in the melting of permafrost during summer. The energy used for the melt of permafrost is therefore not available for increasing surface and soil temperature, thus restricting the increase in *H* or *LE*. It is known that in forested areas, the influence of *G* on the energy partitioning is weak because of the low absolute value. However, the effect of *G* on the energy budget in tundra is not negligible. In tundra regions, for instance, *G* accounted for 25% of R_n during 1986-2004 (Park et al., 2008). Therefore, the increase in *G* in tundra regions caused by the future warming could reduce *H*. The change in the distribution of energy partitioning, especially in the northern region can be explained through an analysis of the reported results (Fig. 3).

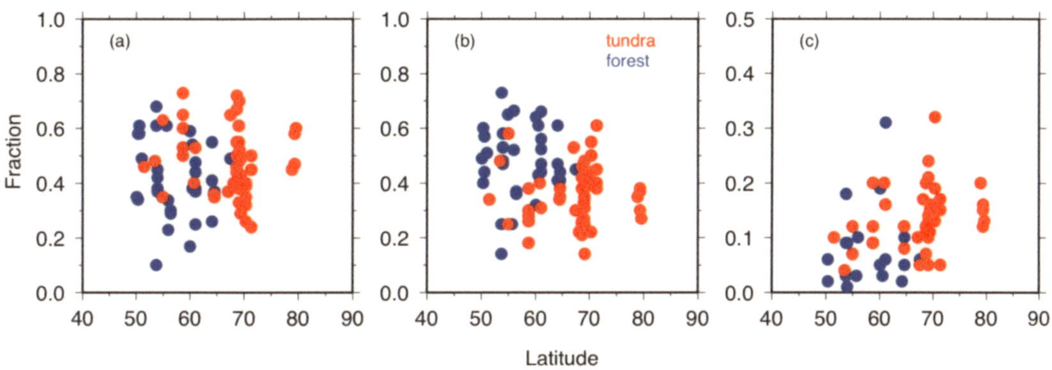

Fig. 3. Distribution of LE/Rn (a), H/Rn (b), and G/Rn (c) of different vegetation types across a latitudinal range. They are arranged based on the data of Eugster et al. (2004), Baldocchi et al. (2000), Beringer et al. (2005), and Bernhofer et al. (2003)

The rate at which R_n is partitioned into energy components is dependent on climatic conditions, species composition, vegetation structure, soil, topography, etc. If vegetation changes, the energy balance of the surface will be greatly changed. Latitudinal energy partitioning rates are plotted in Fig. 3, based on the data observed at various arctic land types. The rate LE/R_n exhibits a large scattering. No specific tendency in the rate is observed along latitudes and between latitudinal vegetation types (Fig. 3a), implying that LE/R_n does not obey simple scale laws. Eugster et al. (2000) suggested that the complex pattern of LE/R_n is associated with moisture gradient, LAI, and plant physiological controls over stomatal conductance. In particular, Eugster et al. (2000) stressed the roles of soil moisture in describing the complex LE/R_n ratio between vegetation types.

The rates of both H and G exhibit a latitudinal trend (Fig. 3). H/R_n tends to decrease with latitude. In the same latitude, H/R_n is higher in boreal forests than in tundra regions. Higher LAI, in addition with tree biomass, contributes disproportionately to roughness length, and therefore, the efficiency of convective exchange. This leads to the conclusion that under the future warming, the migration of forest into tundra regions will cause higher H. Observations over a vegetation transition from tundra to tall shrub and then to forest concluded that the transition could result in an increase in H during the growing season (Beringer et al., 2005). A similar result was obtained in a simulation associated with vegetation change (Chapin et al., 2000). A consistent latitudinal trend is found in G/R_n (Fig. 3c). In particular, higher G/R_n is found in tundra regions. In contrast, the low G/R_n in forests represents the effect of canopy shading. These results also imply that in the future, vegetation shift into tundra can reduce G due to increased soil shading by the canopy. The conclusion drawn from Fig. 3 is that if climatic warming causes the transition of vegetation from tundra to forest, the transition could result in a significant increase in the atmospheric heating due to the increased H.

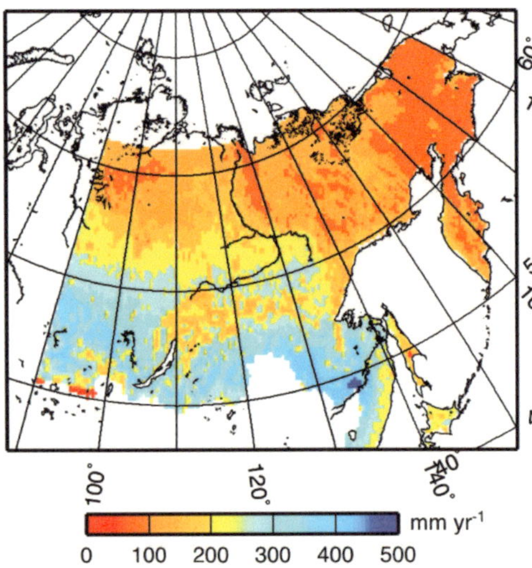

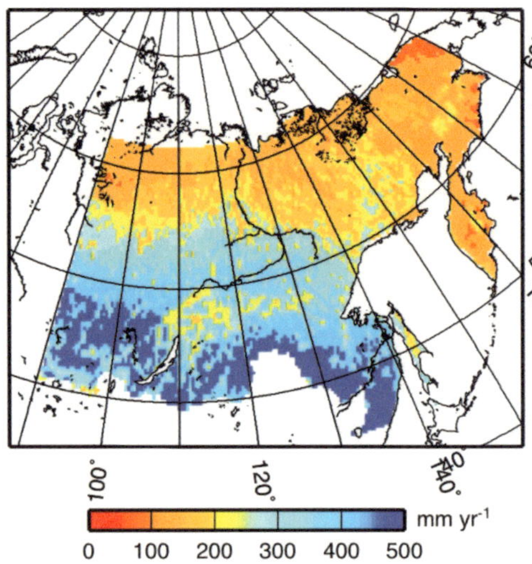

Fig. 4. Comparison of annual mean evapotranspiration in Current (left) and Future (right)

Figure 4 shows the spatial variability of ET in the two periods. In most areas, ET is considerably larger in Future than in Current. The spatial distribution of ET in Future indicates some remarkable characteristics. ET presents a clear latitudinal contrast. The increasing tendency of ET migrates northward as previously mentioned, implying an increase in ET in tundra regions. Finally, the significant increase in ET occurred sporadically in the southern forest area. The increase in ET in Future across all areas of eastern Siberia is closely correlated with the increase in temperatures, precipitation, and LAI. Higher precipitation is likely to be made available for soil moisture and canopy interception. Furthermore, higher LAI increases the absorption of radiation that interactively generates ET.

In Future, the remarkable increase in ET is found in tundra regions. In some of the tundra regions, for instance, ET in Future increased by approximately 50–100 mm when compared to Current. The increase in ET in tundra areas is attributable to higher evaporation rates from the soil surface. Evaporation from the soil decreases nonlinearly with increasing leaf area, whereas soil evaporation is a linear function of available energy. When the soil surface is wet, evaporation rates can exceed available energy (Lindroth, 1985). Therefore, increased precipitation is favourable for high soil evaporation.

5.2 Seasonal variations

5.2.1 Climatic variables

Figure 5 shows mean seasonal variations of air temperature and precipitation in Lena watershed for the two periods (Current and Future). Air temperature exhibits a seasonality, showing a maximum in summer followed by a sharp decrease (Fig. 5a). A significant difference in air temperatures between summer and winter can be observed. The difference in Current was 42.8°C. The air temperature in Future significantly increased as compared to Current, especially during winter. In December, air temperature in Future was on maximum higher 5°C than in Current. IPCC (2007) had projected the increase in air temperature in winters at high latitude under global warming conditions.

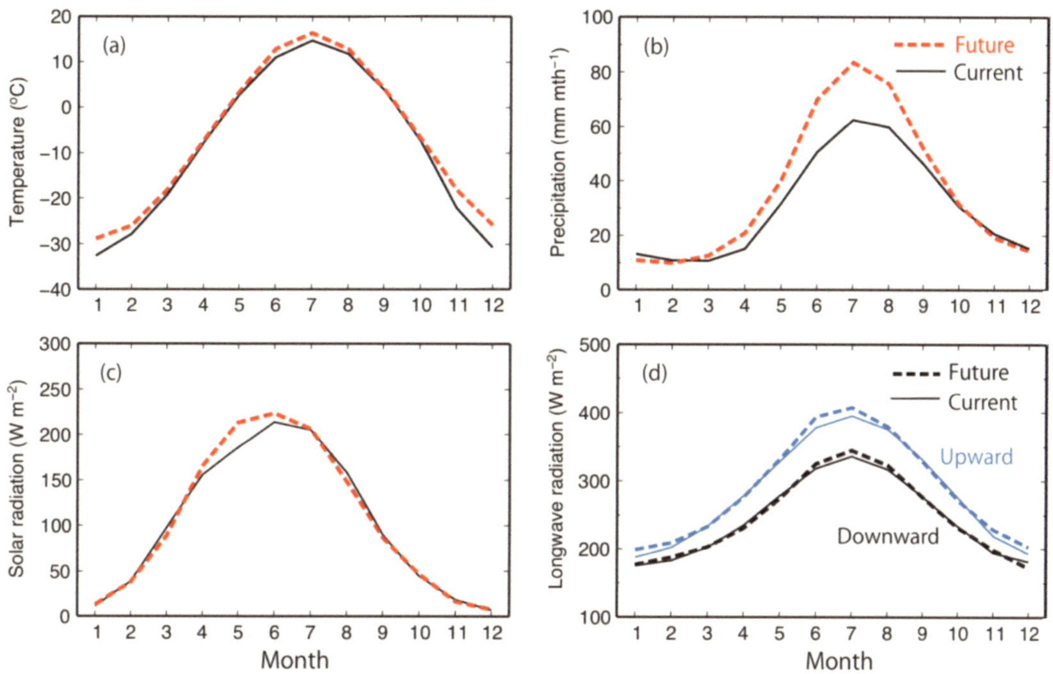

Fig. 5. Monthly variations of temperature (a), precipitation (b), solar radiation (c), and longwave radiation (d) in Lena watershed in Current (straight lines) and Future (dotted lines).

Precipitation also exhibits a clear seasonal pattern in that it is high in summer and low in winter (Fig. 5b). In Current, summer precipitation accounted for 47% of the annual value. As compared to Current, precipitation in Future was higher during spring and summer. The difference in summer precipitation between Current and Future was 59 mm. However, precipitation during October to February was higher in Current than in Future. The low precipitation directly resulted in lower snow depths during the specified period. Annual

mean precipitation was higher by 79 mm in Future than in Current. IPCC (2007) had projected increases in the amount in precipitation in high latitudes of as much as 20% in 2100.
Solar radiation during May through July was higher in Future than in Current (Fig. 5c). However, the solar radiation in the early spring and late summer tended to be rather lower in Future than in Current. Longwave radiations over spring to autumn also exhibited similar seasonal patterns with solar radiation (Fig. 5d). However, longwave radiations during summer and winter were larger in Future than in Current.

5.2.2 Energy budget

Energy exchange at land surface affects the local and seasonal climates through the transfer of heat and water to the atmosphere. The energy exchange is greatly dependent on the surface conditions varying temporarily and spatially. A change in land surface can therefore alter energy exchange ratios, leading to distinctly different local and seasonal climates. Table 1 summarizes the mean seasonal variations of energy components in Lena watershed for the two periods. In both Current and Future, energy exchanges are strongly active in spring and summer. In Current, R_n during spring is almost completely consumed for LE and G. The fraction of G to R_n is higher than that of LE. Prior to leaf-opening, high solar radiation gives rise to higher soil evaporation from the wetted soil surface that is formed with snowmelt. Moreover, frozen soil contributed to the large G by creating a strong thermal gradient between the ground surface and the soil depth. G is therefore not available for increasing the surface temperature, and hence, derives less H. On the other hand, in Future, energy fluxes in the spring exhibit considerably different patterns in comparison with Current. Most noticeably, the increase in R_n is significant, which is strongly associated with the increase in solar radiation (Fig. 5c). Simultaneously, H considerably increased, being 75% of R_n. The increased biomass, especially LAI, is useful for higher H, while reducing G. Evaporation rates over boreal forests and tundra regions are relatively low in the spring, despite higher solar radiation with high evaporative demand. The low evaporation rates during spring are due to cold or frozen soils restricting root uptake of soil moisture (Teskey et al., 1984).

	Current				Future			
	Spring	Summer	Autumn	Winter	Spring	Summer	Autumn	Winter
R_n (W m^{-2})	11.8	104.3	-7.8	-16.3	62.8	111.4	-2.9	-8.4
LE (W m^{-2})	4.9	47.9	2.5	0.1	11.4	58.1	7.7	5.7
H (W m^{-2})	0.9	43.7	-4.0	-9.4	46.8	39.5	-4.7	-8.6
G (W m^{-2})	6.0	12.8	-6.2	-7.0	4.7	13.8	-5.0	-5.5
E_T (mm)	1.8	93.6	6.0	0.0	3.3	94.8	0.2	0.0
E_I (mm)	0.3	33.0	3.0	0.0	20.4	70.1	26.2	8.4
E_S (mm)	11.8	40.0	5.9	0.0	5.3	22.9	2.4	0.0
ET (mm)	13.9	166.6	14.9	0.0	29.0	187.8	28.8	8.4

Table 1. Summary of seasonal variations of averaged energy budget and cumulated evapotranspiration components in Lena watershed in Current and Future. Spring denotes March to May, Summer is June to August, Autumn is September to November, and the remaining months are Winter

In summer, energy fluxes steeply increased in both Current and Future. In particular, there was a marked increase in LE in Future. The increase was accompanied by increased precipitation, deriving higher ET. During summer, soil water deficits can restrict ET by forcing stomata to partially close. Increased precipitation can reduce the restriction of the soil water deficit to ET. In contrast, the presence of clouds and higher atmospheric humidity as fronts passed limited transpiration through their association with low radiative forcing. The autumnal decline in *LE* is covariant with available energy and air temperature, which decrease after the summer. However, the autumnal *LE* in Future is larger than in Current, which can be attributed to the influence of canopy interception. A large increase in winter *LE* in Future is observed, owing to snow interception and sublimation. In dense coniferous canopies, for instance, interception can result in up to 40% of sublimation (Pomeroy and Gray, 1994). The sublimation reduces snow on the ground during the winter, consequently affecting ground heat flux in spring.

Climatic warming resulted in significant increases in both *LE* and *H* during spring and summer. These increases strongly responded to the enhanced available energy and precipitation, in conjunction with biomass. The increase in energy fluxes in summer is a general pattern, although their increasing magnitude depends on a number of factors. A unique change under the Future climatic warming is the promoted atmospheric heating due to greater *H* in spring (Table 1). Above all, the increased solar radiation is primarily associated with the higher *H*. The rise of spring temperature over high latitudes has been observed during the past few decades. The major reason is early snowmelt due to increased air temperature (Groisman et al., 1994; Chapin et al., 2005). As a result, the snow-free period lengthens and sequentially causes summer warming (Chapin et al., 2005). However, snowmelt in Future was somewhat late as compared to Current, which was on average 4 days. Snow depth was lower during October through February in Future than in Current due to lower precipitation and higher sublimation caused by higher temperatures. However, higher precipitation in March and April in Future (Fig. 5b) induced later snow depth with high albedo. After snowmelt, therefore, strong solar radiation and higher air temperature is likely to have caused the higher *H*.

5.2.3 Evapotranspiration

Evapotranspiration in Lena watershed occurred mainly during summer (Table 1). Most of the precipitation in summer is used for ET, and transpiration is a dominant factor of ET in Current and Future. In spring, soil evaporation exceeds transpiration. This is closely correlated with the strong solar radiation observed during spring. Radiation contributes to snowmelt and soil thawing. As a result, increased soil moisture is available for higher soil evaporation. On the other hand, transpiration in spring and autumn is relatively low, despite higher soil moisture. Even though snow is completely melted, the thawing of frozen soil does not occur rapidly. The frozen soil during spring causes the hydraulic conductivity of the roots to be low, reducing leaf turgor and forcing the closing of stomata, thus restricting root uptake of soil moisture (Teskey et al., 1984). The drop in air temperatures below 0°C during spring and autumn results in leaf chilling and freezing. These stresses on the plants affect transpiration through interaction with stomatal conductance and photosynthesis.

When LAI increases, soil evaporation is reduced, but transpiration from the leafy canopy simultaneously increases. The seasonality in interception generally follows the pattern of precipitation, but is closely correlated to LAI. Higher LAI in Future resulted in higher interception. Greater amounts of leaf are aerodynamically rougher. This enhances the transfer of mass and energy to the atmosphere by generating more turbulence and increasing the aerodynamic conductance (Jarvis and McNaughton, 1986). In Future, the increase in interception was significant, especially in summer. Most of the increased precipitation in summer was used for interception, which relatively reduced the contribution of precipitation to transpiration.

Evapotranspiration rates in autumn are low, despite the considerable precipitation. In general, active layer depth in the Arctic reaches the deepest layers between August and October. However, water uptake by trees in these times is relatively low. This is probably related to the decrease in air temperature and available energy that decrease after summer solstice. In autumn, interception in Future is considerably high. ET from wet forests sometimes exceeds available energy. In general, ET rates from wet forests also exceed values over dry forests by factors of 50% and above (Lindroth, 1985). Under climate warming, higher sublimation is a marked change in ET components. The increase in sublimation can decrease snow depth, resulting in a feedback loop that decreases *G* between the snow layer and the soil surface, while heating the atmosphere.

The climate warming greatly changed the partitioning of ET components in Lena watershed. Transpiration was a major contributor to ET in Current and in Future. However, the weight of canopy interception to ET remarkably increased in Future, correlating to the increased precipitation and LAI. In contrast, soil evaporation significantly decreased in Future.

5.3 Inter-annual variations

Annual ET and mean energy budget between Current and Future in Lena watershed is compared in Fig. 6. The comparison shows significant increases in energy budget in Future. R_n increased twofold in Future compared to Current. The increase in R_n caused *LE* and *H* to increase. However, the evapotranspirative fraction (LE/R_n) was higher in Current (0.55) than in Future (0.45). In Lena watershed, transpiration was the major contributor (52%) to ET in Current (Table 1). Baldocchi et al. (2000) also found that across spruce forests, between 50 and 62% of *LE* comes from transpiration. In Future, however, the fraction of transpiration to ET was lower than Current (Table 1). Transpiration is partly controlled by stomatal and boundary resistances to water vapour transport which are in turn related to atmospheric conditions (Oke, 1987). Soil moisture has been considered an important factor in controlling *LE*. Precipitation in Future was higher than in Current (Fig. 5). The increased precipitation in Future contributes to reduce soil water stress and can support transpiration. However, the frequent precipitation can enhance canopy interception with the increased LAI in the Future climate. The increased interception means less contribution of precipitation to the soil moisture. When the canopy is wet, stomata may be temporarily closed, thus limiting transpiration (Park and Hattori, 2004).

The sensible heat fraction (H/R_n) in Future was larger (0.44) than in Current (0.30), which is can be attributed to the contribution of the increased *H* in spring in Future (Table 1). The Bowen ratio (H/LE) is a good factor for comparing the surface energy balance in climates and vegetation types with differing R_n. The Bowen ratio was 0.65 and 0.90 in Current and Future, respectively. According to observations conducted at the same climate zones, the Bowen ratio increased across the vegetation sequence from the low leaf tundra (0.94) to the high leaf forest (1.22) (Beringer et al., 2005). Both *LE* and *H* are sensitive to aerodynamic resistance and surface temperature, but *LE* depends on bulk surface resistance to water vapour flux from the surface. In Future, higher canopy interception reduced soil moisture and higher LAI helped to increase canopy shading. These resulted in higher bulk surface resistance, restricting *LE*. In fact, ET during the dry period was 30 mm lower in Future than in Current (Table 1). Most of this 30 mm was sourced from the decrease in soil evaporation, implying an increase in bulk surface resistance, especially during the dry periods. Boreal woodlands are mostly open, allowing a disproportionate amount of solar radiation to reach the understory, in turn leading to *H* and *LE* exchange at the forest floor. In contrast, the increase in LAI is able to limit this exchange. On the other hand, the increased LAI also enhances aerodynamic conductance. Therefore, the higher Bowen ratio in Future seems to be related to the change in LAI.

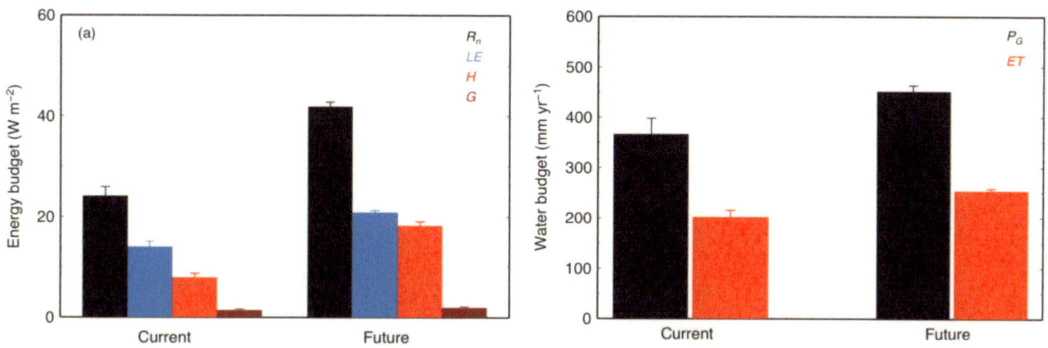

Fig. 6. Comparison of energy (a) and water (b) budgets in Lena watershed between Current and Future

The model specified the ET components over two periods in Lena watershed (Table 1). There have been almost no previous studies that have evaluated ET components on the Arctic watershed scale, making a useful comparison of the estimate ET components impossible. However, Serreze et al. (2003) estimated annual ET of 182 mm in Lena watershed by water budget using measured runoff. Fukutomi et al. (2003) reported annual ET of 160 mm by atmospheric water budget. The values are similar to those estimated by the model for Current ET (196 mm) (Fig. 6b).

The climate warming caused ET to increase by 58 mm in Future (Fig. 6b), which is the value corresponding to the increased precipitation in summer (Table 1). The increase in ET is considerably large, as compared to a projection that ET from the major Arctic watersheds will increase on maximum 10% in the years 2071-2090 relative to 1981-2000 (ACIA, 2005). GCM simulated that in Alpine regions, ET in 2071-2100 will increase by approximately 20% relative that in 1961-1990 (Calanca et al., 2006). Karpechko and Bondarik (2003) also estimated that if warming up of 2°C occurs, ET from a forested catchment in northwest Russia will increase by 15-20 mm per year. The Arctic regions are vulnerable to global warming. Trees may expand into tundra regions, and increased disturbance from fire and logging alters ecosystem structures. Climatic forcing arising from these disturbances may be comparable to that arising from biome shifts (Liu et al., 2005). The most important changes in surface energy partitioning, and hence in the feedback to larger scales, is expected from a combined decrease in precipitation and in fire frequency that would more than double *H* by reducing *LE* to roughly 70% of today's value (Eugster et al., 2000).

5.4 Variables affecting evapotranspiration

Ecoclimatic diversities across the Arctic regions mean that the evaporative rates of the boreal forests present the greatest heterogeneity (Fig. 7). Fig. 7 shows the relationships between annual precipitation (P_G) and annual ET in different types of boreal plants. Classification into 3 regions of continuous permafrost, discontinuous permafrost, and non-permafrost is made. The smallest values of ET are found in plants of continuous permafrost, whereas plants in non-permafrost generally exhibited higher ET. However, the rates of ET to P_G are in contrast the highest in plants of continuous permafrost, in which most of P_G available is used for ET. The rate of ET/P_G tends to decrease with the reduction in permafrost. In non-permafrost regions with low annual P_G (particularly areas with less than 500 mm y^{-1} of P_G), the annual amount of ET has been found to vary linearly with the annual P_G with a very slight deviation (Zhang et al., 2001). The difference between our analysis and the result of Zhang et al. (2001) may represent the specific roles of permafrost on hydrological processes in high latitudes.

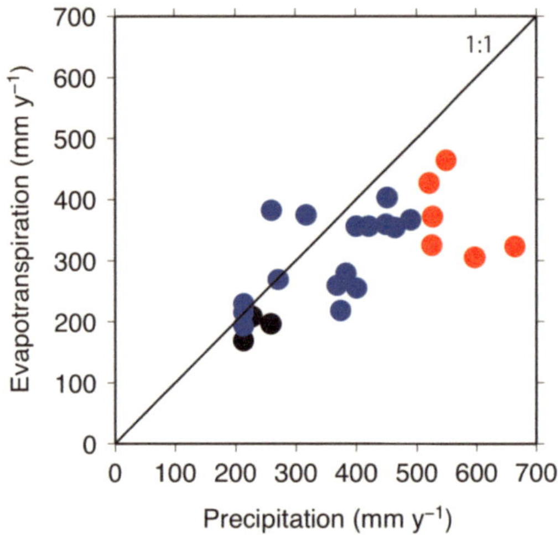

Fig. 7. Relationships between annual precipitation and evapotranspiration in various types of plants in high latitudes. The different colours represent continuous permafrost (black), discontinuous permafrost (blue), and non-permafrost (red). The data are sourced from Amiro (2009), Amiro et al. (2006), Arain et al. (2003), Black et al. (1996), Bond-Lamberty et al. (2009), Grell et al. (1999), Hamada et al. (2004), Nijssen et al. (1997), Nijssen and Lettenmaier (2002), Ohta et al. (2008), Schulze et al. (1999), Wever et al. (2002), and Yuan et al. (2010)

In the Arctic regions, land surface processes are primarily controlled by the presence or absence of permafrost. Land surface processes are also influenced by the thickness of the active layer and the total thickness of the underlying permafrost. As permafrost becomes thinner or decreases in area, the interaction between ET and sub-permafrost groundwater gains in significance. The inability of soil moisture to infiltrate to deeper groundwater zones owing to ice-rich permafrost results in very wet upper soils. The wet soil water is favourable for ET. Even though soil is dry due to small P_G, the melted soil water also helps to alleviate soil water deficit. Sugimoto et al. (2002) examined the water usage of larch forests near Yakutsk, Russia, using a stable isotope of oxygen, δO^{18}. They showed that trees took up precipitation water in wet years, but transpired permafrost melt-water in drought years. Therefore, the active layer plays an important role in the Arctic regions, because most ecological, hydrological, biogeochemical, and pedogenic activity occur within it (Hinzman et al., 1991).
Changes in active layer thickness are influenced by many factors, including surface temperature, physical properties of the surface cover and substrate, soil moisture, and duration and thickness of snow cover (Zhang et al., 2005). In Lena watershed, the average maximum active layer depth was larger in Current (0.94 m) than in Future (0.91 m). The

lower active layer thickness in Future is likely to have affected, in some part, the lower ET in dry period (Table 1). Despite higher air temperature and P_G in Future, the low active layer depth in Future seems to be closely associated with higher LAI (2.8), as compared to LAI of Current (2.0). Higher LAI shades the soil, thus countering the effect of warmer air temperature on the active layer. Walker et al. (2003) also found a similar relationship between LAI and active layer depth when observing various types of vegetation in Alaska. These results indicate that changes in surface conditions can produce significant changes in permafrost, which in turn feeds back to surface processes.

Temperature is one factor influencing ET in the Arctic. In Lena watershed, mean air temperature during the growing season exhibited high correlation with ET (Park et al., 2008). This is because sufficient, but not excessive, heat is a prerequisite for physiological, biological, and biogeochemical reactions. Therefore, boreal vegetation that experiences a short warm summer seems to be more sensitive to low temperatures rather than higher temperatures. The low temperature results in chilling and freezing. Chilling mainly occurs in spring and autumn when ET is relatively low. On the other hand, freezing is often found during night of the growing season. This can cause a non-reversible reduction in photosynthetic capacity and stomatal conductance, consequently limiting transpiration (Schulze et al., 1977).

Soil moisture is an important factor that can greatly influence ET (Ohta et al., 2008). Soil moisture from the previous autumn was highly correlated to ET of the following year (Park et al., unpublished). In the permafrost-covered regions, the soil moisture in autumn is stored as ice until the next melting season. Higher soil moisture in the autumn likely augments soil moisture with the melted snow water during the following spring. The combined higher soil moisture in the spring may largely affect soil evaporation rather than transpiration, because cold or frozen soil in the spring restricts root uptake of soil moisture. However, higher soil water is favourable to soil thawing. Therefore, the speed-up of the soil thawing could enhance root water uptake because considerable amounts of roots in boreal forest are relatively distributed in the upper soil layers.

6. Future model developments

In atmosphere-land systems, vegetation is a dynamic component that exhibits spatiotemporal variations. The representation of vegetation dynamics incorporated thus far is extremely simplified, and most existing LSMs do not consider vegetation as a dynamic component. In most current LSMs, the seasonal variation of LAI is prescribed or defined to be constant. This treatment implies that the effect of climate variability in modifying the structure and physiological properties of vegetation cannot be satisfactorily evaluated. However, this study estimated seasonal LAI in each grid with a semi-empirical formula, based on both prescribed $LAI_{\max}$ and climate variables (eq. 1). The formula is simple, and is dependent on the prescribed $LAI_{\max}$. When actual values of LAI exceed $LAI_{\max}$, eq. (1) can underestimate LAI and can explicitly cause errors in energy and water fluxes.

Plants exposed to elevated levels of CO_2 exhibit increased growth and photosynthesis rates. A number of studies have reported higher soil moisture under elevated CO_2 (e.g., Volk et al., 2000). A decrease in transpiration and an increase in carbon uptake under elevated CO_2 imply increased water use efficiency by plants. Root distribution and maximum rooting depth exhibit seasonal and inter-annual variations depending on plant carbon uptake and soil moisture (Arora and Boer, 2003). The preferential use of moisture from different soil depths by plants highlights the control of vegetation on soil moisture. The spatial structure of soil moisture and its evolution in time are also affected by the root distribution. However, the LSM of this study used an empirical parameterization that roots uptake water from soil layers of 0–22 cm. This parameterization does not consider root dynamics, and thus, it may not be appropriate to simulating the interactions between vegetation and soil moisture, via the control of transpiration by vegetation and the effect of soil moisture on vegetation. Having deeper roots, trees are able to extract water from deep soil layers. When soil is dry, roots tend to grow for extracting water. These processes require a new parameterization for root distribution, or the incorporation of a photosynthesis model into LSM.

7. Summary

This study evaluated the effects of future climate changes on energy and water fluxes over eastern Siberia using a LSM. According to simulations, energy budgets and ET rates in Future considerably increased in comparison with Current. These increases in Future are correlated to the increased temperature and precipitation. The Future climates preferentially yielded higher LAI, increasing the absorption of radiation, in turn leading to higher R_n and consequently to higher LE and H. The increases in LE and H were significant in spring and summer and at tundra in the regional distribution. Considerable increases in ET were found in Future. The increased ET was mainly due to the increase in canopy interception associated with higher LAI and increased precipitation. In contrast, the higher LAI limited soil evaporation. The increase in sublimation during winter in Future is also a remarkable result, caused by higher temperatures and less precipitation. Northward spatial migration of ET was found from the comparison between Current and Future. In conclusion, it is possible to state that tundra regions are susceptible and vulnerable to climate changes.

It has been previously observed that ET is influenced by air temperature during the growing season and date of complete snowmelt (Park et al., 2008). Changes to ET in the Arctic are intimately linked to permafrost levels. As permafrost warms, the active layer thickens. A deeper active layer seems to have the capacity to attenuate soil water deficit owing to increased storage, causing higher ET rates, and vice versa. When permafrost is degraded or absent, ET is completely dependent on precipitation, and thus, the ET rate is decreased.

The Arctic regions are known to be susceptible and vulnerable to climate changes. This study will be useful for decreasing the uncertainties surrounding the susceptibility and vulnerability of the Arctic ecosystems to future climatic changes, especially with regard to water and energy budgets. Further study is needed to examine the details of interactions of vegetation dynamics and climate system and the likely consequences of such changes.

8. References

ACIA. (2005). *Arctic Climate Impact Assessment*, Cambridge University Press, Cambridge.

Amiro, B. (2009). Measuring boréal forest évapotranspiration using the energy balance residual, *J. Hydrol.*, doi:10.1016/j.jhydrol.2008.12.021.

Amiro, B. et al. (2006). Carbon, energy and water fluxes at mature and distrubed forest sites, Saskatchewan, Canada, *Agric. For. Meteorol.*, 136, 237-251.

Arain, M.; Black, T., Barr, A. Griffis, T., Morgenstern, K. & Nesic, Z. (2003). Year-round observations of the energy and water vapour fluxes above a boréal black spruce forest. *Hydrol. Proc.*, 17, 3581-3600.

Arora, V. & Boer, G. (2003). A représentation of variable root distribution in dynamic végétation models. *Earth Interact.*, 7, 1-19.

Baldocchi, D.; Kelliher, F., Black, T. & Jarvis, P. (2000). Climate and végétation controls on boréal zone energy exchange. *Global Change Biol.*, 6, 69-83.

Beringer, J.; Chapin, F., Thompson, C. & McGuire, D. (2005). Surface energy exchange along a tundra-forest transition and feedbacks to climate. *Agric. For. Meteorol.*, 131, 143-161.

Bernhofer, C.; Aubinet, M., Clement, R., Grunwald, G., Ibrom, A., Jarvis, P., Rebmann, C., Schulze, E. & Tenhunen, J. (2003). Spruce forests (Norway and Sitka spruce, including Douglas fir): Carbon and water fluxes and balances, ecological and ecophysiological déterminants, In: *Fluxes of carbon, water and energy of European forests*, R. Valentini (Ed.), 99-123, Springer, Berlin, Germany.

Black, T.; den Hartog, G., Neumann, H. et al. (1996). Annual cycles of CO2 and water vapor fluxes above and within a boréal aspen stand. *Global Change Biol.*, 2, 219-230.

Bonan, G.; Pollard, D. & Thompson, S. (1992). Effects of boréal forest végétation on global climate. *Nature*, 359, 716-718.

Bond-Lamberty, B.; Peckham, S., Gower, S. & Ewers, B. (2009). Effects of fire on régional évapotranspiration in the central Canadian boréal forest. *Global Change Biol.*, doi: 10.1111/j.1365-2486.2008.01776.x

Buermann, W.; Anderson, B., Tucker, C., Dickinson, R., Lucht, W., Potter, C. & Myneni, R. (2003). Interannual covariability in Northern hémisphère air températures and greenness associated with El Nino-Southern oscillation and the Arctic oscillation. *J. Geophys. Res.*, 108(D13), 4396, doi : 10.1029/2002JD002630.

Calanca, P.; Roesch, A., Jasper, K. & Wild, M. (2006). Global warming and the summertime évapotranspiration régime of the Alpine région, *Climate Change*, 79, 65-78.

Chapin, F.; Shaver, G., Giblin, A., Nadelhoffer, K. & Laundre, J. (1995). Response of arctic tundra to expérimental and observed changes in climate. *Ecology*, 76, 694-711.

Chapin, F.; Eugster, W., McFadden, J., Lynch, A. & Walker, D. (2000). Summer différences among Arctic ecosystems in régional climate forcing. *J. Clim.*, 13, 2002-2010.

Chapin, F. et al. (2005). Role of land-surface changes in arctic summer warming. *Science*, 310, 657-660.

Dolman, A.; Maximov, T. & Ohta, T. (2008). Water and energy exchange in East Siberian forest. *Agr. For. Meteorol.*, 148, 1913-1915.

Eugster, W.; Rouse, W. et al. (2000). Land-atmosphere energy exchange in Arctic tundra and boréal forest: available data and feedbacks to climate. *Global Change Biol.*, 2000, 84-115.

Fukutomi, Y.; Igarashi, H., Masuda, K. & Yasunari, T. (2003). Interannual variability of summer water balance components in three major river basins of northern Eurasia. *J. Hydrometeorol.*, 4, 283-296.

Grelle, A.; Lindroth, A. & Molder, M. (1999). Seasonal variation of boréal forest surface conductance and évaporation, *J. Hydrol.*, 98-99, 563-578.

Groisman, P.; Karl, T. & Knight, R. (1994). Observed impact of snow cover on the heat balance and the rise of continental spring températures. *Science*, 263, 198-200.

Hamada, S.; Ohta, T., Hiyama, T., Takahashi, A. & Maximov, T. (2004). Hydrometeorological behaviour of pine and larch forests in eastern Siberia. *Hydrol. Process.*, 18, 23-39.

Hinzman, L. et al. (1991). Hydrologic and thermal properties of the active layer in the Alaskan Arctic. *Cold Regions Sci. Technol.*, 19(2), 95-110.

Hollinger, D.; Kelliher, F., Schulze, E. et al. (1995). Initial assessment of multi-scale measures of CO_2 and H_2O flux in the Siberian taïga. *J. Biogeography*, 22, 425-431.

IPCC. (2007). *Climate Change 2007: The Physical Science Basis. Contribution of Working Grounp I to the Fourth Assessment Report of the Integovernmental Panel on Climate Change.* Cambridge University Press, Cambridge, United Kongdom and New York, NY, USA, 996 pp.

Jarvis, P. (1976). The interprétation of the variations in leaf water potential and stomatal conductance found in canopies in the field. *Philos. Trans. R. Soc. Lond. Ser. B*, 273, 593-610.

Jarvis, P. & McNaughton, K. (1986). Stomatal control of transpiration. *Adv. Ecol. Res.*, 15, 1-49.

Jia, G.; Epstein, H. & Walker, D. (2003). Greening of arctic Alaska. *Geophys. Res. Lett.*, 30, 2067, doi:10.1029/2003GL018268, 2003.

Jolly, W.; Nemani, R. & Running, S. (2005). A generalized bioclimatic index to predict foliar phenology in response to climate. *Global Change Biol.*, 11, 619-632.

Jorgenson, M.; Shur, Y. & Pullman, E. (2006). Abrupt increase in permafrost dégradation in Arctic Alaska. *Geophys. Res. Lett.*, 2, L02503, doi:10.1029/2005GL02496.

Karpechko, Y. & Bondarik, N. (2003). Effect of potential warming on évapotranspiration from forest catchments in Karelia, *Nordic Hydrol.*, 34, 147-160.

Lindroth, A. (1985). Seasonal and diurnal variation of energy budget components in coniferous forests. *J. Hydrol.*, 82, 1-15.

Liu, H.; Randerson, J., Lindfors, J. & Chapin, S. (2005). Changes in the surface energy budget after fire in boréal ecosystems of interior Alaska: An annual perspective, J. Geophys. Res., 110, D13101, doi:10.1029/2004JD005158, 2005.

McNaughton, K. & Spriggs, T. (1989). An évaluation of the Priestley-Taylor équation and the complementary relation using results from a mixed layer model of the convective boundary layer. In: *Estimation of Areal Evaporatranspiration*, Black T. et al. (Eds.), pp. 89-103, IAHS Publications, Wallingford, UK.

Nijssen, B.; Haddeland, I. & Lettenmaier, D. (1997). Point évaluation of a surface hydrology model for BOREAS. *J. Geophys. Res.*, 102, 29367-29378.

Nijssen, B.; O'Donnell, G., Hamlet, A. & Lettenmaier, D. (2001). Hydrologica sensitivity of global rivers to climate change. *Clim. Change.*, 50, 143-175.

Nijssen, B. & Lettenmaier, D. (2002). Water balance dynamics of a boréal forest watershed: White Gull Creek basin, 1994-1996. *Water Resour. Res.*, 38, 1255, doi:10.1029/2001WR000699.

Oechel, W.; Vourlitis, G., Hastings, S., Zulueta, R., Hinzman, L. & Kane, D. (2000). Acclimation of ecosystem CO_2 exchange in the Alaskan Arctic in response to decadal climate warming. *Nature*, 406, 978-981.

Ohta, T.; Maximov, T. et al. (2008). Interannual variation of water balance and summer évapotranspiration in an eastern Siberian larch forest over a 7-year period (1998-2006). *Agric. For. Meteorol.*, doi:10.1016/j.agrformet.2008.04.012.

Oke, T. (1987). *Boundary Layer Climates.* 2nd edn. Methuen, London.

Park, H. & Hattori, S. (2004). Modeling scalar and heat sources, sinks, and fluxes within a forest canopy during and after rainfall évents. *J. Geophys. Res.*, 109, D14301, doi:10.1029/2003JD004360.

Park, H.; Yamazaki, T., Yamamoto, K. & Ohta, T. (2008). Tempo-spatial characteristics of energy budget and évapotranspiration in the eastern Siberia. *Argric. For. Meteorol.*, 148, 1990-2005.

Peterson, B.; Holmes, R., McClelland, R., Vorosmarty, J., Lammers, G., Shiklomanov, R., Shiklomanov, I. & Rahmstorf, S. (2002). Increasing river discharge to the Arctic océan. *Science*, 298, 2171-2173.

Pomeroy, J. & Gray, D. (1994). Sensitivity of snow relocation and sublimation to climate and surface végétation. In: *Snow and Ice Covers; Interactions with the Atmosphere and Ecosystems*, Jones, H. et al. (Eds.), IAHS Publication no. 223, pp. 213-226. IAHS, Wallingford.

Rouse, W. (2000). The energy and water balance of high latitude wetlands: controls and extrapolation. *Global Change Biol.*, 6, 59-68.

Schulze, E.; Fuchs, M & Fuchs, M. (1977). Spatial distribution of photosynthetic capacity and performance in a mountain spruce forest of northern Germany. *Oecologia*, 29, 43-61.

Schulze, E.; Lloyd, J. et al. (1999). Productivity of forests in the Eurosiberian boréal région and their potential to act as a carbon sink - a synthesis. *Global Change Biol.*, 5, 703-722.

Serreze, M.; Walsh, J., Chapin, F., Osterkamp, T., Dyurgerov, M., Romanovsky, V., Oechel, W., Morison, J., Zhang, T. & Barry, G. (2000). Observational évidence of récent change in the northern High-latitude environment. *Clim. Change*, 46, 159-207.

Serreze, M. et al. (2003). Large-scale hydro-climatology of the terrestrial Arctic drainage system. *J. Geophys. Res.* 108(D2), 8160. doi:10.1029/2001JD000919.

Serreze, M. & Barry, R. (2005). *The Arctic Climate System.* Cambridge University Press, Cambridge.

Sturm, M.; Racine, C. & Tape, K. (2001). Increasing shrub abundance in the Arctic. *Nature*, 411, 546-547.

Sugimoto, A.; Yanagisawa, N., Naito, D., Fujita, N. & Maximov, T. (2002). Importance of permafrost as a source of water for plants in east Siberian taiga. *Ecol. Res.*, 17, 493-503.

Teskey, R.; Hinckley, T. & Grier, C. (1984). Temperature induced change in the water relations of *Abies amabilis* (Dougl. Forbes). *Plant Physiology*, 74, 77-80.

Volk, M.; Niklaus, P. & Korner, C. (2000). Soil moisture effects determine CO2 responses of grassland species. *Oecologia*, 125, 380-388.

Walker, D. et al. (2003). Vegetation-soil-thaw-depth relationships along a low-arctic bioclimate gradient, Alaska: synthesis of information from the ATLAS studies. *Permafrost Periglac. Processes*, 14, 103-123.

Wever, L.; Flanagan, L. & Carlson, P. (2002). Seasonal and interannual variation in évapotranspiration, energy balance and surface conductance in a northern temperate grassland, *Agric. For. Meteorol.*, 112, 31-49.

Yamazaki, T. (2001). A one-dimensional land surface model adaptable to intensely cold régions and is application in eastern Siberia. *J. Meteorol. Soc. Jpn.*, 79, 1107-1118.

Yamazaki, T.; Yabuki, H., Ishii, Y., Ohta, T. & Ohata, T. (2004). Water and energy exchanges at forests and a grassland in eastern Siberia evaluated using a one-dimensional land surface model. *J. Hydrometeol.*, 5, 504-515.

Yamazaki, T.; Yabuki, H. & Ohata, T. (2007). Hydrometeorological effects of intercepted snow in an eastern Siberian taïga forest using a land surface model. Hydrol. Process., 21, 1148-1156, doi:10.1002/hyp.6675.

Yuan, W. et al. (2010). Impacts of précipitation seasonality and ecosystem types on évapotranspiration in the Yukon River Basin, Alaska, *Water Resour. Res.*, 46, W02514, doi:10.1029/2009WR008119.

Zhang, L.; Dawes, W. and Walker, G. (2001). Response of mean évapotranspiration to végétation change at catchment scale. *Water Resour. Res.*, 37, 701-708.

Zhang, T. et al. (2005). Spatial and temporal variability in active layer thickness over the Russian Arctic drainage basin. *J. Geophys. Res.*, 110, D16101, doi:10.1029/2004JD005642.

16

Semiarid Riparian Vegetation Water Demand and Its Influence to Compute the Sonora River Basin Water Availability

Ramos J., González F.J., Marrufo L. and Domínguez R.
Instituto de Ingeniería, Universidad Nacional Autónoma de México
Av. Universidad 3000, 04510 Coyoacán, México, D.F.
México

1. Introduction

During the last decade, the riparian zones have received considerable attention in order to restore and manage them since they have an important role to plan water resources and land, among other functions in the ecosystem. The riparian zones are mainly located in the floodplain where the soil is characterised to be alluvial. However, arid zones channels and their floodplain are transitories and subject to frequent and rapid changes (Graf 1988a,b), thus there is not a clear distinction from where the channel finish and the floodplain starts. This condition could be disadvantageous to the vegetation settled in the floodplain which is subject to the floods' force, the wood debris and eroded sediments carried by. Conversely, this vegetation has the advantage to receive organic matter from the debris and minerals from the eroded soil resulting in a straightforward species capable to survive at adversely conditions (Bendix and Hupp, 2000). The adaptation of this vegetation generates species drought and flood resistant. In the first case, plants develop long roots in order to access groundwater and be wet. In the second case, plants can afford inundation since the modification of the hydraulic roughness reduces the flood velocity and spread seeds increasing the moisture and nutrients availability in the area (Bendix and Hupp, 2000).

However, the access of the riparian vegetation to groundwater produces high rates of transpiration and a high evaporative demand of the atmosphere. This is because water requirements are bigger in arid and semiarid regions, where rainfall is less and the vapour pressure deficit on the air is large. This implies great water consumption by this vegetation becoming as one of the components to be considered in the groundwater balance. Particularly, in semiarid sites where groundwater is the main water demand source, since water surface resources are highly compromised, temporal and spatially. Scott et al. (2003) observed that as groundwater is used to provide water among the different users, there is a depletion of the water levels that affects directly the riparian zones and, in consequence, modify ecological and hydrologically the watershed.

Scott et al. (2003) pointed out that in order to compute riparian water requirements it is common to use hydrological models that compute it as the residual discharge resulted after calibrated against known inputs, groundwater levels and discharges. This procedure could over or underestimate the groundwater balance in the basin, thus a better understanding of this component is needed to improve water demand among the users. In order to guarantee a better management of the river basins, Mexico has made different water reforms through the last three decades. Some of these reforms deal with the increasing water over-exploitation and, particularly, there were established aquifers management councils supported by a national water law (Wester et al., 2005). In order to compute precise riparian water requirements, it is necessary to consider both internal (vegetation growth characteristics) and external conditions (atmosphere, plant and soil characteristics) of the vegetation (Ehlers and Goss, 2003). The external conditions can be achieved computing the evapotranspiration (ET), which depends of the solar radiation, vapour pressure, wind speed and direction, stage of the plant development, and soils features as soil moisture, among others. Conventional methods can be used to compute ET, but there are also remote sensing methods that have the advantage to consider its spatial and temporal variability.

Currently, there are not many studies focused on native riparian zones, researches have mainly carried out analysis for cropped income areas. This chapter presents the research performed to establish the water requirement of native vegetation (mesquites and elms) along the main channel and in the floodplain of the Sonora River corresponding to the Pesqueira, Topahue, Ures, La Mesa-Seri and Horcasitas aquifers. To address the water requirements, the ET was estimated using remote sensing techniques based on the energy balance. The study was considered for two hydrological cycles (1996-1997 Spring-summer and 2002-2003 autumn-winter), in particular, for the dry season since the Sonora River tributaries are dry and the flow in the main channel reaches its minimum level. The chapter concludes on the importance to consider a temporal and spatial analysis, to determine the water requirements of the riparian vegetation and its impact on the river basin water budget.

2. Theoretical ET approaches

Evapotranspiration (ET) is the total moisture lost to the atmosphere from the land surface when the vaporisation process starts as function of the input energy received and the vegetation present. Lakshmi and Susskind (2001) described ET as a "useful tool", not only because it provides information that can be applied directly in the water budget, but also because ET has a high sensitivity which can be used to define some biophysical parameters. ET has been studied at different scales (Molden, 1997) providing a wide knowledge of how ET is affected or how it can affect the whole system (Menenti, 2000):

- Macro (e.g. impact of changes in available moisture on cloud formation, radiation budget and precipitation),
- Meso (e.g. depletion of soil moisture, and therefore, crop water requirements of irrigated lands and partitioning of precipitation) and
- Micro (e.g. crop water requirements)

In order to assess ET impacts on the water balance, and in consequence, on the land and water management, ET is defined for a specific crop and land condition. Thus, potential ET (ETp) considers *a reference surface as grass with a crop height of 0.12 m, a fixed surface resistance of 70 s.m^{-1} and an albedo of 0.23*, whereas crop ET (ETc) is *the rate of ET from disease-free, well fertilised crops, grown in large fields under optimum soil water conditions, and achieving full production under the given climatic conditions*. The actual ET (ETa) refers to *the ET from crops grown under management and environmental conditions that differ from the standard conditions* (Allen et al., 1998).

In general, the physics of the ET process is well understood, thus accurate ET values at local level are provided. However, as ET is highly sensitive to various land and atmospheric variables, particularly in their spatially distributed form, it makes regional ET estimations uncertain (Calder, 1998). This uncertainty increases when the ET contribution from riparian vegetation needs to be considered to determine the water budget (Goodrich et al., 2000).
Although some authors have studied the direct and indirect influence of the riparian ET into the water availability in a basin (Goodrich et al., 2000), this vegetation is still poorly understood being a non-easy task to explain its hydrogeomorphological influence since it interacts environmentally at different scales (Bendix and Hupp, 2000).
To compute actual riparian evapotranspiration (ETa) is complex since the vegetation communities are non-uniform linear varying in geometry, altitude and season. Also, the organisation and dynamics of the vegetation are strongly related to the channel river and its floodplain, thus the geomorphological process and forms define the pattern for the different aggregation communities maintained by the fluctuations of water discharge (figure 1). Goodrich et al. (2000) pointed out that this condition limits the application of traditional ET computations since the required fetch conditions are not achieved and in a strict term the definition of potential ET (Allen et al., 1998) did not apply due to differences in the canopy architecture, available energy, water availability and boundary layers differences between the atmosphere and leaf surfaces, among others. Additionally, only in few cases the crop coefficient has been defined for each riparian type (Goodrich et al., 2000).

Fig. 1. Riparian vegetation along the Sonora River

2.1 Traditional ET methods

The relationship between the different ET definitions (potential, crop and actual) is not always determined easily, but an accurate value is required for different uses and users. Favourably, at regional scales whatever the type of ET, these are not independent of each other. For example, during the crop growing period, water needs to be diverted on to the field to meet the ETc demands and to compensate losses by seepage and percolation in order to maintain a saturated root zone. Thus, the estimation of ETp is crucial to obtain ETc

rates and moreover to compute ETa values as the response to different reasons that generate non-standard conditions such as climate, pest, contamination, water shortage or waterlogging.
In the Sonora River Basin (SRB) and others semi-desert basins in Mexico, the Turc equation (1954) have been used to estimate the hydrological water balance and to infer the groundwater balance since very few aquifer data is available. Other condition to apply the Turc equation is that the average ETa is used for longer periods in order to reduce to zero the water retention in the basin. The Turc equation is defined as:

$$\text{if} \quad \left(\frac{P}{L}\right)^2 > 0.1; \qquad \text{ETactual} = \frac{P}{\sqrt{0.9 + \left(\frac{P^2}{L^2}\right)}} \qquad (1)$$

$$\text{or} \quad \left(\frac{P}{L}\right)^2 < 0.1; \qquad \text{ETactual} = P$$

$$L = 300 + 25 * Ta_{prom} + 0.05 * Ta_{prom}^3 \qquad (2)$$

Where P is the precipitation and T is air temperature. Sometimes equation 2 is modified considering T at the power of 2 instead of 3 without any raison. The 300 constant value in L corresponds to the base runoff in the basin. Thus, L could increase above 300 as function of the T and ETa is related directly to rainfall. In 1964, the Truc equation was modified by changing the 0.9 coefficient by 1.0 (Turc-Pike, 1964). Despite to the simplicity of this method a high uncertainty is associated to it. Other methods based on meteorological data have been used such as Blaney-Criddle which is a temperature-based method that can be applied to different climates in a monthly base (Goodrich et al., 2000) or the Hargreaves-Samani equation that considers the air temperature and the extraterrestrial radiation (Allen et al., 1999). However, these types of methods are recommended only when data is not available. However, when data is available it is recommendable to used the Penman-Montheith equation (PM) (Allen et al., 1998) but a crop coefficient accurate is necessary to compute ETc. Gazal et al. (2006) applied the PM method to compute the transpiration of cotton/willow forest. The PM equation was inverted in order to asses the seasonal variation in stomatal resistance, results showed that the spatial and temporal heterogeneity of water availability modified the physiology of these riparian vegetation.

2.2 Advanced ET methods

Although riparian ETa values have been obtained with good accuracy using traditional methods and compared with available ET measurements for local areas, ETa values for large areas are still problematic. Also, it has been demonstrated by some authors such as Schultz and Engman (2000) that studies based only on conventional field data collection are often limited because they cover a specific area. In addition, the lack of available and reliable data is a big constraint in the application of different methods to compute accurate ET values. As riparian vegetation interacts environmentally at different scales, remote sensing techniques have been shown to be a reliable alternative to estimate ET, since some of the main constraints about suitable and available data can be overcome providing a precise spatial representation. One important advantage is that it provides detailed and independent ET estimations on a pixel-by-pixel basis among other data as mapping soil properties based on the reflectance variations, land use and land cover using the spectral signatures of vegetation, water requirements, monitoring water availability, and detecting some properties like water stress effects. This wide range of alternative data offers new possibilities for managing soil, water and land resources efficiently.

Remote sensing techniques have provided accurate estimations of ETa at several scales as a function of the spectral resolution of the satellite sensors. ETa can be estimated using data that describes the conditions in the soil-plant-atmosphere system. Thus, remote sensing provides specific information of some of the parameters involved in the ET estimation such as surface temperature, surface soil moisture, water vapour gradients, surface albedo, vegetative cover and incoming solar radiation.

According to Kite and Droogers (2000), methods and methodologies developed to estimate evapotranspiration (ET) using remote sensing data can be classified into four categories: i) satellite-derived feedback mechanisms, ii) biophysical processes based-model, iii) surface energy balance techniques, and iv) physically based analytical approaches. Qi et al. (1998) noticed that the canopy temperature is lower than the air temperature as water evaporates result of the heat extracted from the canopy leaves and from the air near to the canopy surface. This implies a temperature difference and the surface Energy Balance Techniques (SEB) can be applied because they are based on the spatial and temporal variability of the surface temperature (Ts) and air humidity as a reflection of the partitioning of net radiation available to the surface into soil, sensible and latent heat fluxes release to the atmosphere. Thus, the ET is obtained as the residual component of the energy balance equation:

$$\lambda E = Rn - G - H \tag{3}$$

where Rn is the net surface radiation (Wm^{-2}), G is the soil heat flux (Wm^{-2}), H is the sensible heat flux (Wm^{-2}) and λE is the latent heat flux (Wm^{-2}). These methods estimate each parameter of the surface energy balance equation, which are characterised for a diurnal variation and are subject to significant changes from one day to another. To determine these critical variables, these methods first related the flux parameters of the energy balance in terms of variables such as the soil moisture profile, the surface, ground and near-surface air temperatures, and near-surface humidity. The algorithm employed was MEBES (Surface Energy Balance to Measure Evapotranspiration) that was based on the process theory of the original algorithm SEBAL (Surface Energy Balance Algorithm for Land) developed by Bastiaanssen et al. (1998a;b) applying some modifications to the data and condition of the region. The algorithm was chosen since it presents major advantages to compute ETa values among other SEB techniques (Ramos et al., 2008) using both remote sensing and meteorological-ground data for local and regional areas.

3. Sonora River watershed

The study area is the Sonora River Basin (SRB), located in the north-centre of the Sonora State in Mexico at the extreme latitude: 28°27' to 30°54' N and longitude: 110°06' to 111°03' W coordinates. The Sonora River Basin (SRB) is part of the Hydrological Region No. 9 and it has a total surface of 26,010 Km^2 almost 14.8% of the Sonora state. The SRB limits at the north with the USA and at the south with the Cortes Sea. The main river is the Sonora with 277 Km of longitude, starting at the north in the Cananea City with the union of several streams which down from the Magallanes, Los Ajos and Bacanuchi Mountain ranges. Then the river flows to the south crossing the Bacoachi, Chinipa, Arizpe, Banámichi, Baviácora, Ures and Hermosillo cities. In the last one the San Miguel de Horcasitas and el Zanjon afluents join the main river; both of them from the same basin. Finally, the Sonora River goes to the Coast but due to the sandy soil it disappears before reach the sea (INEGI, 2000). The SRB is divided into the Zanjóna and San Miguel de Horcasitas (8826 Km²), Sonora (11680.6 Km²) and the Coast watersheds (5503.4 Km²) (figure 2).

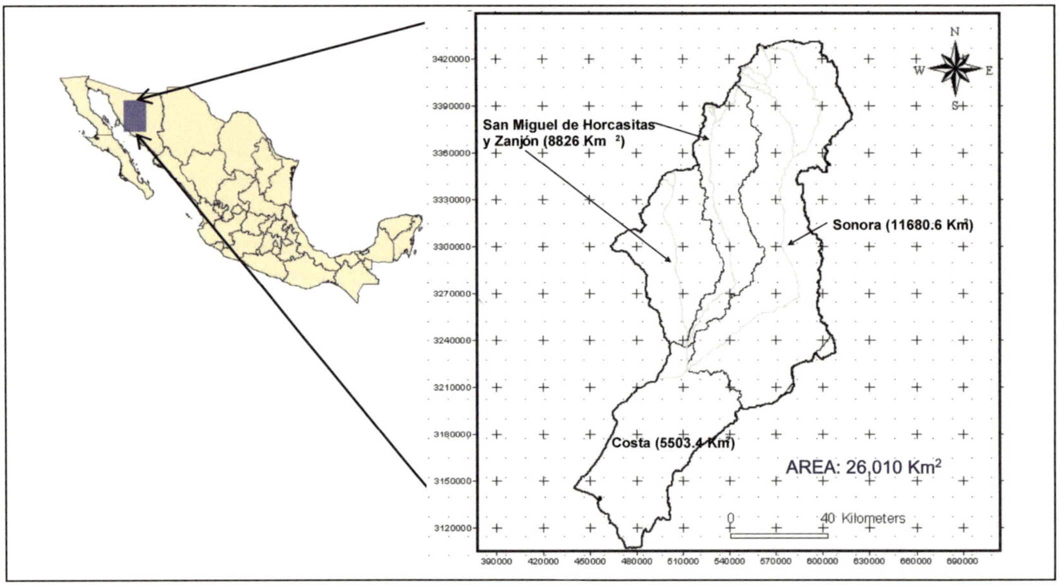

Fig. 2. The Sonora River Basin and its watersheds: Sonora, Zanjón, San Miguel de Horcasitas and the Coast.

The Sonora River has an erratic behaviour and, frequently, the runoff is concentrated into few days during the year. The river flow is controlled by a dam system that includes the Abelardo Rodriguez and Rodolfo Félix Valdés (El Molinito) dams with a total storage capacity of 525 Hm³ (INEGI, 2000). Physiographically, the SRB is divided into two subprovinces (INEGI, 2000):

- Mountain ranges and North Valleys Subprovincie belonging to the Western Mother Mountain Range Province formed in the mountains by volcanic acid rocks with intrusive igneous outcrops and in the valleys by continental sediments. The mountain system is steep from 1000 to 2620 meters above sea level (masl). The weather is dry and semidry and varies as function of the altitude from warm and semi-warm to temperate and semi-cold.
- Sonorenses' Mountain range and plain subprovince belongs to the Sonorense' Desert covering 2/3 of the SRB. This subprovince is characterised ingy lower mountain ranges separate by plains being the mountains narrower than the plains; the plains have 80% of the total area. In the central basin intrusive and lavic igneous rocks predominated as well as metamorphic, ancient limestone and Tertiary conglomerates. In the plains alluvial fans are present with smooth slopes from the near mountains. The weather is very dry semi warm and warm.

There is a high climatological variation observed at the SRB as result of the accidental topography with more than 1000 masl in the mountains and less than 100 masl in the coastline. The climate variation produced two severe droughts (1982-1983 and 1997-1998) which have been related with El Niño event (IPCC, 2007). In fact, during 90s there were three events related to El Niño: 1991-1992, 1993-1994, 1997-1998, and in 2000s two events more in 2002-2003 and 2004-2005 (NOAA, 2008). The mean annual precipitation is 376 mm with a runoff coefficient of 2.8% being the main use the agriculture, followed by domestic, industrial, livestock and recreational. During the drought season it is necessary the groundwater extraction which is used in agriculture (93%), domestic and commercial (4.8%), industry (1.5%) and the rest 0.7% in livestock, recreational and others. The intensive use of groundwater in agriculture has generated an overexploitation of the aquifers present in the SRB. The aquifers are characterized according to the subprovinces in the SRB, thus for the Mountain ranges and North Valleys the non-consolidates material gave a lower infiltration capacity as result of the metamorphic, sedimentary (limestone and conglomerates) and acid extrusive (tuff and riolites) rocks which have a low breaking and porosity reducing the water circulation. Opposite to the Sonorenses' Mountain range and plain subprovince where non-consolidates materials present a high infiltration capacity since the presence of gravel and sand with different sizes and porosity favouring the intercommunication to the water circulation.

The climatic and topographic variation in the SRB provides a diversity of vegetation: pine woods in the north and mesquites (drought tolerant) in the rest of the basin, except irrigated areas aside the rivers and in the coast. The SRB vegetative coverage is quite similar to the state condition due to the extension of the basin, thus almost 50% of vegetal communities correspond to temperate climates under aridity conditions where drought-deciduous low woodlands of thorny and non-thorny trees and bushes are predominant. The thorny shrubs such as Gobernadora (*Larrea tridentata*), Mesquite (*Prosopis laevigata*), Palo Verde (*Parkinsonia aculeate*) and Sangregado (*Jatropha cuneata*) are mainly allocated in hills around the San Miguel de Horcasitas and its affluents, as well as in Bacanuchi, Arizpe and Ures towns (figure 3).

Fig. 3. Mesquite scrubland along the Sonora River and floodplain

The 20% of the vegetation coverage are other type of mesquites as sarcocaulescent scrubland that is a subtype of the xerophyllous scrubland and elms (*Ulmus americana*) (León de la Cruz et al., 2008). These communities are in flat and deep soils, runoff areas, over yermosls, fluvisols or xerosols (figure 4).

Fig. 4. Mesquites as sarcocaulescent scrubland

Grassland represents 13% of the total vegetation and two types can be found in the area: native and introduced. The first one is represented by bluegrass plants and some herbaceous and shrubs used to livestock. The second one is related to the agriculture practise with only 3% of the total (figure 5a). Pine-oak forest and pines with herbaceous secondary vegetation represents 11% of the total vegetation and they can be found in subhumid temperate climates (mountain ranges) (figure 5b).

Fig. 5. (a) Grassland for livestock and (b) pine-oak forest at the SRB

Finally, introduced vegetation is 6.61% from which more or less 4% are agriculture lands where more than 95% of irrigation and the rest is rainfed. Crops are perennial as Lucerne, walnut and sugarcane and seasonal as ryegrass, barley, fodder oats, wheat, garlic and onion for autumn-winter period and fodder sorghum, beans, maize, and vegetables. The winter crops (i.e. wheat) showed a high vulnerability for prolonged drought seasons, thus farmers seemed to change into drought resistance crops or more economically productive crops, especially once the government stopped subsiding grains.

4. Methods

The methodology was divided into three steps: firstly an evaluation of the climate variation, after that a land cover classification was performed in order to identify those areas where the riparian vegetation is present. In this case drought-deciduous low woodlands of thorny and non-thorny trees and bushes were analised, in particular, thorny shrubs such as mesquite (P*rosopis laevigata*) and eml (*Ulmus americana*). Finally, MEBES was applied to estimated ETa and analysed to determine its influence to compute the water availability in the SRB.

4.1 Climatic data analysis

The climatological data available include traditional weather stations (CLICOM database) recording precipitation (P) and air temperature (T) (maximum and minimum), and Observatorios weather stations recording P, T, relative humidity, wind speed and isolation hours. Three Observatorios were analysed: Hermosillo within the SRB, and Nacozari and Empalme at the North and South, respectively, of the neighbour basin at the East.

The CLICOM dataset was updated to 2004 for the Sonora state and it has registered 273 stations. Two filters were applied in order to obtain complete and continuous data for a 20 year period. The results obtained were from 21 stations distributed throughout the entire basin. In order to analyze how the variables related to climate vary in time and space; essential for any study over large areas, data mining techniques were employed. Grouping techniques such as K-means (non-hierarchical) and Ward (hierarchical) methods were evaluated and the result was the basin division into three regions: north, central and south. This definition of homogeneous zones represents the climatic variation in the area accurately.

4.2 Remote sensing analysis

Two hydrological cycles were studied: spring-summer, 1996-1997, and autumn-winter, 2002-2003. Landsat images were acquired (table 1) as well as aerial photographs orthorectified.

The images were used for both classification of the riparian vegetation and estimation of the riparian ETa. For classification, it was considered the spectral response of vegetation to the Visible (VIS) and Infrared (IR) electromagnetic spectrum. A high reflectivity is presented in the near IR as consequence of the leaves structure and in the medium IR where the spectral response is affected by the water content in the surface and a low reflectivity observed in the VIS because the chlorophyll absorption. Also, it was observed the dynamics of the stream interact close with the riparian community characterized by the grow form, size, density, and aerial coverage of the plants (Fischenich and Copeland, 2001).

Sensor	Date of acquisition	Path/row	Day of the Year (DOY)
TM	1997/03/04	35/39 35/40	63
	1997/05/23	35/39 35/40	143
	1997/06/24	35/39 35/40	175
ETM+	2003/03/29	35/39 35/40	88
	2003/04/30	35/39 35/40	120
	2003/01/31	35/39 35/40	31

Table 1. Date and path/row of the Landsat satellite images acquired to the USGS

The images were submitted to a pre-processing analysis to extract useful information from the images, and also to enhance them, to aid in the visual interpretation, and to correct or restore the images if these were subjected to geometric distortion, blurring or degradation. The pre-processing methods used include enhancement of the image, radiometric, atmospheric and geometric corrections, and the georeference of the scene for a chosen map scale and coordinate system.

In order to estimate the ETa, the first part involves the determination of the land surface physical parameters from spectral reflectance and radiance. This stage estimates surface albedo (αo), emissivity (εo), surface temperature (Ts), vegetation indices (NDVI and SAVI), fractional vegetation coverage (Pv) and leaf area index (LAI), and the roughness height (or height of the vegetation, zo). Here ground data are required, however, if the data are not available, it can be replaced using the vegetation indices and considering a standard crop height of 1.0 m. The second stage includes the introduction of meteorological data such as air temperature, humidity, and wind speed at a reference height. The reference height is the measurement height at the weather station (2 m). The last stage includes the estimation of the energy flux parameters and obtains ET as the residual form of the energy balance equation (Bastiaanssen and Bandara, 2001).

5. Results and discussion

5.1 Classification

To achieve the classification, vegetation indices were used since they represent a direct relation to the vegetation health linking the biomass and the leaf area index (LAI) in a spatial basis. In particular, the normalised difference vegetation index (NDVI) was used as well as the soil and wetness indices in the temporal analysis. Additionally, bands 5 and 7 of Landsat were applied to separate areas with high soil moisture, in order to differentiate between irrigated lands and riparian vegetation. The reason to use bands corresponding to the medium IR is to cover a major reflective spectrum allowing the observation of watered overages. Finally, LAI was also monitored, this parameter offers advantages since the geometrical, size and other conditions for the riparian vegetation is very characteristic.

A supervised classification was performed using the NDVI values for each image available, then other classification was done to a multitemporal image generated with the NDVI values. This classification allowed the identification of temporal homogenous areas. Finally, a multitemporal image was produced using a combination of the vegetation, soil and wetness indices, and the ratio between bands 5 and 7. The number of classes was established into 6, as the main interest was the aggregation of riparian vegetation thus mesquites, elms, palo verde, grass, water and bare soil were selected and the agriculture zones were removed from the images. The error matrix showed a confident level of 12% associated to a mix between palo verde and mesquite, and to a very small surfaces. Figure 6 illustrates the aggregation of elms, mesquites and some herbaceous vegetation in the Ures and Topahue aquifer.

The number of hectares obtained for elms was 68% bigger than the mesquites one within the 2000 ha in the Topahue and Ures region. The mesquite tree is a drought resistance obligate phreatophyte and indicator of the water table, and soil and nitrogen fixed, thus it controls the erosion and the soil fertility. The mesquite tree can be 10 m tall and its roots can go up to 50 m deep reaching the water table, also it has lateral roots that can be extended to 15 m. However, as riparian areas are characterised by the growth form, size, density, and aerial coverage of the plants aboveground, it made elms the predominant species since they are a taller facultative phreatophyte usually with 12 to 15 m in height, although they can reach 20 to 30 m or more (figure 7). As well as mesquites, the roots of elms have the ability to go deep

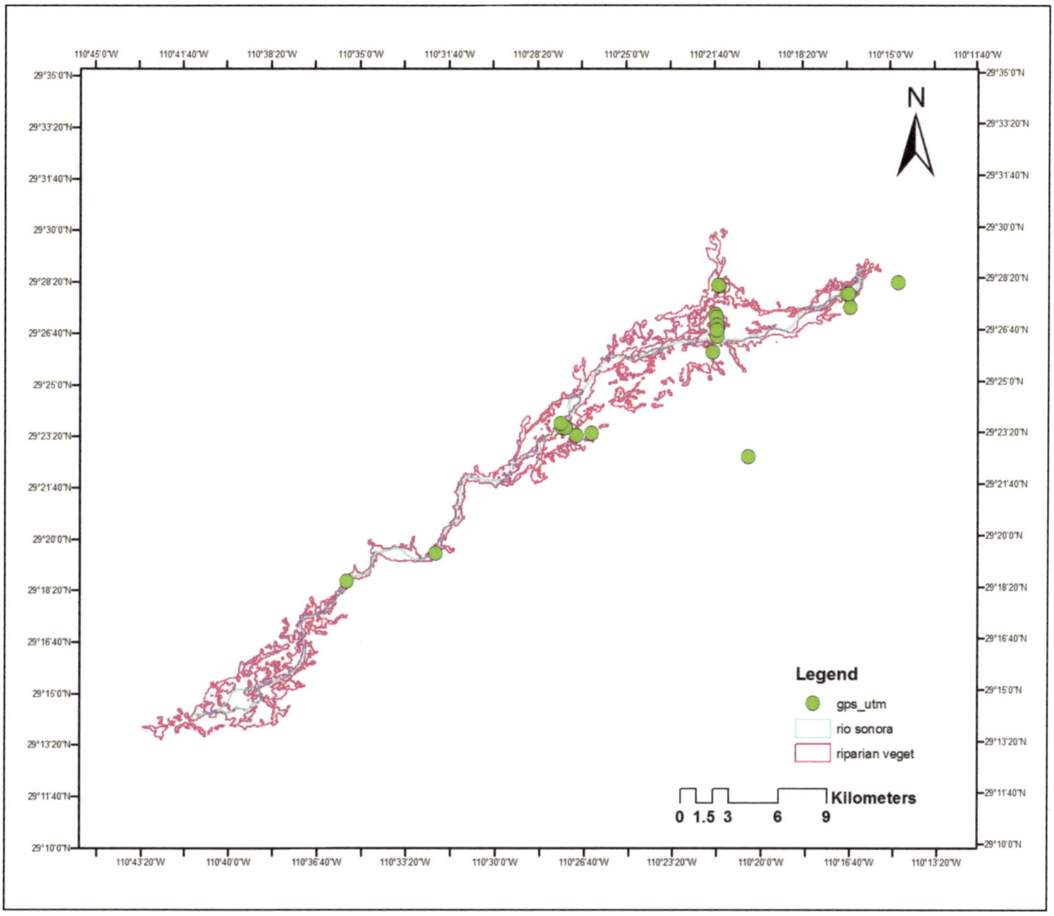

Fig. 6. Elms and mesquites distribution following the river channel and the floodplain and the gps points used in the supervised classification.

to get water, thus the elms can survive during the drought period. The elms posses lateral surface roots that go deep as they need to gather nutrients for the photosynthesis process, approximately from 5 to 20 cm. As elms growth well in moisture and drainage soils, the first precipitation or runoff is enough to obtain the water they need. Elms seeds can be disseminated by wind and water.

The grass area or other secondary herbaceous vegetation was also important for the number of hectares covered (1090 ha). It was observed in the floodplain as a groundcover of the riparian vegetation and their presence is related to the amount of flooding and the radiation allowed by the trees.

Elms LAI values for DOY143 were from 0.7 until 2.0 being the lowest ones during the drought season, whereas for mesquites were from 1.2 to 2.0, these values agreed with Kiniry (1998). LAI for grass or herbaceous vegetation growing around the riparian corridor was between 0.2 and 0.4, and in well irrigated orchards was between 3 and 4. Gazal et al. (2006) found that LAI is more sensitive in intermittent streams sites than in perennial ones due to the depth water table, thus plants modified their canopy structure in order to cope with water scarcity as it was observed in the study site.

Fig. 7. Elms predominance over mesquites and other thorny and non-thorny shrubs

5.2 ETa estimations

Average ETa values in DOY63 (after winter) showed for elms and mesquites 2.1 $mm \cdot d^{-1}$ and 1.5 $mm \cdot d^{-1}$ for groundcover grass of the riparian vegetation, respectively. It was observed a similar ETa rate for elms and mesquites and irrigated orchard. At this time the orchard is less irrigated since the fruit development is not impacted thus important water saving is made. For DOY143 and DOY120 during spring-summer where dry conditions are presented having more sunshine hours during the day, the average ETa values were from 5 to 8 $mm \cdot d^{-1}$ for riparian, and less than 1 $mm \cdot d^{-1}$ for their groundcover grass. Also, ETa values observed in mature orchard (more than 5 years old) and very well irrigated crops such as Lucerne were quite similar to riparian. The water evaporation was around 28 $mm \cdot d^{-1}$ in these DOYs (figure 8).

A 20 year climatological analysis at the centre of the SRB showed that temperature stars to rise from March until June where it reaches its maximum peak (28°C) as well as the wind speed with values of 5.2 ± 0.2 $km \cdot h^{-1}$, contrary to the precipitation that reach its minimum value in May (11 mm). The monsoon season starts late July and August increasing the relative humidity to 50 ± 5%. As the monsoon rains did not wet the saturated zone for DOY143, the climatic conditions imply a high vapour exchange between the surface and the atmosphere , thus the water necessities of the plants need to be covered to prevent wilting and the only source of water is from the aquifer. Due to the elms are located along the main channel their facultative conditions help them to obtain water at bit deep, thus drought conditions impact less although they are more exposed to radiation. This condition is not the same for mesquites where the main disadvantage is the size competence with elms receiving less energy and being protected from the strong winds. Mesquites as obligate phreatophytes needs to go deep to reach the water table establishing a high soil water dependence and, in consequence, reducing their hydraulic capacity to carried water from the roots to the stomata cavities as Gazal et al. (2006) noticed for cottonwood.

6. Conclusions

The riparian vegetation has a very important role in the ecosystems not only because they are a habitat to flora and fauna spices but also because they provide some hydrological control, particularly when flood and drought are present. The vegetation structure is highly related to the geomorphological process of the region being determinant in the valleys where the fluvial soils supply adequate conditions to the development of these plants. Also,

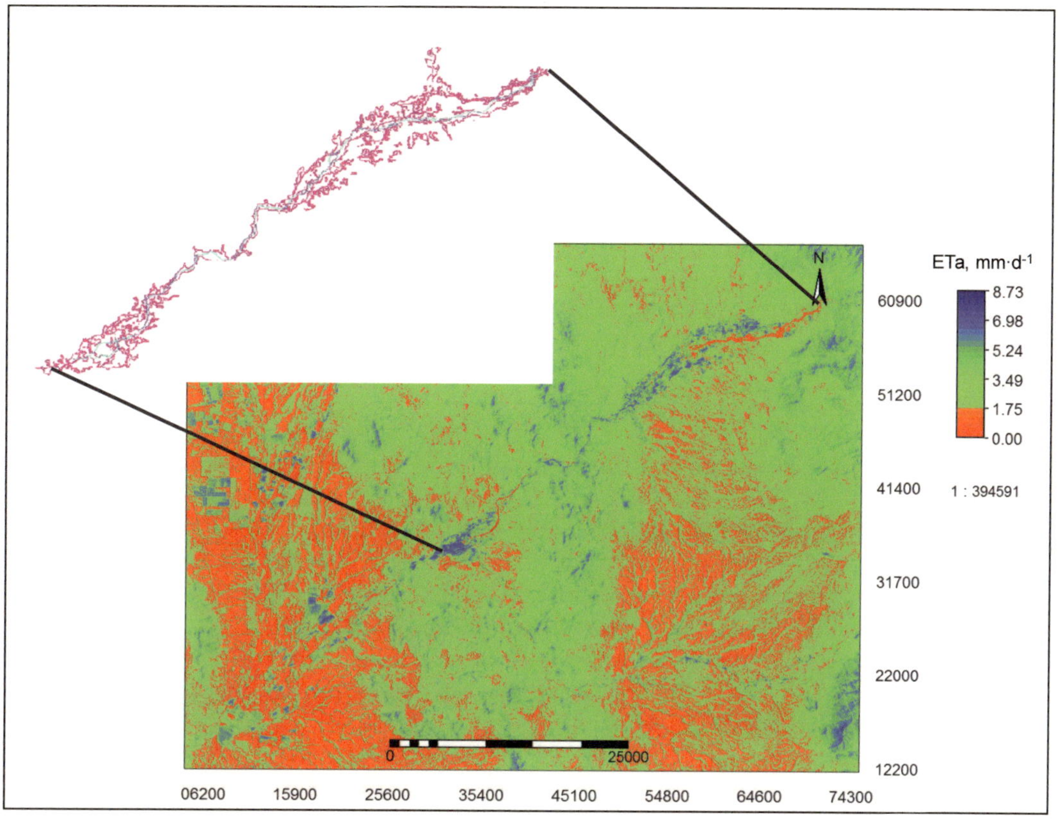

Fig. 8. ETa values for DOY143

the environmental and climatological conditions provide important characteristics since in arid and semi-arid zones the main river channel and the floodplain are undistinguished as result of the erosion and sediments deposition allowing the growth of obligate and facultative phreatophytes as well as herbaceous groundcover vegetation. This resulted in a hydrogeomorphological influence on the vegetation being severely affected but at the same time benefit by flooding and developing the capacity to survive under drought conditions.

In early stages the hydrological regime and the available energy are the main factors to control ET from riparian corridors, although these conditions change as the vegetation growth. Also, as the dry conditions is a limitation, the plant's capacity to extract water from the aquifer is determinant during their mature lives.

The similitude between the irrigated land and the riparian corridor implies a good water resource, in the first case irrigation is the explanation but in the second the only one possibility is the adaptation of plants to obtain water from the aquifers reaching their roots the water table after the depletion of the soil moisture in the saturated zones.

In the SRB, 95% of irrigation is provided by groundwater extraction with an average volume of 5 $m^3 \cdot s^{-1}$ for a total cropped area of 12,300 ha in 2003. In the Ures and Topahue region the cropped lands were almost 2000 ha in 2003 with a total volume of 3.5 $m^3 \cdot s^{-1}$, as ETa values were similar to very well irrigated lands, this would imply a volume of 6.6 $m^3 \cdot s^{-1}$ for riparian corridors considering elms, mesquites and grass. This raw estimation pointed out the amount required from the aquifer to cope with irrigation and riparian corridors without consider other uses. This amount needs to be carefully used since there is not still enough data to provide confident recharge values to these aquifers (Ures and Topahue), however, it is clear the high influence of the ETa to compute water requirements in arid zones. Further work is required in order to compute accurately each component of the water balance paying also attention to the precipitation and interception effect on the riparian corridor since its rapid recovering after the first precipitation during the monsoon months.

7. Acknowledgments

This project was founded by the Sonora State Water Commission (CEA) in 2005 and by the Instituto de Ingeniería, UNAM, in 2008.

8. References

Allen Jr. L.H. (1999). Evaporation Responses of Plants and Crops to carbon dioxide and temperature. 37-70. Kirkham M.B.Editor, Water Use in Crop Production. The Harworth Press, pp. 385, ISBN 1-5 6022-068-6. NY, USA.

Allen R.G., Pereira L.S., Raes D. and Smith M. (1998). Crop Evapotranspiration. Guidelines for Computing Crop Water Requirements. Irrigation and Drainage, Paper No. 56, pp.300. UN Food and Agriculture Organisation. Rome, Italy.

Bastiaanssen W.G.M., Menenti M., Feddes R.A.and Holtslag A.A.M (1998a). A remote sensing surface energy balance algorithm for land (SEBAL), Part 1: Formulation. J. of Hydrology, 212-213, 198-212.

Bastiaanssen W.G.M., Pelgrum H., Wang J., Ma Y., Moreno J., Roerink G.J. and van der Wal T. (1998b). The Surface Energy Balance Algorithm for Land (SEBAL): Part 2 validation, J. of Hydrology, 212/213: 213-229.

Bastiaanssen, W. G. M., and Bandara, K. M. P. S. (2001). "Evaporative depletion assessments for irrigated watersheds in Sri Lanka." *Irrig. Sci.*, 21, 1–15.

Bendix, J. and Hupp, C. (2000). Hydrological and geomorphological impacts on riparian plant communities, *Hydrol. Process.*, 14, 2977-2990

Calder, I. R. (1998). Water-resource and land-use issues. SWIM Paper 3. International Water Management Institute. ISBN: 92-9090-361-9. Colombo, Sri Lanka.

Ehlers W. and Goss M. (2003). Water Dynamics in Plant Production CABI Publishing, ISBN 0-85199-694-9 273 pages, Wallingford, UK

Fischenich J.C. and Copeland R.R. (2001). Flood Damage Reduction Research Program Environmental Considerations for Vegetation in Flood Control Channels, Flood Damage Reduction Research Program, U.S. Army Corps of Engineers, pp. 70, Washington

Gazal R.M., Scott R.L., Goodrich D.C. and Williams D.G. (2006). Controls on transpiration in a semiarid riparian cottonwood forest, *Agricultural and Forest Meteorology*, 137, 56-67

Goodrich D.C:, Scott R., Qi J., Goff B., Unkrich C.L., Moran M.S., Williams D., Schaeffer S., Snyder K., MacNish R., Maddock T., Pool D., Chehbouni A., Cooper D.I., Eichinger W.E., Shuttleworth W.J., Kerr Y., Marsett R. and Ni W. (2000). Seasonal estimates of riparian evapotranspiration using remote and in situ measurements, *Agricultural and Forest Meteorology*, 105, 281-309

Graf, W.L. (1988a). Fluvial processes in dryland rivers, Springer-Verlag, New York

Graf, W.L. (1988b). Definition of flood plains along arid regions rivers, Flood Geomorphology, Baker V.R., Kochel R.C., Patton P.C. (eds), Wiley and Sons, pp. 231-242, New York

Instituto Nacional de Estadística, Geografía e Informática (INEGI). 2000. Síntesis de Información Geográfica del estado de Sonora. Editado por el Instituto Nacional de Estadística, Geografía e Informática. México.

IPCC (2007). Climate Change: Synthesis Report, IPCC, Valencia, Spain, November 2007.

Kiniry J.R. (1998). Biomass accumulation and radiation use efficiency of honey mesquite and eastern red cedar, *Biomass and Bioenergy*, 15, 6, pp. 467-473

Kite G. and Droogers P. (2000). Comparing estimates of actual evapotranspiration from satellites, hydrological models, and field data: A case study from Western Turkey. Research Report No. 42, International Water Management Institute, pp. 32. Colombo, Sri Lanka.

Lakshmi V. and Susskind J. (2001). Utilization of satellite data inland surface hydrology: sensitivity and assimilation. *Hydrol. Process*, 15, 877-892.

Leon de la Cruz J.L., Rebman J., Domínguez-Leòn M. and Domínguez-Cadena R. (2008). The vascular flora and floristic relationship of the Sierra de la Garganta in Baja California, *Revista Mexicana de Biodiversidad*, 79, 29-65

Menenti M. (2000) Evaporation. Chapter 8, 156-196. Gert A. Schultz and Edwin T. Engman, Eds. Springer, ISBN 3-540-64075-4. Heidelberg, Germany

Molden D. (1997). Accounting for water use and productivity. International Irrigation Management Institute, SWIM Paper 1, pp.16, Colombo, Sri Lanka.

NOAA (2008). Oceanic Niño Index: Cold and Warm Episodes by Season, Climate Prediction Centre Web page consulted on May, 2008 http://www.cpc.ncep.noaa.gov/products/analysis_monitoring/ensos tuff/ensoyears.shtml

Pike, J. G. 1964. "The estimation of annual run-off from metrological data in tropical climate." *Journal of Hydrology* 2: 116-123.

Qi J., Moran M.S., Goodrich D.C., MArsett R., Scoot R., Chehbouni A. and Schaeffer S. (1998). Estimation of evapotranspiration over the San Pedro riparian area with remote and in situ measurements, American Meteorological Society Symposium on Hydrology, Phoenix, Arizona, 11-16 January 1998, Session 1: Paper 1.13

Ramos J.G., Cratchley C.R., Kay J.A., Casterad M.A., Martı´nez-Cob A., Domı´nguez R. (2009). Evaluation of satellite evapotranspiration estimates using ground-meteorological data available for the Flumen District into the Ebro Valley of N.E. Spain, *Agricultural water management*, 96, 638 – 652

Schultz G. A. and Engman E.T. (2000). Remote Sensing in Hydrology and Water Management. Chapter 1, 3-14. Gert A. Schultz and Edwin T Engman Eds. Remote Sensing in Hydrology and Water Management, 484pp. Springer. ISBN 3-540-64075-4. Heidelberg, Germany.

Scott, R.L., Goodrich, D.C., Levick, L.R. (2003). A GIS-based management tool to quantify riparian vegetation groundwater use. Proc. 1st Interagency Conf. on Research in the Watersheds, K.G. Renard, S. McElroy, W. Gburek, E. Canfield, and R.L. Scott (eds.), Oct. 27-30, Benson, AZ, pp. 222-227. (112 kb PDF)

Turc, L. (1954). "Le bilan d'eau des sols: relation entre les précipitations, l'évapotranspiration et l'écoulement." Annales agronomiques Série A: 491-595

Wester, P., Scott, C. A. and Burton, M. (2005). River basin closure and institutional change in Mexico's Lerma-Chapala Basin, in: M. Svendsen (Ed.) Irrigation and River Basin Management: Options for Governance and Institutions, pp. 125–144, Chapter 8 (Wallingford, UK: CABI Publishing).

17

Evapo-transpiration in Ecological Engineering

Andrzej Białowiec[1], Irena Wojnowska-Baryła[1] and Peter Randerson[2]
[1]*University of Warmia and Mazury in Olsztyn, Department of Environmental Biotechnology, ul. Sloneczna 45 G, 10-900 Olsztyn,*
[2]*Cardiff University, School of Biosciences, Cardiff, CF10 3AX,*
[1]*Poland*
[2]*U.K.*

1. Introduction

1.1 Constructed wetlands – ecological engineering technologies

Odum (1971) described ecological engineering as half science and half engineering: techniques of designing and operating the economy with nature. Ecological engineering or ecotechnology are defined by Guterstam (1996) as the design of human society with its natural environment for the benefit of both. One of the technologies included in ecological engineering is using constructed wetlands for wastewater treatment.

Constructed wetlands (vegetation filters or treatment wetlands) are artificial complexes of water, matrix, vegetation and the associated invertebrate and microbial communities designed to simulate the ability of natural wetlands to remove pollutants from water (Brix, 1987; Kangas, 2004). They provide an inexpensive and reliable method for treating a variety of waste waters such as sewage, landfill leachate, mine leachate, agricultural run-off, and are comparatively simple to construct, operate and maintain (Randerson, 2006, Randerson *et al.*, 2007).

Based on hydraulic regime CWs may be divided into two main groups: surface flow systems (SF), and subsurface flow systems (SSF) (Kadlec *et al.*, 2000). The latter may employ horizontal flow (HF), vertical flow (VF) or tidal flow (TF) hydraulic regimes and these may be combined in hybrid systems to optimize pollutant removal (Randerson, 2006). Both aerobic and anaerobic processes are involved, but degradation of carbonaceous matter to CO_2 and nitrification requires availability of oxygen. That may be achieved most efficiently in compact VF systems; as the surface is flooded, air is forced into the bed, while effluent percolates downwards through the matrix. HF beds typically achieve lower oxygen transfer rates but, with largely anaerobic conditions, they are effective in removing nitrogen to atmosphere *via* denitrification. SF wetlands most resemble natural wetlands, as the water level is typically above the soil surface (Randerson, 2006). With HF, water flows laterally below the surface, through a gravel bed. Oxygen is consumed by microbial activity, and oxygenation of the bed is limited by surface diffusion and transport via aerenchyma tissue to the root zone, so that anaerobic conditions predominate. Hence nitrification is limited by oxygen, and denitrification is limited by the supply of nitrate and usable carbon compounds. In the VF system, pulses of water flow downwards through layers of increasing particle size. Air is drawn into the bed between each pulse of water. Removal of BOD and nitrification is very efficient, for a given surface area, but nitrate and phosphate are typically high in the outflow (Cooper, 1999).

In subsurface systems the root zone is limited mostly to an upper layer of 30-40 cm, below which the influence of plants is reduced. Shallow HSSF beds (0.2–0.5 m in depth and with a water table close to the top substratum) removed organic matter and nitrogen at very high rates and higher than deeper beds (Garcia *et al.*, 2004, Albuquerque *et al.*, 2009). Those beds are normally more oxidized, possibly due to the shallow water and enhanced oxygen diffusion at the air–water interface.

1.2 Importance of plants in constructed wetlands

Plants commonly used in constructed wetlands include: cattail (*Typha latifolia* L.), reed (*Phragmites australis* (Cav.) Trin ex Steudel), rush (*Juncus effusus* L.), yellow flag (*Iris pseudacorus* L.), and manna grass (*Glyceria maxima* (Hartm.) Holmb.). As well as these typical natural wetland plant species, willow (*Salix sp.*) may be used in constructed wetlands with high efficiency. Willow vegetation filters have been shown to be effective in cleaning polluted drainage water from agricultural land, and as a final stage in wastewater treatment (Aronsson, 2000; Aronsson & Perttu, 2001; Elowson, 1999; Kowalik & Randerson, 1994; Perttu & Kowalik, 1997). Willows have been used successfully in the treatment of agricultural runoff and leachate from landfill sites (Bialowiec *et al.*, 2007; Duggan, 2005; Randerson, 2006) and are especially successful at removing high levels of ammonia and nitrogen from solution. A review of the potential for the use of willow filter beds and short rotation willow coppice (SRC) to treat landfill leachate by Duggan (2005) concluded that a number of studies showed success of willow filter beds in treating leachate and that the leachate treatment improved with the number of the willows.

In a CW, the plant root-soil interface plays a significant role in the removal of pollutants. Oxygen released from roots creates aerobic conditions in the otherwise anaerobic rhizosphere, which induces growth of both heterotrophic and autotrophic aerobic bacteria (nitrifiers) and the aerobic breakdown of organic material (Brix, 1997). Reddy *et al.* (1989) reported that approximately 90% of the oxygen available in the rhizosphere was transported there by aquatic vegetation. The other 10% arrives by soil surface diffusion (Lloyd *et al.*, 1998).

Plant uptake also plays an important role in the enhancement of N removal, especially in treatment wetlands containing fast-growing plants such as willows (Randerson, 2006). Release of oxygen into the rhizosphere and uptake of nutrients are not the only ways that the presence of plants can facilitate the removal of pollutants.

Most macrophytes release carbon compounds into the rhizosphere (*e.g.* polysaccharides, sugars, amino acids, organic acids and fatty acids) that may be removed through aerobic oxidation, fermentation and denitrification pathways (Pinton *et al.*, 2007), especially the ones with low molecular weight. Other compounds may be converted into recalcitrant organics (*e.g.* humins). Organic rhizodeposition includes lysate, root border cells and root exudates, released passively (diffusates) or actively (secretions). The main mechanisms of root exudation are diffusion, ion-channels, vesicle transport and exocytosis. In addition these exudates, which make up 5-25% of photosynthetically fixed carbon, assist the degradation of toxic organic chemicals (Brix, 1987).

Aerenchyma also plays a role in the removal of nitrogen (N) from wetlands through releasing N_2 and N_2O produced by anaerobic denitrification of NO_3^- into the atmosphere (Reddy *et al.*, 1989). Plants are also largely responsible for the release of CO_2 and CH_4 from the soil, with up to 90% of the net efflux of methane from peat lands being attributed to gas transport by plants (Beckmann and Lloyd, 2001).

As well as the influence of plant species, the release of oxygen into the rhizosphere is related to some extent to photosynthesis, light intensity, stomatal aperture, and temperature (Stein and Hook, 2005). The amount of oxygen in the rhizosphere therefore fluctuates over diurnal periods and varies between seasons (Williams *et al.*, 2010). There is higher oxygen release in periods of illumination and even in periods of relatively low light intensity the amount of oxygen released into the rhizosphere can meet the respiratory oxygen demand of the roots and micro-organisms in the rhizosphere (Sheppard and Lloyd, 2002). In view of the differences between plant species in the release of oxygen into the rhizosphere, assessing the oxidizing/reducing capability of the plant system may improve understanding of the effect of this flux on the remediation of LL in treatment wetlands. The majority of microbial processing that occurs in wetlands is attributed to biofilms made up of communities of algae, bacteria, protozoa and invertebrates. It has been shown that up to 90% of organic and inorganic N can be removed from LL by biofilms (Welander *et al.*, 1998).

Other studies have shown the importance of evapotranspiration during hot periods in natural wetlands and also in constructed wetlands. Evapotranspiration may be defined as water evaporation from soil or/and open water surface, supported by transpiration of plants (Allen *et al.*, 1998).

Tchobanoglous (1987) informs that water plants are more productive than terrestrial ones, because of higher resistance to environmental changes and high photosynthetic efficiency. Annual transpiration of water plants (hydrophytes) is in the range of 1000-1400 mm/year (Siuta, 1996). Williams et al. (1987) compiled a water balance for Jack Valley wetland (155 ha) in USA. Water loss due to evaporation was about 1524 mm/year. Reeds, cattails and, near banks, willows and poplars were the most common colonizing plants. Raisin (1999) assessed the annual evapotranspiration to be 1100 mm/year at the Reids Wetland in NE Victoria, Australia, where *Phragmites australis, Juncus usitatus* L., and *Carex fascicularis* Sol. Ex Boot dominated. Also, Wieüner et al. (1999) observed the loss of water at wetlands planted with reed for industrial sewage treatment. The measured annual evapotranspiration was about 1500 mm/year. Białowiec at al. (2006), reported the total lack of effluent from a hydrophyte system planted with reed designed for wastewater treatment.

Evapotranspiration as a process might be used in constructed wetlands for decreasing the volume of treated wastewater or landfill leachate (Dobson & Moffat, 1995). Such soil-plant systems are planted with macrophytes like willows, poplars and reeds, as these are the plants that are recommended by the US EPA (2000) for landfill leachate evapotranspiration. Systems where evapotranspiration dominates might be called evapotranspirative soil-plant systems (Bialowiec et al., 2007) or zero-discharge systems (Bialowiec & Randerson, 2010).

In this chapter the functioning of a zero discharge system located in Poland is described: an evapotranspirative system (ES) with reed at the landfill in Zakurzewo near Grudziądz. Additionally, the possibilities of employing transpiration measurements as a tool in phytotoxicity of landfill leachate assessment are presented.

2. Evapotranspirative system for landfill leachate treatment

2.1 Landfill leachate

One of the most important problems associated with landfill sites is the leachate, that may cause significant environmental impacts. Landfill leachate (LL) is formed when rain water

filters through a landfill becoming contaminated with pollutants (Kjeldsen *et al.*, 2002). The composition of LL varies from site to site and includes various organic and inorganic compounds that may be either dissolved or suspended, but is generally characterised by high levels of total nitrogen (N), ammonia (NH_4^+ or NH_3, depending on pH), chemical oxygen demand (COD), sodium (Na), chloride (Cl) and a low ratio of BOD/COD (Jones *et al.*, 2006). A high concentration of ammonia in the range from 50 to 5000 mg/dm^3 (Yildiz and Unlu, 2003) is the main problem causing toxicity of LL, but also oxygen depletion and changes in the fauna and flora of surface water bodies (Christensen *et al.*, 1992, Kjeldsen *et al.*, 2002). It was shown that low phosphorus concentration, usually not exceeding 60 mg/dm^3, can act as a limiting factor in applying LL as fertilizer (Hasselgren, 1998).
The most commonly used method of treatment is to transfer landfill leachate directly to a wastewater treatment plant. In this case, in practice the major treatment process is dilution. Another way of leachate treatment is its recirculation on to the original waste heap. Unfortunately, when this technique is poorly operated, it causes water overloading of the landfill system. There are several technologies of landfill leachate treatment on-site. However, because of high variation in the production and composition of leachate, treatment using just one process is not efficient. Therefore the combined systems of physical, biological and chemical technologies are needed, increasing the cost of treatment.

2.2 Using plants for landfill leachate treatment

Despite differing views on leachate treatment, many experts agree that on-site treatment is desirable, since it is easier to operate, and more economical in terms of cost and energy (Bulc, 2006). Owing to low investment and maintenance costs, there is a growing interest in applying plants in LL disposal (Ashbee & Fletcher, 1993). On-site treatment using constructed wetlands (CW) is one of the low cost methods of LL treatment which has been widely practiced in several countries for many years (Kadlec *et al.*, 2000, Schnoor *et al.*, 1995, Wallace and Knight, 2006, Vymazal & Kropfelova, 2008). CW with reed (Wojciechowska *et al.* 2009, Wojciechowska & Obarska-Pempkowiak, 2008) as well as with willow (Bialowiec *et al.*, 2007, Randerson, 2006) have been shown to be effective in treating leachate from landfill sites and are especially successful at removing high levels of nitrogen from solution.
The use of plants in landfill leachate evapotranspiration has been widely tested. Two willow species: *Salix alba* and *Salix nigra* were used for landfill leachate disposal by Alker *et al.* (2003). Evapotranspiration values ranging from 8 to 9 mm/d and 5 mm/d obtained from willow stands, were determined by Persson & Lindroth (1994), and Elowson (1999) respectively. In *Salix viminalis* L. and *Salix aquatica* Sm. stands, located at municipal and industrial landfill sites, evapotranspiration was significantly higher than the annual precipitation rate in Finland (Ettala, 1992). Transpiration by willow stands may be high enough to affect significantly the ground water table level (Dulohery *et al.*, 2000) and exceed annual precipitation (Persson, 1995). Research on young willow sprouts revealed that shortly after planting, evapotranspiration was effective at reducing landfill leachate volume. Agopsowicz (1994) determined that evapotranspiration of 3-month willow sprouts was 1.6-1.8 times higher than an average precipitation rate in Poland, which is about 600 mm/y. Białowiec *et al.* (2003) confirmed that transpiration of 3-month old sprouts of *Salix amygdalina* L. resulted in evapotranspiration ranging from 80 to 90 % of the supplied volume.
The decrease in leachate volume due to evapotranspiration from soil-plant systems provides a great advantage in landfill leachate disposal. Evapotranspiration technology might therefore be used in soil-plant systems for landfill leachate treatment.

In Polish climatic conditions (mean precipitation about 550 mm/year), it is possible to achieve a negative water balance for the landfill. One of the most important advantages of this technology is the low cost of investment and operation in comparison to other ways of leachate treatment.

Currently, plant species resistant to pollutants in leachate and characterized by high transpiration and adaptation to a wide range of pollutant concentrations are sought. So far, the use of different species of willows for leachate evapotranspiration has been tested mostly in Sweden (Hasselgren, 1992, Dimitriou & Aronsson, 2003, 2005, Dimitriou *et al.*, 2006), Finland (Ettala, 1992), Great Britain (Alker *et al.*, 2003) and also in Poland (Agopsowicz, 1994, Białowiec et al, 2003, Białowiec 2005, Białowiec *et al.* 2007). However, numerous reports concerning transpiration ability, gross biomass growth and resistance to salinity of halophytes, point to the suitability of common reed– *Phragmites australis* in soil-plant systems for landfill leachate evapotranspiration.

Reed can by characterized by high transpiration ability, gross biomass growth, resistance to high concentrations of pollutants in soil solution, resistance to waterlogged soil and anaerobic conditions in soil. There is also a simple method of breeding, planting and cultivation. Reed was found to be the most suitable plant for landfill leachate evapotranspiration, because of very high evapotranspiration about 1800 mm/year (5 mm/d), biomass yield about 12 Mg d.m./ha/y and high resistance to pollutant loads supplied with leachate up to 6.7 MgN/ha and 1.5 $MgCl^{-}$/ha (Białowiec, 2005).

On the basis of recent results, soil-plant systems with reed - *Phragmites australis* have been implemented at the landfill in Zakurzewo near Grudziądz, Poland. This chapter presents the design and two years operational experiences with this system.

2.3 Landfill leachate treatment in evapotranspirative system – case study

2.3.1 Landfill site

The municipal landfill is located at Zakurzewo, about 15 km north of Grudziądz. Total area of the site is 13.5 ha, and over 2.5 ha are reserved for piles of waste material (Agopsowicz *et al.* 2006). Zakurzewo landfill was built in 1997 and is operated by SWECO company based on waste disposal in "energetic piles". This was the second landfill in Poland involving waste disposal using this technology for energy production. Design guidelines were based on the results from the pilot installation in Sweden (Hagby) (Trzosińska & Podkański, 2000). Economic analysis of the investment assumes biogas generation and recovery, and selling various recyclable materials (plastic, metal, glass, textiles), as well as the likely sale of part of the waste as a fuel (Trzosińska & Podkański, 2000).

Firstly, four waste piles were constructed according to design guidelines. A rectangular basin (70 x 38 m), was prepared with a height of 3 m and a vertical section as an inverted trapezium, enclosed by a 5 m high embankment. The basin was sealed by a 50 cm layer of clay. The predicted capacity of each pile was 20 000 Mg of waste (in practice the mass of collected waste in the piles ranged from 19 154 Mg to 23 269 Mg, comprising 96-116 % of predictable capacity). Currently the piles are covered by an isolation layer, and are excluded from biogas recovery. Two bigger waste piles (90 x 90 x 10 m), working alternately from 2002, are now almost filled. The predictable capacity of these piles was 84 000 Mg of waste (in practice the collected waste mass ranged from 62 869 Mg to 89 333 Mg, comprising 75-106 % of predictable capacity). During the period of landfilling, degassing wells with a diameter of 1 m were installed, consisting of broken stone inside which was a 100 mm

diameter perforated pipe made of PE. In five pipes there are heads joined to a gas collector. According to the landfill operator, in the years 2002 and 2003, 6 700 Mg of sewage sludge with moisture about 80 % was deposited in one of the bigger piles.
Leachate is collected by a system of drainage pipes, and flows by gravity to a collection well, and then pumped to the main collector. From there, leachate passes to a concrete tank with a volume of 76 m³ that is joined to a newly built tank with a working volume of about 2 500 m³. SWECO technology recommends heating the waste to 40 °C before pumping under pressure into the waste piles, using elastic hoses inserted near the surface of the waste layer. This allows irrigation of the waste independently of the degree of filling of the pile. Currently this system is not working (leachate is not heated and not forced into the piles). Hence about 700 m³ of leachate per month were taken for treatment to the wastewater treatment plant, at a total cost of about 29 000 $/year.

2.3.2 Evapotranspirative system construction and leachate characteristics

On the top of the one of the big waste piles, a soil-plant system with area of 2 500 m² was built. The base of the soil-plant system consists of two liners: 0.5 m of compacted clay soil, overlain with stabilized sewage sludge (dosage 250 Mg/ha). The soil-plant system is surrounded by a 0.5 m high clay bank with a gradient of 1:3 (H:L) (figure 1). Existing gas wells inside the soil-plant systems have been isolated by additional clay banks to a height of 0.5 m to reduce water percolation into the gas wells (Fig. 1) (Białowiec *et al.*, 2007).

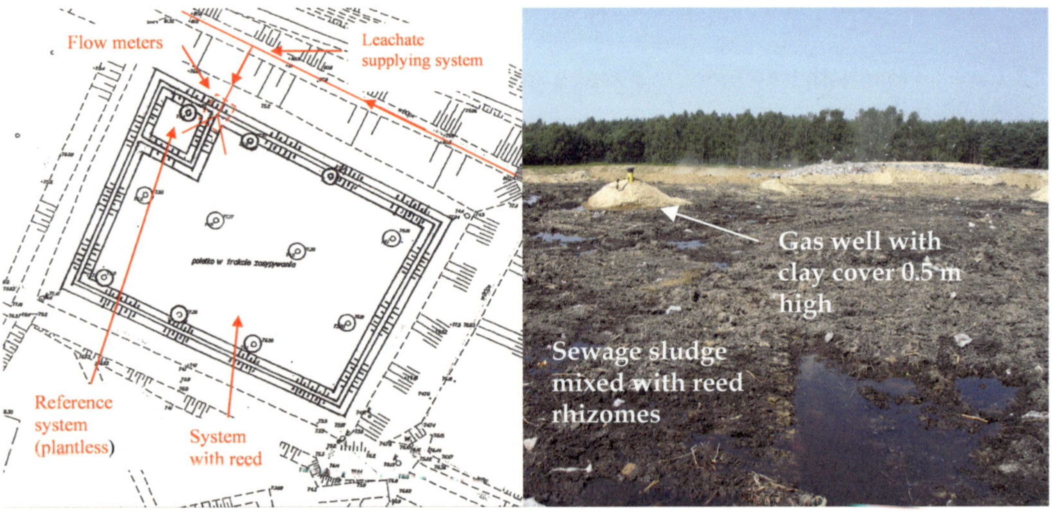

Fig. 1. Soil-plant system construction.

Seedlings of reed (rhizomes) were placed into the sewage sludge liner and irrigation of the soil-plant system started from that time (1 July 2006). Firstly, clean rain water, gathered in a chamber, was supplied. After the reed had established (5 weeks), irrigation of leachate started, using the existing leachate recirculation system.
Reed was flooded by both incoming rain water and leachate in accordance with the design. From the beginning of the project the amount of irrigated clean water, leachate, and also precipitation were measured.

2.3.3 Field measurements

The efficiency of leachate treatment is evaluated in respect of water balance, based on the amount of leachate pumped into the soil-plant system, precipitation measurements, and meteorological data of evaporation measured with a Peach Evaporometer.

Treated landfill leachate was characterized by low concentration of organic compounds (COD - 997.0 mgO_2/dm^3) which consisted of about 20 % dissolved solids. The nitrogen concentration was very high (ammonium nitrogen 576 mg/dm^3). It caused a slightly alkaline reaction about 8.5 pH.

During the second year, in the peak of the vegetative season, the degree of leaf stomata opening (transpiration), was measured on 100 reed plants, using a Porometer AP-4-UM-3 Delta-T Devices [$mmolH_2O.m^{-2}.s^{-1}$]. Measurements were made every day during July. These results were used to assess the behavior of reeds in the toxic conditions of landfill leachate supply. During recent laboratory studies, the parameter - degree of leaf stomata opening (transpiration) had been used to determine the resistance of reeds to pollutants contained in leachate. The lowest effective concentration causing a toxic effect (LOEC) was estimated for leachate from Zakurzewo, to be at a level of 9 % share of leachate in water supplied to plants (Białowiec & Kasiński, 2008). Direct measurements of the degree of leaf stomata opening were compared to the value obtained for LOEC.

2.3.4 Leachate treatment efficiency

In the period (152 days) from 1 July 2006 to 30 November 2006 (end of vegetation season), the following amount of water had been supplied to the soil-plant system: clean rain water 231.0 m^3 (collected and stored during July in the unused waste deposition cell), precipitation 648.1 m^3 and leachate 283.0 m^3. The share of landfill leachate in the total amount of water supplied to the soil-plant system was 24.4 % (Tab. 1). During that period the hydraulic loading rate (HLR) of the soil-plant system by leachate was low, about 0.7 mm/d. In next year the share of leachate increased to 38 %, higher HLR (about 2.4 mm/d), and a much higher efficiency of water loss (about 4200 m^3 compared to about 1100 m^3), was obtained. Recent experience indicates that hydraulic loading rate of reed stands by leachate may be higher than the value required to balance transpiration loss (5 mm/d) (Białowiec & Wojnowska-Baryła, 2007). We decided to use a low HLR because we wanted to avoid the toxic influence of leachate on young, developing plants.

Year of operation	The share of leachate in total water supplied into soil-plant system [%]	Leachate hydraulic loading rate [mm/d]	Evapotranspiration efficiency [m^3]
2006 (initial 35 days)	0.0	0.00	270
2006 (total 152 days)	24.4	0.74	1100
2007 (270 days)	37.6	2.36	4241
Recommended values	9.0*	<5.00**	-

Table 1. Parameters of soil-plant system with reed in Zakurzewo. * LOEC – (Białowiec, Ksiński, 2008), ** (Białowiec & Wojnowska-Baryła, 2007)

The calculation of water balance showed that evapotranspiration from the soil-plant system did not exceed the total hydraulic loading rate; the difference was 84 mm after the end of first year (Fig. 2), and about 60 mm at the end of second year of operation (Fig. 3). However, the retention volume of the soil-plant system (500 mm) was high enough to accommodate all incoming water within the system, and evapotranspiration rate was high enough to reduce level of water gathered in the system during the second year (Fig. 3).
Measurements of the degree of leaf stomata opening indicated the mean value of transpiration rate to be 261.3 [mmolH_2O.m^{-2}.s^{-1}]. Comparison with laboratory results (Białowiec & Kasiński, 2008), indicated a strong capacity for adaptation by reed growing at the landfill. The electrical conductivity (EC) of the leachate collected in the soil-plant system at the landfill was 10.8 mS/cm. During laboratory tests a similar degree of reed leaf stomata opening, 250.3 [mmolH_2O.m^{-2}.s^{-1}] was measured for a solution with EC of only 1.1 mS/cm (Fig. 4). This indicates that during one year of operation of the system, the plants' resistance to higher salinity had increased about ten times. This is a promising aspect for implementation of this technology.
In addition the influence of the soil-plant system on biogas production was examined. Static measurements of CH_4 content in biogas gathered from the waste pile with the soil-plant system (biogas sample taken from a gas well) showed the CH_4 concentration to be 60 %. Dynamic measurements gave similar results (biogas sample taken from the pump station) during 0.5 h of pumping, indicating that the soil-plant system had no negative influence on fermentation processes inside the pile.

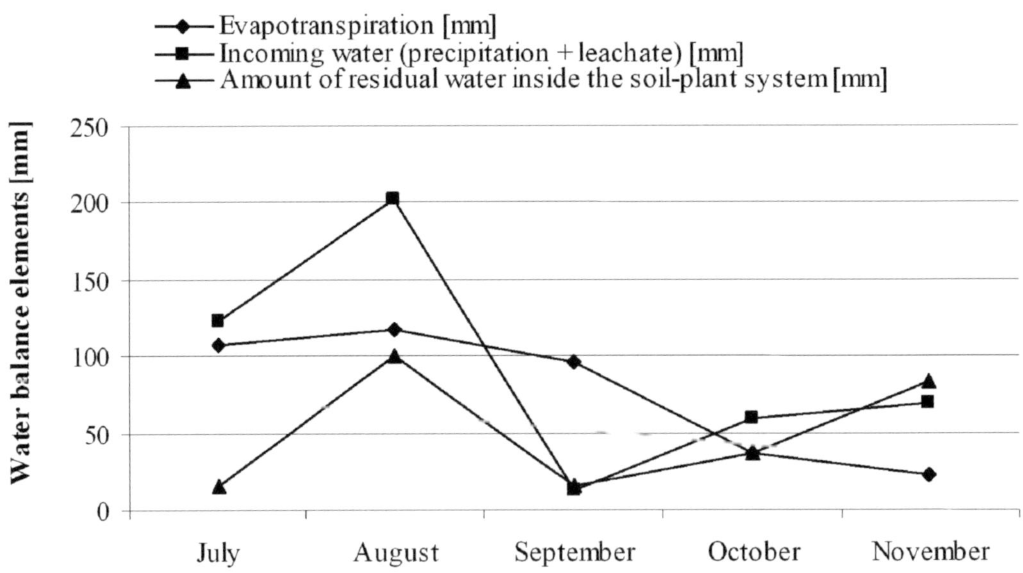

Fig. 2. Water balance elements of the soil-plant system supplied by landfill leachate during 152 days period of the year 2006.

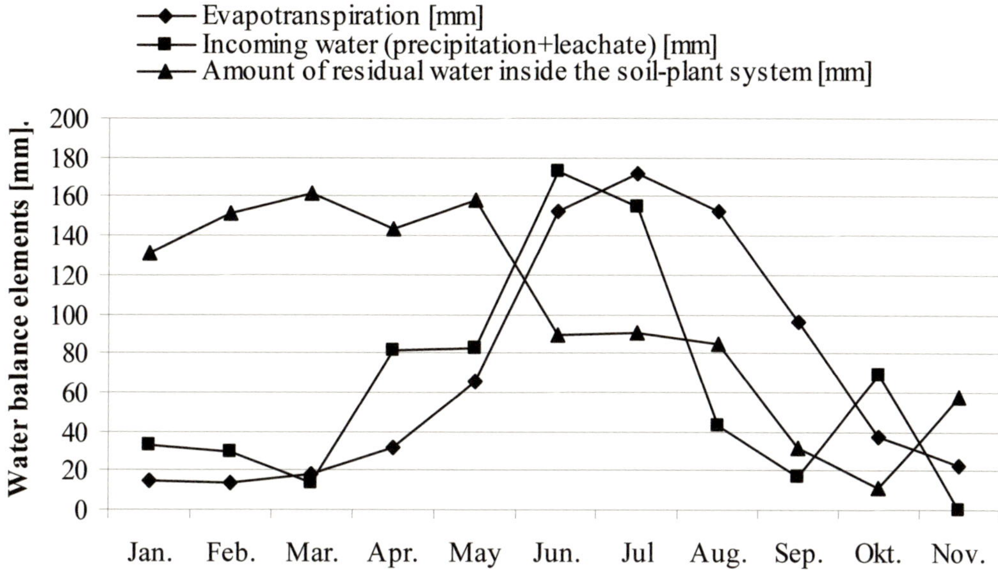

Fig. 3. Water balance elements of the soil-plant system supplied by landfill leachate during 270 days period of the year 2007.

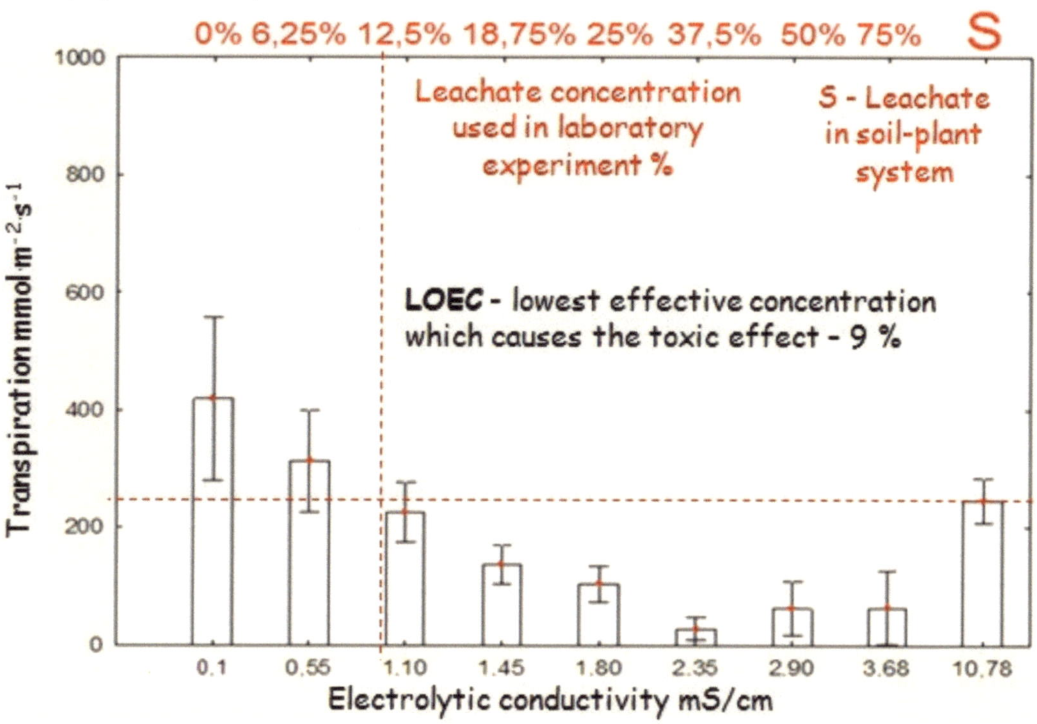

Fig. 4. Comparison of degree of leaf stomata opening (transpiration) in laboratory tests with measurements at the soil-plant system, Zakurzewo landfill, during July 2007.

3. Transpiration – a useful tool for landfill leachate phytotoxicity evaluation

3.1 Phytotoxicology in ecological engineering

Phyto-indicative methods for characterization of the environment have recently become popular. In such methods, observations of plant performance provide information about environmental conditions. Algae have been used as active components in phytotoxicological tests (Baun *et al.*, 1998; Baun *et al.*, 2002; Fargasowa *et al.*, 1999; Halling-Sorensen, 2000; LeBlond & Duffy, 2001; Okamura *et al.*, 2002; Rojickowa–Padrtowa & Marsalek, 1999; Sepic *et al.*, 2003; Toricelli *et al.*, 2002; Wundram *et al.*, 1996). Multi-celled plants are also in use, mostly floating plants like *Lemna minor* L. but also rooted aquatic and terrestrial plants. Usually, such plants are used for toxicity assessment of pesticides, polycyclic aromatic hydrocarbons, and heavy metals as well as for landfill leachate toxicity determination. Similar studies have been conducted with willows: *Salix amygdalina* L., *Salix viminalis* L., *Salix purpurea* L. (Bialowiec & Agopsowicz, 2007), *Salix viminalis* (Dimitriou *et al.*, 2006), and reed *Phragmites australis* (Cav.) Trin. ex Steud. (Bialowiec & Kasinski, 2008). Phytotoxicological tests may be important in environmental engineering in designing phytoremediation technologies for treatment of harmful substances (wastewater, landfill leachate) or for remediation of polluted soils. Such tests can determine the maximal dose of treated pollutant which causes no negative effect on the plants. In the case of phytoremediation using higher plants, with complex tissue and organ structures, there is a question of selecting the best parameter to indicate plant response to the toxicant. Different parameters have been chosen individually for each case of plant and phytotoxic compound. Fairchild *et al.* (1999) counted number of leaves and total biomass of *L. minor* L. Marwood *et al.* (2001) measured the concentration of chlorophyll a and b in cells of *Lemna gibba* L. Fairchild *et al.* (1998) measured biomass growth of *Ceratophyllum demersum* L., *Elodea canadensis* Michx., and *Myriophyllum heterophyllum* Michx. whereas growth inhibition of *Lolium perenne* L., and *Raphanus sativum* L. was employed by Prokop *et al.* (2003). Moreover, Dimitriou *et al.* (2006) examined the usefulness of leaf length and asymmetry measurements for determination of tolerance of five willow clones to landfill leachate. Dimitriou *et al.* (2006) used relative growth rate (RGR) based on measurements of shoot, leaf, root and total plant dry weights, leaf area and shoot length. Changes caused by toxicants are usually not visible in short term experiments, and size measurement with a ruler is not accurate enough to indicate small changes in the width, length, shape and symmetry of leaves. Selecting the most appropriate parameter from such a list as an indicator of phytoxicity may lead to confusion. The parameter which may include all possible changes in leaf morphology is the fractal dimension of the leaf (Kopik & Bialowiec, 2007).

In ecological engineering, phytotoxicological tests are used in soil remediation, waste management and wastewater treatment in addition to chemical analysis (Wilson *et al.*, 2002) and they may be used for landfill leachate toxicity assessment. Many studies indicate the utility of macrophytes in toxicological tests (Barbero *et al.*, 2001; Kirk *et al.* 2002; Wang, 1987; Wang & Williams, 1988), and involve measuring basic parameters such as seed survival (germination), biomass growth, stem and root elongation. From a technological point of view, the most important parameters to determine the landfill leachate dose rate are transpiration and biomass growth, these being processes with the greatest impact on treatment efficiency. Successful implementation of a soil-plant (evapotranspirative) system with willow irrigated by leachate depends on an appropriate dose rate, especially in the initial phase of plant growth. In the next section measurements of biomass growth and transpiration of three species of willow are used to determine landfill leachate phytotoxicity.

3.2 Materials and methods

The experiment was designed to determine the optimum duration of testing of willow tolerance to landfill leachate, using two easily measured physiological parameters, as indicators of plant response: stem elongation and transpiration.

Three willow species marked as W_1, W_2 and W_3 were tested in the greenhouse located at the University of Warmia and Mazury in Olsztyn, Poland. Individual willow cuttings were placed in 90 transparent plastic bottles with volume 1 dm^3 (30 bottles for each willow species). Bottles containing willow cuttings (Fig. 5), were filled with the following solutions of landfill leachate, clean tap water and water taken from a river: 0% (clean tap water), 0% (river water), 12.5 %, 25 % 50 % and 100 % concentrations of leachate. Tap water and river water were used as a reference. The bottles were filled to the same level, and the volume of solution used was in the range 955 to 990 cm^3/per bottle. Every variant of the experiment was replicated 5 times. Figure 6 shows the experimental configuration.

The willow cuttings were taken from an experimental plantation owned by UWM. The diameter and length of each cutting was in the range 0.8 to 1.9 cm and 31 to 33 cm. Three quarters of the cutting's length was inside the bottle.

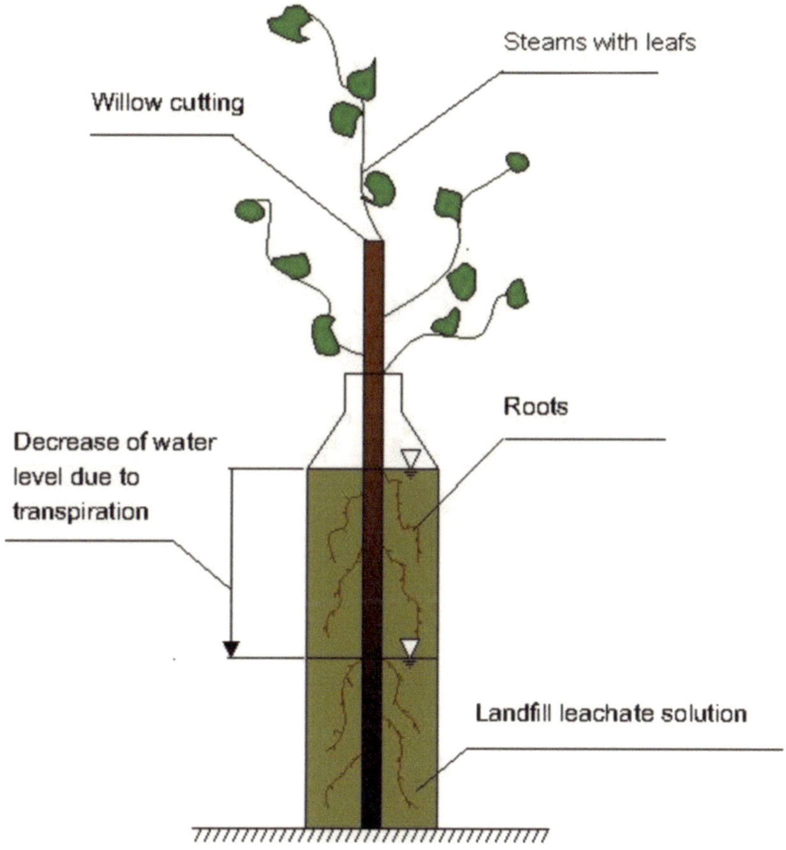

Fig. 5. The construction of experimental stand.

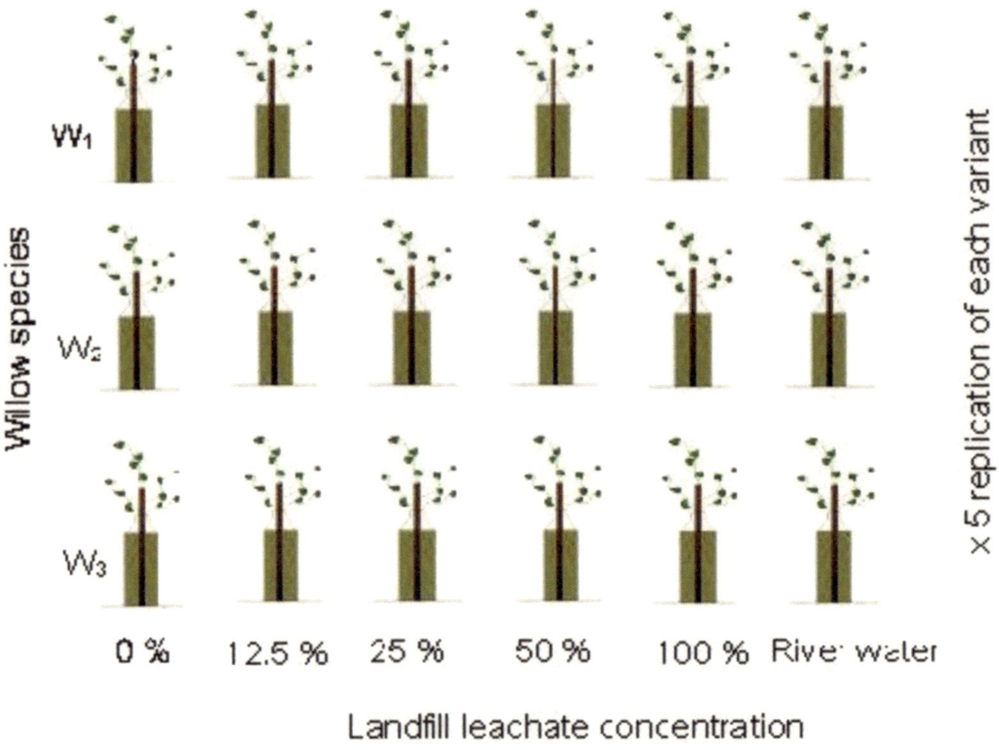

Fig. 6. Configuration of the experiment.

Tap water used in the experiment is typical for water supplied in Olsztyn municipality. The river water was taken for Kortowka River passing through UWM campus. Landfill leachate came from landfill in Zakurzewo near Grudziadz, Poland. The chemical characteristics of landfill leachate and its dilutions are shown in table 2.

Parameter	Unit	Leachate solutions [%]			
		100.0	50.0	25.0	12.5
Reaction	pH	8.74	8.65	8.58	8.4
Conductivity	mS/cm	4.46	2.9	1.8	1.1
COD	mgO_2/dm^3	997.0	592.0	229.0	170.0
Chlorides	mg/dm^3	131.0	63.1	32.6	16.9
Kjeldahl nitrogen	mg/dm^3	680.0	339.0	168.3	85.4
Ammonium nitrogen	mg/dm^3	576.0	286.0	143.0	85.4
Dissolved solids	mg/dm^3	6040	2570	1515	982
Residual after ignition	mg/dm^3	4805	2010	1170	727
	%	79.5	78.2	77.2	74.0
Lost at ignition	mg/dm^3	1235	560	345	255
	%	20.5	21.8	22.8	26.0

Table 2. The chemical characteristics of landfill leachate and solutions used in the experiment

The duration of the experiment was 35 days (5 weeks), during which time the average temperature inside the greenhouse was 22.5 °C.
The tolerance of plants to the prepared leachate concentration was determined by measurements of biomass growth and transpiration rate. Biomass growth was determined by stem elongation measured with 1 mm accuracy. Transpiration was measured as the decrease of water in the bottles, determined gravimetrically with 0.1 g accuracy. Measurements were made every week.
The minimum duration for future experiments to indicate the tolerance of willows to leachate was assessed on the basis of ANOVA analysis of mean values of stem length (significance $p<0.05$). The toxic effect of leachate on willows was evaluated by analysis of plant biomass growth curves (stem length) and transpiration cumulative curves. Coefficients of growth and transpiration were determined by regression analysis (significance $p<0.05$). The no-effect concentration of leachate on plants was calculated using DEBTox software tool (body growth model).

3.3 Results and discussion

Optimal duration for the toxicity test (long enough to obtain significant differences between variants), was evaluated using ANOVA with post-hoc Tukey's test (significance level $p<0.05$) (Fig. 7).
It is clear that during two weeks in most cases there is lack of differences between variants. After a third week differences started to be observed and from the 28th day of experiment differences had stabilized. Extending the experiment beyond 28 days is not necessary, because it did not change the configuration of differences (Fig. 7).
Linear regressions of biomass growth rate during the experiment are shown as cumulative curves of stem elongation (Fig. 8), and are described by equation 1:

$$y = a \cdot t - b \qquad (1)$$

where:
y - stem length [cm],
t - time [days],
a - regression coefficient (the stem elongation ratio) [cm/day],
b - initial stem length [cm].
The calculated regression coefficients (a and b) and also the determination coefficients (R^2) are shown in table 3.
The high values of R^2 obtained (significance $p<0.05$) indicate a good fit of the estimated regression parameters to the results. Comparing between species, stem elongation ratios obtained as regression slopes (a) in leachate concentration 0% solution (control) the highest was in willow W1 2.53 cm/day, and the lowest in willow W3 1.46 cm/day. Increasing the leachate concentration to the 12.5 % level did not affect stem elongation ratio. Further increase in leachate concentration caused a slight decrease (of 22 to 53 %) in the stem elongation ratio. Increasing leachate concentration to 50 and 100 % caused stem elongation ratio to decrease dramatically to values in the range 0.02 - 0.31 (Tab. 3). In the case of willow W2 with concentration of leachate 100 % biomass growth did not occur.
Values of stem elongation ratio obtained for willows supplied with river water (similar to values for plants growing in tap water), show no negative effect of its constituents on biomass growth.

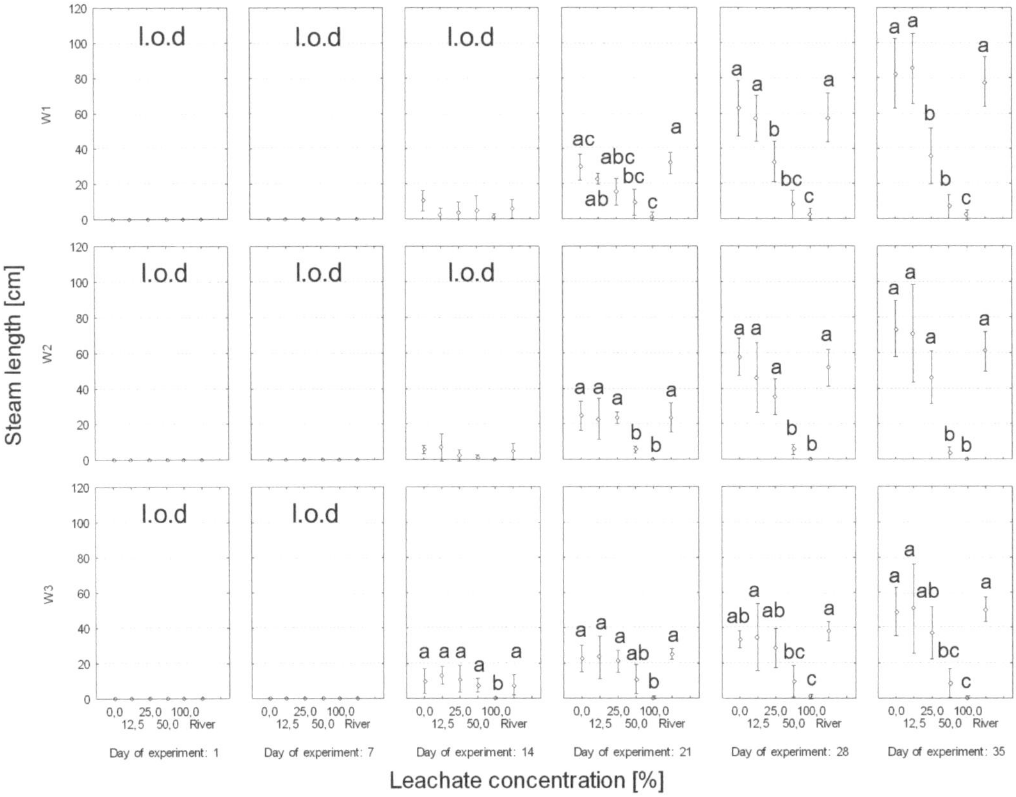

Fig. 7. Mean stem length differentiation during experiment. l.o.d. – lack of significant ($p<0.05$) differences between means. Letters (a, b, c and combinations) show significant ($p<0.05$) differences between means.

Parameters	Experiment variants																	
	W1						W2						W3					
	0	12.5	25	50	100	Riv.	0	12.5	25	50	100	Riv.	0	12.5	25	50	100	Riv.
a	2.53	2.52	1.18	.25	.07	2.40	2.28	2.08	1.46	.15	-	1.95	1.46	1.55	1.14	.31	.02	1.56
b	13.4	16.3	6.1	-.5	.2	13.2	13.0	11.9	7.7	.1	-	10.7	6.4	6.1	3.7	-.8	.04	7.2
R^2	.913	.855	.898	.676	.900	.909	.890	.894	.907	.622	-	.895	.950	.961	.970	.679	.703	.944

Table 3. Linear estimation parameters of the willow stem elongation during experiment.

The no-effect concentration of leachate on plants (NEC) is the concentration of toxicant in the medium that will not induce effects after prolonged exposure. NEC values, calculated using DEBTox software tool (body growth model), are in range 20.74 (W1) - 24.21 (W2) % (Tab. 4). Those results are similar to observations of changes of stem elongation ratio. This indicates the usefulness of the DEBtox tool for selection of leachate dose for treatment by plants.

Bialowiec at al. (2003) in earlier research on leachate with a lower concentration of nitrogen, 148 mg/dm^3, observed that increasing the leachate concentration up to 50 % stimulated biomass growth, but further increase in leachate concentration to 100 % slowed it down.

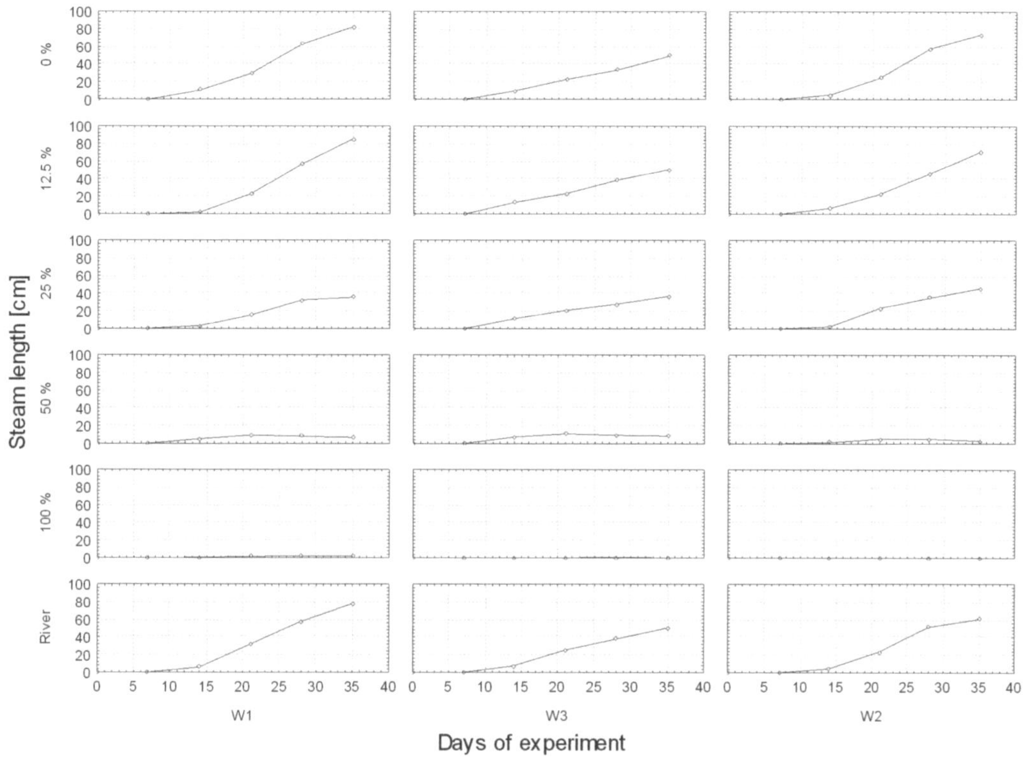

Fig. 8. Stem elongation cumulative curves during experiment.

Parameter	Unit	Willow species		
		W1	W2	W3
NEC	%	20.74	24.21	23.04
The ultimate length*	cm	217.9	189.3	141.3
Mean deviation	cm	10.2	9.1	4.5

Table 4. The parameters of NEC calculation. * The ultimate length is the body length after a very long time without distortion of the growth process.

This shows the necessity of performing individual tests for each kind of leachate, because of different possible responses of plants to variable leachate properties.

There is an influence of leachate concentration on transpiration. Non-linear regression of transpiration during the experiment is shown as cumulative curves of the mass of transpired water [g H_2O per plant] (Fig. 9).

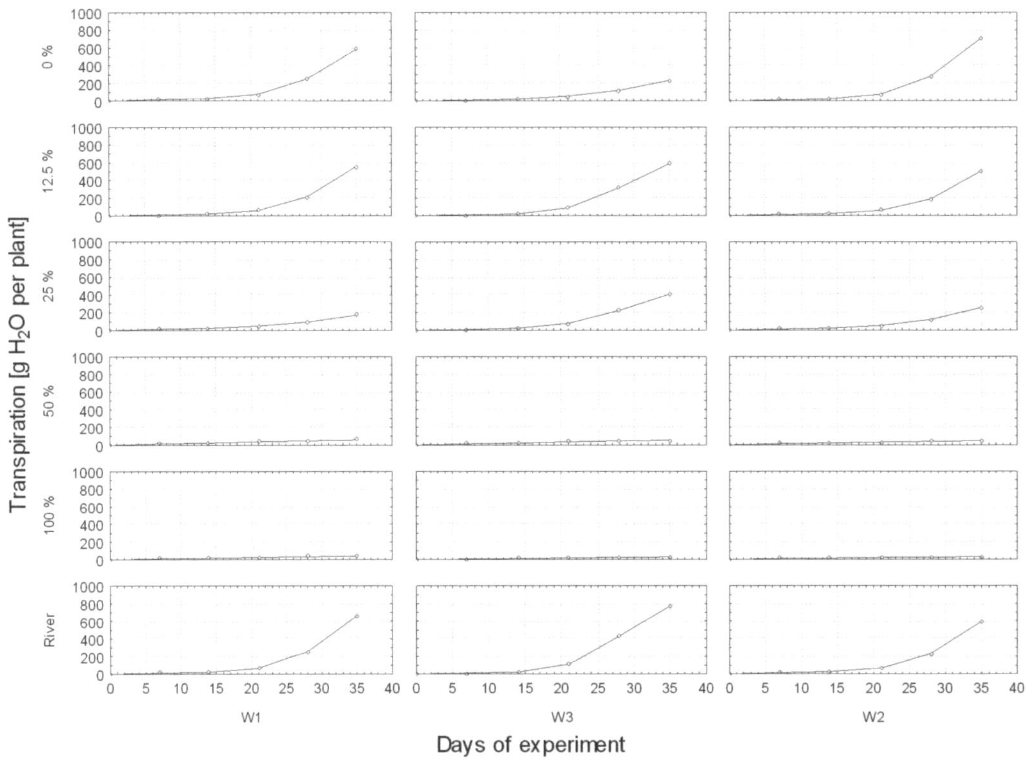

Fig. 9. Transpiration cumulative curves during the experiment.

This dependence is described by equation 2:

$$T = T_0 \cdot e^{k \cdot t} \qquad (2)$$

where:
T - transpiration [g H_2O per plant],
T_0 - initial transpiration [g H_2O per plant]
t - time [days],
k - regression coefficient (transpiration rate constant) [1/day],
The calculated regression coefficients (k and T_0) and also the determination coefficient (R^2) are shown in table 5.

Parameters	Experiment variants																	
	W1						W2						W3					
	0	12.5	25	50	100	Riv.	0	12.5	25	50	100	Riv.	0	12.5	25	50	100	Riv.
k	.131	.145	.090	.047	.046	.146	.144	.144	.111	.045	.041	.141	.103	.116	.109	.047	.048	.115
T_0	5.89	3.49	7.29	12.0	9.00	4.06	4.54	3.29	5.27	10.7	8.65	4.27	6.51	10.4	9.26	11.3	6.29	14.0
R^2	.998	.999	.997	.901	.896	.998	.997	.999	.999	.890	.850	.998	997	.982	.987	.883	.892	.975

Table 5. Non-linear estimation parameters of the willow transpiration during experiment.

The high values of R^2 obtained (significance $p<0.05$) indicate a good fit of the estimated regression parameters to the results. Comparing between species, transpiration rate constants obtained as regression slopes (k) in leachate concentration 0% solution (control) the highest was in willow W2 0.144 1/day, and the lowest in willow W3 0.103 1/day. Increasing the leachate concentration to 12.5 % did not affect transpiration. Similarly, as in the case of stem elongation ratio, increasing leachate concentration from 12.5 to 25 % caused a slight decrease in transpiration rate constant. Increasing leachate concentration to 50 and 100 % caused intense decrease of transpiration rate constant to values in the range 0.041 - 0.048 (Tab. 5).

Similarly, Bialowiec *at al.* (2003) in earlier research, determined that leachate concentrations exceeding 25 % had a negative influence on willow transpiration. Leachate used in the cited experiment was characterized by a much higher concentration of chlorides (1540 mg/dm^3) than in the results presented here. This may show that different pollutants influence each physiological process in different ways. The presence of dissolved compounds in water may affect transpiration in two ways; (i) it may pose toxic threats to plants, (ii) it may decrease the difference between soil water tension and soil water tension at the point of air entry, and hence decrease water flux (Persson, 1995). It may shown that transpiration is a more sensitive physiological process than biomass growth in the presence of toxicants. This parameter (qualitative differences between toxicants), should be included in the test procedure, because lack of knowledge about the influence of pollutants on transpiration my lead to further mistakes during dose selection of landfill leachate for treatment by plants. A possible mistake is to obtain good biomass growth but with low transpiration. This should be taken into consideration because the main aim of leachate treatment in soil-plant systems is not biomass production but leachate volume decrease due to evapotranspiration.

4. Summary

Early operational results of landfill leachate treatment in soil-plant systems show that this method may be used as an alternative technology in contrast to treatment in a wastewater treatment plant. In the initial phase (July 2006 to the end of November 2006), about 1100 m^3, and during the year 2007, 4241 m^3 water were lost by evapotranspiration.

Reeds growing in a soil plant system had higher resistance to pollutants present in leachate than plants exposed to leachate in the laboratory. Despite high concentration of ions in water collected from the soil-plant system (10.78 mS/cm] reeds showed high transpiration rate, about 250 mmol m^{-2} s^{-1}. During the phytotoxicological tests the same transpiration rate value was observed with plants watered by a solution with 10-fold lower ionic concentration (1.10 mS/cm).

Experiences with the first pilot scale evapotranspirative system for landfill leachate allowed us to propose a formula to calculate the minimum size of such zero discharge systems (Białowiec, 2009):

$$As = \frac{-A_l(Qp - Ev)}{(Ev - E)} \tag{3}$$

where:

Qp - annual precipitation [m/y],
Ev - annual evaporation [m/y],
E - annual evapotranspiration [m/y],
A_l - landfill site area (sealed) [m^2],
As - evapotranspirative system area (located within the landfill site) [m^2]

Before implementation of a soil-plant system for evapotranspiration of landfill leachate the LL dose rate should be assessed on the basis of phytotoxicological test. The proposed procedure to determine willow tolerance to landfill leachate is easy to perform, proposed indicators of plant response to leachate (stem elongation and transpiration) are easy to measure and allow information about possible leachate dose for willow irrigation to be obtained. The duration of tests does not need to be longer than 28 days (4 weeks). Both stem elongation and transpiration should be measured during the experiment, because of different sensitivity of those physiological processes to leachate constituents. Because of different leachate chemical properties and landfill conditions the proposed testing procedure for the assessment of leachate dose for willow irrigation should be repeated every time a new plantation is planned. Using data from another experiment with different leachate may lead to failure of the whole system. Experiments performed here showed that the willow plantations should not be irrigated by solutions of this kind of leachate with a share higher than 25 % of the total amount of water supplied to the plantation (leachate + precipitation).

5. Acknowledgements

This experimental work was done with financial support of Polish Ministry of Science and Higher Education, grant No. N3313/T02/2007/32

The Authors wish to thank Mrs. Monika Sosińska and Mr. Sławomir Kasiński for their valuable help in the analytical work.

6. References

Agopsowicz, M. (1994). Badania przydatności wierzb *„Salix"* do unieszkodliwiania odcieków z wysypisk komunalnych (Research on usefulness of willow - „Salix" to municipal landfill leachate treatment). *V Polski Kongres Oczyszczania Miast*, Szczecin 19-22 września, 1994 *(In polish).*

Agopsowicz, M.; Białowiec, A. & Radziemska, M. (2006). Municipal waste disposal in energetic piles in SWECO technology - seven years of operation and what now? *Archives of Environmental Protection*, vol. 32, 3.

Albuquerque, A., Oliveira, J.; Semitela, S. & Amaral, L. (2009). Influence of bed media characteristics on ammonia and nitrate removal in shallow horizontal subsurface flow constructed wetlands. *Bioresource Technology*, 100, 6269–6277.

Alker, G.R.; Godley, A.R. & Hallett, J.E. (2003). Landfill leachate management by application to short rotation willow coppice. *Proceedings of the Ninth International Waste Management and Landfill Symposium in Sardinia*, 6 - 10, October, 2003.

Allen, R.G.; Pereira, L.S.; Raes, D. & Smith, M. (1998). *Crop evapotranspiration. Guidelines for computing crop water requirements*. FFAO Irrigation and Drainage Paper No 56, Food and Agriculture Organization of the United Nations, Rome, pp. 300.

Aronsson, P. (2000). *Nitrogen retention in vegetation filters of short rotation willow coppice*. PhD Thesis, Swedish University of Agricultural Sciences, Uppsala, Sweden.

Aronsson, P. & Perttu, K.L. (2001). Willow vegetation filters for waste-water treatment and soil remediation combined with biomass production. *Forest Chronicle, 77*, 293-299.

Ashbee, E. & Fletcher, I. (1993). Reviewing the options for leachate treatment, *Wastes Management*, 8, 32-33.

Barbero, P.; Beltrami, M.; Baudo, R., & Rossi, D. (2001). Assessment of Lake Orta sediments phytotoxicity after limiting treatment. *Journal of Limoolgy*, 60 (2), 269 - 276.

Baun, A.; Bussarawit, N. & Nyholm, N. (1998). Screening of pesticide toxicity in surface water from an agricultural area at Phuket Island (Thailand). *Environmental Pollution*, 102 (2–3), 185–190.

Baun, A.; Justesen, K.B. & Nyholm, N. (2002). Algal tests with soil suspensions and elutriates: A comparative evaluation for PAH - contaminated soils. *Chemosphere*, 46 (2), 251–258.

Beckmann, M. & Lloyd, D. (2001). Mass spectrometric monitoring of gases (CO_2, CH_4, O_2) in a mesotrophic peat core from Kopparås Mire, Sweden. *Global Change Biology*, 7: 171-180

Białowiec, A.; Wojnowska-Baryła, I. & Agopsowicz, M. (2003). Effectiveness of leachate disposal by the young willow sprouts *Salix amygdalina*. *Waste Management Research*, 21 (6), 21, 557-566.

Białowiec, A. (2005). *Landfill leachate treatment in soil-plant systems*. Ph.D. dissertation, The University of Warmia and Mazury in Olsztyn, Poland.

Białowiec, A.; Zieliński, M. & Dębowski, M. (2006). Problemy eksploatacyjne hydrofitowych systemów oczyszczających ścieki (Operational problems constructed Welland for wastewater treatment). *Prace Naukowe Instytutu Ochrony Środowiska Politechniki Wrocławskiej*, 82, Seria: Studia i Materiały 22: 26-39 *(In polish)*.

Bialowiec, A. & Agopsowicz, M. (2007). Using phytotoxicological test for landfill leachate dose selection in willow short rotation plantations. *Proceeding of the Eleventh International Waste Management and Landfill Symposium in Sardinia*, 1–5 October, 2007.

Białowiec, A. & Wojnowska-Baryła, I., (2007). The efficiency of landfill leachate evapotranspiration in soil-plant system with reed Phragmites australis. *Ecohydrology and Hydrobiology*, Vol 7, No 3-4.

Białowiec, A.; Wojnowska-Baryła, I. & Agopsowicz, M. (2007). The efficiency of evapotranspiration of landfill leachate in the soil–plant system with willow *Salix amygdalina* L. *Ecological Engineering*, 30 (4), 356-361.

Białowiec, A. & Kasiński, S. (2008). Transpiration, an indicator of reed (*Phragmites australis*) behavior during landfill leachate treatment. *Seminar constructed wetlands for wastewater treatment and reuse* (hosted by the EVAWET project - PTDC/AMB/73081/2006) Kobyla Gora, Poland 19-20 September 2008.

Białowiec, A. (2009). Systemy roślinno-gruntowe do odparowania odcieków składowiskowych. (Soil-plant system for landfill leachate evapotranspiration). *Przegląd Komunalny*, ABRYS 11(218)/2009 *(In Polish)*.

Białowiec, A., Randerson, P., (2010). Zero Discharge Systems - a case study. *CWA 6th Annual Conference, Wastewater Management and the Application of Constructed Wetlands*, Stoneleigh Park, Warwick, 22-24 June 2010.

Brix, H. (1987). Treatment of wastewater in the rhizosphere of wetland plants - the root zone method. *Water Science and Technology*, 19, 107-119.

Bulc, T.G. (2006). Long term performance of constructed wetlands for landfill leachate treatment. *Ecological Engineering*, 26, 365-374.

Christensen, T.H.; Cossu, R. & Stegmann, R. (1992). Landfill leachate: an introduction, *in: Christensen, T.H., Cossu, R., Stegmann, R., (Eds.), Landfilling of Waste: Leachate. Elsevier*, Barking, UK, pp. 1-14.

Cooper, P.F. (1999). Areviewof the design and performance of vertical flow and hybrid reed bed treatment systems. *Water Science and Technology*, 40 (3), 1–9.

Dimitriou, I. & Aronsson, P. (2003). Evaluation of stress factors after irrigation of willow with landfill leachate. *Proceedings of the Ninth International Waste Management and Landfill Symposium in Sardinia*, 6 - 10 October, 2003.

Dimitriou, I. & Aronsson, P. (2005). Willows for energy and phytoremediation in Sweden. *Unasylva*, 221, Vol. 56.

Dimitriou, I., Aronsson, P. & Weih, M. (2006). Stress tolerance of five willow clones after irrigation with different amounts of landfill leachate. *Bioresource Technology*, 97, 150–157.

Dobson, M.C. & Moffat, A.J. (1995). A re-evaluation of objections to tree planting on containment landfills. *Waste Management and Research*, 13, 6.

Duggan, J. (2005). The potential for landfill leachate treatment using willows in the UK - A critical review. *Resource Conservation and Recycling*, 45, 97-113.

Dulohery, C.J.; Kolka, R.K. & McKevlin, M.R. (2000). Effects of willow overstory on planted seedlings in bottomland restoration. *Ecological Engineering*, 15, 57-66.

Elowson, S. (1999). Willow as a vegetation filter for cleaning of polluted drainage water from agricultural land. *Biomass and Bioenergy*, 16, 281-290.

Ettala, M. (1992). Effect of vegetation on landfill hydrology, *in Christensen, T.H., Cossu, R., Stegmann, R. (Eds.), Landfilling of Wastes, Leachate.* E&FN Spon, and imprint of Chapman&Hall, London, UK, 53-64.

Fairchild, J.F.; Ruessler, D.S. & Carlson, A.R. (1998). Comparative sensitivity of five species of macrophytes and six species of algae to atrazine, matribuzin, alachlor and metolachlor. *Environmental Toxicology and Chemistry*, 17 (9), 1830–1839.

Fairchild, J.F.; Sappington, L.C. & Ruessler, D.S. (1999). An ecological risk assessment of the potential for herbicide impacts on primary productivity of the lower Missouri River. *Pages 323-330 in U.S. Geological Survey Toxic Substances Hydrology Program – Volume 2 of 3 – Contamination of Hydrologic Systems and Related Ecosystems, D. Morganwalp and H. Buxton, (eds.)*, U.S. Geol. Surv. Water-Resour. Invest. Rep. 99-4018B.

Fargasowa, A.; Bumbalowa, A. & Havranek, E. (1999). Ecotoxicological effects and uptake of metals (Cu, Cu^{2+}, Mn^{2+}, Mo^{6+}, Ni^{2+}, V^{5+}) in freshwater alga Scenedesmus quadricauda. *Chemosphere*, 38 (5), 1165–1173.

Garcıa, J.; Aguirre, P., Mujeriego, R., Huang, Y., Ortiz, L. & Bayona, J. (2004). Initial contaminant removal performance factors in horizontal flow reed beds used for treating urban wastewater. *Water Research*, 38, 1669–1678.

Guterstam, B. (1996). Demonstrating ecological engineering for wastewater treatment in a Nordic climate using aquaculture principles in a greenhouse mesocosm. *Ecological Engineering*, 6, 73 - 97.

Halling-Sorensen, B. (2000). Algal toxicity of antibacterial agents used in intensive farming. *Chemosphere*, 40 (7), 731–739.

Hasselgren, K. (1992). Soil-plant treatment system. *in Christensen, T.H., Cossu, R., Stegmann, R. (Eds.), Landfilling of Wastes, Leachate.* E&FN Spon, and imprint of Chapman&Hall, London, UK, 361-380.

Hasselgren, K. (1998). Use of municipal waste products in energy forestry: highlights from 15 years of experience. *Biomass and Bioenergy*, 15, 71–74.

Jones, D.L.; Williamson, K.L. & Owen, A.G. (2006). Phytoremediation of landfill leachate. *Waste Management*, 26, 8, 825-837.

Kadlec, R.; Knight, R.; Vymazal, J.; Brix, H.; Cooper, P. & Haberl, R. (2000). *Constructed wetlands for pollution control: processes, performance, design and operation.* Report No. 8, IWA Publishing, London, UK.

Kangas, P.C. (2004). *Ecological Engineering: Principles and Practice.* CRC Press LLC, Florida.

Kirk, J.L.; Klironomos, J.N.; Lee, H. & Trevors, J.T. (2002). Phytotoxicity assay to assess plant species for phytoremediation of petroleum - contaminated soil. *Bioremediation Journal*, 6 (1), 57–63.

Kjeldsen, P.; Barlaz, M.A.; Rooker, A.P.; Baun, A.; Ledin, A. & Christensen, T.H. (2002). Present and Long-Term Composition of MSW Landfill Leachate: a Review. *Critical Reviews in Environmental Science and Technology*, 32, 297-336.

Kopik, M. & Bialowiec, A. (2007). Sezonowa zmienność wymiaru fraktalnego liści brzozy *Betula pendula* Roth. (Seasonal changes of fractal dimension of birch *Betula pendula Roth.*). *Proceedings of the Environmental biotechnology Conference*, Wisla, 6-8 December. *(In Polish).*

Kowalik, P.J. & Randerson, P.F. (1994). Nitrogen and Phosphorus removal by willow stands irrigated with municipal wastewater - a review of the Polish experience. *Biomass and Bioenergy*, 6, 133-139.

LeBlond, J.B. & Duffy, L.K. (2001). Toxicity assessment of total dissolved solids in effluent of Alaskan mines using 22 - h chronic Microtox® and Selenastrum capricornatum assays. *Science of the Total Environment*, 271 (1-3), 49–59.

Lloyd, D.; Thomas, K.L.; Benstead, J.; Davies, K.L.; Lloyd, S.H., Arah, J.R.M. & Stephen, K.D. (1998). Methanogenesis and CO2 exchange in an ombrotrophic peat bog. *Atmospheric Environment*, 32, 3229–3238.

Marwood, Ch.A.; Solomon, K.R. & Greenberg, B.M. (2001). Chlorophyll fluorescence as a bioindicator of effects on growth in aquatic macrophytes from mixtures of polycyclic aromatic hydrocarbons. *Environmental Toxicology and Chemistry*, 20 (4), 890–898.

Odum H.T. (1971). *Environment, power and society.* Wiley-Interscience, John Wiley & sons, Inc., New York.

Okamura, H.; Piao, M.; Aoyama, I.; Sudo, M.; Okubo, T. & Nakamura, M. (2002). Algal growth inhibition by river water pollutants in the agricultural area around Lake Biwa, Japan. *Environmental Pollution,* 117 (3), 411–419.

Persson, G. & Lindroth, A. (1994). Simulating evaporation from short rotation forest - variation within and between seasons. *Journal of Hydrology.* 156, pp 21-45.

Persson, G. (1995). Willow stand evapotranspiration simulated for Swedish soils. *Agricultural Water Management,* 28, 99, pp. 271-293.

Perttu, K.L. & Kowalik, P.J. (1997). Salix vegetation filters for purification of waters and soils. *Biomass and Bioenergy,* 12(1), 9-19.

Pinton, R.; Varanini, Z. & Nannipieri, P. (2007). *The rhizosphere: biochemistry and organic substances at the soil-plant interface.* 2nd edition, CRC Press, 447 pp.

Prokop, Z.; Vangheluwe, M.L.; Van Sprang, P.A.; Janssen, C.R. & Holoubek, I. (2003). Mobility and toxicity tests into sandy sediments deposited on land. *Ecotoxicology and Environmental Safety,* 54 (1), 65–73.

Raisin, G.; Bartley, J. & Croome, R. (1999). Groundwater influence on the water balance and nutrient budget of a small natural wetland in Northeastern Victoria, Australia. *Ecological Engineering,* 12, 133–147.

Randerson, P.F. (2006). Constructed wetlands and vegetation filters: an ecological approach to wastewater treatment. *Environmental Biotechnology.* 2(2), 78-79.

Randerson, P.F.; Jordan, G. & Williams, H.G. (2007). The role of willow roots in sub-surface oxygenation of vegetation filter beds - mass spectrometer investigations in Wales, U.K. *Ecology & Hydrobiology,* 7(3-4), 255-260.

Reddy, K.R.; Patrick, W.H. & Lindau, C.W. (1989). Nitrification-denitrification at the plant-root sediment interface in wetlands. *Limnology and Oceanography,* 34, 1004-1013.

Rojickowa-Padrtowa, R. & Marsalek, B. (1999). Selection and sensitivity comparisons of algal species for toxicity testing. *Chemosphere,* 38 (14), 3329–3338.

Schnoor, J.L.; Licht, L.A.; McCutcheon, S.C.; Wolfe, N.L. & Carreira, L.H. (1995). Phytoremediation of organic and nutrient contaminants. *Environmental Science and Technology,* 29, 318-323.

Sepic, E.; Bricelj, M. & Leskovsek, H. (2003). Toxicity of fluoranthene and its biodegradation metabolites to aquatic organisms. *Chemosphere,* 52 (7), 1125 - 1133.

Sheppard, S.K. & Lloyd, D. (2002). Diurnal Oscillations in Gas Production (O_2, CO_2, CH_4 and N_2) in Soil Monoliths. *Biological rhythm research,* 33: 577-597.

Siuta, J. (1996). Roślinna sanitacja gruntów zanieczyszczonych odpadami (Plants in soil remediation). *Ogólnopolska XIII Konferencja Naukowa (Tom 2), Unieszkodliwianie i utylizacja odpadów płynnych w środowisku naturalnym, Zeszyty Naukowe Akademii Rolniczej we Wrocławiu,* 293, 57-67 *(In polish).*

Stein, O.R. & Hook, P.B. (2005). Temperature, plants and oxygen: how does season affect constructed wetlands performance. *Journal of Environmental Science and Health,* 40, 1331-1342.

Tchobanoglous, G., (1987). Aquatic systems for wastewater treatment: engineering considerations. *Aquatic plants for wastewater treatment and resource recovery, edited by Reede K.R., Smith W.H.,* Magnolia Publishing Inc. Orlando, Florida.

Toricelli, L. ; Manzo, S.; Accornero, A. & Manfra, L. (2002). Biomonitoring of marine waters by the use of microalgal tests: results from the Campania coastal zone (south Tyrrhenian Sea). *Fresenius Environmental Bulletin* 11.

Trzosińska, M. & Podkański, W. (2000). Projekt modernizacji wysypiska odpadów dla Grudziądza w miejscowości Zakurzewo (Project of landfill modernization in Zakurzewo near Grudziądz). Przedsiębiorstwo Robót Inżynieryjno - Melioracyjnych „Melbud" Sp. z o.o., Grudziądz, *(In Polish)*.

Vymazal, J. & Kropfelova, L. (2008). Wastewater treatment in constructed wetlands with horizontal sub-surface flow. *in Series of Environmental Pollution, vol. 14, Springer, Germany, pp. 566. Vymazal, J., Kropfelova, L., 2008. Wastewater treatment in constructed wetlands with horizontal sub-surface flow, in Series of Environmental Pollution, vol. 14,* Springer, Germany, pp. 566.

USEPA, (2000). *Introduction to phytoremediation*. EPA/600/R-99/107.

Wallace, S.D. & Knight R.L. (2006). *Small-scale constructed wetland treatment systems - feasibility, design criteria, and O&M requirements*. Water Environment Research Federation, ISBN 1-84339-728-5.

Wang, W. (1987). Root elongation method for toxicity testing of organic and inorganic pollutants. *Environmental Toxicology and Chemistry*, 6, 409 - 414.

Wang, W. & Wiliams, J.M. (1988). Screening and monitoring of industrial effluents using phytotoxicity tests. *Environmental Toxicology and Chemistry*, 7, 645 - 652.

Welander, U., Henrysson, T. & Welander, T. (1998). Biological nitrogen removal from municipal landfill leachate in a pilot scale suspended carrier biofilm process. *Water Research*, 32, 1564-1570.

Wieûner, A.; Stottmeister, K.; Struckmann, N. & Jank, M. (1999). Treating a lignite pyrolysis wastewater in a constructed subsurface flow wetland. *Water Research,*, 33, 5, 1296-1302.

Williams, R.B.; Borgerding, J.; Richey, D. & Kadlec, R.K. (1987). Start-up and operation of an evaporative wetlands facility, *Aquatic plants for wastewater treatment and resource recovery, edited by Reddy K.R., Smith W.H.*, Magnolia Publishing Inc. Orlando, Florida.

Williams, H.G.; Białowiec, A.; Slater, F. & Randerson, P.F. (2010). Diurnal cycling of dissolved gas concentrations in a willow vegetation filter treating landfill leachate. *Ecological Engineering*, 30, 1680-1685.

Wilson, J.J.; Hatcher, J.F. & Goudey, J.S. (2002). Ecotoxicological endpoints for contaminated site remediation. *Ann Ist Super Sanita*, 38 (2), 143 - 147.

Wojciechowska, E.; Gajewska, M.; Waara, S.; Obarska-Pempkowiak, H.; Kowalik, A.; Albuquerque, A. & Randerson, P. (2009). Leachate from sanitary landfills treated by constructed wetlands *Proc. 12th International Waste Management and Landfill Symposium*, S. Margherita di Pula (Cagliari), Sardinia, Italy, 5-9 Oct 2009.

Wojciechowska, E. & Obarska-Pempkowiak, H. 2008. Performance of Reed Beds Supplied with Municipal Landfill Leachate. *In: Wastewater Treatment, Plant Dynamics and Management in Constructed and Natural Wetlands, Ed. Vymazal J.*, Springer, Netherlands, 251-265.

Wundram, M.; Selmar, D. & Bahardi, M., (1996). The Chlamydomonas Test: A new phytotoxicity test based on the inhibition of algal photosynthesis disposal on salt mines. *Chemosphere,* 32 (8), 1623–1631.

Yildiz, E.D. & Unlu, K. (2003). Effects of landfill development on leachate characteristics. *Proceedings of the Ninth International Waste Management and Landfill Symposium.* S. Margherita di Pula, Cagliari, Italy; 6 - 10 Oct 2003.

18

Turfgrass Growth, Quality, and Reflective Heat Load in Response to Deficit Irrigation Practices

Benjamin Wherley
Texas AgriLife Research Urban Solutions Center
United States

1. Introduction

Turfgrass irrigation practices have come under intense scrutiny in recent years due to concerns over increasing population growth and diminishing water availability. Municipal water restrictions have become commonplace and more recently, the U.S. Environmental Protection Agency has developed guidelines that would restrict irrigation and/or amount of turf within the landscape (Watersense Single Family New Home Specification, 2009). Thus, turfgrass managers are increasingly faced with the challenge of maintaining acceptable turfgrass quality using less water. Understanding the minimal irrigation requirements and extent of water stress that a particular turfgrass species can tolerate while exhibiting acceptable quality is therefore highly valuable information for turfgrass managers and homeowners.

Deficit irrigation is the practice of intentionally under-irrigating of a plant to below its maximum water demand. This practice has been long used in crop production, where it often culminates in overall reductions in growth, development, and yield. Turfgrass systems are perhaps uniquely adapted for deficit irrigation because reductions in shoot growth are perceived to be beneficial, as long as visual and functional quality are not significantly sacrificed. Deficit irrigation has been practiced across a number of species, although the particular level of irrigation needed to maintain acceptable quality appears to vary among species. Using minilysimeters in the field, DaCosta & Huang (2006) determined that bentgrass species required ≥60% ET_a for maintaining acceptable summer quality, but that irrigating at only 40% ET_a was sufficient for maintaining acceptable quality during fall months. Qian and Engelke (1999) found that minimal irrigation requirements for grasses grown along a linear gradient of irrigation ranged from 26% to 68% of Class A pan evaporation (Ep) in a study of five turfgrass species along a linear gradient of irrigation. Feldhake et al. (1984) studied deficit irrigation of three turfgrass species grown in lysimeters and determined that irrigation deficits up to ~27% only decreased growth, but greater deficits resulted in significant loss of quality for Kentucky bluegrass and tall fescue.

Zoysiagrass is a warm-season (C_4) turfgrass native to Southeast Asia, but has become an increasingly popular turfgrass for use on lawns and golf courses throughout the southern half of the United States and many other tropical, subtropical, and temperate regions of the world (Turgeon, 2002). Whereas some turfgrass species are capable of avoiding drought through production of a deep root system, physiological studies indicate that zoysiagrass tolerates drought largely through osmotic adjustment (Qian & Fry, 1997). Very little information is available regarding the minimal irrigation requirements or response of this species to deficit irrigation practices. Therefore, the objective of this study was to determine the response of 'Empire' zoysiagrass to four levels of deficit irrigation and to identify the maximally acceptable irrigation deficit at which acceptable turf quality could be maintained in this species.

2. Materials and methods

This study was carried out from August 27 through October 15, 2008 in a greenhouse at the University of Florida campus in Gainesville, FL. 'Empire' zoysiagrass (*Zoysia japonica* Steud.) was grown in pots constructed from 10 cm diameter, 20 cm tall PVC pipes fitted with a flat end cap. A small hole was drilled into the center of each end cap for drainage. The top, open end of the pipes was fitted with a toilet flange for attaching a photosynthesis chambers. Four weeks prior to deficit irrigation studies, zoysiagrass sod pieces (2.5 cm depth) were removed from established, sand-based research plots at the University of Florida G.C. Horn Memorial Turfgrass Research Field Laboratory, Citra, FL, using a 10-cm diameter golf cup cutter. The sod was washed free of soil and established atop medium-coarse textured sand in PVC pots. A complete slow-release fertilizer (24-5-11, Turfgro Professional, Sanford, FL) was applied and grasses were grown in the greenhouse for four weeks to fully root in the soil prior to deficit irrigation experiments. Greenhouse temperatures during the study period were controlled at 32/22 °C (day/night). During this time, grasses were clipped weekly at a height of 6.4 cm. Visual observations confirmed that grass roots had reached container bottoms at the start of the experiment. A full cover of turf was also present on all pots at the start of the experiment, so that water loss was primarily a function of transpiration.

The study was initiated by fully saturating all pots. Following 24- hour period, when drainage had ceased, holes in pot bottoms were plugged to prevent drainage for the duration of the six-week study. At this time, four pots were randomly selected to be fully watered controls; and the initial, well-watered weights of these pots were measured. These pots were kept fully watered throughout the study by adding water daily in an amount equivalent to 100% of ET_a, measured gravimetrically. In order to more rapidly attain the desired water stress levels within the irrigation deficit treatments, drought-stress treatments were allowed to dry down the soil as a result of transpirational water loss, as described by Sinclair and Ludlow, 1986. Four plants were assigned to each of four deficit irrigation treatments (80%, 60%, 40%, and 20% of ET_a), which were initiated by allowing soil water content to decrease below that of fully watered controls prior to the beginning of the experiment. For example, in the 80% ET_a treatment, plants were allowed to decrease soil water content until the daily transpiration was measured to be 80% of that for well-watered plants. Upon reaching 80% relative transpiration (RT) rate, this stress level was permanently maintained by replenishing the pots daily with 80% of ET_a. The number of days of soil dry down required to reach targeted stress levels varied from 2 days for 0.8 RT (80% ET_a) to 6 days for the 0.2 RT (20% ET_a) stress treatment. As each of the stress treatments were reached, irrigation was returned to pots daily at 100% (controls), 80%, 60%, 40%, or 20% of ET_a. ET rates within the water stress treatments were also measured each afternoon to determine water loss, and corresponded highly to the prescribed irrigation amounts added to the respective treatments (data not shown). As such, ET of deficit treatments maintained a steady state proportional to fully irrigated controls.

Over the course of the six weeks, data were collected including daily ET rates, turfgrass visual quality, clipping dry weights, degree of leaf wilt and firing, photosynthetic rates, and reflective heat load in response to deficit irrigation. Turfgrass visual quality was visually estimated on a 1-9 scale, with 6 representing minimally acceptable turf (Morris & Shearman, 2006). For measuring clipping dry weights, grasses were clipped to 6.4 cm weekly, with clippings oven dried at 65 C for 72 hrs. Percent leaf wilt and firing were determined weekly by visually estimating the percentage of leaves within the 10 cm diameter plug that were either wilted of firing during the afternoon on a clear day. A portable chamber system modified from that described by Pickering et al., 1993, was used to measure canopy photosynthetic rates within the treatments at the conclusion of the study. This involved a portable photosynthesis system (LI-6200, LI-Cor Inc., Lincoln, NE) with an open leaf chamber mounted inside the translucent canopy chamber (figure 9). Chambers caused a 20% reduction in incoming photosynthetic active radiation (PAR). Three readings were recorded for each pot on a cloudless day during the afternoon hours, during which PAR levels within chambers always exceeded 1200 μmol m^{-2} s^{-1}. Temperatures inside chambers during the measurement period averaged 37.7 +/- 0.2 C. Canopy net photosynthetic rate (Pn) was expressed as CO_2 uptake per unit turf canopy area (m^{-2}). Reflective heat load from within the treatments was determined with a handheld Crop Trak Mini IR thermometer (Spectrum Technology, Plainfield IL) by measuring leaf canopy temperatures during the afternoon on a clear day.

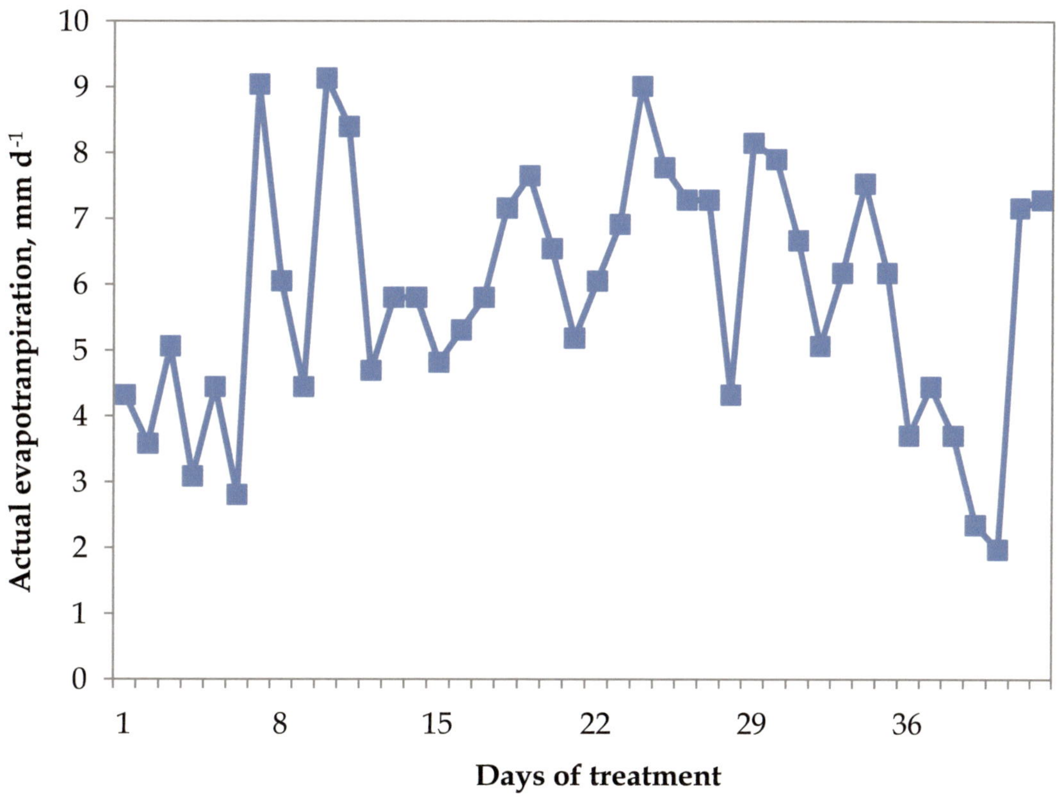

Fig. 1. Mean evapotranspiration (ET_a) rates for 100% ET control plants over the six week study. Amounts are equivalent to the daily irrigation supplied to 100% ET control plants, with deficit ET treatments receiving a fraction (80, 60, 40, or 20%) of this amount

3. Results and discussion

The objective of this study was to determine the response of 'Empire' zoysiagrass to four levels of deficit irrigation, provided daily. Actual evapotranspiration (ETa) and (equivalent to irrigation requirements) for control plants over the 42-day study ranged from 2 to 9 mm d^{-1}, with total irrigation volume applied of 246 mm (Figure 1). Deficit irrigated treatments received a total of 197 mm (80% ETa), 157 mm (60% ETa), 126 mm (40% ETa), and 101 mm (20% ETa).

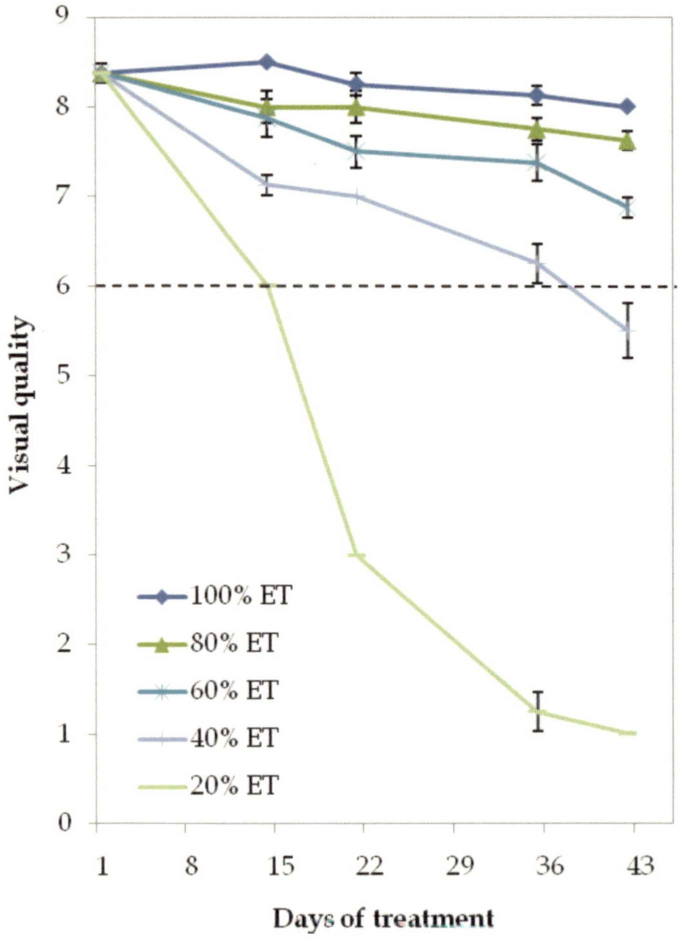

Fig. 2. Visual quality of turfgrass plants over six-week experiment, based on a 0-9 scale, with 0 = brown, dead turf, 6 = minimally acceptable and 9 = optimal color, density, and uniformity. Error bars denote standard error of the mean (n=4)

Visual Quality

Our results demonstrated that irrigating zoysiagrass at > 60% of ET_a was sufficient to sustain acceptable turfgrass quality over the six-week study (Figures 2 and 8). Conversely, visual quality of 40% ET_a treatments gradually declined over the study, but fell to unacceptable levels after five weeks. Within two weeks of initiating treatments, turf receiving 20% ET_a declined rapidly to unacceptable levels characterized by rapid wilt and significant leaf firing throughout the entire turf canopy. Previous work has shown that irrigating to 80% ET_a twice weekly was necessary to maintain acceptable zoysiagrass quality (Fu et al., 2004) and irrigating above 73% ET_a three times weekly was necessary for maintaining adequate Kentucky bluegrass quality (Feldhake et al., 1984). Our results indicate that acceptable zoysiagrass quality may be achievable at even greater irrigation deficits if irrigation is supplied with greater frequency.

Photosynthesis and Shoot Growth

Canopy photosynthetic rates of fully watered (100% ET_a) plants exceeded that of all deficit irrigated treatments and decreasing irrigation amounts resulted in proportional reductions in photosynthetic rates (Figure 3). Rates of photosynthesis ranged from an average of 6.1 μmol m^{-2} s^{-1} at 100% ET_a to 0.8 μmol m^{-2} s^{-1} at 20% ET_a. This is not surprising, given that dry matter accumulation and transpiration have been shown to be intimately linked to leaf gas exchange through stomata (Sinclair et al., 1984).

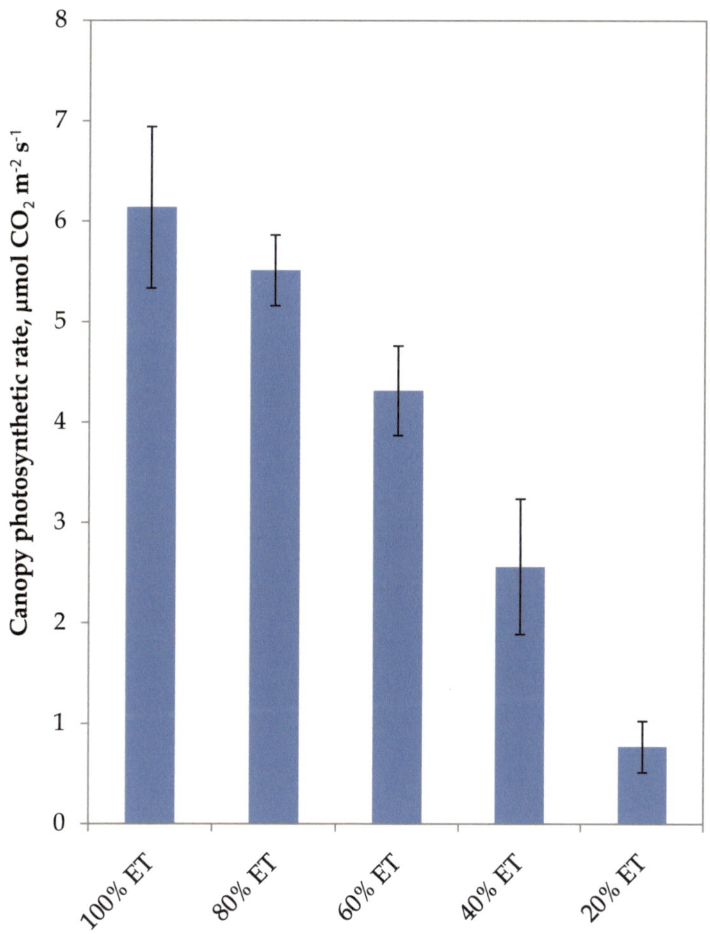

Fig. 3. Canopy photosynthesis rates measured at the conclusion of the six-week study. Error bars denote standard error of the mean (n=4)

Shoot growth is a process that is dependent on increases in cell volume, and thus, is generally highly sensitive to water deficit. Turfgrass shoot growth in this study was measured by weekly clipping collections, and was found to be progressively reduced with decreasing irrigation (Figure 4). Our data show that reducing irrigation to levels of 60% of ET_a, while causing little change in overall quality, led to as much as 25% reductions in shoot growth. Increased rates of turfgrass shoot growth have been correlated with higher turfgrass ET in a number of other warm and cool-season turfgrasses (Bowman & Macaulay, 1991; Shearman, 1986; Kim, 1983). Interestingly, while there was generally a proportional decrease in transpiration and shoot growth as deficits increased; shoot dry weights did not change between the 60% and 80% ET_a treatments. Thus, in terms of shoot growth, transpirational water use efficiency was greatest at 60% ET_a with this species (data not shown). From a practical standpoint, whereas soil water deficits may negatively affect field crop yields, moderate growth reductions in turfgrass sytems may actually be desirable, as they could result in less mowing requirements.

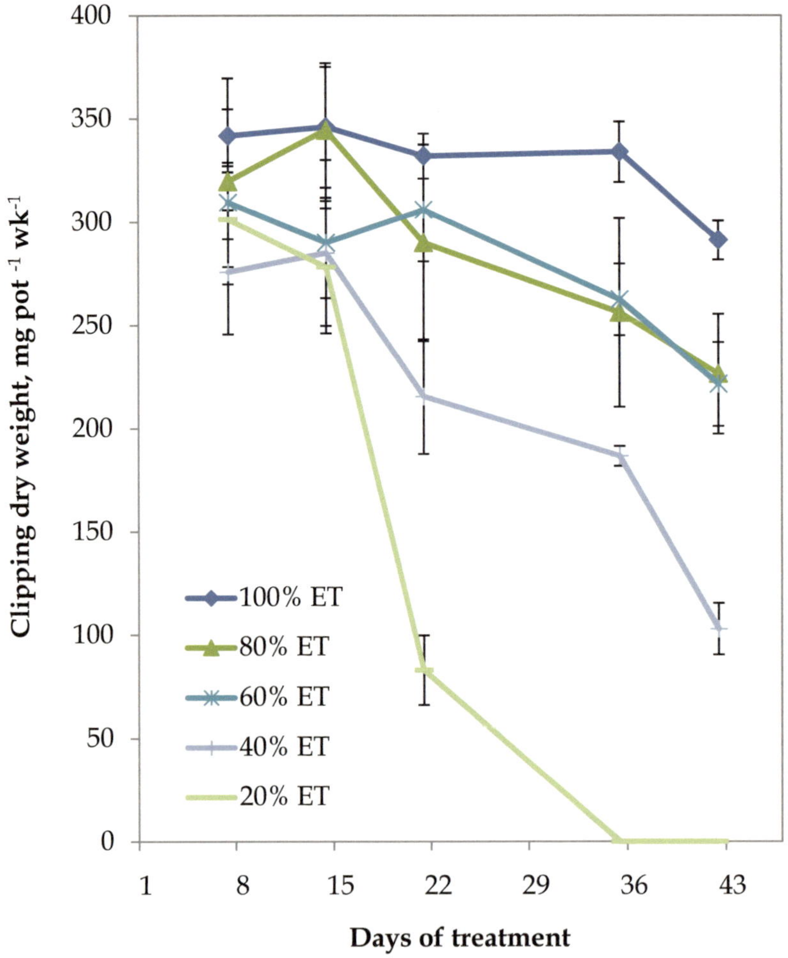

Fig. 4. Clipping dry weights collected weekly from 81 cm^2 pots. Error bars denote standard error of the mean (n=4)

Leaf Wilt and Firing

Leaf wilt and firing were the primary symptoms of plant water stress leading to quality loss in grasses receiving ≤ 60% of ET_a in the study (Figures 5 and 6). Leaf canopies of 20% ET_a plants declined most rapidly, with half of the canopy wilted within two weeks of initiating treatments, and becoming almost entirely fired by week six. 40% ET_a plants were nearly 50% wilted, with 30% of the canopy firing by week six. Only 5-10% of the canopy in 60% ET_a plants showed signs of wilt or firing; thus, quality was deemed acceptable throughout the study. Zoysiagrass irrigated at ≥80% ET_a never showed signs of wilt during any measurement period, indicating that soil moisture was sufficient to meet transpirational demand.

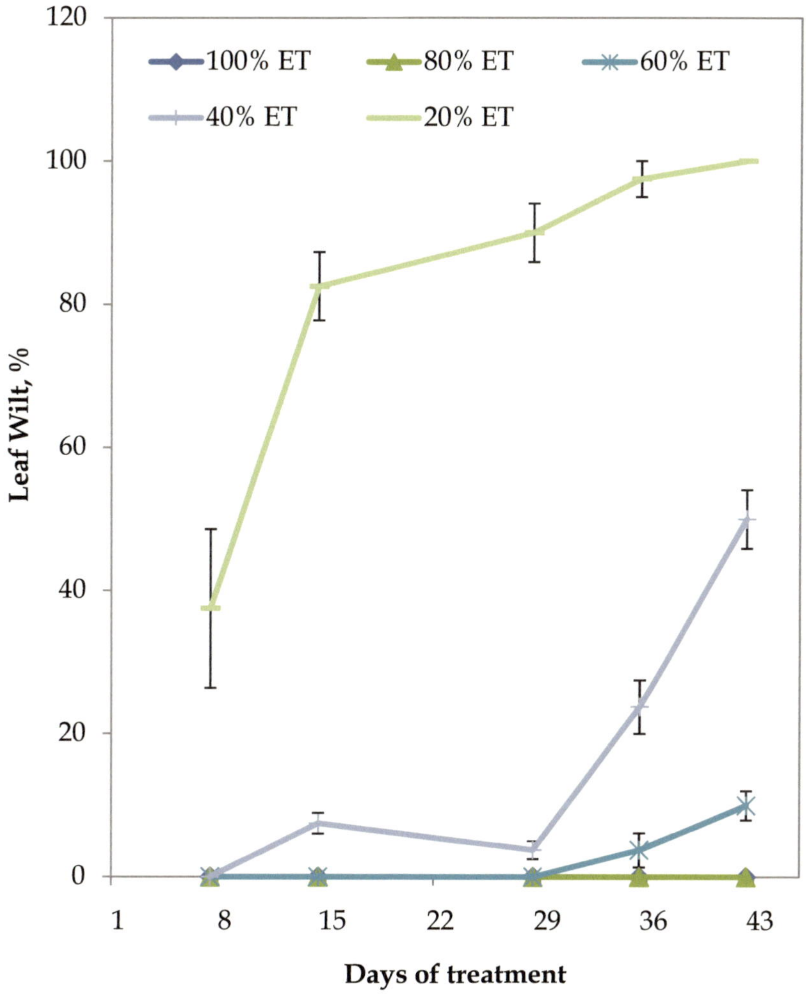

Fig. 5. Percentage of leaf wilt visible within the 81 cm^2 turfgrass canopy. Measurements were obtained during mid-afternoon hours on cloudless days. Error bars denote standard error of the mean (n=4)

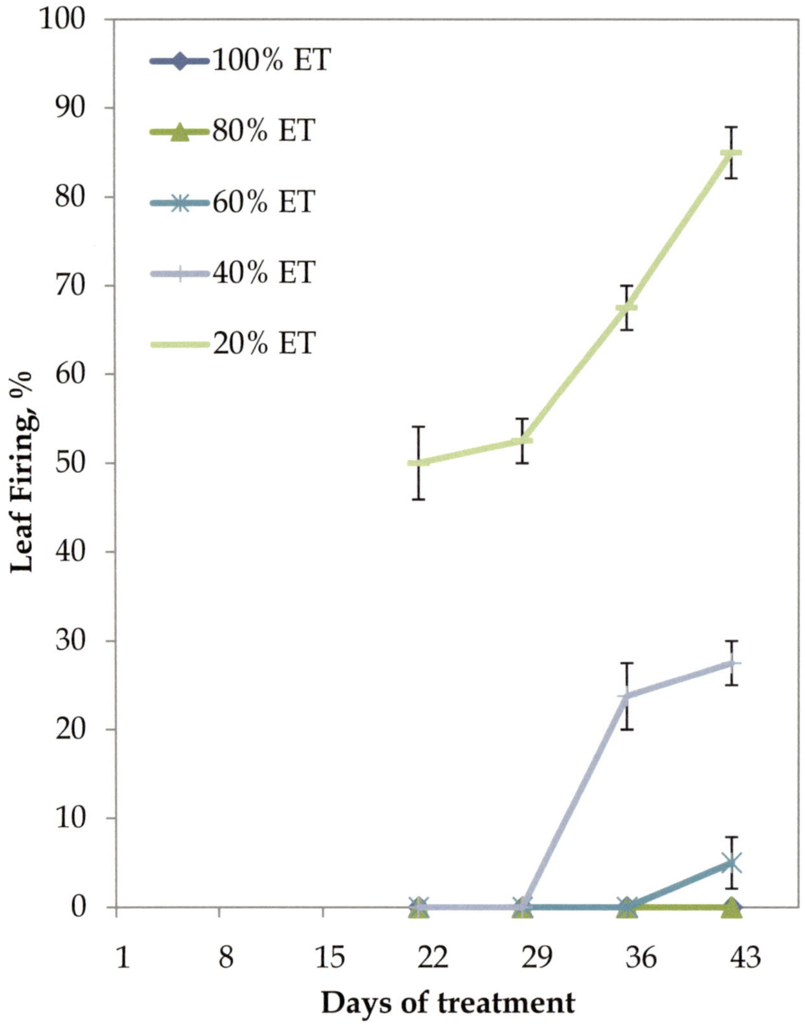

Fig. 6. Percentage of leaf firing visible within the 81 cm^2 turfgrass canopy. Measurements were obtained during mid-afternoon hours on cloudless days. Error bars denote standard error of the mean (n=4)

Reflective Heat Load

A substantial amount of incident solar radiation is converted to latent heat during the transpiration process. Therefore, vegetated surfaces such as turfgrass systems possess significant cooling capacity and play a critical role in heat dissipation within urban areas. It has been reported that green turfs and landscapes can save energy by reducing the energy input required for interior mechanical cooling of adjacent homes and buildings (Johns & Beard, 1985). Decreasing irrigation levels for water conservation conserves resources and maintenance requirements, but it also is likely to substantially impact reflective heat loads. Over the five sampling dates, we observed up to a 16 °C increase in canopy temperatures from fully irrigated to 20% ET_a irrigated turf (Figure 7). Interestingly, this heat load difference was reduced by half when irrigation was only slightly increased to 40% ET_a, due to the presence of significantly more green vegetation (Figure 8). Our data for zoysiagrass are similar to those of Feldhake et al. (1984), who reported a 1.7 °C increase in Kentucky bluegrass canopy temperature for each 10% reduction in ET. While a non-irrigated control treatment was not used in this study, it seems likely that even greater heat loads would result from turf receiving no irrigation, as no living vegetation would likely be present. From a practical standpoint, these data suggest that large-scale landscape deficit irrigation practices, such as those mandated through municipal water restrictions, could contribute to significantly increased surface temperatures.

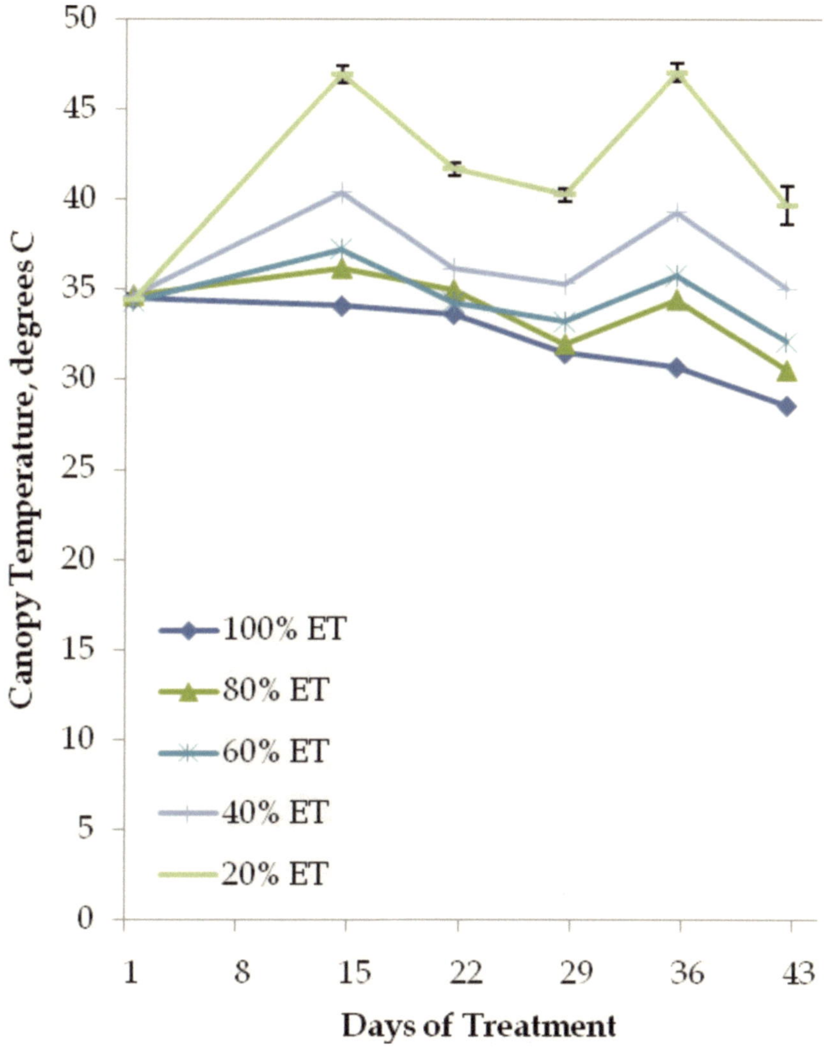

Fig. 7. Canopy temperatures within the irrigation treatments. Measurements were obtained during mid-afternoon hours on cloudless days. Error bars denote standard error of the mean (n=4)

Fig. 8. Visual appearance of 'Empire' zoysiagrass following six weeks of deficit irrigation, applied daily. (From left to right: 20% ET_a, 40% ET_a, 60% ET_a, 80% ET_a, and 100% ET_a)

Fig. 9. Portable chamber system for measuring canopy gas exchange

4. Conclusions

On the basis of our results, deficit irrigation can be a useful means of conserving water in the management of turfgrass. Irrigating zoysiagrass at up to a 40% deficit (60% ET_a) was sufficient to maintain acceptable turfgrass quality over a six-week period. Zoysiagrass response to the deficit irrigation included a reduction in evapotranspiration and photosynthetic rates, as well as shoot growth reductions. Such effects on shoot growth may be viewed as beneficial, as they would likely result in fewer mowing requirements. Our results suggest that irrigating this species at >40% irrigation deficits would not be advisable, as significant leaf wilt and firing occurred which negatively affected the appearance of the turf canopy and produced less-than-acceptable visual quality of the turf. While we were able to maintain acceptable quality at these levels with daily watering, it is likely that less frequent irrigation scheduling (2 or 3 days per week) might result in greater loss of quality at similar levels of ET replacement, due to longer periods of water stress encountered. An important consideration with deficit irrigation practices is the increasing reflective heat loads that are generated with diminishing amounts of irrigation. Canopy temperatures in this study progressively increased with greater deficits to a maximum increase of 16° C between 100% and 20% ET_a plants. This is an important consideration that could have large-scale implications on the surface temperatures and human comfort levels within urban environments.

5. References

Bowman, D.C. & Macaulay, L. (1991). Comparative evapotranspiration rates of tall fescue cultivars. *HortScience 26:122-123*

DaCosta, M. & Huang, B. (2006). Minimum water requirements for creeping, colonial, and velvet bentgrasses under fairway conditions. *Crop Science* 46: 81-89

Feldhake, C.M.; Danielson, R.E., & Butler, J.D. (1984). Turfgrass evapotranspiration. II. Responses to deficit irrigation. *Agronomy Journal* 76:85-89

Fu, J.; Fry, J., & Huang, B. (2004). Minimum water requirements of four turfgrasses in the transition zone. *HortScience* 39: 1740-1744

Johns, D. & Beard, J.B. (1985). A quantitative assessment of the benefits from irrigated turf on environmental cooling and energy saving in urban areas. In: *Texas Turfgrass Research.* p.134-142. Texas Agric. Exp. Stn. PR-4330. College Station

Kim, K.S. (1983). Comparative evapotranspiration rates of thirteen turfgrasses grown under both non-limiting soil moisture and progressive water stress conditions. p. 64. M.S. thesis, Texas A&M University, Colege Station, Texas

Morris, K.N., & Shearman, R.C. (2010). NTEP Evaluation Guidelines. National Turfgrass Evaluation Program. Available at: http://www.ntep.org/pdf/ratings.pdf Accessed 6 September 2010

Pickering, N.B.; Jones, J.W., & Boote, K.J. (1993). Evaluation of the portable chamber technique for measuring canopy gas exchange by crops. *Agricultural and Forest Meteorology* 63:239-254

Qian, Y.L., & Engelke, M.C. (1999). Performance of five turfgrasses under linear gradient irrigation. *HortScience* 34(5):893-896

Qian, Y.L., & Fry, J.D. (1997). Water relations and drought tolerance of four turfgrasses. *Journal of the American Society for Horticultural Science* 122: 129-133

Shearman, R.C. (1986). Kentucky bluegrass cultivar evapotranspiration rates. *HortScience* 21:455-457

Sinclair, T.R., & M.M. Ludlow. (1986). Influence of soil water supply on the plant water balance of four tropical grain legumes. *Australian Journal of Plant Physiology* 13:329-341

Sinclair, T.R.; Tanner, C.B., & Bennet, J.M. (1984). Water-use efficiency in crop production. *BioScience* 34:36-40

Turgeon, A.J. (2002). *Turfgrass Management*. Prentice Hall, ISBN 0-13-027823-8, Upper Saddle River, NJ, USA

WaterSense® Single Family New Home Specification. (2009). http://www.epa.gov/WaterSense/docs/home_finalspec508.pdf

19

Understanding the Effects of Fires on Surface Evapotranspiration Patterns Using Satellite Remote Sensing in Combination with an Energy Balance Model

Juan M. Sánchez[1], Vicente Caselles[2] and Eva Rubio[1]
[1]*University of Castilla-La Mancha*
[2]*University of Valencia*
Spain

1. Introduction

Forest fires are highly destructive for nature, affecting the landscape, the natural cicle of the vegetation, and the structure and functioning of ecosystems. Beyond that, they also provoke changes in the local and regional meteorology, and particularly in the surface energy flux patterns. In a fire-affected area, changes in the ecosystem structure and species composition modify the evapotranspiration (*LE*) and the rest of the terms involved in the energy balance equation. Besides, these changes in the local energy balance may persist for decades (Randerson et al., 2006). There is an increasing concern among the scientific community about the effect of forest fires on climate change at this point (Randerson et al. 2006). In this work we focus on the study of the changes in the energy flux patterns after a forest fire, with particular emphasis on the evapotranspiration, which effect on the global system should be further analyzed by the radiative forcing models.

The physical characterization of the hydrological processes plays a very important role in the framework of the activities for the management of hydrological resources. Particularly, the soil-vegetation-atmosphere energy exchanges are the basis of an appropriate hydrological balance, and thus, of an appropriate planning of the hydrological resources. The fusion of physical models for estimating the hydrological balance, and particularly the evapotranspiration, with technological advances for the characterization of hydrological, hydro-geological, and atmospheric issues, is of great utility. Although there are several surface-based methods that can accurately measure surface heat fluxes at point locations, it is not feasible to use a network of these systems to create spatially distributed flux maps because of the high variability of real landscapes. As stated by Scott et al. (2000), micrometeorological approaches can only realistically provide measurements representative of a particular type of vegetation cover when there is a reasonably extensive, uniform area of that vegetation immediately upwind of the instruments. The use of remote sensing techniques supplies the frequent lack of ground-measured variables and parameters required to apply the local models at a regional scale. Modelling evapotranspiration is very sensitive to the surface features and conditions. For this reason, a regional model must account for the surface variability. In this context, satellite remote sensing has become a basic tool since it allows us the regular monitoring of extensive areas. Different surface variables and parameters can be extracted from the combination of the multi-spectral information contained in a satellite image. The surface can be characterized with a detail depending on the spatial resolution of the sensor used. One of the goals of remote sensing is to provide us with data on barely accessible areas. This includes some regions such as, for example, forest areas where evapotranspiration retrieval is more difficult due to evident limitations on experiments.

The basis of the remote sensing models using thermal infrared data for determining evapotranspiration is the direct relationship between surface thermodynamic temperature and energy balance, which has long been recognized by meteorologists and hydrologists. The surface thermodynamic temperature can be obtained from the brightness temperature after atmospheric and emissivity corrections. Monteith (1973) proposed a single source model based on a convection analogue to Ohm´s Law to obtain the sensible heat flux. However, the evaluation of this single source model showed important limitations for partial canopy cover conditions (Hall et al. 1992). Two-source (soil + vegetation) layer models have been developed to accommodate partial canopy cover conditions considering energy exchange between soil and canopy components (Shuttleworth & Wallace, 1985; Choudhury & Monteith, 1988, Norman et al., 1995, Sánchez et al., 2008).

Inoue and Moran (1997) proposed a simple method to estimate daily values of actual canopy transpiration. The method utilizes instantaneous differences of canopy and air temperature around mid-day as a major input. Results were found to be well correlated to those measured by sap-flow heat balance method in soybean canopies. Anderson et al. (1997) presented an operational two-source (soil+vegetation) model for evaluating the surface energy balance given measurements of the time rate of change in radiometric surface temperature during morning hours. Using this model, the need for ancillary measurements of near-surface air temperature is eliminated. The performance of this model was evaluated in comparison with data collected during the first International Satellite Land Surface Climatology Project field experiment, in Kansas, and the Monsoon ´90 experiment, conducted in southern Arizona. Comparisons yielded uncertainties comparable to measurement errors typical of standard micrometeorological methods for flux estimation. Chehbouni et al. (2001) used dual angle observations of radiative surface temperature in conjunction with a two-layer model to derive sensible heat flux over the Semi Arid Land Surface Atmosphere program (SALSA) in Mexico. The average error was about 23%. Moran et al. (1994) introduced a water deficit index for evaluating evapotranspiration rates of both full cover and partially vegetated sites. This index can be computed using remotely sensed measurements of surface temperature and reflectance with limited on-site meteorological data. Comparison with simulations of a two source energy balance model showed accurate estimates of field evapotranspiration rates. French et al. (2005) used data from ASTER collected over an experimental site in central Iowa, in the framework of the Soil moisture Atmosphere Coupling Experiment (SMACEX), to retrieve surface energy fluxes. Two different approaches, designed to account for the spatial variability, were considered: the Two-Source Energy Balance model (TSEB) and the Surface Energy Balance Algorithm for Land model (SEBAL). Comparison of the results with eddy covariance measurements showed better agreement using the TSEB model with average deviations lower than 20 W m^{-2}. These results were also supported by Li et al. (2005). These authors compared local model output using two different versions of the TSEB (series and parallel) with tower-based flux observations. Root mean square differences ranged from 20 to 50 W m^{-2}. In this case, land surface temperatures were derived from high resolution Landsat Thematic Mapper (TM) and Enhanced Thematic Mapper (ETM+) scenes and aircraft imagery. Su et al. (2005) also used SMACEX data to evaluate the Surface Energy Balance System (SEBS) model using both high-quality local-scale data and high-resolution remote sensing data from the Landsat ETM+.

Despite the wide variety of remote sensing-based works and proposed models on evapotranspiration retrieval, there is not a generalized agreement about the most appropriate model depending on the application area. In this work, we present a methodology focused on the *LE* retrieval from high-resolution satellite data. The bases of this method are the energy balance equation and a Simplified Two-Source Energy Balance (STSEB) model proposed by Sánchez et al. (2008) for estimating instantaneous surface fluxes. This model was validated over a variety of surface conditions with good results. Also, an exhaustive analysis of sensitivity to typical uncertainties (assumed for a regional variability) in the required inputs was performed by these authors (Sánchez et al., 2008, 2009). The combination of the STSEB model with remote sensing techniques results in an operational methodology to retrieve evapotranspiration at a regional scale from remote sensing and local meteorological data. In this work we show the application of this methodology to Landsat imagery and a particular region in Spain, but it could be extended to other high-resolution sensors, and what it is even more interesting, to any other regions in the world.

In this work we focus on an area, located in Almodovar del Pinar, Cuenca (Spain), affected by a forest fire in the summer of 2001. This fire destroyed 172 ha of pine forest and shrubland. After the fire this area became an intensive study site. Our objective is to quantify the effect of this forest fire in terms of net radiation, soil and sensible heat fluxes, and evapotranspiration in the two land cover classes dominant in the area, mature pine forests and shrublands. With this aim we applied the STSEB model (Sánchez et al., 2008) to a set of 5 images Landsat 5 Thematic-Mapper (TM) corresponding to the period July 2007-July 2008.

2. Methodology

The governing equation is the Energy Balance Equation (EBE) of the land surface, which models a system formed by vegetation, surrounding soil, and atmosphere:

$$R_n = G + H + LE \tag{1}$$

where R_n is the net radiation flux (W m^{-2}), G is the soil heat flux (W m^{-2}), H is the sensible heat flux (W m^{-2}) and LE is the latent heat flux in the atmosphere boundary layer (W m^{-2}). According to Seguin and Itier (1983):

$$\frac{H_d}{R_{nd}} = \frac{H_i}{R_{ni}} \tag{2}$$

where the subscripts i and d refer to instantaneous and daily fluxes, respectively.

On diurnal timescales, G can constitute an important contribution to the EBE (Choudhury, 1987; Santanello and Friedl, 2003). However, at a daily scale G can be neglected in equation (1) (Seguin and Itier, 1983; Lagouarde and McAneney, 1992; Sánchez et al., 2007), and LE can be obtained from equations (1) and (2) as:

$$LE_d = \frac{R_{nd}}{R_{ni}}(R_{ni} - H_i) \tag{3}$$

Using equation (3), LE_d can be obtained from the instantaneous values of R_n and H at a particular time of day, and the relative net radiation contribution at that time when global radiative exchange is integrated, R_{nd}/R_{ni}.
The instantaneous net radiation is estimated by establishing a balance between the long-wave and the short-wave radiation:

$$R_n = (1-\alpha)S + \varepsilon L_{sky} - \varepsilon\sigma T_R^4 \tag{4}$$

where S is the solar global radiation (W m^{-2}), T_R is the radiometric land surface temperature, a is the surface albedo, ε is the surface effective emissivity, and σ is the Stefan-Boltzmann constant. L_{sky} is the incident long-wave radiation (W m^{-2}).
In the STSEB approach proposed by Sánchez et al. (2008), the ground surface is divided in two components, canopy and soil surrounding (Fig. 1). According to this configuration, the total sensible heat flux is obtained by the addition between the soil and canopy contributions, H_s and H_c, respectively:

$$H = P_v H_c + (1-P_v)H_s \tag{5}$$

In this equation, P_v is the vegetation cover, and H_s and H_c are expressed as:

$$H_c = \rho C_p \frac{T_c - T_a}{r_a^h} \tag{6a}$$

$$H_s = \rho C_p \frac{T_s - T_a}{r_a^a + r_a^s} \tag{6b}$$

where ρC_p is the volumetric heat capacity of air (J K^{-1}m^{-3}), T_a is the air temperature at a reference height (K), T_c and T_s are the canopy and soil radiometric temperatures, respectively, r_a^h is the aerodynamic resistance to heat transfer between the canopy and the reference height (m s^{-1}), r_a^a is the aerodynamic resistance to heat transfer between the point z_0+d (z_0: roughness length, d: displacement height) and the reference height (m s^{-1}), r_a^s is the aerodynamic resistance to heat flow in the boundary layer immediately above the soil surface (m s^{-1}). A summary of the expressions to estimate these resistances using the wind speed and crop height, and more details about the STSEB model can be seen in Sánchez et al. (2008).
Finally, the instantaneous soil heat flux can be obtained as a fraction of the net radiation:

$$G = C_G(1-P_v)R_n \tag{7}$$

In this work we used a value of C_G=0.275 (mean value of the usual range 0.15-0.40).

2.1 Surface temperature and emissivity

Landsat-TM and ETM+ sensors possess a unique thermal band with a spectral range of 10.4-12.5 µm, and an effective wavelength of 11.457 µm. This limitation does not allow to apply split-window methods neither Temperature/Emissivity Separation (TES) methods.

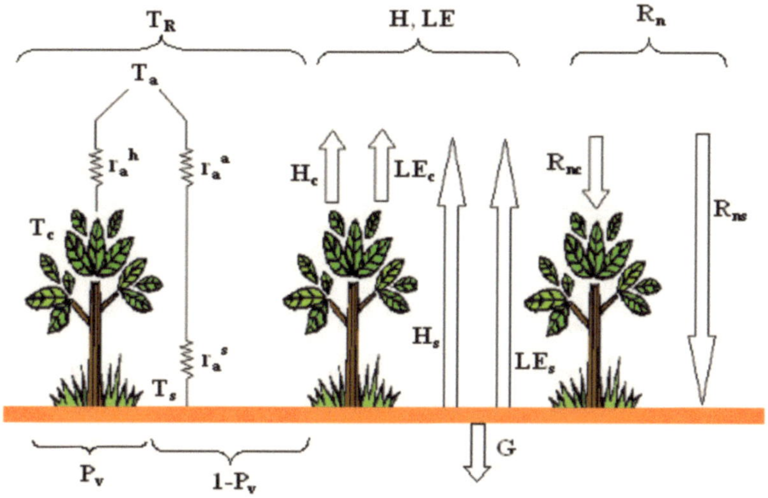

Fig. 1. Scheme of resistances and flux partitioning between soil and canopy, corresponding to the STSEB approach. Symbols are defined in the text.

Therefore, a single-channel method, based on the radiative transfer equation, was used. The remotely measured channel radiance, R_i, consist of two main contributions: (1) the radiance at surface level, which is attenuated by the absorption of the atmosphere between the surface and the instrument, characterized by the atmospheric transmittance, τ_i, and (2) the upwelling sky radiance emitted by the atmosphere in the viewing direction, $L_{i\ atm}^{\Uparrow}(\theta)$, so that R_i, in agreement with the radiative transfer equation, is stated as:

$$R_i = \left[\varepsilon_i B_i(T_R) + \left[1 - \varepsilon_i\right] L_{i\ atm\ hem}^{\Downarrow} \right] \tau_i + L_{i\ atm}^{\Uparrow} \tag{8}$$

where $B_i(T_R)$ is Planck´s function for a temperature T_R, ε_i is the surface emissivity and $L_{i\ atm\ hem}^{\Downarrow}$ is the hemispheric downwelling sky irradiance divided by π (Lambertian reflection assumed). Radiosounding data were introduced into the MODTRAN 4.0 code (Berk et al., 1999) to get estimates of τ_i, $L_{i\ atm}^{\Uparrow}(\theta)$ and $L_{i\ atm\ hem}^{\Downarrow}$. Altough equation (8) depends on the observation angle (θ), the nadir view provides good results for Landsat-TM and ETM+. A simple and operational equation proposed by Valor & Caselles (2005) was used to estimate the surface emissivity from the knowledge of the vegetation cover, P_v, and the emissivities of the soil and canopy components, ε_s and ε_c, respectively.

$$\varepsilon = \varepsilon_c P_v + \varepsilon_s (1 - P_v)(1 - 1.74 P_v) + 1.7372 P_v (1 - P_v) \tag{9}$$

Equation (9) includes the effect of the geometrical distribution of the vegetated surface and also the internal reflections. Typical emissivity values can be assumed for ε_c and ε_s (Rubio et al., 1997, 2003).

2.2 Vegetation cover

Bands 3 (0.63-0.69 µm) and 4 (0.76-0.90 µm) of TM and ETM+ were used to estimate NDVI. Previously, visible and near-infrared bands were corrected of atmospheric effects using the radiosounding data and the MODTRAN 4.0 code. For this purpose, the at-surface channel reflectivity, ρ_i, is calculated with the following equation:

$$\rho_i = \frac{\pi(R_i - L_{i\ atm}^{\Uparrow})d^2}{\tau_i(ESUN_i \cos(\alpha)\tau(\alpha) + L_{i\ atm\ hem}^{\Downarrow})} \tag{10}$$

where $\tau(a)$ is the atmospheric transmissivity between the sun and the surface, a is the zenithal solar angle, $ESUN_i$ is the spectral solar irradiance on the top of the atmosphere, and d is the Earth-Sun distance.
Vegetation cover was obtained through the expression (Valor & Caselles, 1996):

$$P_v = \frac{\left(1 - \frac{NDVI}{NDVI_s}\right)}{\left(1 - \frac{NDVI}{NDVI_s}\right) - K\left(1 - \frac{NDVI}{NDVI_v}\right)} \tag{11}$$

where the coefficient K is obtained by:

$$K = \frac{R_{NIR_v} - R_{RED_v}}{R_{NIR_s} - R_{RED_s}} \tag{12}$$

where R_{NIR} is the near infrared reflectivity, and R_{RED} is the red visible reflectivity. The subscript v and s correspond to completely vegetated and unvegetated areas, respectively, selected by looking at the spectral contrast among bands 3-5. These selected areas were also used to estimate T_c and T_s, required in equations (6a) and (6b), respectively, from the land surface temperature maps generated.

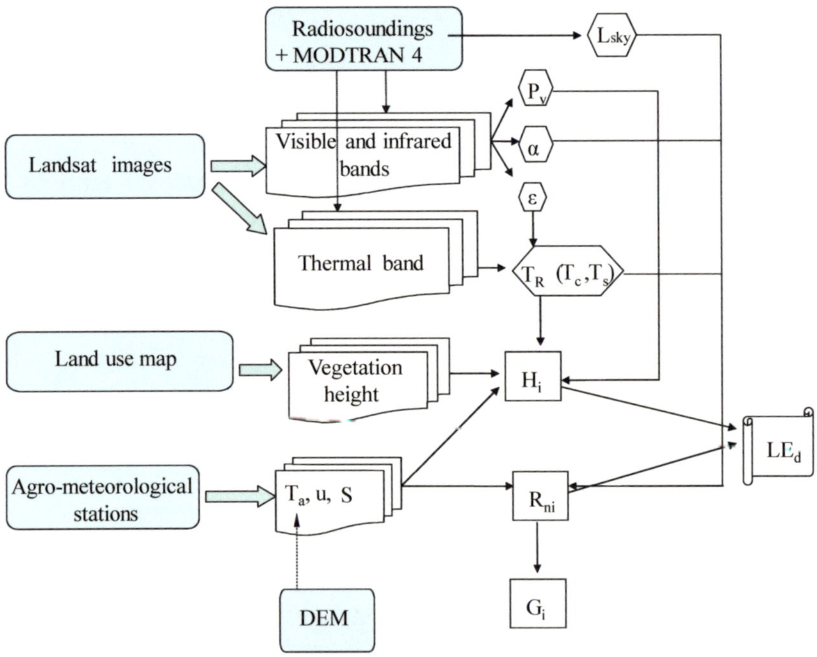

Fig. 2. Scheme of the methodology proposed to retrieve actual daily evapotranspiration at a regional scale.

2.3 Albedo

Some authors such as Dubayah (1992), Duguay (1992), etc, divided the spectral region from 0.3 to 3.0 µm into 10 spectral bands to estimate surface albedo. Each band has a different integrating weight according to the typical vegetation spectral reflectance pattern. In this paper, the surface albedo is integrated by using the equation (Dubayah, 1992):

$$\alpha = 0.221\rho_1 + 0.162\rho_2 + 0.102\rho_3 + 0.354\rho_4 + 0.059\rho_5 + 0.0195\rho_7 \tag{13}$$

where ρ_i is the corrected reflectivity for the i band of TM or ETM+. The average error of the calculated surface net radiation, using equation (12) for estimating the albedo, is around 2% when comparing to field measurements (Dubayah, 1992).

2.4 Meteorological variables

Ancillary meteorological data are required to complete the set of variables and parameters involved in the previously shown scheme of equations. Air temperature, T_a, is necessary in equations (6a) and (6b) to estimate the exchange of sensible heat flux between the surface and the atmospheric boundary layer. Wind speed, u (m s^{-1}), is required in the expressions to calculate the aerodynamic and soil resistances taking part in the STSEB model (Sánchez et al., 2008). The global solar radiation, S, and the incident long-wave radiation, L_{sky}, are necessary in the net radiation balance (Eq. 4). All these variables, except L_{sky}, are continuously registered in typical agro-meteorological stations, regional maps can be created by interpolating the data registered in a network of stations distributed within the study area. Regarding L_{sky}, due to its known spatial homogeneity across a relative extensive area (Humes et al., 2002), a single value of this variable can be used for each image, and it can be obtained from launched radiosoundings. Sánchez et al. (2008) showed that particular care must be taken with the air temperature when the STSEB model is applied since an uncertainty in T_a can lead to a significant error in the evapotranspiration retrieval. For this reason a Digital Elevation Model (DEM) can be considered in order to obtain more reliable maps of this meteorological variable.
A scheme of the methodology exposed is shown in Figure 2.

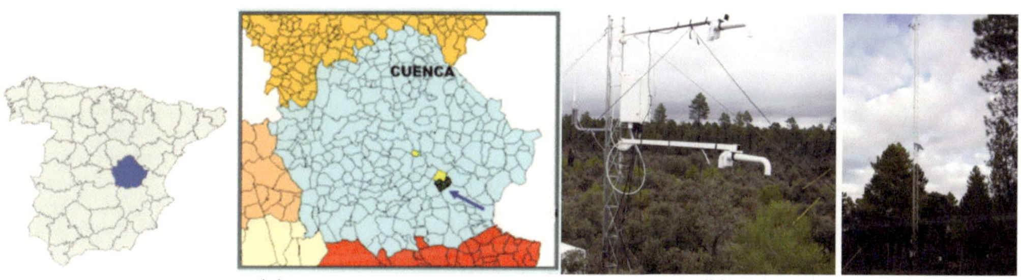

(a) (b) (c)

Fig. 3. (a) Location of the study site. (b) Overview of the burned area, and the Bowen station placed in site. (c) Overview of the forested_c areas, and the meteorological tower placed in site.

3. Study site

The study site is a forest area, with some inserted crop fields, located in Almodóvar del Pinar, Cuenca (39° 40´N, 1° 50´W, 950 m above sea level) (Fig. 3a). Climate is mediterranean, with warm and dry summers, and cool winters. The dominant tree species is *Pinus pinaster Ait.*, but many other species coexist. In the summer of 2001, a fire affected a total of 172 ha of which 113 ha were covered by pines and 59 ha by shrubs (Fig. 4). After the fire, the species *Quercus ilex L.* was occupying the burned area as a consequence of a natural regeneration process. Four 125×125 m test sites, two inside and two outside the burned area perimeter, were selected for this study, as samples of both pine areas and shrublands. Test sites outside the fire perimeter were called control sites (_c). Environmental conditions in these control sites mimic those in the two test sites inside the burned perimeter in case the fire had never happened. A meteorological tower was placed in the *Forest_c* area (Fig. 3c) with the instrumentation necessary to measure air temperature and wind speed, at several heights, as well as incident solar radiation, precipitation, etc. Also, a Bowen station was set up in the *Forest* site in september 2007 (Fig. 3b). Moreover, we selected a nearby area representative of a mature holm oak forest to analyze the effect of the fire on a future scenario in which pines are replaced by holm oaks as the dominant species. For this work we have used a set of 5 Landsat 5-TM scenes (19 July 2007, 4 August 2007, 28 September 2007, 2 May 2008, 21 July 2008) with a spatial resolution of 30 m for the visible and near infrared bands, and 120 m for the thermal band (all bands were rescaled to 25 m). Also, the atmospheric profiles required for the atmospheric correction of the images were obtained from the website http://atm-corr.gsfc.nasa.gov/.

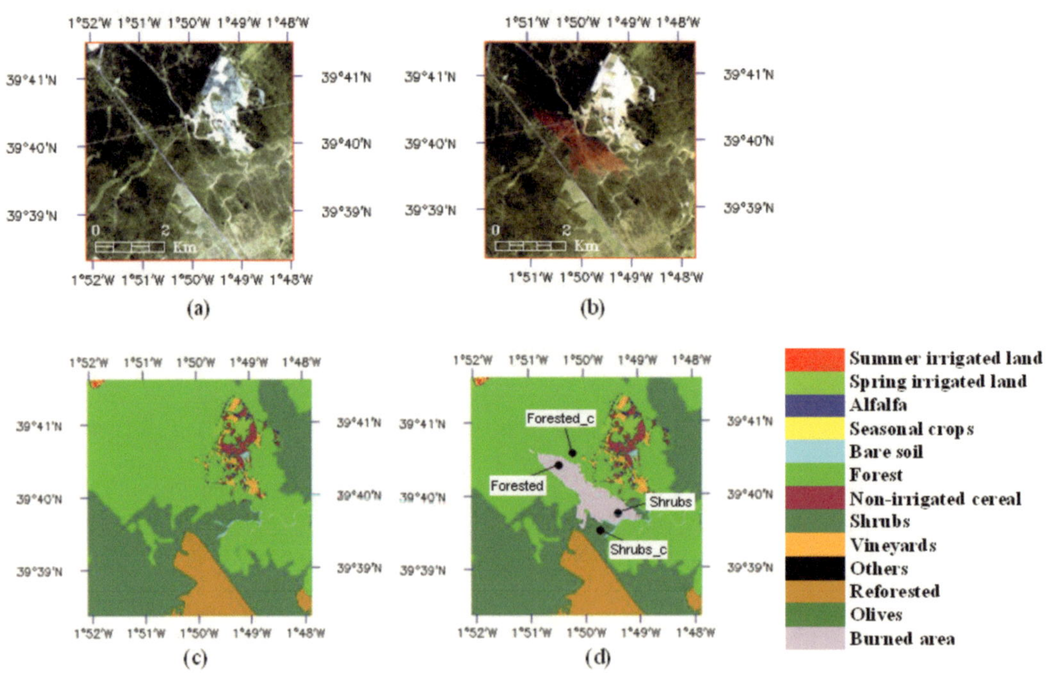

Fig. 4. Study site: (a) False color composition (7,5,3) from a L7-ETM+ image for the 8 June 2001 (prior fire), (b) Idem for the 26 July 2001 (post- fire), (c) Land use map before the fire, (d) Idem after the fire.

4. Results

4.1 Comparison with control areas

Firstly, the similarity between the control sites and the fire-affected areas was tested. For this, two Landst 7-ETM+ were used, before and after the fire date (8 June 2001 and 26 July 2001). Using digital count data, without further processing, we compared values obtained in both sites for each one of the 7 Landsat channels. Figure 5a shows this comparison prior the fire. Differences between the control areas and those affected by the fire a few days afterwards are negligible. However, these differences are evident after the fire (Fig. 5b). A significant increase is observed in values of bands 6 (thermal infrared) and 7, whereas values decrease in band 4 (near infrared), consequence of the vegetation disappearance.

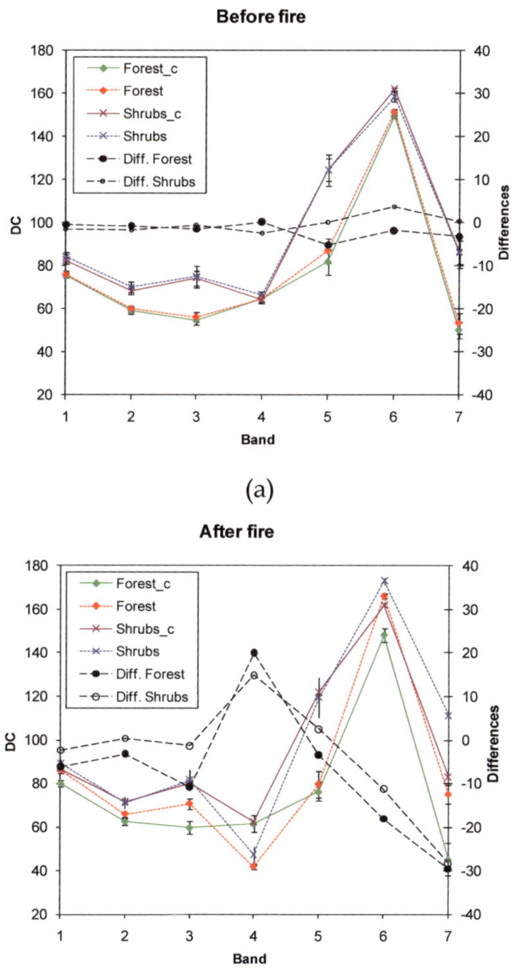

(b)

Fig. 5. Average digital count values, and their corresponding error bars, for each one of the selected areas, and differences with their corresponding control areas, (a) before the fire, (b) after the fire. Spectral ranges for the different bands: (1) 0.45-0.52 µm, (2) 0.52-0.60 µm, (3) 0.63-0.69 µm, (4) 0.76-0.9 µm, (5) 1.55-1.75 µm, (6) 10.4-12.5 µm, (7) 2.08-2.35 µm.

4.2 Comparison with observed fluxes

Once processed with MODTRAN the atmospheric profiles for each one of the Landsat images, we obtained the atmospheric parameters required for the atmospheric correction of the different bands. Inputs required in equations (6), (11), and (12), were extracted after selecting a full covered area and a bare soil area. Merging the satellite information with values of air temperature, wind speed and global radiation, the data set we need is completed. Figure 6 shows, as an example, maps generated from the 28 September 2007 image. Three of the five scenes in the present study are concurrent with ground flux measurements, which allows us to compare flux results with values registered in the Bowen station located in the *forest* site. Table 1 lists the values estimated and those ground-measured. A good agreement is shown between predicted and observed fluxes. To sum up, relative errors of 7% (40 W m^{-2}), 12% (10 W m^{-2}), 19% (70 W m^{-2}), and 21% (21 W m^{-2}) are obtained for R_n, G, H, and LE, respectively, at a instantaneous scale, while at a daily scale errors are 6% (11 W m^{-2}), 3% (4 W m^{-2}), and 22% (10 W m^{-2}) for R_n, H, and LE, respectively. Note that this is not a robust validation, however the model has been previously tested under different coverages (Sánchez et al., 2008, 2009).

Date	Instantaneous		Daily	
	Observed	/ Estimated	Observed	/ Estimated
R_n				
28 Sep.2007	492	544±7	118	130.6±1.7
2 May 2008	633	640±6	187	192.0±1.8
21 Jun.2008	653	626±7	203	194±2
G				
28 Sep.2007	123	98±6		
2 May 2008	134	125±6		
21 Jun.2008	101	105±7		
H				
28 Sep.2007	305	390±30	92	91±7
2 May 2008	386	365±19	118	110±6
21 Jun.2008	427	375±25	122	116±8
LE				
28 Sep.2007	64	60±40	25	38±8
2 May 2008	113	150±30	51	83±7
21 Jun.2008	126	150±40	61	78±9

Table 1. Estimated and observed values of the surface energy fluxes (Wm^{-2}) at the forest site.

4.3 Analysis of the fire effect

Average values of the fluxes and most significant parameters of the model were obtained for each one of the test sites and dates. Note that the fire event occurred in 2001, and 6 years is enough time for the shrublands to recover its original stage (prior the fire); however, it is a very short time period for the forested area. This can be seen in the average values of NDVI and P_v listed in Table 2. A similar effect can be observed in terms of surface temperature. In the forest site T_R values are, in average, 6 °C higher in the burned area, whereas in the shrubs site, temperature is 0.6 °C lower in the burned area. These differences are in agreement with those observed by other authors in similar studies (Amiro et al., 1999). After those 6 years, the effect of the fire on the energy flux pattern has almost disappeared in the shrub sites, while it is still significant in the forested sites. Figure 7 shows the plots with the average values (for the 5 study dates) of the differences in terms of energy fluxes between calculated values for the forested and shrub test sites and their respective control sites. The increase in albedo and T_R produces a decrease in both shortwave and longwave net radiation, yielding an average net decrease in R_{ni} of 54±5 W m^{-2}. The opposite effect is observed in G_i, with an average increase of 43±10 W m^{-2}. At a daily scale the average decrese in R_{nd} results 15±3 W m^{-2}. Sensible heat flux is higher in the burned area, in average 160±40 W m^{-2} at the time of the satellite overpass, and 45±9 W m^{-2} at daily scale, whereas latent heat flux is lower, 250±50 W m^{-2} at the time of the satellite overpass, and 61±7 W m^{-2} (2.1±0.2 mm/day) at daily scale. Therefore, even though the effect of the fire on the total net radiation is not very important, it is significant the increase in the Bowen ratio (*H/LE*), and the drastic decrease in the evapotranspiration in forested areas.

Date	Forest		Shrub		Mature holm oak forest
	Control	Burned	Control	Burned	
Albedo					
19 Jul. 2007	0.117	0.140	0.124	0.139	0.144
4 Aug.2007	0.119	0.134	0.126	0.138	0.143
28 Sep.2007	0.095	0.115	0.098	0.12	0.132
2 May 2008	0.116	0.127	0.119	0.128	0.135
21 Jun.2008	0.114	0.131	0.120	0.134	0.138
NDVI					
19 Jul. 2007	0.43	0.29	0.32	0.31	0.35
4 Aug.2007	0.40	0.28	0.29	0.29	0.33
28 Sep.2007	0.63	0.38	0.42	0.38	0.45
2 May 2008	0.46	0.28	0.29	0.29	0.37
21 Jun.2008	0.44	0.33	0.35	0.35	0.39
P_v					
19 Jul. 2007	0.58	0.33	0.37	0.36	0.44
4 Aug.2007	0.52	0.29	0.31	0.31	0.39
28 Sep.2007	0.74	0.34	0.41	0.35	0.46
2 May 2008	0.63	0.29	0.33	0.32	0.48
21 Jun.2008	0.61	0.39	0.43	0.44	0.51
T_R (°C)					
19 Jul. 2007	36.8	42.9	42.1	41.2	39.5
4 Aug.2007	37.5	43.5	43.3	42.8	40.6
28 Sep.2007	21.4	27.1	26.0	25.8	24.5
2 May 2008	18.6	25.3	22.1	21.4	22.0
21 Jun.2008	26.2	33.1	31.6	30.7	30.0

Table 2. Average values of the main radiometric parameters for all study sites and dates.

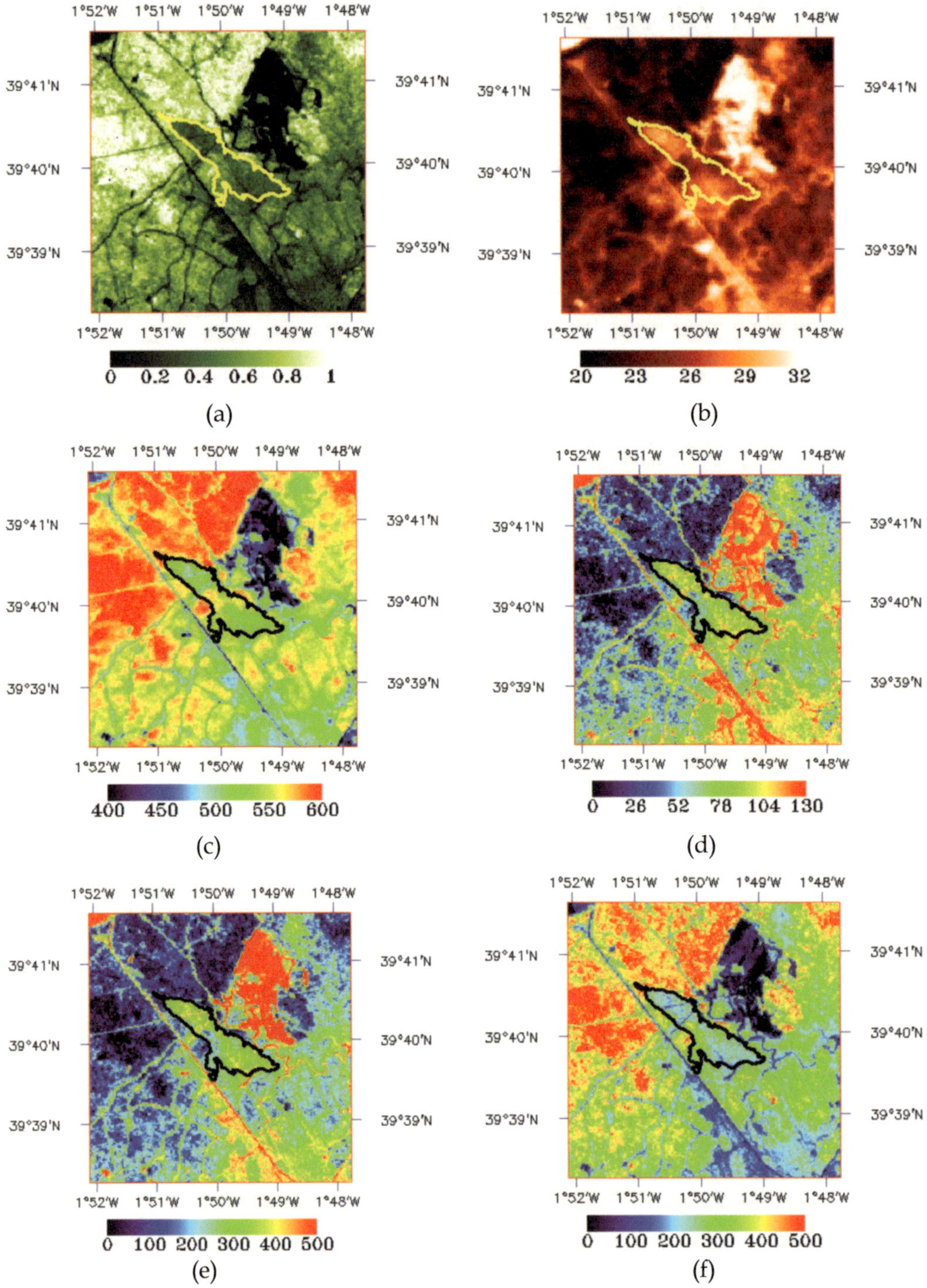

Fig. 6. Instantaneous values of: (a) P_v, (b) T_R (°C), (c) R_n (W m^{-2}), (d) G (W m^{-2}), (e) H (W m^{-2}) and (f) LE (W m^{-2}), obtained from the L5-TM image corresponding to the date 28 September 2007.

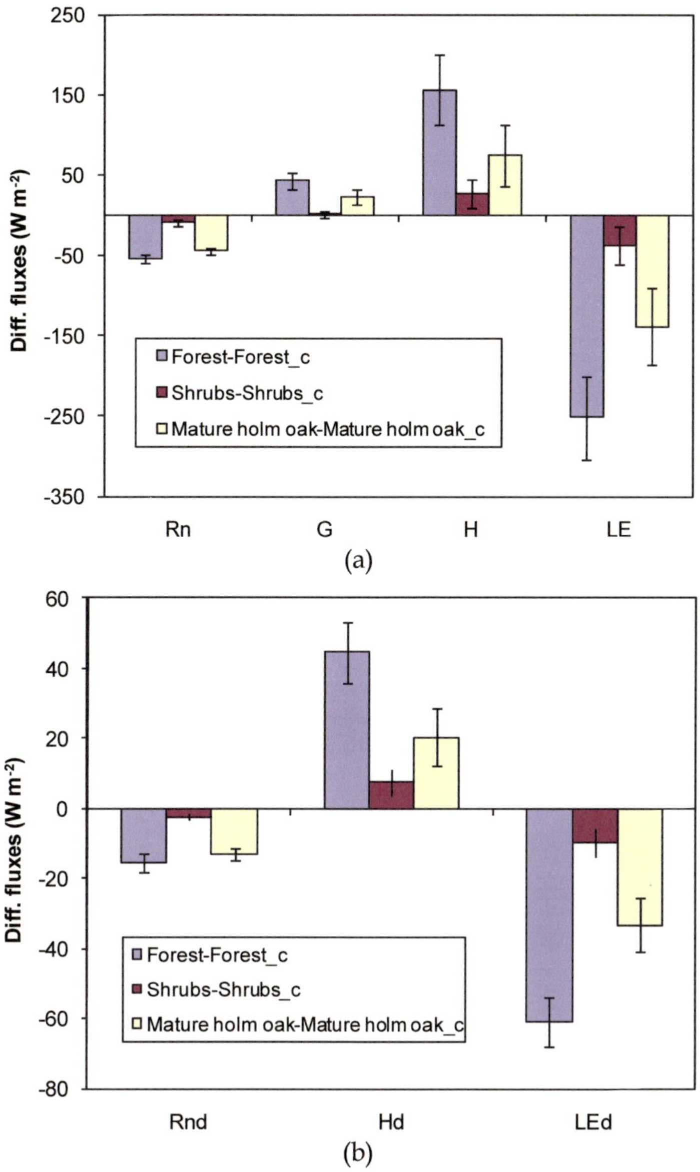

Fig. 7. Average values of the differences between burned and non-burned areas in terms of R_n, G, H, and LE: (a) Instantaneous fluxes; (b) Daily fluxes. Error bars represent the standard deviations of the averages.

Figure 7 shows also the effect of the fire in a future scenario in which the burned area has been naturally reforested and a mature holm oak forest is occupying the area. With this aim, differences between the forest_c area and a selected area representative of the mature holm oak forest are shown. Note that the net radiation is barely affected by the pass of the years. However, differences in terms of *G*, *H*, and *LE* would be half reduced at both instantaneous and daily scales. Despite this reduction significant differences in the flux patterns remain when replacing the pine with the holm oak as the dominant species. Thus we might conclude a irreversible reduction above 1 mm/day in the fire-affected area, even many years after the event. It would be desirable to expand this study to coming years and perform a more detailed temporal monitoring of the flux patterns in the area.

5. Conclusions

We have analyzed the effect of a fire on the energy flux pattern in a mediterranean forest area affected by a fire in the summer of 2001, located in a central Spain region. High resolution satellite imagery has been used, and two different ecosystems, a pines area and a shrublands area, have been studied. Maps of the different fluxes have been obtained for each one of the 5 Landsat 5-TM images, applying the described methodology. Validation with ground measurements shows relative errors of 7, 12, 19, and 21% for R_n, G, H, and LE, respectively, at a instantaneous scale, and of 6, 3, and 22% for R_n, H, and LE, respectively, at a daily scale. These results are in agreement with those obtained in recent validations of the presented methodology under different land cover types. The effect of the fire in the shrubland test site is negligible after 6 years. However, in the forested test site, an increase in H over 150 W m^{-2}, and a decrease in LE over 250 W m^{-2}, still remain around midday. At a daily scale the increase in H results 40 W m^{-2}, and the decrease in LE over 60 W m^{-2} (2.1 mm/day). This is mainly due to fire effects such as the decrease in P_v of almost 30%, or the increase in T_R of 6 °C approximately. Additional comparison with a nearby area covered by mature holm oak allows the analysis of the effect of the replacement of the pine with the holm oak as the dominant species, consequence of a natural post-fire regeneration process. The main result is the irreversible reduction of more than 1 mm/day produced in terms of daily evapotranspiration, that might have an impact on the local hydrological cycle and also on the local, or even regional, meteorology.

6. Acknowledgements

This work has been funded by the Spanish Science and Innovation Ministry (Projects CGL2007-64666/CLI, CGL2010-17577/CLI, AGL2009-13124, CSD2008-00040, CGL2008-04047, and Juan de la Cierva contract of J. M. Sánchez), Generalitat Valenciana (project PROMETEO/2009/086) and the JCCM (project ECOFLUX II, Ref: PCC08-0109).

7. References

Amiro, B. D., Macpherson, J. I., & Desjardins, R. L. (1999). BOREAS flight measurements of forest-fire effects on carbon dioxide and energy fluxes. *Agricultural and Forest Meteorology*, 96, 199-208.

Anderson, M. C., Norman, J. M., Diak, G. R., Kustas, W. P., & Mecikalski, J. R. (1997). A two-source time integrated model for estimating surface fluxes using thermal infrared remote sensing. *Remote Sensing of Environment*, 60, 195-216.

Berk, A., Anderson, G. P., Acharya, P. K., Chetwynd, J. H., Bernstein, L. S., Shettle, E. P., Matthew, M. W., & Adler-Golden, S. M. (1999). MODTRAN 4 user´s manual. In: *Air Force Research Laboratory, Space Vehicles Directorate* (Hascom AFB, MA: Air Force Materiel Command).

Chehbouni, A., Nouvellon, Y., Lhomme, J.-P., Watts, C., Boulet, G., Kerr, Y. H., Moran, M.S., & Goodrich, D. C. (2001). Estimation of surface sensible heat flux using dual angle observations of radiative surface temperature. *Agricultural and Forest Meteorology*, 108, 55-65.

Choudhury, B. J., Idso, S. B., & Reginato, R. J. (1987). Analysis of an empirical model for soil heat flux under a growing wheat crop for estimating evaporation by an infrared-temperature based energy balance equation. *Agricultural and Forest Meteorology*, 39, 283-297.

Choudhury, B., & Monteith, J. (1988). A four-layer model for the heat budget of homogeneous land surfaces. *Quarterly Journal of the Royal Meteorological Society*, 114, 373-398.

Dubayah, R. (1992). Estimating net solar radiation using Landsat Thematic Mapper and digital elevation data. *Water Resource Research*, 28, 2469-2484.

Duguay, C. R. (1992). Estimating surface reflectance and albedo from Landsat-5 thematic Mapper over rugged terrain. *Photogrammetric Enginnering & Remote Sensing*, 58, 551-558.

French, A. N., Jacob, F., Anderson, M. C., Kustas, W. P., Timmermans, W., Gieske, A., Su, Z., Su, H., McCabe, M. F., Li, F., Prueger, J., & Brunsell, N. (2005). Surface energy fluxes with the Advanced Spaceborne Thermal Emission and Reflection radiometer (ASTER) at the Iowa 2002 SMACEX site (USA). *Remote Sensing of Environment*, 99 (1-2), 55-65.

Hall, F., Huemmrich, K., Goetz, S., Sellers, P., & Nickerson, J. (1992). Satellite remote sensing of surface energy balance: Success, failures and unresolved issues in FIFE. *Journal of Geophysical Research*, 97, 19061-19089.

Humes, K., Ardí, R., Kustas, W.P., Prueger, J., & Starks, P. (2002). High spatial resolution mapping of surface energy balance components with remotely sensed data. In: *Thermal Remote Sensing in Land Surface Processes*, New York, CRC Press. pp. 110-132.

Inoue, Y., & Moran, M. S. (1997). A simplified method for remote sensing of daily canopy transpiration-a case study with direct measurements of canopy transpiration in soybean canopies. *International Journal of Remote Sensing*, 18 (1), 139-152.

Lagouarde, J.-P., & McAneney, K. J. (1992). Daily sensible heat flux estimation from a single measurement of surface temperature and maximum air temperature. *Boundary-Layer Meteorology*, 59, 341-362.

Li, F., Kustas, W. P., Prueger, J. H., Neale, C. M. U., & Jackson, T. J. (2005). Utility of Remote Sensing based two-source energy balance model Ander low and high vegetation cover conditions. *Journal of Hydrometeorology*, 6 (6), 878-891.

Monteith, J. L. (1973). Principles of environmental physics, London, Edward Arnold Publisher, 241 pp.

Moran, M. S., Clarke, T. R., Inoue, Y., & Vidal, A. (1994). Estimating crop water deficit using the relation between surface-air temperature and spectral vegetation index. *Remote Sensing of Environment*, 49, 246-263.

Norman, J. M., Kustas, W., & Humes, K. (1995). A two-source approach for estimating soil and vegetation energy fluxes from observations of directional radiometric surface temperature. *Agricultural and Forest Meteorology*, 77, 263-293.

Randerson, J. T., Liu, H., Flanner, M. G., Chambers, S. D., Jin, Y., Hess, P. G., Pfister, G., Mack, M. C., Treseder, K. K., Welp, L. P., Chapin, F. S., Harden, J. W., Goulden, M. L., Lyons, E., Neff, J. C., Schuur, E. A. G., & Zender, C. S. (2006). The impact of Boreal Forest Fire on Climate Warming. *Science*, 17, 1130-1132.

Rubio, E., Caselles, & Badenas, C. (1997). Emissivity measurements of several soils and vegetation types in the 8-14 µm wave band: Analysis of two field methods. *Remote Sensing of Environment*, 59, 490-521.

Rubio, E., Caselles, V., Coll, C., Valor, E., & Sospedra, F. (2003). Thermal infrared emissivities of natural surfaces: Improvements on the experimental set-up and new measurements. *International Journal of Remote Sensing*, 20(24), 5379-5390.

Sánchez, J. M., Caselles, V., Niclòs, R., Valor, E., Coll, C., & Laurila, T. (2007). Evaluation of the B-method for determining actual evapotranspiration in a boreal forest from MODIS data. *International Journal of Remote Sensing*, 28 (5-6), 1231-1250.

Sánchez, J. M., Kustas, W. P., Caselles, V., and Anderson, M. (2008). Modelling surface energy fluxes over maize using a two-source patch model and radiometric soil and canopy temperature observations. *Remote Sensing of Environment*, 112, 1130-1143.

Sánchez, J. M., Caselles, V., Niclòs, R., Coll, C., & Kustas, W. P. (2009). Estimating energy balance fluxes above a boreal forest from radiometric temperature observations. *Agricultural and Forest Meteorology*, 149, 1037-1049.

Santanello, J. A., & Friedl, M. A. (2003). Diurnal covariation in soil heat flux and net radiation. *Journal of Applied Meteorology*, 42, 851-862.

Scott, R. L., Shuttleworth, W. J., Goodrich, D. C., & Maddock III, T. (2000). The water use of two dominant vegetation communities in a semiarid riparian ecosystem. *Agricultural and Forest Meteorology*, 105, 241-256.

Seguin, B. & Itier, B. (1983). Using midday surface temperature to estimate daily evaporation from satellite thermal IR data. *International Journal of Remote Sensing*, 4 (2), 371-383.

Shuttleworth, W., & Wallace, J. (1985). Evaporation from sparse crops: an energy combination theory. *Quarterly Journal of the Royal Meteorological Society*, 111, 1143-1162.

Su, H., McCabe, M. F., Wood, E. F., Su, Z., & Prueger, J. H. (2005). Modelling evapotranspiration during SMACEX02: Comparing two approaches for local and regional scale prediction. *Journal of Hydrometeorology*, 6 (6), 910-922.

Valor, E. & Caselles, V. (1996). Mapping Land Surface Emissivity from NDVI: Application to European, African, and South American Areas. *Remote Sensing of Environment*, 57, 167-184.

Valor, E. & Caselles, V. (2005): Validation of the vegetation cover method for land surface emissivity estimation. In: *Recent Research Developments in Thermal Remote Sensing*, Kerala (India), Research Signpost, pp. 1-20.